INTRODUCTION TO ELECTRICAL ENGINEERING LABORATORIES

Circuits, Electronics, and Digital Logic

Elliot B. Slutsky

*California State
Polytechnic University, Pomona
(formerly of AT&T Bell Laboratories
Reading, Pennsylvania)*

David W. Messaros

Hewlett-Packard Corporation

PRENTICE HALL, Englewood Cliffs, NJ 07632

Editorial/production supervision: *Sharen Levine and Colleen Brosnan*
Cover designer: *Lundgren Graphics*
Manufacturing buyers: *Linda Behrens/Patrice Fraccio/Dave Dickey*

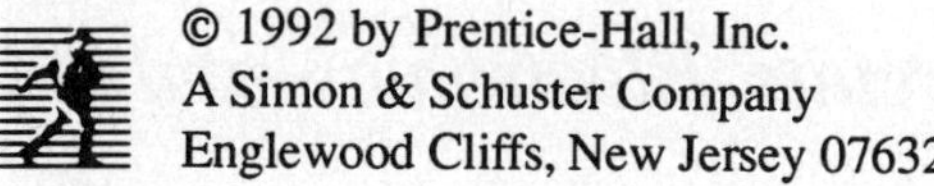

© 1992 by Prentice-Hall, Inc.
A Simon & Schuster Company
Englewood Cliffs, New Jersey 07632

All rights reserved. No part of this book may be reproduced, in any form
or by any means, without permission in writing from the publisher.

Printed in the United States of America

10 9 8 7 6 5 4 3 2

ISBN 0-13-488750-6

Prentice-Hall International (UK) Limited, *London*
Prentice-Hall of Australia Pty. Limited, *Sydney*
Prentice-Hall Canada Inc., *Toronto*
Prentice Hall Hispanoamericana, S.A., *Mexico*
Prentice-Hall of India Private Limited, *New Delhi*
Prentice-Hall of Japan, Inc., *Tokyo*
Simon & Schuster Asia Pte. Ltd., *Singapore*
Editora Prentice-Hall do Brasil, Ltda., *Rio de Janeiro*

To Our Wives

Bonnie and Tiffany

CONTENTS

CHAPTER 12   **DIGITAL SYSTEMS AND DIGITAL ELECTRONICS**   *352*

PREFACE

This text is an outgrowth of material developed by the authors while teaching laboratory courses as part of our assignments as Visiting Professors. The authors have over ten years of college level teaching experience, while participating the AT&T (EBS) and Hewlett-Packard and (DWM) Visiting Professor Programs.

This book is designed for students who are taking electrical engineering laboratory for the first time. These students include those in electrical engineering and in other disciplines, such as chemical engineering, computer science, mechanical engineering, physics, and chemistry who may be required to take an introductory electrical engineering laboratory. Because of gradations of difficulty of the laboratory experiments, this text would also be appropriate for community college and technical school use.

The text is divided into four parts. Part I (Chapters 1-4) covers introductory concepts, Part II (Chapters 5-8) covers circuits, Part II (Chapters 9-11) covers electronics, and Part IV (Chapter 12) covers digital systems and digital electronics.

Chapters 1 thru 3 cover laboratory report writing, data and error analysis, and the basic (generic) instruments for electrical measurements. Chapter 4, Laboratory Practice, uses fundamental electrical engineering theorems to develop simple models that show why certain approaches to circuit implementation are considered good laboratory practice and why others are not. For example, circuit grounds are modeled as disturbance inputs.

In Chapter 5, we discuss Ohm and Kirchhoff's Laws. The operation amplifier is introduced as an example of a dependent voltage source. In Chapter 6 we cover such linear network theorems as superposition, Thevenin and Norton's Theorems and Tellegan's Theorem. In Chapters 7 and 8 we discuss Frequency and Transient Response and AC Steady State Power, respectively. In the next three chapters we study the fundamental electronic families of devices: diodes in Chapter 9, bipolar transistors, junction field effect transistors and metal oxide semiconductor transistors in Chapter 10 and nonideal operational amplifiers in Chapter 11.

In the digital systems section of Chapter 12 we cover such topics as adders, subtractors, decoders, multiplexers and flip-flops in the digital system section. In the digital electronics section such topics as emitter-coupled logic, transistor-transistor logic, and NMOS and CMOS gates are introduced.

In our text, each of the laboratory chapters consists of a theory section with five or six experiments, using generic measurement equipment, to illustrate each of the major concepts. The experiments represent a spectrum of difficulty; each experiment has a prelaboratory associated with it. The prelab links the theory section of the chapter to the laboratory section so that the student, in performing the experiment, will be verifying the results of the prelab calculations.

The lab report is based on a series of questions at the end of each of the labs. The instructor can change the laboratory experiments each semester, based on the level of his/her class, or a desire to simply alter the experiments each year. Also, the amount of material presented in each of the laboratories is not absolute. The instructor should feel free to alter the amount of material used from each laboratory experiment to suit the level of his/her class. The book is self-contained: sufficient material is included in the theory sections so that outside references may be useful, but are not necessary.

In writing this laboratory book we wanted to make the material compatible with the topics covered in the popular texts used in electrical engineering. Therefore, we based the material of Chapter 3 on Wolf's *Guide to Electronic Measurements and Laboratory Practice*. For Chapters 5, 6, 7 and 8, we used *Electric Circuits* by Nilsson, *Basic Engineering Circuit Analysis* by Irwin, *Engineering Circuit Analysis* by Hayt and Kemmerly, and *Basic Electric Circuit Analysis* by Johnson, Hilburn and Johnson. The topics covered in Chapters 9, 10, and 11 were based on the material of Sedra and Smith (*Microelectronic Circuits*), Gray and Meyer (*Analysis and Design of Analog Integrated Circuits*), Savant, Roden and Carpenter (*Electronic Circuit Design*), and Burns and Bond (*Principles of Electronic Circuits*). For the digital chapter, we found Mano's books to be especially useful for the digital systems sections and Sedra and Smith and Hodges and Jackson (*Analysis and Design of Digital Integrated Circuits*) to be excellent references for the digital electronics sections of Chapter 12.

Many of the new circuits and electronics texts have sections, appendices and even entire chapters on SPICE. We used the SPICE manual distributed by the University of California at Berkeley. In addition, we found PSPICE, SPICE for PCs, developed by MicroSim Corporation to be very useful. Micro-CAP II, developed by Spectrum Software and distributed by Addison-Wesley Publishing Company, is another electric circuit analysis program.

ACKNOWLEDGMENTS

The authors would like to offer much thanks and appreciation to Sandra Cacciacarne of AT&T Microelectronics, Reading, Pennsylvania, for entering the manuscript on the UNIX™ system, supervising the drafting work, and contacting the production editors, and for her patience and good humor which allowed us to meet our writing schedules. We also want to acknowledge Robert Schneider and Richard Cronan of the drafting department at AT&T Bell Laboratories, Reading, Pennsylvania, for all the plots, schematics and tables contained in the manual. The authors would also like to express their thanks to Phyllis Fegley of the Text Processing Department of AT&T Bell Laboratories, Reading, Pennsylvania for her work on the manuscript.

We also want to acknowledge a number of people at AT&T Bell Laboratories and Hewlett Packard for their encouragement and support, not only during the course of this project, but also for their support of the Visiting Professor Program. Thanks to M. J. Thompson, R. A. Herrman, F. Junger, J. M. Goldey, G. F. Foxhall, and William C. Ballamy. Special thanks to L. E. Miller, A. J. Kerin, L. Heckman A. E. Jones and L. Aikens, all of Bell Laboratories, and B. Recchia, W. Buffington, T. Roohe, M. Biles, P. Crilly and C. L. Teng of Hewlett Packard Corporation.

Chapter 1

REPORT WRITING

1.1 Introduction

The engineer is called on to write memoranda, technical notebooks, proposals, professional papers, magazine articles, patent disclosures, letters, specifications, technical bulletins, instruction manuals, handbooks and even textbooks. Communications, both written and oral, are an essential part of an engineer's responsibilities. Good ideas have to be explained to others, and that requires good communication skills. In a survey entitled "Goals of Engineering Education," published in January 1968 by the American Society for Engineering Education, the opinions from about 4,000 engineers in 129 companies were obtained regarding subjects taken in college.[1] Both English composition and speech were ranked very high as subjects to be emphasized for future engineers and as subjects used by these engineers either in the past month or during their careers.

Just as mastery of circuit design is attained by actually designing circuits, mastery of communication skills requires practice in writing reports or giving oral presentations. Although many concepts are common to both written and oral forms, in this book we shall limit ourselves to written communication; we refer the reader to other sources for oral communication.[1],[2],[3],[4]

The academic laboratory report not only presents the engineering student with the opportunity to reinforce classroom concepts, but gives the student practice in presenting technical data and communicating observations. However, before the engineer can write a technical report, experiments must be performed, data recorded, and observations made. This information is recorded in some form of notebook; we shall make suggestions as to the form this notebook might take.

1.2 *Laboratory Notebook*

Before entering the laboratory, it is essential to review the theory behind the experiment to be performed, to sketch anticipated graphical results, to identify appropriate scales, and to determine the number of experimental points to be taken and the range of values within which the points lie. Circuit connection diagrams should also be prepared before coming to the laboratory.

The laboratory notebook is usually specified by the student's department, but notebooks often used are the Computation Notebook, Vernon McMillian, Inc., No. 09-9890; and the National, No. 43-648, both usually available in campus bookstores. The purpose of the laboratory notebook is to train the student to keep complete and accurate records of his or her work so that the habit will carry over into the industrial or academic research laboratory or in fieldwork. In the notebook, the engineer is documenting, in a legally recognized way, the development of inventions that could be used in patent enforcement or field investigations. Any legal problems these inventions run into later may require reference to the original information kept in the notebook.

The following information should be recorded in the laboratory notebook:

a. Descriptive title of the experiment.

b. Objective of the experiment.

c. Circuit diagram, including measurement connections.

d. Complete list of equipment, including type, model number, serial number, and corresponding symbol on the circuit diagram.

e. Data obtained in tabular form, including instrument range.

f. Graphs of the data.

g. Any observations made during the report write-up that might be helpful.

h. Calculations.

i. Discussion of the results.

j. Conclusions.

All of the entries in the notebook must be in ink and must be written consecutively on sequentially numbered pages, without leaving blank spaces or pages. Pages must not be torn from the notebook; if mistakes are made, they may be crossed out. Sketches, drawings, photographs, and equipment and device data sheets should be glued to blank pages of the notebook.

During the laboratory period, the student working with one or two partners should identify the test equipment and take experimental data in a clear, organized format. Before the circuit is dismantled, it is a good idea to plot the data to reveal points that are incorrect or, if enough data points have been taken, to show important details of the curve.

After the experiment is performed, but still in the laboratory, the student should work up all required results and computations and complete the tables. Scales and parameters on all curves should be clearly marked with variable names, as well as with units and dimensions. After the laboratory session, the student's objective is to put in discussion form how well the experiment went, the accuracy of the data, and the limitations of both the equipment and the test circuit. The student should finally include a summary of the experiment and its results. The laboratory information report should be complete enough so that another experimenter could enter the laboratory and duplicate the results.

1.3 Data Presentation[5],[6]

A graph is a good deal more than simply a plot of experimental points; it is an effective analysis tool. Graphs are an efficient and convenient way of portraying and analyzing data and are used to help visualize analytic expressions, to interpolate data, and to discuss errors.

Graphs should contain a title, the date the data were taken, and axes that are adequately labeled and scaled. Convention dictates that the vertical or ordinate axis contains the dependent variable or the measured response to the controlled parameter, while the horizontal or abscissa axis contains the controlled independent parameter. The scale should be convenient and readable so that interpolation can be easily accomplished; it is preferable to use units of 1, 2, or 5 for a division. The scales of a graph should provide a rough indication of the accuracy of the measurements and should not imply a greater accuracy than that of the measurements. The scales also should be selected so that the graph is spread to a convenient and readable extent, but not overspread. This requires some judgment. For multiple curves, use different symbols and explain the meaning of each by including a key within the graph. A sharp-pointed pen, straightedge, or French curve should be used to draw smooth curves through the data points. Plotting data on a graph as the data are taken allows unexpected data points to be rechecked before an experimental setup is dismantled. If a mathematically derived curve is plotted, it should be drawn smoothly and without any of the plotted points shown discretely. If the graph is derived from experimental results, the experimental points should be plotted. Curves should not contain discontinuities or other peculiarities, especially if there are theoretical reasons for expecting a smooth result. It is not necessary for the curve to pass through all the experimental points, but there should be an equal number of points falling on both sides of the curve. In order to separate several curves that are plotted on the same graph, different symbols can be used, such as small circles, triangles, and squares. A curve that connects the experimental points in a smooth way can be drawn in, but it should stop at the edges of the points (Figure 1.1). Figures and Tables should be arranged to be readable without having to rotate the page. If the figure has to be turned, it should be rotated 90° clockwise.

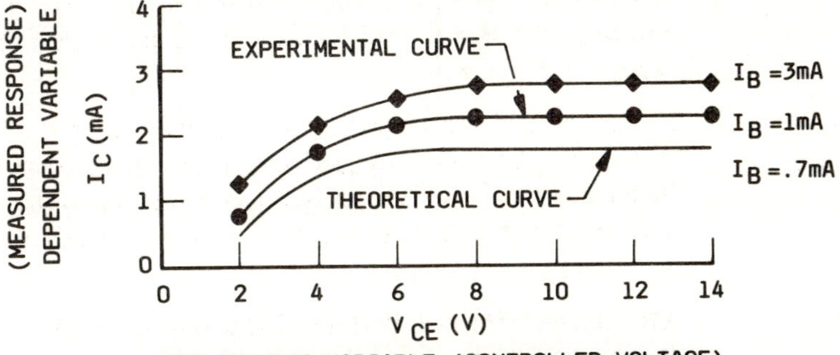

Figure 1.1: Current-voltage family of curves
for a bipolar junction transistor

Three of the most popular types of graph paper are Cartesian (Figure 1.2), semilog (Figure 1.3), and log-log (Figure 1.4). In Cartesian, or linear, graph paper the abscissa and ordinate scales are linear.

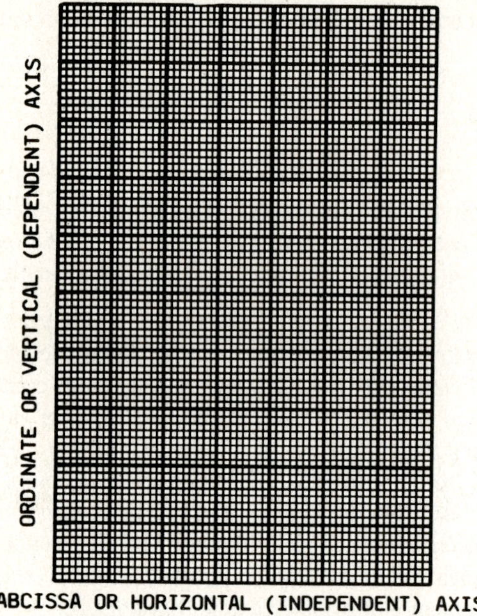

Figure 1.2: Cartesian graph paper

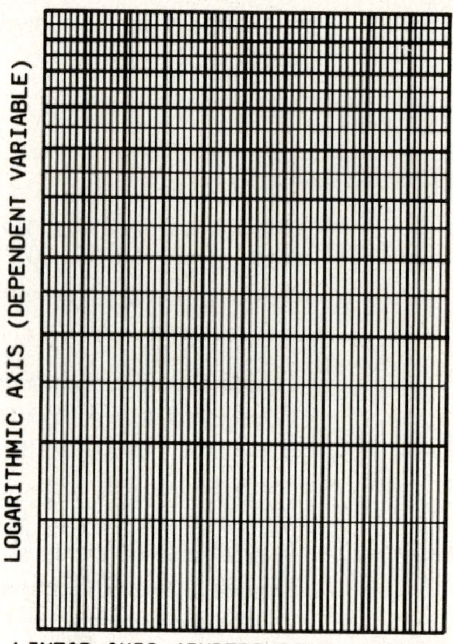

Figure 1.3: Semilog graph paper

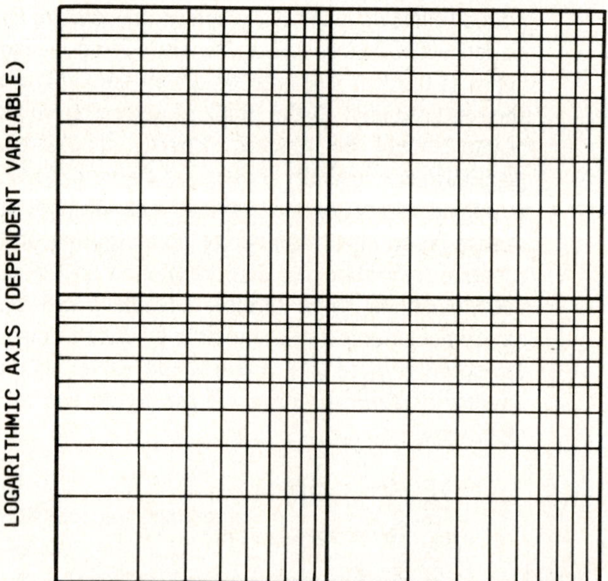

Figure 1.4: Log-log graph paper

By contrast, semilog paper contains one logarithmic axis and one linear axis. Semilog paper is convenient for plotting exponential functions or functions that have one widely ranging parameter. For example, in the equation

$$y = Ae^{x/\alpha} \tag{1.1}$$

the dependent parameter y would be plotted on the log scale and the independent variable x on the linear scale. On semilog paper the plot of this equation would be a straight line, with α determined from the slope of the line and A found from the value of y at x=0.

The equation

$$y = Bx^{\beta} \tag{1.2}$$

is conveniently plotted on log-log paper when both the independent and dependent parameters vary over a wide range. The plot of this equation is a straight line. The value of the constant B is determined by setting $x = 1$, and β is the slope of the straight line.

Tables should be neat. Columns should be headed with the name or symbol of the quantity being measured and the units used. It is often convenient to include a column for remarks and a column marked "computed data."

If a photograph, plotter printout, or sketch of an oscilloscope display is included, it should capture those aspects of the trace that are relevant to the experiment. If the gradicule lines are drawn, scale factors are required.

We shall continue the discussion of data presentation in Chapter 2.

1.4 Form of the Report

In this section, we discuss two types of reports: the academic laboratory report and the formal report. The two reports have a number of principles in common, namely, that (1) they should be written for a specific person, i.e., an intelligent but uninformed reader; (2) their purpose should be made clear to the reader with every sentence and word; (3) their language should be simple and concise; and (4) they should follow the simple rule that you should (a) prepare the reader for what you are going to say, (b) tell the reader the information, and (c) review what you told the reader.

There are a number of report forms or styles. An informal laboratory report with emphasis on the data section might conform to the following style:

Informal Laboratory Report[7]

A. Cover Sheet

1. Name

2. Lab partner

3. Course number and section

4. Experiment number and title

5. Date of experiment

B. Data Section

1. Schematic of the circuit

2. Equipment

a. Manufacturer's name

b. Model number

      c. Serial number

   3. Data table

   4. Graphs on graph paper

   5. Sample calculations

C. Analysis Section

   1. Technical discussion of your data

   2. Comparison with expected results

   3. Causes of errors

D. Conclusions

A more formal report might take the following form:

Formal Laboratory Report

A. Title Page or Folder Cover

   1. Name

   2. Lab partner's name

   3. Course number and section

   4. Experiment number and title

   5. Date of the experiment

B. Introduction Section
   Statement of the objective of the experiment

C. Theory Section
   Concise and pertinent discussion of the theory

D. Method of Investigation

   1. Brief outline of the experimental procedure

   2. Schematic of circuit

E. Equipment List

   1. Manufacturer's name

   2. Model number

   3. Serial number

F. Data Section

G. Sample Calculation

H. Analysis Section

   1. Comparison of experimental data and expected results

   2. Probable causes of any errors

I. Conclusions

J. Rough Data Section

1.5 Summary

In this chapter, we have emphasized the importance of clean, concise, well-organized report writing in an engineering environment. Because this goal is obtained by practice, the academic laboratory serves not only as a medium for verifying classroom theory, but also as a means to practice both organizing a laboratory notebook and presenting technical observations in written form. Two different report forms were presented and details of their constituent sections were explained.

Data presentation in the form of tables and graphs was also discussed. The IEEE publishes an "information for authors" document with which you should become familiar. This document contains standards of practice for documents published by journals of the IEEE. Units, symbols and abbreviations, mathematical notation, references, and more are specified.

REFERENCES

[1] J. D. Kemper, *Introduction to the Engineering Profession* (New York: Holt, Rinehart and Winston, 1985), Chapter 8.

[2] E. Cohen, *Speaking the Speech,* 2d ed. (New York: Holt, Rinehart and Winston, 1983).

[3] D. E. Zimmerman and D. G. Clark, *The Random House Guide to Technical and Scientific Communication* (New York: Random House, 1987), Chapter 15.

[4] G. H. Mills and J. A. Walter, *Technical Writing,* 5th ed. (New York: Holt, Rinehart and Winston, 1986), Chapter 16.

[5] S. Wolf, *Guide to Electronic Measurements and Laboratory Practice,* 2d ed. (Englewood Cliffs, NJ: Prentice-Hall, 1983), Chapter 2.

[6] Kemper, Chapter 6.

[7] L. Jones and A. F. Chin, *Electronic Instruments and Measurements* (New York: Wiley, 1983), pp. 371-3.

Chapter 2

DATA
AND
ERROR ANALYSIS

2.1 Introduction

No measurement or observation is ever made with absolute accuracy. There is always a definite limit to the accuracy with which the scale of an electrical meter, meter stick, thermometer, etc., can be read. Although instruments with accurate scales are available and the uncertainty in the position of the measuring mark is minimized, a limit to the accuracy with which the scales can be read always exists in every instrument.

The accuracies of digital meters are usually greater than those of analog meters, but there are still inaccuracies. Therefore, it is important to study and understand the manufacturer's specifications. This will be done in Chapter 3, where we discuss analog and digital electronic meters.

The **accuracy** of a measurement specifies the difference between the measured and the true value. Deviation from the true value is an indication of how accurately a reading has been made. The **precision** of a measurement specifies the repeatability of a set of readings, with each reading made independently using the same instrument. An estimate of precision is determined by the deviation of a reading from the average value of all the readings.

A physical measurement, then, is not complete until its "error" or, more precisely, its experimental uncertainty has been determined.

Studying errors is important because it provides a means of reducing, if not altogether eliminating errors.

The sources of errors can be divided into several varieties:

1. **Human error** represents true mistakes or blunders either in measurement or in computation. These errors can be eliminated by double-checking readings and making rough calculations of expected results before completing the measurements.

2. **Scale uncertainty** represents the smallest quantity that can be estimated on the scale of an instrument or tool and establishes a lower limit to the accuracy of the measurement.

3. **Systematic errors** are due to faulty calibration or use of the equipment, biased observers, or external influences such as temperature and humidity. These errors can be minimized by understanding the manufacturers' specifications for all the equipment used, by knowing the limitation and accuracy of each piece of equipment, and by understanding the environmental effects on the measurement system. Systematic errors are not reduced by successive measurements.

4. **Statistical errors, random errors**, and **residual errors** are three names given to errors that arise from a wide variety of sources. These errors are often caused by the erratic combinations of a large number of small effects, some of which are known and some of which are not. Usually, statistical or random errors are considered after the systematic errors are subtracted or taken into account in some other way. Some characteristics of statistical errors are:[1]

   a. They are distributed in a random manner.

   b. Positive and negative errors are equally likely.

   c. The probability of a small error is greater than the probability of a large error.

   d. Extremely large errors are very improbable.

   e. They may be reduced by additional measurements.

Although systematic errors are not reduced by successive measurements, residual or statistical errors may be reduced. This is because the laws of probability utilized by statistics operate on random errors but not on systematic errors. Therefore, systematic errors must be small compared to random errors if the results of statistical evaluation are to be meaningful.

Often, the experimenter contributes to the error in measurement, depending on habit and observational abilities. Errors may arise through estimating fractions of a division in the reading of an instrument or peculiarities in the observer's sense of timing (i.e., making a reading too early or too late). If you cock your head to one side while reading the scale of an electrical meter, you may introduce a parallax error if the line of sight differs from a direction perpendicular to the plane of the scale.

When there is time or when a data measurement is crucial, you should repeat the measurement several times. The scattering of values about the average shows how large the random error will be. You should look through the data for obvious mistakes immediately after measuring and you should repeat a set of measurements if necessary. You are allowed to throw out an odd or grossly inconsistent reading only if you can explain its occurrence and if the explanation can be checked.

2.2 Significant Figures and Scale Uncertainty[1],[2],[3]

A number representing a measured quantity has at least three distinct features: significant figures, the location of the decimal point, and units or dimensions. **Significant figures** are numbers that convey actual information regarding the magnitude of a quantity. As an example, if a resistance has been determined to three significant figures, the value might be written in any of several ways, including the following:

(a) 105×10^2 ohms
(b) 10.5 kilohms
(c) 0.0105 megohms
(d) 10,500 ohms

Specifying 105×10^2 ohms implies that the value is closer to 105×10^2 ohms than to 104×10^2 ohms or 106×10^2 ohms. Writing 105.0×10^2 ohms implies four significant figures, but also that the resistance is closer to 105.0×10^2 ohms than it is to 104.9×10^2 ohms or 105.1×10^2 ohms. The conclusion is that the extra digits should not be added unless they can be justified.

The following are some conventions concerning significant figures:

a. The last digit represents the point of uncertainty.

b. It is understood that there is a total uncertainty of one unit in the last digit. That is, if a resistance is closer to 105×10^2 ohms than it is to 104×10^2 ohms or 106×10^2 ohms, then the resistance lies between 104.5×10^2 ohms and 105.5×10^2 ohms.

c. To avoid the appearance of zeros beyond the uncertain digit, an appropriate power of 10 should be used.

Conventions for rounding off to a specified number of significant figures are as follows:

a. If the first digit to be discarded is less than 5, the preceding digit is left unaltered.

b. If the first digit to be discarded is greater than 5, the preceding digit is increased by 1.

c. If the first digit to be discarded is equal to 5 and is increased by digits greater than zero, the digit preceding the 5 is increased by 1.

d. If the first digit to be discarded is equal to 5 and is followed by either zeros or no further digits, the digit preceding the 5 is rounded to its nearest even value. (The choice of even over odd is arbitrary). As examples of the preceding, 10.449 becomes 10.4, 10.46 becomes 10.5, 10.4501 becomes 10.5, 10.45 becomes 10.4, and 10.35 becomes 10.4.

In the operations of addition, subtraction, multiplication, and division, it is normal to carry more digits than ultimately will be used. The extra digits are dropped in the final result, because the figures have no meaning beyond the accuracy of the original data. In the addition or subtraction of two numbers, no digit should be retained that is more than one place beyond the least significant figure of the other number. For example, if $R_1 = 24.3$ ohms and $R_2 = 0.516$ ohm, the sum $R_1 + R_2$ would be 24.8 ohms. In this operation, numbers 24.3 and 0.516 are added as 24.3 and 0.52 = 24.82 and this number is rounded to 24.8.

For analog meters, the scale uncertainty of an instrument gives you the precision with which you can read the scale. For example, a scale is shown in Figure 2.1 that can be read to about 0.01 mA. In the figure,

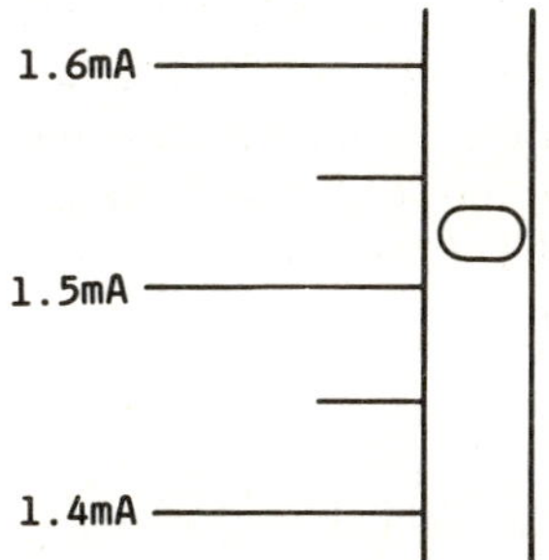

Figure 2.1: Scale uncertainty of an ammeter

the reading may be between 1.52 mA and 1.54 mA, or 1.53 ± 0.01 mA. The value 0.01 gives the precision with which you can read the scale. Even if the readings are repeated and the pointer goes to the same place shown, the precision of the measurement is still only ± 0.01 mA. Therefore, you must first examine each measuring instrument and write down what both the limit of your accuracy and the precision of the reading will be. These establish the error inherent in the process of reading the scale. The precision of any one reading gives the lower limit of accuracy of your experiment.

2.3 Accuracy

Accuracy is defined as the degree of conformity of a measurement to a standard or assumed true value. In our case, the assumed true value will come from the prelaboratory calculations. The task is to compare the theory or prelab calculations with the results obtained from the experiment and to determine whether the experimental data are "close enough" to the theoretically predicted values. The phrase "close enough" must be quantified so that the predicted value of a data point is, for example, not simply x volts but x volts ± y volts. The variation of this voltage, ± y volts, is not due to deviations in the same reading or voltage, but depends on the tolerances of the circuit components and the accuracies of the measuring instruments.

2.3.1 Example

You are told to build and verify a circuit with a nominal output of 1 volt (1 V), one-half the input voltage. The circuit is shown in Figure 2.2. Resistors R_1 and R_2 have nominal resistances of 10 kΩ.

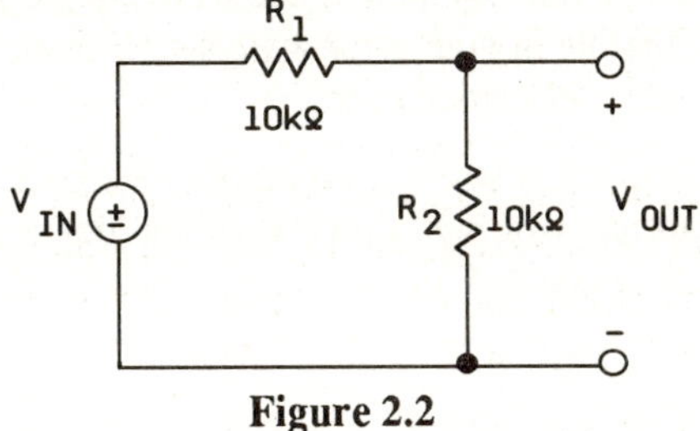

Figure 2.2

First, we need to analyze the circuit and determine the output voltage. To do so, we write the node equation summing the current flowing out of node V_{out}.

$$\frac{V_{out} - V_{in}}{R_1} + \frac{V_{out}}{R_2} = 0 \tag{2.1}$$

This simplified to

$$V_{out} = \frac{R_2}{R_1 + R_2} V_{in} \tag{2.2}$$

Using the nominal values for R_1, R_2, and V_{in}, we obtain

$$V_{out} = \frac{10 \text{ k}\Omega}{10 \text{ k}\Omega + 10 \text{ k}\Omega} (2 \text{ volts}) = 1 \text{ volt} \tag{2.3}$$

After construction of the circuit, V_{out} was measured and found to be 0.95 volts instead of 1.0 volt. An error analysis will reveal if the results are "close enough"; that is, analysis will determine the maximum error the components can produce. The error analysis is, then, simply an accounting of all the errors in the circuit, determining how they affect the variable of interest, the output voltage V_{out}.

The first step in the error analysis is to determine the tolerance of all of the components used in the experiment. For this experiment, assume the following tolerances:

Component	Tolerance
R_1, R_2	$\pm\,5\%$
Voltmeter	$\pm\,1\%$

This means that the actual value of R_1 or R_2 could be as high as 10,500 Ω or as low as 9,500 Ω. The tolerance or accuracy given for the voltmeter implies that any voltage reading is within plus or minus one percent of the actual voltage. In some meters, the percent error is based on the full scale value. For example, if our hypothetical meter were used to set the 2-V value of V_{in}, the actual value of V_{in} could be anywhere between 2.02 V and 1.98 V if the meter indicated 2.00 V. If we could only set V_{in} to 1.90 V, then the actual voltage could be anywhere between 1.92 and 1.88 V.

Another source of error is the temperature coefficients associated with the meter and resistors in the circuit. That is, the accuracy or tolerance of a device or component may vary with temperature. Yet another source of error is the input impedance (resistance) of the voltmeter. Ideally, this will be infinite and not cause a change to a node voltage when the meter is connected. For the preceding example, these effects are neglected. But knowing whether an error is insignificant or within the limits of the measurement equipment is not an easy task. That is why it is important to know as much as possible about your equipment and its loading effect and to include as many sources of error as possible in your error analysis. If some source generates a negligible error, this error can be discarded later.

In the next section, we use two methods for quantifying the variation of V_{out} given the accuracy of our components and devices. We assume that the major sources of error are accounted for.

2.3.2 Method I: Worst Case

In this method, we examine the mathematical relationship between V_{out} and all parameters that affect V_{out}. We vary the parameters in such a way as to maximize V_{out} and then we repeat this process to minimize V_{out}. If, during the process, a parameter is found in both the numerator and the denominator of the relationship (R_2 is an example) we repeat the calculation using both the maximum and minimum value of this parameter to determine the largest error. In equation (2.2), we find the maximum value of V_{out} by maximizing the numerator and minimizing the denominator. Since R_2 occurs in both the numerator and denominator, we repeat this process for $R_{2_{max}}$ and $R_{2_{min}}$. We use a bar over a parameter to indicate a maximum and a bar under a parameter to indicate a minimum. We have

$$\overline{V}_{out}= \frac{\overline{R}_2\,\overline{V}_{in}}{\underline{R}_1 + \overline{R}_2} = \frac{10,500\ \Omega\ (2.02\ V)}{10,500\ \Omega + 9,500\ \Omega} = 1.06\ V \qquad (2.4)$$

or

$$\overline{V}_{out}= \frac{\underline{R}_2\,\overline{V}_{in}}{\underline{R}_1 + \underline{R}_2} = \frac{9,500\ \Omega\ (2.02\ V)}{9,500\ \Omega + 9,500\ \Omega} = 1.01\ V \qquad (2.5)$$

Therefore the maximum value of $\overline{V}_{out} = 1.06$ V.

Now, we find the minimum value of V_{out}, $\underline{V}_{out}$, by minimizing the numerator and maximizing the denominator. The calculation must be done twice for R_2 to determine $\underline{V}_{out}$.

$$\underline{V}_{out}= \frac{\underline{R}_2\,\underline{V}_{in}}{\overline{R}_1 + \underline{R}_2} = \frac{9,500\ \Omega\ (1.98\ V)}{10,500\ \Omega + 9,500\ \Omega} = 0.94\ V \qquad (2.6)$$

or

$$\underline{V}_{out}= \frac{\overline{R}_2\,\underline{V}_{in}}{\overline{R}_1 + \overline{R}_2} = \frac{(10,500\ \Omega)\ (1.98\ V)}{10,500\ \Omega + 10,500\ \Omega} = 0.99\ V \qquad (2.7)$$

Therefore, the minimum value of V_{out} is 0.94 volt, and

$$V_{out}\ (actual) = 1 \pm 0.06\ V \qquad (2.8)$$

This means that the actual output voltage could be as high as 1.06 V or as low as 0.94 V. Therefore, the measured value of 0.95 V for V_{out} is acceptable.

In measuring V_{out}, it must be realized that the meter is $\pm1\%$ accurate, so an additional 1% error must be considered when reading V_{out}. Thus,

$$V_{out}\ (reading) = 1 \pm 0.07\ V \qquad (2.9)$$

2.3.3 Method II: Total Derivative

A more systematic approach is to use the concept of the total derivative. In this method, we take the derivative of V_{out} with respect to each parameter and add the variations. This procedure yields the total variation of V_{out} due to all possible variations in the circuit.

In equation (2.2) again, we take the partial derivative of V_{out} with respect to each parameter. Then we multiply each partial derivative by the variation in the nonfixed parameter. Finally, we add all the variations in V_{out}. This gives the total derivative,

$$dV_{out} = \left[\frac{\partial V_{out}}{\partial R_1}\right]\partial R_1 + \left[\frac{\partial V_{out}}{\partial R_2}\right]\partial R_2 + \left[\frac{\partial V_{out}}{\partial V_{in}}\right]\partial V_{in} \qquad (2.10)$$

$$= \left[\frac{(-R_2)V_{in}}{(R_1 + R_2)^2}\right]\partial R_1 + \left[\frac{R_1 V_{in}}{(R_1 + R_2)^2}\right]\partial R_2 + \left[\frac{R_2}{R_1 + R_2}\right]\partial V_{in}$$

∂V_{in}, ∂R_1, and ∂R_2 are the maximum variation in each component. Since $\overline{V}_{in}$ equals 2.02 and $\underline{V}_{in}$ equals 1.98, the maximum variation in V_{in}, ∂V_{in}, is 0.04 V. R_1 can vary $\pm\,500\ \Omega$ resulting in a ∂R_1 equal to 1000 Ω. R_2 is the same, but the maximum dV_{out} is obtained by letting ∂R_2 equal -1000 Ω. This is valid because the value of R_2 can go from 9,500 Ω to 10,500 Ω, as well as from 10,500 Ω to 9,500 Ω. The rest of the parameters are taken at their nominal or mean value. That is, $R_1 = R_2 = 10\ k\Omega$ and $V_{in} = 2.0$ V. Using these parameters in equation (2.10), we obtain

$$dV_{out} = \frac{(10\ k\Omega)(2\ V)(1\ k\Omega)}{(20\ k\Omega)^2} + \frac{(10\ k\Omega)(2\ V)(1\ k\Omega)}{(20\ k\Omega)^2} + \frac{10\ k\Omega}{10\ k\Omega}(0.04) \qquad (2.11)$$

so that

$$dV_{out} = 50\ mV + 50\ mV + 20\ mV = 120\ mV$$

and

$$V_{out}\ (actual) = 1 \pm 0.06\ V \qquad (2.12)$$

This result is the circuit variation. In addition, the meter has a $\pm 1\%$ error, which translates to a $\pm 7\%$ total variation in the measured value of V_{out}. The results of the two methods agree.

2.4 Summary

In this chapter, we have learned that you must understand the theory behind the experiment and have a thorough grasp of the characteristics and limitations of the equipment being used. You must also minimize and correct for external factors that could adversely affect the results of the experiment. When the larger factors have been eliminated, the residual factors must be accounted for.

We described and classified the sources of measurement error and learned that there is a fundamental limitation to our ability to make a precise measurement. We discussed the concept of significant figures and scale uncertainty and how to combine measurement uncertainties when we perform simple mathematical manipulations. The concept of accuracy was discussed and illustrated.

The concepts of data and error analysis give us the background to make intelligent use of the measuring equipment to be discussed in the next two chapters and set the stage for the measurements we shall be making starting with Chapter 5.

QUESTIONS

1. In Figure 2.3, find the voltage V_o expressed as a function of the circuit's components.

    a. Take the total derivative of this equation for voltage.

    b. For the component values given find the deviation possible for V_o.

    $I_1 = 1\ A \pm 5\%$, $V_1 = 10\ V \pm 1\%$,
    $R_1 = R_2 = 10\ k\Omega \pm 10\%$, and $R_3 = 1\ k\Omega \pm 5\%$

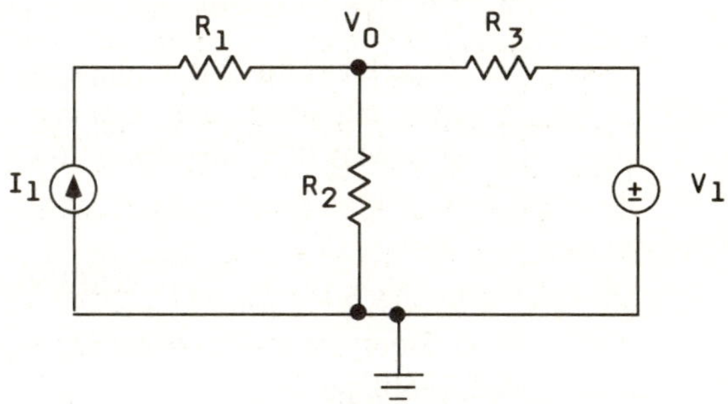

Figure 2.3

2. Show that if two numbers x and y have a 10% error associated with each of them, then if you multiply x and y or divide x by y or y by x, the resulting number has a 20% error.

3. Show that the addition or subtraction of two numbers x and y with the same percent error results in a number z with the same percent error. What is the percent error of z if x has half the percent error of y?

4. For the circuit of Figure 2.4 to act as a current source of 100 µA with less than a $\pm 1\%$ error for a range of load resistance from 10 Ω to 1 MΩ, what must be the values R_{source} and V_{source}

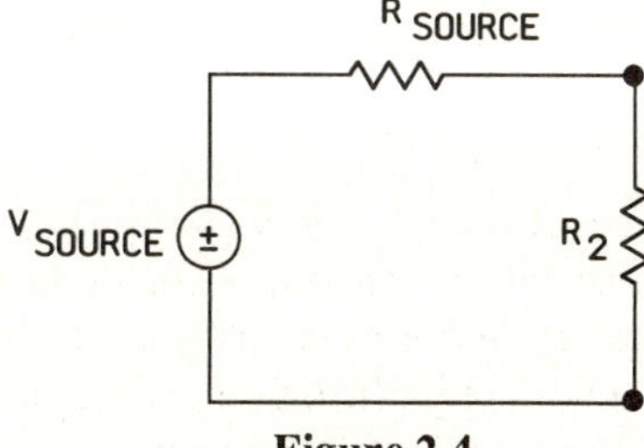

Figure 2.4

5. The circuit of Figure 2.5 has the following component values and tolerances:

$$
\begin{array}{lll}
R_1 = R_2 \ = \ 10\ k\Omega & \pm 5\% \\
R_3 \ = \ 1\ k\Omega & \pm 1\% \\
R_4 \ = \ 5.11\ k\Omega & \pm 10\% \\
V_{in} \ = \ 10\ V & \pm 1\% \ \ (\text{set by the} \pm 1\% \text{ voltmeter).} \\
\text{Voltmeter:} & \pm 1\%
\end{array}
$$

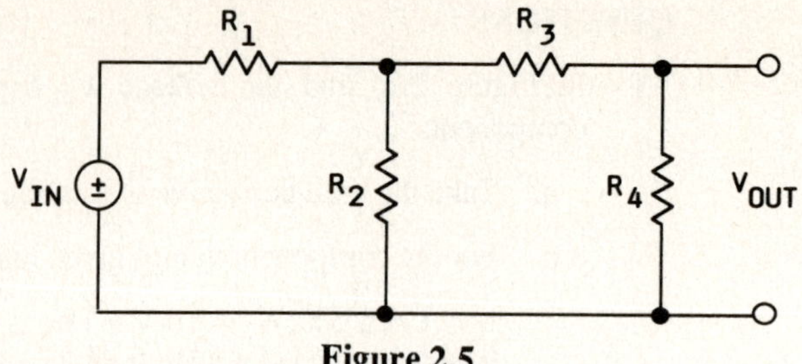

Figure 2.5

a. Find the maximum and minimum values of V_{out}.

b. If each resistor has a temperature coefficient of 1,000 ppm,* what are the new maximum and minimum values for V_{out} when the temperature increases 20°C? decreases 20°C?

REFERENCES

[1] M. L. Stout, *Basic Electrical Measurements* (Englewood Cliffs, NJ: Prentice-Hall, 1950), pp. 16-26.

[2] E. Frank, *Electrical Measurement Analysis* (New York: McGraw-Hill, 1959), pp. 392-4.

[3] IEEE, "Recommended Practice for Units in Published Scientific and Technical Work," IEEE Spectrum, March 1966, Appendix B.

---

*ppm means "parts per million" and implies "per degree Celsius." So, for example, a 10 kΩ resistor with a temperature coefficient of 100 ppm means that the resistor changes 100 ohms per million ohms or 10 Ω per 100 kΩ or 1 ohm per 10 kΩ, all per one degree Celsius increase in temperature.

Chapter 3

BASIC INSTRUMENTS FOR ELECTRICAL MEASUREMENT

3.1 Introduction

In this chapter we discuss some of the instruments students find in the typical electrical engineering laboratory. We offer a brief discussion of the theory of operation for some of these instruments.

The chapter serves as a guide to the instruments and measurement techniques that the student will use to perform the experiments, starting with Chapter 5. It is not the purpose of the chapter to substitute for such excellent texts as Wolf's *Guide to Electronic Measurements and Laboratory Practice*[1] *or* Jones and Chin's *Electronic Instruments and Measurement.*[2] The student is also encouraged to read the manufacturer's instruction manual for each instrument used.

3.2 Analog DC Meters

Current and voltage meters can be grouped into two general categories: analog meters and digital meters. Analog meters use electromechanical movements and pointers to display the quantities being measured along a continuous scale. Digital meters will be discussed in Section 3.4.

3.2.1 D'Arsonval Movement

The basic mechanism of the analog dc meter is the **D'Arsonval meter movement**. The mechanism detects current by using the force resulting from the interaction of a fixed magnetic field with the magnetic field caused by the current flowing through the fixed field. The force is used to generate a mechanical displacement, which can be measured on a calibrated scale. Many references discuss the D'Arsonval movement in detail, so our discussion will be limited. It is important to note that the basic quantity measured by the D'Arsonval based instrument is current, regardless of the units in which the scale is calibrated. Since the D'Arsonval meter movement uses very fine wire in its winding, the basic movement is modified by placing a low resistance in parallel with the basic meter movement resistance, R_m. For ammeters, the purpose of the additional resistance R_{SC} is to provide an alternative path for the total current, I, around the meter movement. The basic circuit for the meter is shown in Figure 3.1.

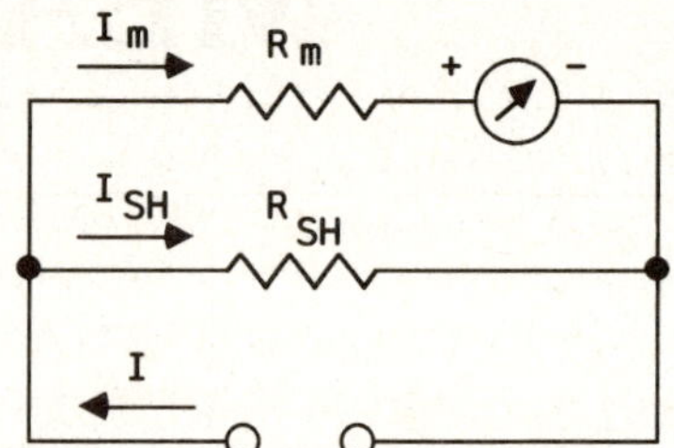

Figure 3.1: Equivalent circuit for the D'Arsonval meter movement as an ammeter

An ammeter is always connected in series with a circuit branch and measures the current flowing in the branch. Real ammeters, as we see from Figure 3.1, always possess some internal resistance, causing the current in the branch to change because of the addition of the meter. By using the model of Figure 3.1, it is possible to calculate the error caused by introducing an ammeter into a circuit.

3.2.2 DC Voltmeter

The D'Arsonval meter movement can be used to make a dc voltmeter by connecting a multiplier resistor R_S in series with the meter movement (see Figure 3.2).

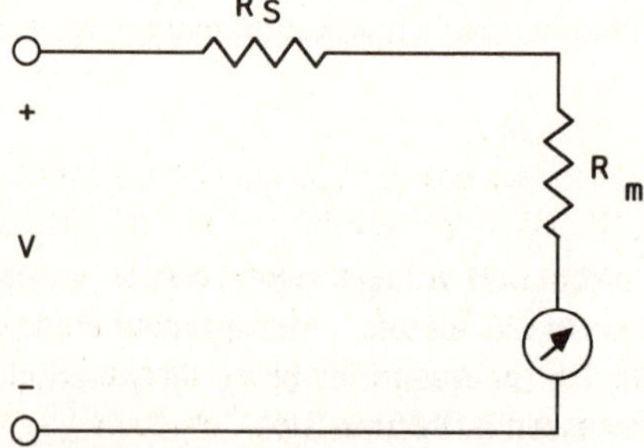

Figure 3.2: DC voltmeter using a D'Arsonval movement

The resistor R_S increases the voltage that can be measured by the meter. The purpose of the multiplier resistance is to extend the voltage range of the meter and to limit current through the D'Arsonval movement to a maximum full-scale deflection current. The value of the multiplier resistor can be found by determining the sensitivity of the meter movement. The voltmeter's sensitivity is usually found by taking the reciprocal of the full-scale deflection current, or

$$S = \frac{1}{I_{fs}} \ (\Omega/V) \tag{3.1}$$

The sensitivity is an indication of how well the behavior of an actual voltmeter approaches that behavior of an ideal voltmeter. An ideal voltmeter would have an infinite Ω/V ratio and would appear as an open circuit to the circuit under test. Typical dc laboratory voltmeters have a 20,000 Ω/V rating.

3.2.3 Voltmeter Loading

When a voltmeter is used to measure the voltage across a circuit component, the voltmeter is placed in parallel with the component. The parallel combination of two resistors is less than either resistor alone. Therefore, the voltage across the component is less whenever the voltmeter is connected. The effect of the voltmeter depends on its sensitivity. Remember, sensitivity is inversely proportional to the current drawn by the voltmeter. That is, the less current drawn, the more the voltmeter appears as an open circuit to the circuit under test. The disturbance caused by current drawn by the voltmeter is called loading.

An example of loading is shown in Figure 3.3.

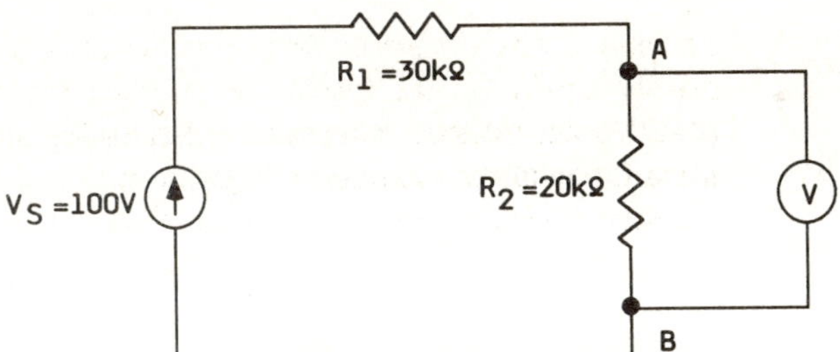

Figure 3.3: Voltmeter loading

The true voltage reading across resistor R_2 is

$$V_T = \frac{V_S}{R_1 + R_2} R_2 = 40 \ V \tag{3.2}$$

Now, measure the voltage with voltmeters. Let voltmeter 1 have a sensitivity of 1,000 Ω/V and voltmeter 2 have a sensitivity of 20,000 Ω/V. Use the 50-V scale. Voltmeter 1 then has an equivalent resistance of 50 kΩ while voltmeter 2 has a equivalent resistance of 1 MΩ. The parallel combination of R_2 and 50 kΩ is approximately 14.3 kΩ, but the parallel combination of R_2 and 1 MΩ is approximately 19.5 kΩ. Then the circuit with the less sensitive meter gives

$$V_{AB} = V_S \ \frac{R_{AB}}{R_1 + R_{AB}} = 32.3 \ V \tag{3.3}$$

The error is

$$error = \frac{40\text{ V}-32.3\text{ V}}{40\text{V}} \times 100\% = 19.2\% \tag{3.4}$$

Using the more sensitive voltmeter gives

$$V_{AB} = 100\frac{19.6\text{ K}}{30\text{ K} + 19.6\text{ K}} = 39.5\text{ V} \tag{3.5}$$

The error for the more sensitive meter is then

$$error = \frac{40\text{ V} - 39.5\text{ V}}{40\text{ V}} \times 100 = 1.25\% \tag{3.6}$$

Clearly, the meter with the higher sensitivity rating will yield the more reliable reading in terms of loading.

Inserting an ammeter in a circuit increases the resistance of the circuit and thus reduces the current. The error caused by the meter is a function of the resistance of the circuit and the ammeter resistance. Ammeter loading can be determined by the ratio

$$\frac{I_m}{I_e} = \frac{R_1}{R_1 + R_m} \tag{3.7}$$

$$\begin{aligned}
\text{where} \quad I_m &= \text{current with the ammeter present} \\
I_e &= \text{current without the ammeter present} \\
R_m &= \text{meter resistance} \\
R_1 &= \text{circuit resistance}
\end{aligned}$$

3.2.4 Ohmmeter

The D'Arsonval meter movement with a battery and a resistor can be used to construct a simple ohmmeter circuit (see Figure 3.4).

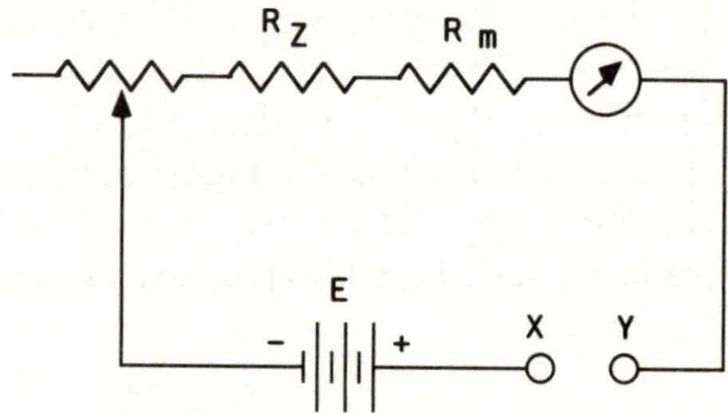

Figure 3.4: Ohmmeter circuit

If points X and Y are connected, the result is a simple series circuit with current through the meter due to the source E. The amplitude of the current is limited by R_Z and R_m. Connecting X and Y allows the user to "zero" the meter before use. The "zero control" on the ohmmeter allows the user to eliminate the effect of the test leads so that the resistance read by the meter is that at the end of the leads. The variable part of resistor R_Z is adjusted to obtain exactly full-scale deflection on the meter movement. The variable part of R_Z is often used as a means of adjusting for battery aging.

Combining the ammeter, voltmeter, and ohmmeter in a single instrument with the proper switching arrangement results in an instrument called the **analog multimeter** or **Volt-Ohm-Milliammeter** (VOM).

This device is capable of measuring voltages (dc and ac), currents (dc), and resistance. For dc and ac measurements, the D'Arsonval movement is used. In the case of ac, a rectifier circuit is added. Resistance is measured using an ohmmeter circuit. The ohmmeter circuit applies the voltage from a battery across a series connection consisting of a known and an unknown resistance. The D'Arsonval movement is used to determine the value of the resistors being tested by measuring the fraction of the battery voltage drop across the known resistor. The various sensitivities and full-scale ranges of the VOM are obtained from the specification sheets and the operating instruction manual for the instrument used.

Among the guidelines for using basic meters and taking account of meter errors are the following:[3]

1. Ammeters should always be connected in series while voltmeters should always be connected in parallel.

2. "Zero" the meter first.

3. DC meters must be connected so that meter terminals are connected to test points with matching polarities.

4. Start all readings by setting the meter on its highest scale. Take the final reading on the scale that yields a deflection that is nearest the full scale.

5. Correct the readings for loading effect.

6. To prevent battery drain in VOMs, have the function selector switch set to a high dc-volts scale.

3.3 Analog AC Meters

AC meters are capable of responding to the average, peak, or rms value of an applied periodic ac signal. It is best if the meter being used responds to the characteristic value of the waveform being sought. Otherwise, a correction factor between the actual reading and the desired value of the characteristic of the signal must be determined.

The D'Arsonval meter movement is frequently used in ac meters, but it cannot be used directly to measure ac current or voltage because it responds to the average value of the current and, of course, the average value of a sine wave is zero. In order to measure ac current with this meter movement, the ac signal must be rectified by a diode rectifier to produce unidirectional current flow. The meter responds to the rectified average, but the meter scale is calibrated in rms for a sine wave. The simplest circuit is shown in Figure 3.5.

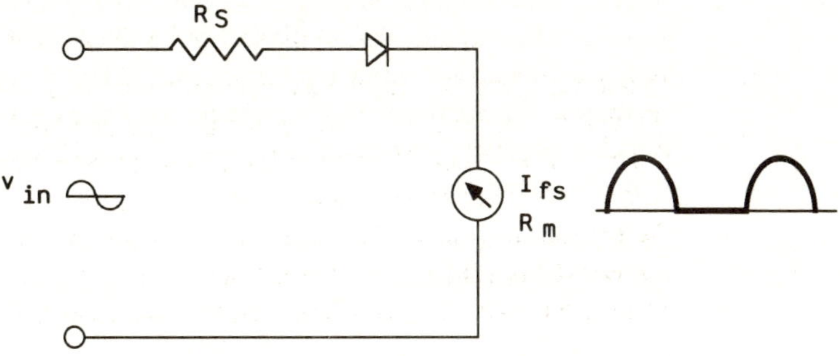

Figure 3.5: Half-wave rectifier ac voltmeter

It can be shown[4] that the multiplier resistor is given by

$$R_S = \frac{0.45V_{in}}{I_{dc}} - R_m \qquad (3.8)$$

And, for a half-wave rectifier, the sensitivity of the meter movement is

$$S_{ac} = 0.45\, S_{dc} = 0.45\frac{1}{I_{fs}} \qquad (3.9)$$

Improvement in sensitivity can be obtained by using a full-wave rather than a half-wave rectifier in ac voltmeters (see Figure 3.6).

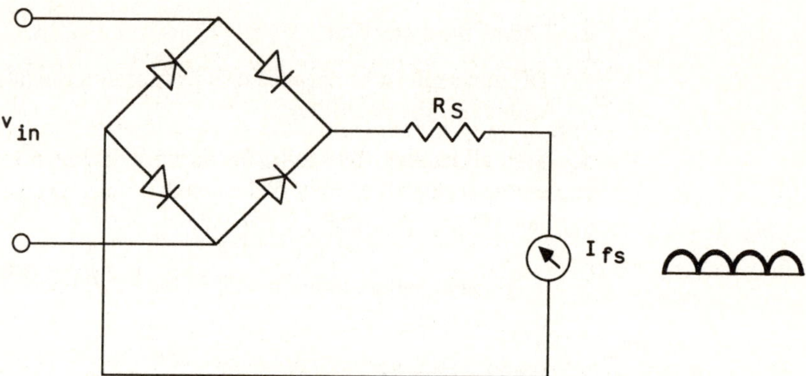

Figure 3.6: AC voltmeter using full-wave rectification

For a full-wave rectifier, the sensitivity obtained is [5]

$$S_{av} = 0.9S_{dc} \qquad (3.10)$$

The sensitivity of dc voltmeters using half-wave or full-wave rectification is lower than the sensitivity of dc voltmeters that do not use either of these techniques. From the concept of sensitivity, the loading effect of an ac voltmeter is greater than that for a dc voltmeter.

Voltmeters using half-wave or full-wave rectification are useful for measuring only sinusoidal ac voltages, and the equations are not valid for nonsinusoidal waveforms such as squares, triangular, or sawtooth waves.

Another source of error in basic ac meters is that a forward bias of 0.6 to 0.7 is required for conduction by the silicon diodes. As a result of this diode voltage drop, only that portion of the source voltage in excess of 0.6-0.7 V is effective in causing a current in the meter. To overcome this drawback, amplifiers can be incorporated into the meter. With the amplifier connected in the meter circuit before the rectifier, ac voltages down to the microvolt range can be measured.

Limitations in ac meters include frequency response limitations due to coil inductance and stray capacitances.

3.4 Digital Meters[6],[7]

Digital meters have a number of advantages over analog meters, including greater speed, increased accuracy and resolution, and increased repeatability or precision. One of the significant features of the digital meter is the readability of the measurement. With this instrument, the user is eliminated as part of the measurement process. Digital meters provide better resolution than their analog counterparts, and this greater resolution reduces the number of ranges required to cover a given range of voltages. The digital meter output can be fed directly to recorders and then processed by computer.

In addition to the full digits that the digital meter displays, an overrange digit is added to allow the user to read beyond the full scale. The overrange digit is called a "1/2" digit because it displays only the digits 0 and 1. The digital meter specification states X and 1/2 digits, with X specifying the number of full digits and the 1/2 digit the overrange digit.

Overranging extends the usefulness of digital multimeters by maintaining resolution up to and beyond the full scale. For example, a four-digit digital voltmeter without overranging would measure a number that changed from 9.999 to 10.015 as 9.999. To read the new value of 10.015 would require a range change, and the number would be read as 10.01. The remaining .005 would be lost. With overranging, the measurement would be 10.015. The half digit for overranging is the leftmost digit in the display.

The most important part of the digital meter is the circuitry that converts the measured analog signal into a digital signal. This circuitry is called the **analog-to-digital (A/D) converter.** Several techniques, including the staircase ramp, successive approximation, the dual slope, voltage-to-frequency conversion and parallel circuitry, implement the conversion. These techniques are explained in detail in Wolf.[8]

The input voltage measured by a digital voltmeter must either be a dc voltage or an ac voltage that is changed by an ac converter circuit into an equivalent dc form, i.e., average, rms, or peak voltage.

The output of the A/D converter is a string of pulses. These pulses are counted for a prescribed time interval by a binary-coded decimal counter. The result of the count represents the value of the input signal reading in digital form. The contents of the counter can then either be transmitted to other digital storage devices or, with some additional processing, be used to drive a display mechanism.[9],[10]

The resolution of the meter indicates the number of digits in the display. The accuracy of a digital voltmeter can be expressed in terms of a combination of constant and proportional errors.[11] Constant errors are errors that remain constant over the full range of the meter. Such errors are expressed in terms of the number of digits or the percentage of the full-scale reading. Proportional errors are proportional to the magnitude of the digital indication and are expressed in terms of a percentage of the reading. As an example, if 7.000 V are measured with a four-digit meter whose accuracy is 0.01% of the reading plus one digit, the maximum error is 0.01% x 7 V + 0.001 V, or a total of 0.0017 V.

Digital meters can load down a circuit. However, the input impedance of digital voltmeters is quite high and usually does not present a loading problem. A rule of thumb is that the input impedance of the digital voltmeter should exceed the measured source impedance by at least a factor of 10^n, where n is the number of digits in the display.[12]

Additional characteristics of the digital voltmeter include a meter reading speed of one reading per second. The range selection for the digital voltmeter is manual or automatic. When a large number of measurements over a wide range needs to be made, the automatic range condition is useful.

If additional circuitry is added to the digital voltmeter, the meter becomes a digital multimeter (DMM) capable of measuring dc and ac voltage, and currents and resistance.[13]

3.5 The Oscilloscope[14],[15],[16]

3.5.1 Introduction

The cathode-ray oscilloscope (CRO) is the most versatile of all the laboratory measuring instruments. The CRO provides a visual display of waveform versus time of the signal under investigation. Interpretation of the display on the cathode-ray screen can reveal the current, time, frequency, and phase difference of the signal under study. The oscilloscope is very well suited for the study of repetitive signals; it is capable of displaying transients as well as random waveforms.

The oscilloscope is capable of measuring amplitudes as low as microvolts with frequencies from dc to the GHz range. Oscilloscopes can be divided into three groups: (a) low-frequency scopes, i.e., those measuring dc to ~1 MHz; (b) high-frequency scopes, i.e., dc to ~50 MHz (many can go even higher); and (c) special purpose scopes, such as sampling and storage oscilloscopes.

3.5.2 Operation

The CRO is made up of the following subsystems:

1. Cathode-ray tube

2. Vertical deflection subsystem

3. Horizontal deflection subsystem (including timing circuit)

4. Power supplies

5. Calibration circuits

6. Probes

A block diagram of the oscilloscope is shown in Figure 3.7.

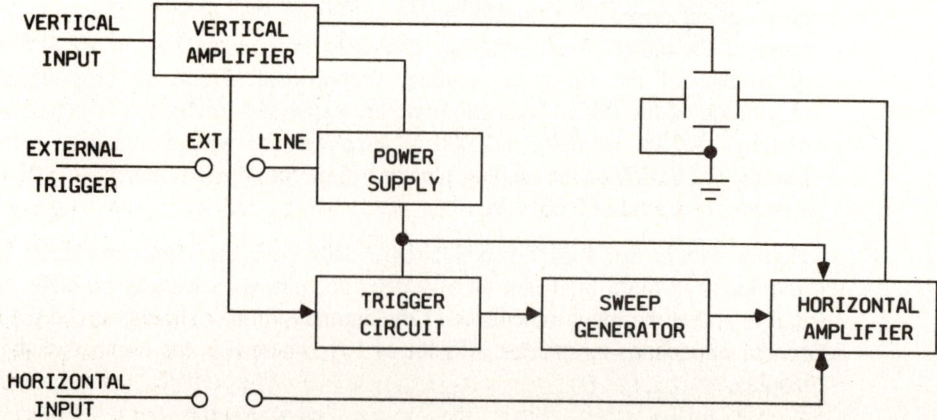

Figure 3.7: Block diagram of a cathode-ray oscilloscope

The signal, which is sensed by the scope probes, can be a voltage or a nonvoltage that is converted to a voltage. It is transmitted through the use of cables to the scope input terminals. If the signal is too small to activate the display subsystem it is amplified by the vertical deflection system and then applied to the vertical deflection plates of the cathode-ray tube (CRT). The electron beam is focused and strikes the fluorescent screen, creating a spot of light. The beam is deflected vertically in proportion to the amplitude of the voltage applied to the CRT vertical deflection plates. The horizontal deflection subsystem sweeps the electron beam horizontally across the screen at a uniform rate. Simultaneous deflection of the electron beam in the vertical and horizontal directions causes the spot of light produced by the electron beam to trace a path across the CRT screen. When the input is periodic and the time-based circuitry properly synchronizes the horizontal sweep with the vertical deflection, the spot of light will trace the same path on the screen repeatedly. For a complete discussion of the various subsections that make up the CRO, see Wolf.

3.5.3 Input Coupling Selector

An input signal may be either a dc signal, an ac signal, or a signal consisting of an ac component superimposed on a dc component (ac + dc signal). Sometimes only the ac component of the (ac + dc) signal is of interest. At other times, both the ac and dc components are to be displayed. The input coupling selector gives the option of which of the signal components will be coupled to the amplifier circuitry for display. The input coupling selector functions as a three-position switch, with the switch positions ac, gnd, and dc. When the dc position is selected, the input terminal connects the entire signal to the vertical deflection system. If the ac position is chosen, a capacitor is placed in series with the input branch to the amplifier. The capacitor blocks dc components from entering the vertical amplifier (see Figure 3.8).[17]

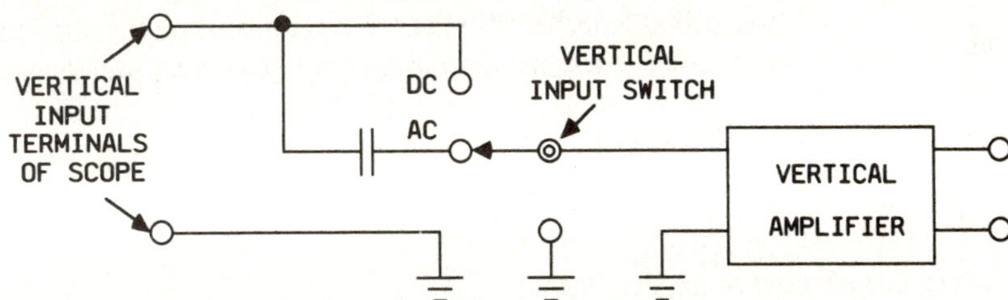

Figure 3.8: Input coupling selector (Reprinted by permission from
Guide to Electronic Measurements and Laboratory Practice,
Second Edition, by Stanley Wolf. Copyright © 1983 by Prentice-Hall.)

The two types of inputs through which a signal can be connected to an oscilloscope are the single-ended input and the differential input.

The single-ended input is the most common form of input. It consists of only one input terminal besides the ground terminal. Only voltage values relative to ground can be measured with a single-ended input, unless the scope is "floated" above earth ground with the use of a three-to-two wire adapter.

A differential input (an input in which the scope is equipped with a differential amplifier) has three terminals, including the ground terminal at each amplifier channel. With a differential input, the voltage between two nongrounded points in a circuit can be measured without floating the scope. The differential amplifier electronically subtracts the voltage levels applied at the two terminals and displays the difference on the screen.

3.5.4 Dual Trace Oscilloscope[18]

Another feature of the CRO is the dual-trace oscilloscope. Here, two signals can be displayed simultaneously on a CRO screen. An obvious use of this device is to display the input and output signals of an amplifier simultaneously. The dual-trace scope produces the dual image by switching the two separate input signals electronically.

3.5.5 Scope Loading[19]

Oscilloscopes, like voltmeters, can load the circuits on which they perform measurements. The input impedance of the scope provides a measure of how much the scope will load a circuit that is being tested. A typical scope equivalent circuit is shown in Figure 3.9.

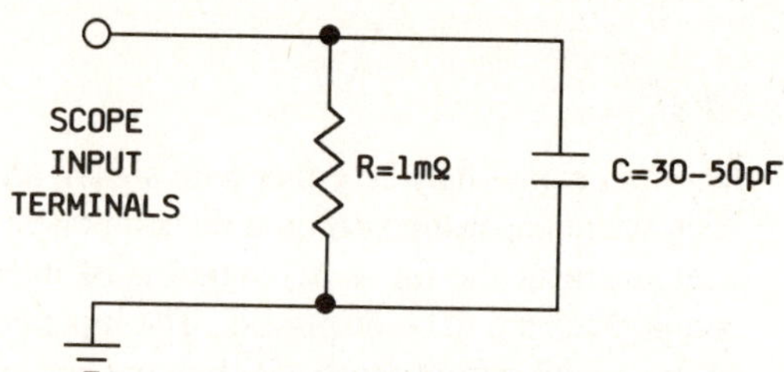

Figure 3.9: Typical scope equivalent circuit (Reprinted by permission from *Guide to Electronic Measurements and Laboratory Practice,* Second Edition, by Stanley Wolf. Copyright © 1983 by Prentice-Hall.)

In measuring a dc voltage, the input impedance is 1 MΩ because the capacitor is an open circuit to dc currents. Therefore, loading effects become important if the test resistance approaches 100 kΩ. When an ac signal is applied to a scope, the capacitive reactance of the scope amplifier affects the input impedance. As the frequency of the input signal increases, the capacitance makes the effective input impedance of the scope decrease.

3.5.6 Making Connections to an Oscilloscope[20]

In testing a circuit with an oscilloscope, the connection to the oscilloscope must be done properly. If the voltage of the point being measured is relative to ground and if the input is single ended, then connect the probe to the test point in the circuit and to the nonground input terminal of the oscilloscope. If the input to the scope is a differential input, then connect the probe to the test point in the circuit and to the plus input terminal of the differential input. Ground the minus terminal of the input by setting its input coupling switch to the ground position or by ensuring that the grounding strap to that terminal is connected.

If the voltage being measured is between two nongrounded points in a circuit and if a differential input is available, connect one point to the plus input terminal of the differential input and the other to the minus input. (Each voltage input is referred to ground.) The scope electronically subtracts one voltage from the other and displays the difference.

3.6 Time and Frequency Measurements[21]

One method of measuring time is through the use of electronic timers, a principle which involves the charging of a capacitor in an RC circuit. It takes a finite time to charge a capacitor when a dc voltage is applied across a series RC connection. We can define the time it takes for the voltage across the capacitor to reach 63% of its full charge (see Figure 3.10) as

$$\tau(s) = R(\Omega) \times C(\text{farads}) \tag{3.10}$$

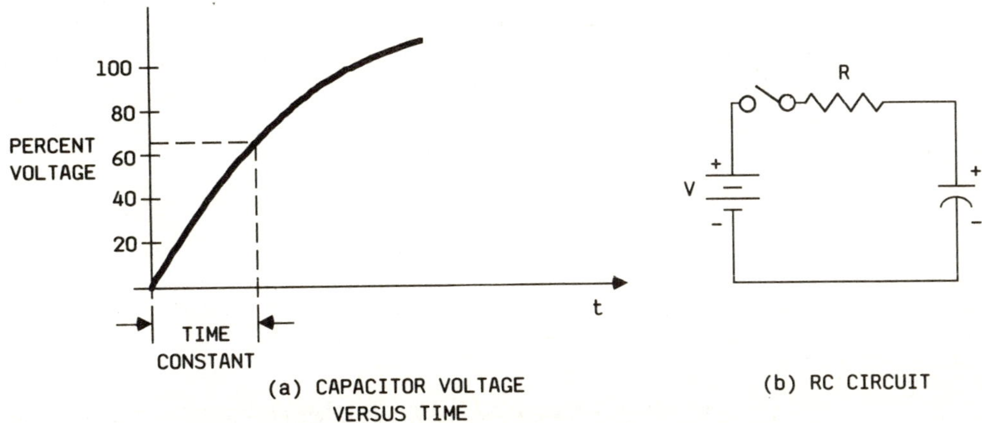

Figure 3.10: RC circuit as a timing circuit

Frequency measurements can be made using a number of instruments. The oscilloscope can measure frequencies by using the triggered sweep and the time base, or Lissajous pattern method, but the accuracy is not always acceptable.[21]

The digital frequency counter is the most accurate instrument for measuring unknown frequencies. The counter can measure frequencies from dc to the mega- or gigahertz range. Its subsections include the digital counting and display assembly, the time-based generator, a pulse-forming circuit such as a Schmitt trigger, and a gate circuit.

An instrument similar to a digital frequency counter is a universal time counter. This instrument measures both frequency and time intervals. It can also be used to measure low frequencies accurately. Typically, a universal time counter has an input impedance of 1 MΩ in parallel with 25 pF. The input sensitivity is approximately 100 mV.

The frequency composition or frequency spectrum of a signal can be analyzed through the use of a spectrum analyzer, a Fourier analyzer, or a wave analyzer. The spectrum analyzer is a real-time instrument that simultaneously displays the amplitudes of all signals in the frequency range of the analyzer. The instrument operates by using a filter that rejects all but a very narrow band of frequencies. The central frequency of the narrow band of the filter is swept over the range of interest by the spectrum analyzer. The frequency components of the signal being analyzed are passed through to the display only when their frequency matches that of the swept-filter frequency. A cathode-ray tube is used as the display device and shows the amplitude of each harmonic of the signal versus frequency over the swept frequency range. Spectrum analyzers can display signals over the range from 5 Hz to 40 GHz.

The Fourier analyzer uses digital signal processing techniques to provide measurements that are an improvement over the capabilities of the spectrum analyzer. The Fourier analyzer is capable of measuring very low frequencies or very closely spaced signals, as well as random signals obscured by noise. The instrument is based on the calculation of the discrete Fourier transformer using an algorithm called the Fast

Fourier Transform (FFT). The FFT calculates the amplitude and phase of each signal component from a set of time domain samples of the input signal. Because the analysis of the signal can be performed in a simultaneous parallel procedure, the frequency spectrum of the signal can be displayed very rapidly. The results can then be displayed on a CRT. The Fourier analyzer is basically a digital device; therefore, it can be interfaced to a computer rather easily.

3.7 Resistance Measurement[22]

The ohmmeter and voltmeter-ammeter methods are used for rapid but rough estimates of resistance values. The voltmeter-ammeter method is a simple method for rapidly determining resistance when only voltmeters and ammeters are available. The use of a high-input impedance digital voltmeter and digital ammeter will reduce the loading effect (see Figure 3.11).

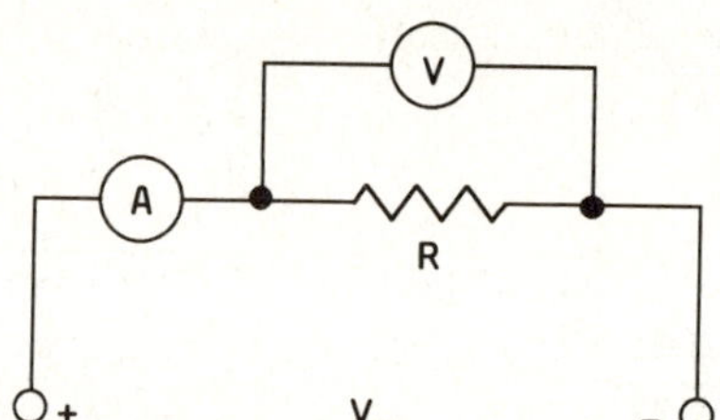

Figure 3.11: Resistance measurement
using the voltmeter-ammeter method

An ohmmeter applies the fixed voltage of a battery across two resistors in series. One resistor is known, and the other is the resistor being tested. The voltage across the known resistor is measured by a dc voltmeter. This, however, is calibrated to indicate the value of the test resistor. Ohmmeters are capable of measuring resistance values from milliohms to 50 MΩ. The accuracy of an ohmmeter is about ±2%, so ohmmeters are generally not used in applications where high accuracy is needed. Digital ohmmeters eliminate the reading errors associated with analog meter scales. Also, precision-regulated power supplies aid the accuracy of the ohmmeter.

Some tips for using the ohmmeter include the following:[23]

1. If autoranging is not being used and the resistance is completely unknown, use the highest resistance scale.

2. "Zero" the meter. For an ohmmeter, this is equivalent to full-scale deflection.

3. Make sure that the circuit power to the resistor is disconnected. Put the probes across the resistor being tested and switch to a scale that gives half-scale deflection.

4. Even if the power is off, if the resistor is part of a circuit that contains capacitors, make sure that the capacitors have been discharged.

5. The ohmmeter contains a dc power supply that could damage voltage-sensitive circuit components.

6. In-circuit resistance measurements indicate the parallel combination of the resistor and all other dc conducting paths in parallel.

The Wheatstone bridge is used when a very accurate determination of a resistance value is necessary. The principle for this technique is that no current will flow through the D'Arsonval galvanometer if there is no potential difference across it. When no current flows, the bridge is balanced (see Figure 3.12).

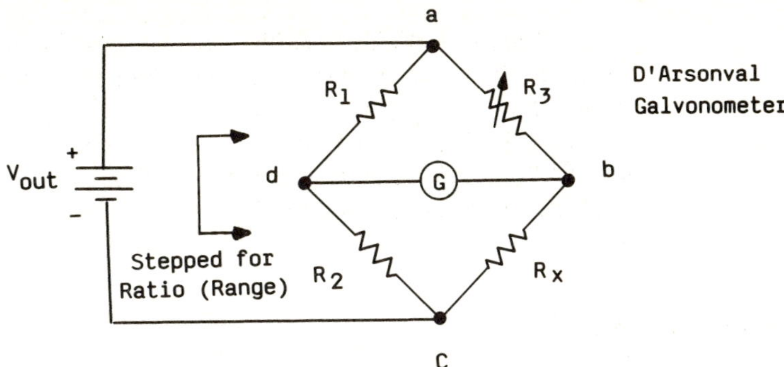

Figure 3.12: Wheatstone bridge circuit

When balanced is achieved, the voltage V_{out} is divided in the path adc by resistors R_1 and R_2 in the same ratio as it is in the path abc by resistors R_3 and R_x. Since points d and b are at the same potential, no current flows through the galvanometer, which results in the equation

$$\frac{R_x}{R_3} = \frac{R_2}{R_1} \tag{3.12}$$

Since R_1, R_2, and R_3 are known, R_x can be found.

The Kelvin bridge method is a procedure used to determine the value of very small resistances, i.e., below 1 Ω. For very small resistance values, the contact resistance and the resistance of the connecting leads are all the same order of magnitude. The Kelvin bridge is able to separate out the contact resistance and the wiring resistance so that the unknown resistance can be determined.[24]

For very high resistance measurements, a megohm bridge is used. This bridge employs a high applied voltage (~1,000 V) and has high sensitivity to avoid circuit loading. This bridge is capable of measuring resistances up to $10^{15}\Omega$ with accuracies of 0.1 to 1.0%.

The megohmmeter can also measure very high resistances, but the accuracy of this instrument is only 3 to 10%. Its advantage is that it is easier to use and faster than a megohm bridge. The megohmmeter functions by applying a very high voltage across an unknown resistance. The voltage is measured across a high resistance standard using a high input impedance voltmeter.

3.8 Capacitance, Inductance, and Impedance Measurements[25]

Capacitance (C) and inductance (L) measurements can be made using bridge-type instruments or special impedance meters. However, direct inductance- and capacitance-indicating instruments are being used more frequently because balancing bridges is difficult, balancing takes time, and the bridges are limited in frequency.

If the resistances of the Wheatstone bridge are replaced by complex impedances (Figure 3.13) and an ac voltage is applied, the balance equation is

$$Z_x Z_1 = Z_2 Z_3 \tag{3.13}$$
$$\text{where } Z = R + jX$$

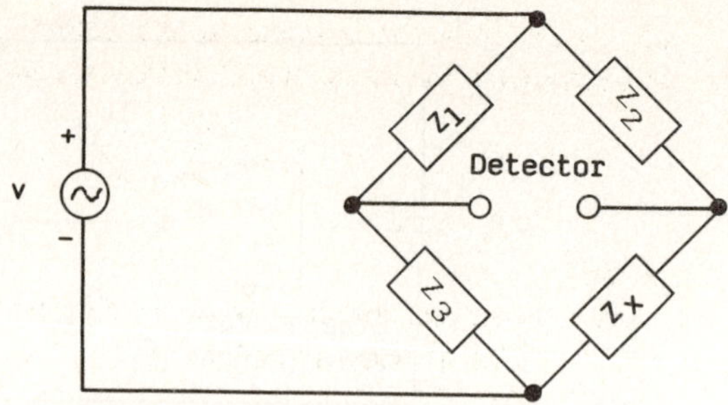

Figure 3.13: AC impedance bridge circuit

For a null condition to be achieved,

$$Z_X = R_X + jX_x = \text{Re}\left[\frac{Z_2 Z_3}{Z_1}\right] + I_m\left[\frac{Z_2 Z_3}{Z_1}\right] \tag{3.14}$$

where Re and Im stand for the real and imaginary parts of Z_x, respectively.

A universal impedance bridge[26] is designed to measure R, L, and C over wide ranges. This type of instrument has five or six built-in bridge circuits. Usually these bridges include a Wheatstone bridge, the series and parallel capacitance bridges, the Maxwell bridge, and the Hay bridge. Both internal fixed-frequency signals and external ac sources are included in the universal impedance bridge.

Digital capacitance meters are also available. These meters determine capacitance by measuring the discharge time of a capacitor through a known resistor. The unknown capacitor is connected to the meter and is charged to a known voltage. When the reference voltage is reached, the capacitor discharges through a known resistor R to a second voltage value, and the discharge time is measured. If the time constant of RC_X is large compared to the discharge time measured, the capacitance is found by dividing the change in the charge on the capacitor by the change in the voltage.

Quick but less accurate (>10% error) capacitance measurements can be made using a high-impedance ac voltmeter.[27] The unknown capacitor (usually >0.001 µF) is connected in series with a resistor, then connected across the power line 115 V, 60 Hz (see Figure 3.14). The result is

$$C = \frac{I}{2\pi f V_C} \tag{3.15}$$

where
V_C is the rms voltage across the capacitor
$I = V_R/R$ is the rms current through the resistor
$f = 60$ Hz

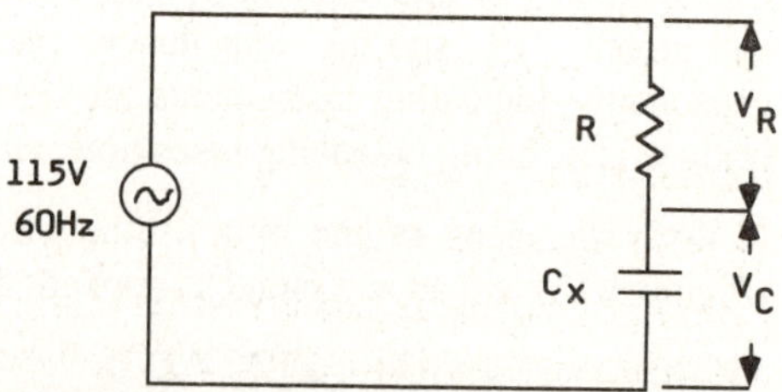

Figure 3.14: AC voltmeter measurement of C_X

Quick measurements of inductance can be made in a manner similar to that of determining capacitance. However, in the inductance procedure, a variable resistance is placed in series with the inductor and the resistor value is varied until the voltage across the inductor is equal to the voltage across the resistor. At this point, the impedances of the inductor and resistor are equal; thus, the inductance is found to be

$$L = \frac{\sqrt{R^2 - r^2}}{2\pi f} \tag{3.16}$$

where

    R   is the resistor value when the circuit is balanced
    r   is the dc resistance of the coil
    f   is 60 Hz

The quality factor Q of an inductor can be measured with a Q-meter. A continuously variable oscillator with a low-output impedance acts as a source. Q-meters also contain a high-impedance voltmeter and a variable capacitor. The scale of the voltmeter is calibrated to indicate Q directly.

3.9 AC Signal Generators[28]

In this section, ac signal sources such as oscillators, and sweep-frequency, pulse, and function generators are discussed.

An oscillator is an instrument that generates sinusoidal output signals. Functionally, it is an amplifier with positive feedback. The circuit model for the oscillator can be represented by an ideal ac voltage source in series with an impedance, often a resistor R_G (see Figure 3.15).

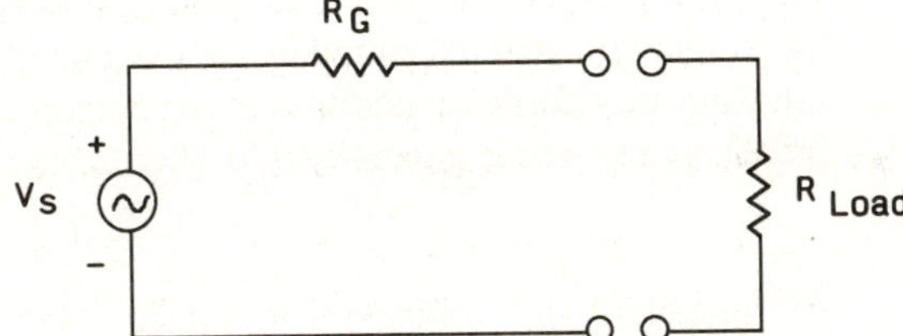

Figure 3.15: Circuit model of an oscillator

For an audio frequency oscillator, R_G is usually 600 Ω, while for radio frequency, signal generators, and pulse generators the value of R_G is usually 50 Ω. To transfer the maximum power from the generator to the load, the impedances of the generator and the load must be matched.

Sweep-frequency generators are instruments that provide a sine wave output whose frequency is automatically swept between two chosen frequencies. The rate at which the frequency is varied can be linear or logarithmic. The amplitude stays constant over the entire frequency range of the sweep. Sweep-frequency generators are basically used to measure the responses of amplifiers, filters, and electrical components over some frequency bands.

The RF oscillator has several operating ranges, with the frequency of the signal generator's output signal varied by either a mechanical or electronic process. In the electronically tuned oscillator, the frequency of the main oscillator is kept fixed, but a second oscillator is used to provide a varying frequency signal. The second oscillator contains an element whose capacitance value depends on the voltage applied across it and is called a voltage-controlled oscillator (VCO). The output of the VCO is combined with the output of the RF oscillator in the mixer. The output of the mixer is a sine wave whose frequency depends on the difference between the frequencies of the two applied signals.

Pulse generators are instruments that produce a periodic train of equal amplitude pulses (see Figure 3.16). The duration of the *on* time of a pulse may be independent of the time between pulses. The **duty-cycle** is defined as the ratio of the pulse width to the pulse period (in percentage) or

$$\text{Duty Cycle} = \frac{\text{pulse width}}{\text{pulse period}} \times 100 \qquad (3.17)$$

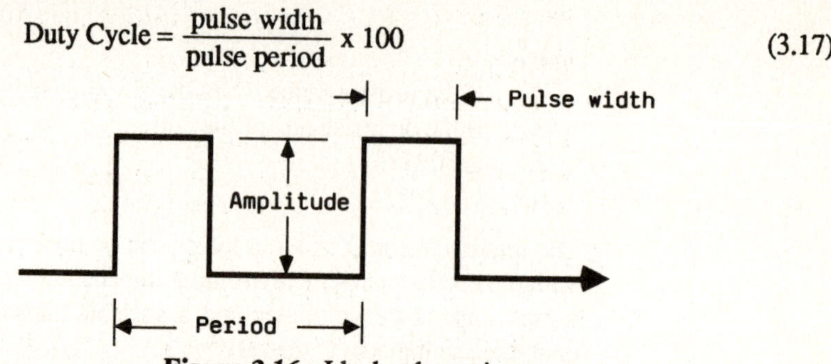

Figure 3.16: Ideal pulse train

A square wave has a 50% duty cycle, but the duty cycle of a pulse generally varies from 5% to 95%.

The generator impedance of pulse generators is 50 Ω. This value allows pulse generators to be matched to the cables that transmit the pulses. Any mismatching results in part of the pulse being reflected back to the pulse generator. This reflected signal is completely absorbed by the generator.

A function generator is a signal source that has the capability of producing several different types of waveform as its output signal. Most function generators can generate sine waves, square waves, and triangular waves over a wide range of frequencies. Some function generator models are capable of generating pulse, ramp, trapezoidal, or sawtooth waveforms in addition to the common waveforms. The frequency range of a function generator is generally 0.01 Hz to about 10 to 20 MHz.

3.10 Summary

All ammeters and voltmeters introduce some impedance into a circuit being tested because of the instruments' loading effect. The loading effect may be reduced by using a meter with a high sensitivity rating.

For analog ac measurements, the D'Arsonval meter movement, which is the heart of the analog dc meter, can be used with the addition of either a half-wave or a full-wave rectifier.

The digital electronic meter indicates the quantity being measured by a numerical display rather than a pointer and scale. Digital meters are more accurate and more convenient than analog meters and present less loading to the circuit being tested.

The oscilloscope is a versatile instrument that is capable of displaying or measuring a wide variety of signals. The scope consists of a number of subsystems, including the cathode-ray tube, the horizontal and vertical deflection sections, power supplies, calibration circuits, and probes.

Signal generators include oscillators, function generators, pulse generators, and sweep-frequency generators. An oscillator is an instrument that generates sinusoidal output signals. Function generators provide sine, square, and triangular waveforms. Pulse generators produce a rectangular output waveform that can be varied both in frequency and duty cycle. Sweep-frequency generators can provide a sine wave output that can be varied over an entire frequency band.

Time measurements can be made using electronic timers, while frequency measurements can be made using a variety of instruments. These include the oscilloscope, the digital frequency counter, and the universal timer counter. The frequency spectrum of a signal can be analyzed by a number of instruments, including the spectrum analyzer, the Fourier analyzer, and the wave analyzer.

The ohmmeter and Wheatsone bridge are two instruments used to make resistor measurements. Capacitance and inductance measurements are made using bridge circuits. Digital meters that determine capacitance by measuring the discharge time of a capacitor through a known resistor are commercially available. Capacitors and inductors can also be measured using an ac voltmeter.

A vector impedance meter is an instrument that can measure the magnitude and phase of an impedance at the frequency of interest.

In all measurements, the frequency and amplitude characteristics of the instruments must be considered as they relate to the circuit or component being tested.

REFERENCES

[1]   Stanley Wolf, *Guide to Electronic Measurements and Laboratory Practice*, 2d ed. (Englewood Cliffs, NJ: Prentice-Hall, 1983).

[2]   Larry Jones and A. Foster Chin, *Electronic Instruments and Measurements*, (New York: John Wiley & Sons, 1983).

[3]   Wolf, pp. 97-8.

[4]   Jones and Chin, pp. 56-7.

[5]   Jones and Chin, p. 61.

[6]   Wolf, Chapter 5.

[7]   Jones and Chin, Chapter 13.

[8]   Wolf, pp. 106-14.

[9]   Wolf, pp. 114-5.

[10]   Jones and Chin, pp. 309-13.

[11]   Wolf, pp. 117-18.

[12]   Wolf, p. 118.

[13]   Wolf, pp. 119-20.

[14]   Wolf, Chapter 6.

[15]   Jones and Chin, Chapter 6.

[16]   *The XYZs of Using an Oscilloscope* (2215A), Tektronix Education Products, Request 46-AX-47588-2.

[17]   Wolf, pp. 133-4.

[18]   Wolf, pp. 135-7.

[19]   Wolf, pp. 145-7.

[20]   Wolf, pp. 153-4.

[21]   Wolf, Chapter 8.

[22]   Wolf, Chapter 10.

[23]   Wolf, p. 247.

[24]   Wolf, Chapter 11.

[25]   Wolf, p. 282.

[26]   Wolf, pp. 285-6.

[27]   Wolf, pp. 288-9.

[28]   Wolf, Chapter 13.

Chapter 4

LABORATORY PRACTICES

4.1 Introduction

Many of the laboratory techniques and strategies a student learns are through experience or trial and error. This experience is indirect or intuitive, unlike the formal training of the classroom. But laboratory techniques are -- for example, Ohm's law -- the result of the application of physical principles. These principles may not be evident to the student, but they work.

What we hope to accomplish in this chapter is to give some idea of the basic physical principles underlying good laboratory techniques. The techniques or topics covered include electrical safety, both for students and equipment, grounding techniques, troubleshooting, input and output impedance, power supply decoupling, and circuit layout techniques.

It is impossible to cover all of the electrical techniques that can be used in the laboratory. What we hope to cover are the basic problems the student might encounter in the laboratory, based on our experience.

Many students will not as yet have the proper background to understand much of the material in this chapter. However, the topics covered are too important or useful to be left for the end of the text. This chapter attempts to sensitize the reader to the importance of such topics as safety, circuit layout, proper grounding, and troubleshooting. As you use the laboratory sections of the text, refer to this chapter.

Hazards in the laboratory environment include electric shock, moving machinery, and hot soldering irons.

Use common sense when dealing with power tools and soldering irons. Return the iron to its holder between soldering operations, and turn the iron off when you are finished using it. Wear proper protection, including eye protection, when using power tools. Take proper precautions with hair and clothing: long hair and loose clothing can get caught in power machinery.

The severity of electric shock is a function of the amount of <u>current</u> that flows through the human body. Table 4.1 documents the effects of various current levels on the human body.[1]

Current intensity, 1-second contact	Effect
1 milliampere	Threshold of perception.
5 milliamperes	Accepted as maximum harmless current intensity.
10-20 milliamperes	"Let-go" current before sustained muscular contraction occurs.
50 milliamperes	Pain. Possible fainting, exhaustion, mechanical injury. Heart and respiratory functions continue.
100-300 milliamperes	Ventricular fibrillation will start, but respiratory center remains intact.
6 amperes	Sustained myocardial contraction followed by normal heart rhythm. Temporary respiratory paralysis. Burns if current density is high.

Table 4.1: Effects of 60-Hz electric shock current
on an average human through the body trunk
(© 1990 Hewlett-Packard Company. Reproduced with permission)

From the table, the threshold of perception of current is about 1 ma. At that level of current, the sensation is an unpleasant heating at the point of contact. At currents above 10 ma, the victim loses the ability to control muscles, so, even though the pain is severe, the victim is unable to let go of the electrical conductor. The lethal current range is between 100 and 300 ma.

Wet skin has a far lower resistance than dry skin; therefore, a given voltage allows higher current through the wet skin. In addition, body fluids offer less resistance to current conduction, so once current enters the body, a low resistance path is encountered; further skin resistance breaks down with current conduction. These facts make it important to work around potentially dangerous electrical conditions with dry hands and to break contact with the conductor as rapidly as possible before skin resistance has a chance to break down.

The following safety rules should be observed:

1. Never work alone.

2. Use three-wire power grounded power cords.

3. Three-phase circuits use four wires and one ground.

4. Always shut off power before handling live conductors.

5. Look for damaged cords and do not use until repaired.

6. Do not become a low-resistance path to ground. Wear dry shoes and avoid standing on conducting flooring. Avoid handling conducting wiring.

7. Connect the "hot" side of a conductor last. Use control switches or circuit breakers to energize systems when they are ready for testing.

4.3 Grounding

4.3.1 Safety Ground

Grounding is an important consideration in safety, and it is also one of the important ways to minimize noise. In Figure 4.1, an instrument's metal chassis that is not connected to ground is shown.[2]

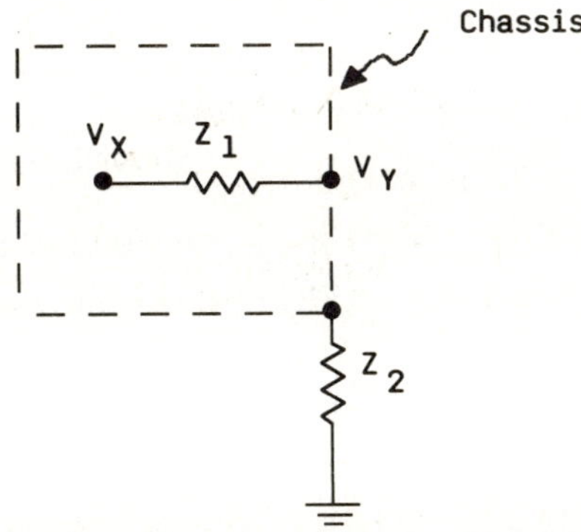

Figure 4.1: Ungrounded chassis (Reprinted from *Noise Reduction Techniques in Electronic Systems* by H. W. Ott. Copyright © 1976 by John Wiley & Sons, Inc.)

If an uninsulated wire comes into contact with the ungrounded chassis represented by Z_1, a voltage V_Y is placed on the surface of the chassis. There then exists a voltage drop of V_Y between the chassis and ground. The size of V_Y is a function of the relative values of Z_1 and Z_2, where Z_2 is a high impedance from the chassis to ground. The voltage could be relatively high and be a potential shock hazard for the user. If the chassis is grounded, however, its potential is zero since Z_2 becomes zero.

4.3.2 Power Ground

Figure 4.2 shows the standard three-wire system that is normally used for electrical equipment.

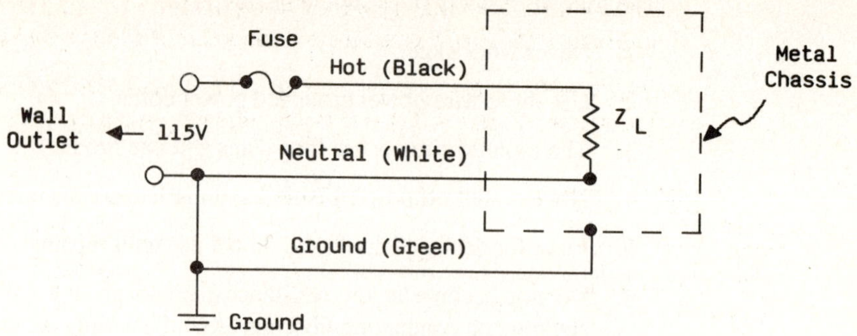

Figure 4.2: Three-lead distribution system

Current flows through the black wire and the fuse and returns through the white wire. The green wire is connected between the chassis and ground, and normally, current does not flow through this wire. The neutral and ground wires connect at only one point so that none of the neutral wire's current flows through the ground wire. A four-wire 115/230-V system is shown in Figure 4.3.

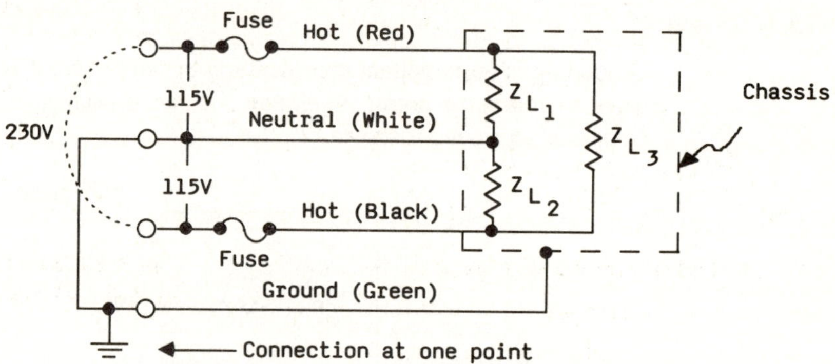

Figure 4.3: Four-wire power distribution system

4.3.3 Signal Ground

There are a number of ways that ground connections can be made.

For all methods shown in Figure 4.4, the conductors have a finite impedance consisting of resistances and inductances, and there is a voltage drop between two physically separated ground points. For power grounds, the voltage between two points can range from several hundred millivolts to several volts. Small signal voltages are only in the millivolt range, so the power ground and the signal ground should be kept separate. However, for safety reasons, the two grounds should be connected at one point.

The easiest ground system to implement is shown in Figure 4.4(a). But this is the least desirable system, because large ground currents can flow and thus produce a noisy ground system. A ground loop is obtained by grounding circuit 3 at point (2) in the figure.

The ground system of Figure 4.4(b) is a good system at low frequencies because there is no cross coupling between ground currents from different circuits. However, this system can be complicated to implement. At higher frequencies, there is inductive and capacitive coupling between the grounds, as shown in Figure 4.4(e).

The ground system of Figure 4.4(c) is used at high frequencies to reduce the ground impedance. Here, circuits are connected to the nearest ground plane, usually the chassis. To keep impedances low, the ground leads should be kept as short as possible.

This type of grounding system is not good for low frequencies, since all of the ground currents from all of the circuits will flow through the ground plane (see Figure 4.4(d)).

Figure 4.4(c) shows a parallel ground system that provides good low frequency grounding. Figure 4.4(f) shows a multipoint ground system used at high frequencies.

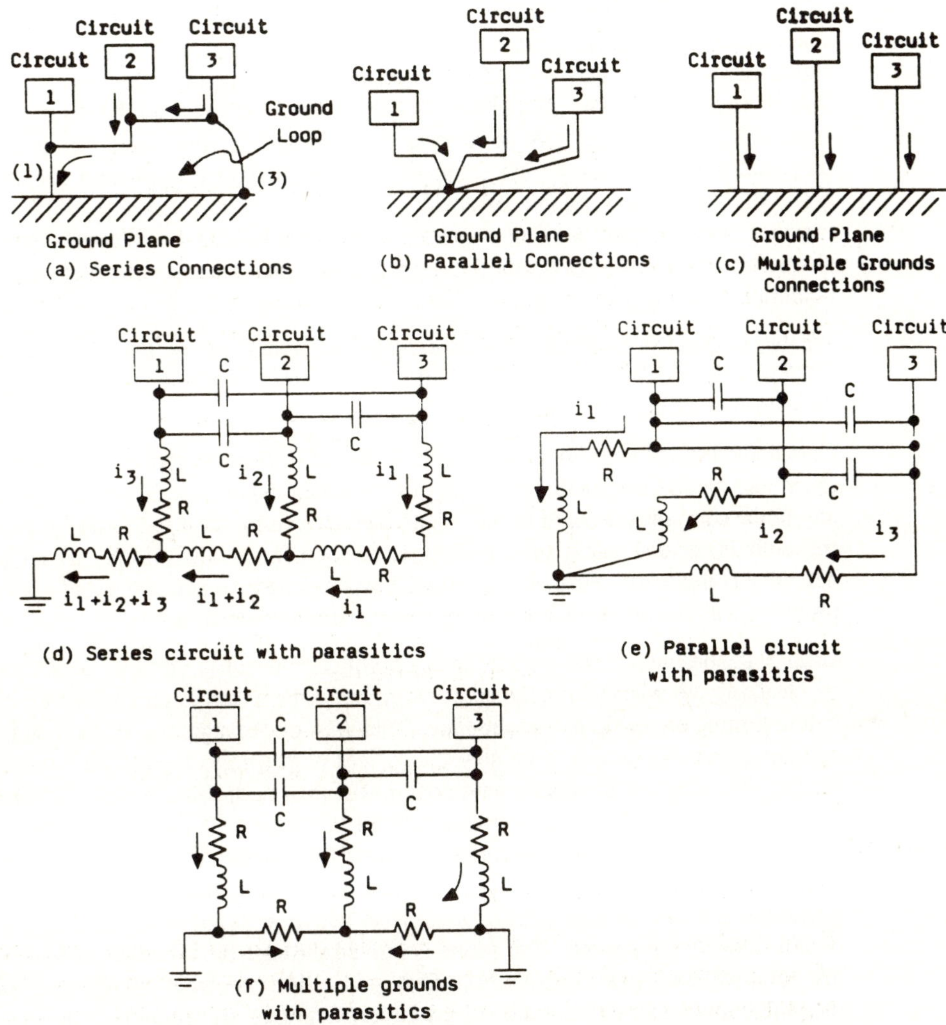

Figure 4.4: Grounding methods (Reprinted by permission from *Noise Reduction Techniques in Electronic Systems* by H. W. Ott. Copyright © 1976 by John Wiley & Sons, Inc.)

4.4 Troubleshooting

It has been our observation that students, particularly in introductory laboratory classes, will put an experiment together and then find that their setup doesn't work. Then, either they complain that they did it right, but it doesn't work, or they will completely rebuild the experiment. What students do not understand is that often experiments, circuit designs, or integrated circuit layouts don't work the first time and that patience and good troubleshooting techniques are needed to determine what went wrong. First, we briefly discuss some principles of a good circuit layout.

Neatly drawn circuit diagrams can minimize errors that occur when constructing and analyzing circuits. The circuit diagram can be converted to a wiring diagram to show how the components are actually arranged on the circuit board. Sometimes it is possible to insert components in a circuit so that their placement matches their location in the circuit diagram. This arrangement is helpful in trouble-shooting wiring errors when a circuit isn't working.

The circuit should have proper power supply decoupling, and if bypass elecrolytic capacitors are used, they should be bypassed with small high-frequency capacitors. Where necessary, there should be separate signal, noise, and hardware grounds. Ground leads should be kept as short as possible, and circuits should be grounded at only one point, except at high frequencies. The student should take care to avoid ground loops or questionable or accidental grounds. Noisy leads should be twisted together, and all leads leaving a noisy environment should be filtered. The student should also be aware of whether the test equipment being used is floating or at ground potential.

We begin our troubleshooting with a preliminary analysis of the symptoms of the problem and possible causes. Tests and measurements should then be made, to isolate the incorrect wiring or the defective component or device.

One method of locating faults is through the use of the human senses, i.e., sight, sound, touch and smell. For example, do any of the components feel hot to the touch or do any of the components smell burnt? Can you hear humming or high voltage arcing? If the fault is located using this method, it is not a good idea simply to replace the defective component. Further inspection should be made to detect the real cause of the problem, which may be another component or circuit connections.

Usually the problem is more subtle than a burnt component. Start your troubleshooting by checking the wiring to make sure that it conforms to the circuit or wiring diagram before turning on the dc power supplies. Then turn on the supplies. If the circuit is not operating properly, check to be sure that power is getting to the circuit by measuring the supply voltage at the component node rather than at the supply node. There could be a defective wire from the power supply to the circuit. If the supplies are present, then trace the signal voltage through the circuit. A basic troubleshooting method is signal tracing with the use of an oscilloscope. This procedure is used to measure quantities such as voltage gain to determine where a signal is blocked or to check circuit frequency response. The probe is moved through the circuit from test point to test point while a signal is applied at a fixed point. The procedure allows you to isolate the problem in a particular section of the circuit. Another procedure is to inject a signal from point to point, keeping the test point fixed. It is, of course, important to have a circuit diagram or system schematic in front of you with the expected results while the circuit is being tested. If an abnormal or distorted waveform is found, it should be analyzed to determine its amplitude, duration, phase, and shape. This type of evaluation will help in determining where in the circuit the fault occurred.

For dc circuits, or after performing waveform analysis, dc voltage measurements should be made with the circuit dc biased but with no applied signal. The object is to check node voltages against the expected values calculated from circuit analysis and listed on a circuit schematic or data sheet. It is good practice to check the nodes with the highest voltage first.

Resistance measurements are an aid in detecting short and open circuits and forward and reverse states for diodes and transistors. These measurements are made with the power supplies turned off. If shunting conditions exist -- that is, if two or more circuit components are in parallel -- one terminal of the component being tested should be disconnected from the circuit to isolate it from the rest of the circuit. Any capacitors in the circuit should be discharged to protect the ohmmeter.

4.5 Input and Output Impedance

It is often useful to know the input and output impedance of an electrical circuit, device, or instrument so that the effects of loading can be determined. The input impedance can be found by connecting a voltage source across the input terminals and measuring the current that flows for a particular voltage setting (see Figure 4.5).

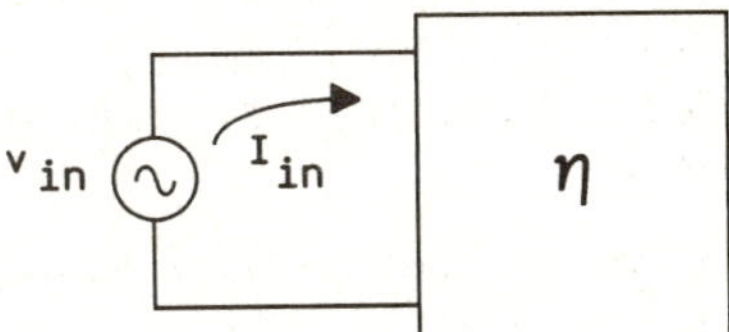

Figure 4.5: Input impedance measurement

In this arrangement, the input impedance is given by

$$Z_{in} = V_{in} / I_{in} \tag{4.1}$$

where V_{in} is the voltage setting and I_{in} is the measured current. The value of Z_{out} can be determined by replacing any external input signal source by its Thevenin equivalent resistance and applying a test voltage V_{test} to the output terminals of the circuit. The current corresponding to V_{test} is determined, and the ratio of V_{test}/I_{test} is defined as Z_{out} (see Figure 4.6). That is,

$$Z_{out} = V_{test} / I_{test} \tag{4.2}$$

Figure 4.7 illustrates the technique for determining the input and output impedances when a circuit contains active devices.

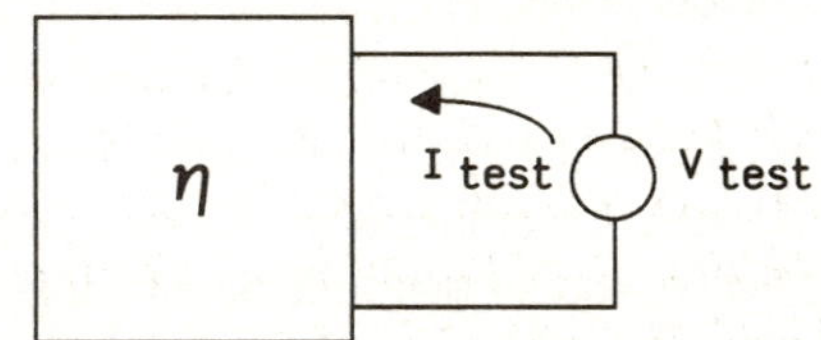

Figure 4.6: Output impedance measurement

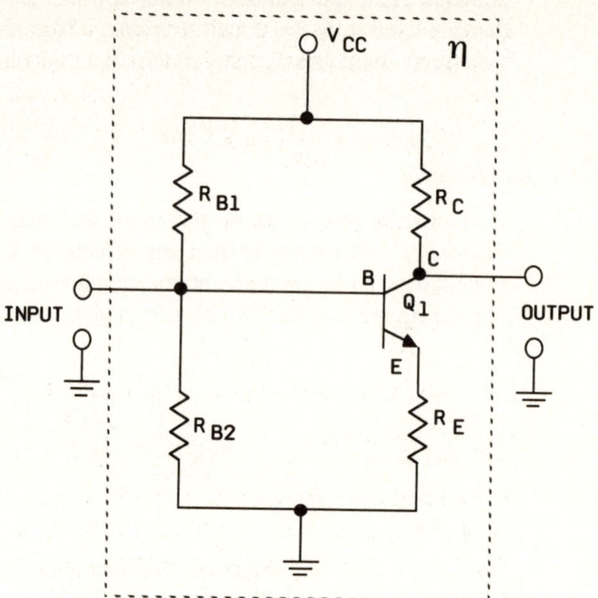

Figure 4.7: Example for input and output impedance measurements

Input Impedance

The transistor requires specific dc voltages applied to the collector, base, and emitter (labeled C, B, and E, respectively) to bias it in the active region and to have it operate linearly as an amplifier. The impedance to be measured is the small signal input impedance. To measure such a small impedance, the applied ac voltage v_{in} should be no larger than several hundred millivolts. As a rule of thumb, the applied peak-to-peak voltage should be less than 10% of the dc bias voltage. In order to avoid altering the dc voltages, the applied voltage should be connected to the network through a capacitor (see Figure 4.8).

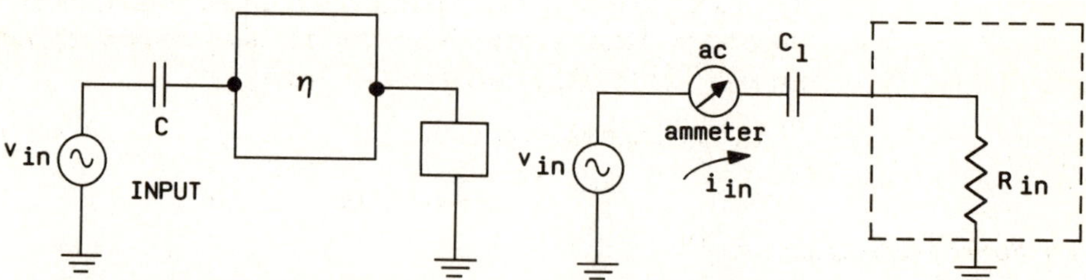

Figure 4.8: AC input impedance measurement

Figure 4.9: Model for input impedance of network η in Figure 4.8

The value of the capacitor is set up so that it looks like a short circuit at the frequency of interest. Experimentally, this set up can be confirmed by setting the frequency of the applied voltage to twice the initial value and determining that the impedance does not change. If it does not change, the capacitor must look like a short circuit at both frequencies. Note that the impedance of a capacitor is $1/j\omega C$.

In order that the ratio $|v_{in} / i_{in}|$ resemble R_{in} and not be affected by C_1 (as far as magnitude goes), the frequency of v_{in} must be greater than the RC time constant, Figure 4.9, by at least a factor of 10.

R_{in} is found by increasing the frequency of v_{in}; the amplitude of i_{in} will increase until the frequency of v_{IN} is at least a factor of 10 above the RC time constant. At this point i_{in} will remain constant, and the ratio of $|\,v_{in}/i_{in}\,|$ will give R_{in}. Mathematically,

$$|\,Z_{in}\,| = |\,\frac{1}{j\omega C_1} + R_{in}\,| = |\,\sqrt{\frac{1+(\omega C_1 R_{in})^2}{(\omega C_1)^2}}\,| \tag{4.3}$$

and for $\omega > 10/R_{in}C_1$,

$$|\,Z_{in}\,| \approx |\,\sqrt{\frac{(\omega C_1 R_{in})^2}{(\omega C_1)^2}}\,| = R_{in} \tag{4.4}$$

Note that in Equation (4.4), B_{in} is independent of frequency.

Output Impedance
____

The output impedance can be similarly measured in a manner similar to measuring the input impedance. We capacitively couple the input to ground and capacitively couple V_{test} to the output, as shown in Figure 4.10. The value of C_2 is chosen such that it acts as a short circuit for the frequency of interest.

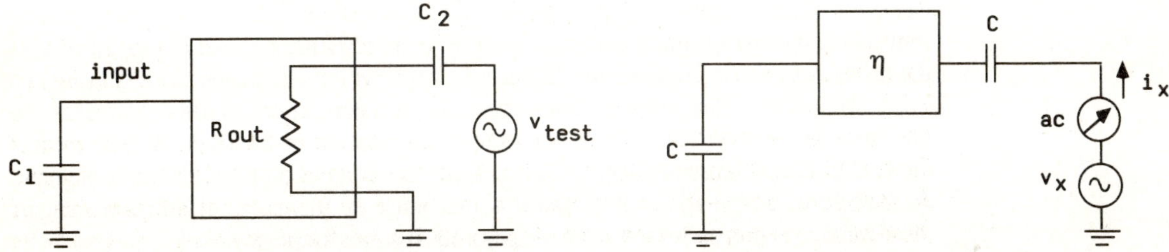

<table>
<tr><td>Figure 4.10: Model for output impedance</td><td>Figure 4.11: Circuit for measuring Z_{out}</td></tr>
</table>

The output impedance of a two-part amplifier is found by first shorting the signal source, i.e., by capacitively coupling the input of the amplifier to ground, Figure 4.11. The value of the coupling capacitor can be the same as that used to determine the input impedance. Next, a capacitively coupled voltage is applied to the output terminals of the amplifier and the current produced is monitored.

The output impedance will be

$$Z_{out} = v_x/i_x \tag{4.5}$$

4.6 Power Supply Decoupling[3]

The purpose of a power supply is to distribute a nearly constant dc voltage to all loads under conditions of varying load currents. Any variation in a load resistance for the circuit of Figure 4.12(a) will cause a variation in the node voltages due to the resistances of the connecting wires. Because of the series connections in this circuit, the disturbance is coupled to all loads. This is not the case for Figure 4.12(b), where the only coupling is through the nonzero output impedance of the power supply.

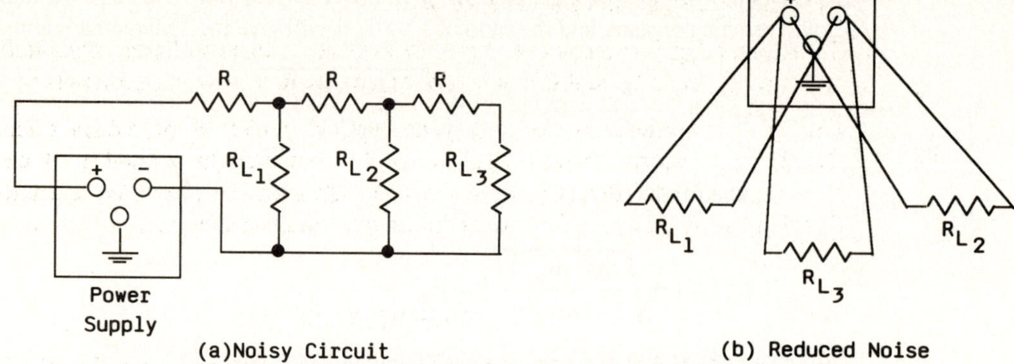

Figure 4.12: Power supply connections to loads (Reprinted by permission from *Guide to Electronic Measurements and Laboratory Practice*, Second Edition, by Stanley Wolf. Copyright © 1983 by Prentice-Hall.)

Capacitor, resistor-capacitor, and inductor-capacitor decoupling networks can be used to isolate circuits from power supply noise, to eliminate coupling between circuits, and to keep the power supply noise from entering the circuit (see Figure 4.13). With a capacitor filter (Figure 4.13(a)), decoupling depends on the power supply source impedance (normally small) and the capacitor to supply filtering. An improved filter is the resistor-capacitor combination (Figure 4.13(b)), which attenuates noise starting at a lower frequency. The voltage drop across the resistor-capacitor filter decreases the power supply voltage; thus, there is a limit to the amount of filtering. A third filter is the L-C circuit (Figure 4.13(c)), which reduces the voltage drop in the series element. An additional advantage to this circuit is that noise is attenuated at a higher rate. A disadvantage is that series resonance is possible at some frequency.

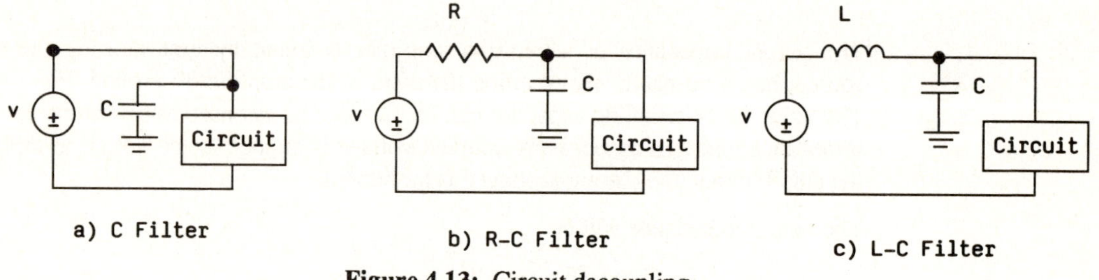

Figure 4.13: Circuit decoupling

Care must be taken with the capacitor-inductor filter that the resonant frequency of the filter is well below the band of frequencies used by the circuit connected to the filter.

4.7 Summary

This chapter has introduced students to safety and good laboratory practice before they have had a chance to develop bad laboratory habits. We saw that the lethal current level is only about 100-300 ma. To minimize accidents in the laboratory, a list of safety rules was presented.

After presenting some guidelines regarding good circuit layout, we examined the topic of grounding. Grounding is divided into three areas: grounding for safety, power grounds, and signal grounds. For safety, a three-wire system should be used and the chassis of electrical equipment should be connected to ground through a separate wire.

To minimize noise, signal grounds and power grounds should be kept separated until connected at a single point at earth ground.

Troubleshooting is a technique too often neglected. In this chapter, troubleshooting approaches were offered that will help a student determine why a circuit fails to work.

Finally, we discussed power supply decoupling, a method of reducing unwanted signals in a system. Power supplies can be connected to a number of electrical systems; therefore, unwanted signals in one system can be channeled to another. In this chapter, we discussed how to minimize this unwanted coupling.

REFERENCES

[1] S. Wolf, *Guide to Electronic Measurements and Laboratory Practice*, 2d ed. (Englewood Cliffs, NJ: Prentice-Hall, 1983), Chapter 3.

[2] H. W. Ott, *Noise Reduction Techniques in Electronic Systems*, (New York: Wiley, 1976), Chapter 3.

[3] Ott, Chapter 4.

Chapter 5

OHM'S AND KIRCHHOFF'S LAWS

5.1 Introduction

Chapters 1 through 4 provide the background for the remaining chapters of this book. With Chapter 5, we are ready to perform procedures or laboratory experiments that will both reinforce our knowledge of electrical engineering concepts and introduce us to new concepts.

In this chapter, we study the basic techniques of circuit analysis using the experimental laws of Ohm and Kirchhoff. We also introduce the concepts of the dependent voltage and dependent current source. These sources are very important in circuit theory, particularly in electronic circuits. The ideal operational amplifier is a device that may be used to obtain dependent sources.

5.2 Ohm's Law

Ohm's law says that the voltage across many types of conducting material is directly proportional to the current flowing through the material. When the material has a constant conductivity, the relationship simplifies to

$$v = Ri \tag{5.1}$$

where the constant of proportionality R is called the **resistance**.

Equation (5.1) results from a plot of v versus i and represents a straight line passing through the origin. The equation is linear and can be considered the definition of a linear resistor. A **linear resistor** is an idealized circuit element and is a mathematical model of a physical device. The current-voltage ratio of this physical device is

46

constant only within certain ranges of current, voltage, or power and depends also on temperature and other environmental factors.

Figure 5.1 shows the circuit symbol for the resistor.

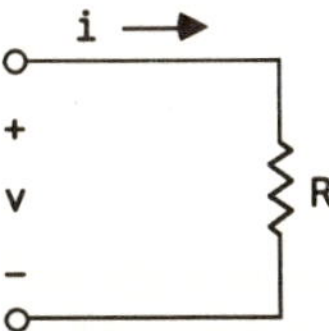

Figure 5.1: Circuit symbol for the resistor

Since charges are transported from a higher to a lower potential in passing through a resistor, the energy qv lost by a charge q is absorbed by the resistor in the form of heat. The rate at which energy is dissipated is

$$p(t) = v(t)i(t) = Ri^2(t) = v^2(t) / R \tag{5.2}$$

A plot of equation (5.2) is given in Figure 5.2.

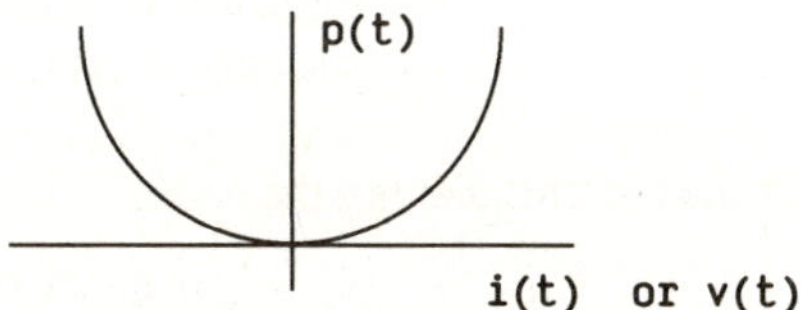

Figure 5.2: Instantaneous power

The figure shows that p(t) is a parabolic function of v(t) or i(t) and is always positive.

Another important quantity useful in circuit analysis is **conductance**, defined by

$$G = 1/R \tag{5.3}$$

The unit of conductance is the mho (A/V), and the international unit for conductance is siemens (S). Alternative expressions for Ohm's law and instantaneous power are, respectively,

$$i = Gv \tag{5.4}$$

and

$$p(t) = i^2(t)/G = Gv^2(t) \tag{5.5}$$

5.3 Kirchhoff's Laws

A circuit can be completely described when we know the voltage across and current through every element. Kirchhoff's laws together with the terminal characteristics of the circuit elements give solutions for any electrical network.

Kirchhoff's current law (KCL) states that

The algebraic sum of all the currents at any node in a circuit equals zero.

This means that

The algebraic sum of the currents entering any node is zero,

and

The algebraic sum of the currents leaving any node is zero.

Stated in a third way;

The sum of the currents entering any node equals the sum of the currents
leaving the node.

KCL simply says that charge cannot accumulate at a node. Since a node is not a circuit
element and it cannot store, destroy, or generate charge, there cannot be an
accumulation of charge at a node. Hence, the currents at a node must sum to zero.

Consider the node shown in Figure 5.3. By KCL, the algebraic sum of the four
currents entering the node must be zero. That is,

$$i_A + i_B - i_C - i_D = 0 \tag{5.6}$$

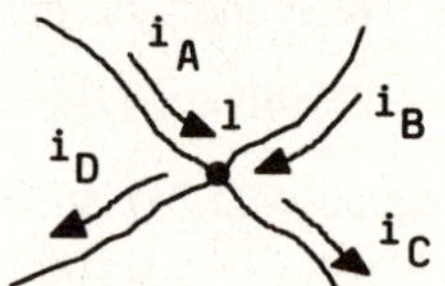

Figure 5.3: The algebraic sum of the four
currents entering a node

For the currents leaving the node,

$$-i_A - i_B + i_C + i_D = 0 \tag{5.7}$$

If we multiply both sides of Equation (5.7) by -1, we get

$$i_A + i_B - i_C - i_D = 0 \tag{5.8}$$

which is identical to Equation (5.6). In general, a mathematical expression of KCL is

$$\sum_{n=1}^{N} i_n = 0 \tag{5.9}$$

where i_n is the nth current entering (or leaving) the node and N is the number of the
node currents.

Kirchhoff's voltage law (KVL) states that

The algebraic sum of the voltages around any closed path is zero.

Kirchhoff's voltage law is a consequence of the conservation of energy and the
conservative property of the electric circuit. In a conservative electric field the
summation of potential around a closed loop is zero.

Application of KVL to the closed path ACBA of Figure 5.4 gives

$$-v_1 + v_2 - v_3 = 0 \tag{5.10}$$

where the algebraic sign for each voltage is taken as positive when moving from + to -
(higher to lower potentials) and negative when moving from - to + (lower to higher
potential) in going across the element. Here we are equating the sum of the voltage
drops around the loop to zero. We could just as well equate the voltage rises to zero.

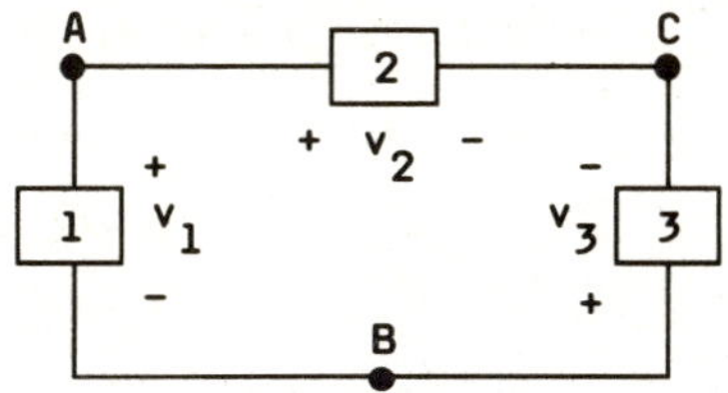

Figure 5.4: Voltage around a closed path

The application of KVL is independent of the direction in which the path is traversed. Thus, for the path ABCA, we get

$$v_3 - v_2 + v_1 = 0 \tag{5.11}$$

which is equivalent to Equation (5.10).

In general, a mathematical expression for KVL is

$$\sum_{n=1}^{N} v_n = 0 \tag{5.12}$$

where v_n is the nth voltage in a loop of N voltages. The sign of each voltage is chosen to be positive either for a voltage drop or for a voltage rise, but the convention used must be consistent.

5.4 Series Elements and Voltage Division

Circuit elements that carry the same current are said to be **connected in series**. For Figure 5.5, it can be shown that the resistors are connected in series by applying KCL to each node in the circuit. Doing so, we obtain

$$i_s = i_1 = -i_2 = i_3 = i_4 = -i_5 = i_6 = -i_7 \tag{5.13}$$

From this equation, if any one of the seven currents is known, then all the currents are known. Figure 5.5 can then be redrawn with only the unknown current i_s, as in Figure 5.6.

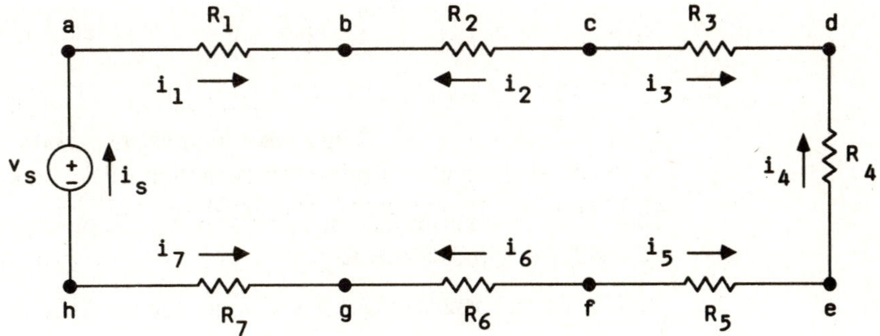

Figure 5.5: Resistors in series

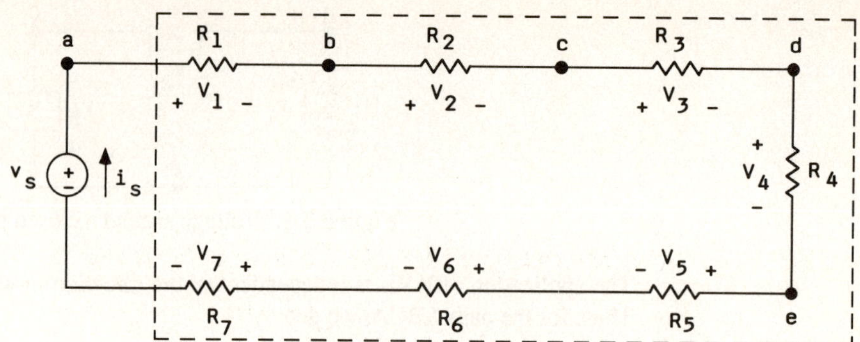

Figure 5.6: Series resistors with an unknown current i_s

KVL applied around the closed loop of Figure 5.6 gives

$$-v_s + v_1 + v_2 + v_3 + v_4 + v_5 + v_6 + v_7 = 0 \tag{5.14}$$

Applying Ohm's law gives

$$-v_s + i_s R_1 + i_s R_2 + i_s R_3 + i_s R_4 + i_s R_5 + i_s R_6 + i_s R_7 = 0 \tag{5.15}$$

where voltage drops are taken as positive and voltage rises as negative. Factoring out i_s gives

$$-v_s = i_s(R_1 + R_2 + R_3 + R_4 + R_5 + R_6 + R_7) \tag{5.16}$$

The seven resistors may then be replaced by a single resistor equal to their sum, and Equation (5.16) can then be represented by

$$v_s = i_s R_{eq} \text{ or } i_s = v_s / R_{eq} \tag{5.17}$$

where $R_{eq} = R_1 + R_2 + R_3 + R_4 + \cdots R_7$.

Figure 5.6 can then be redrawn as shown in Figure 5.7.

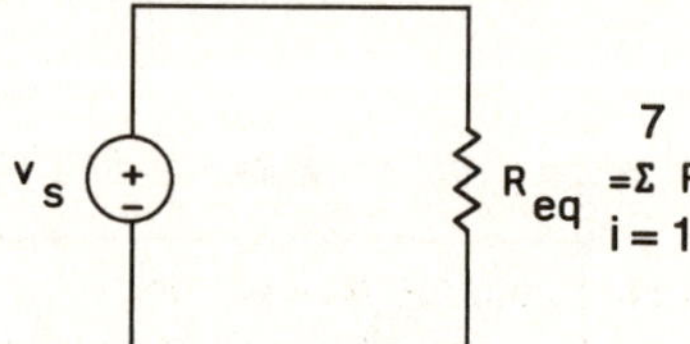

Figure 5.7: Equivalent version
of the circuit of Figure 5.6

This equivalence means that the current-voltage relationship at terminals a-h in Figures 5.6 and 5.7 would be identical.

For N resistors connected in series, the equivalent resistor is

$$R_{eq} = \sum_{n=1}^{N} R_i = R_1 + R_2 + \cdots + R_N \tag{5.18}$$

From Equations (5.14) and (5.15),

$$v_1 = R_1 i_s$$
$$v_2 = R_2 i_s$$
$$v_3 = R_3 i_s$$
$$\cdot$$
$$\cdot$$
$$\cdot$$
$$v_7 = R_4 i_s \tag{5.19}$$

Substituting Equation (5.17) into Equation (5.19) yields

$$v_1 = (R_1 / R_{eq}) v_s$$
$$v_2 = (R_2 / R_{eq}) v_s$$
$$v_3 = (R_3 / R_{eq}) v_s$$
$$\cdot$$
$$\cdot$$
$$\cdot$$
$$v_7 = (R_7 / R_{eq}) v_s \tag{5.20}$$

Thus, voltage divides in direct proportion to the resistance, and the voltage appearing across one of the series resistors is the total voltage times the ratio of its resistance to the total resistance.

For the two-resistor voltage divider circuit of Figure 5.8, the output voltage is easily obtained from Equation (5.20) as

$$v_o = \left[R_2 / (R_1 + R_2) \right] v_s \tag{5.21}$$

Note that the output is open circuited.

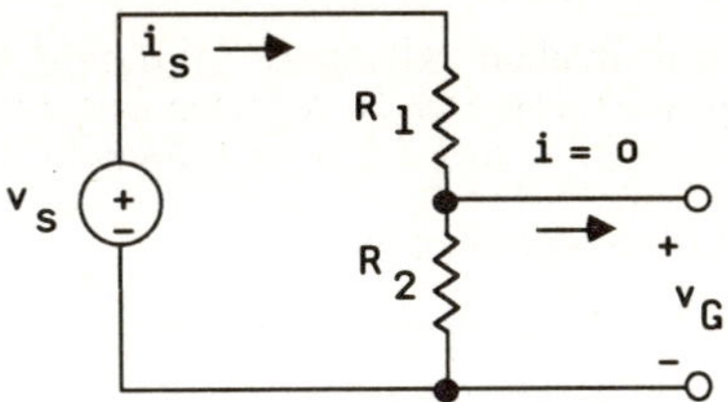

Figure 5.8: Two-resistor voltage divider

Capacitors connected in series (Figure 5.9(a)) can be reduced to a single equivalent capacitor (Figure 5.9(b)). The reciprocal of the equivalent capacitance is equal to the sum of the reciprocals of the individual capacitances. If each capacitor carries its own initial voltage, the initial voltage on the equivalent capacitor will be the algebraic sum of the initial voltages on the individual capacitors. The relevant equations are

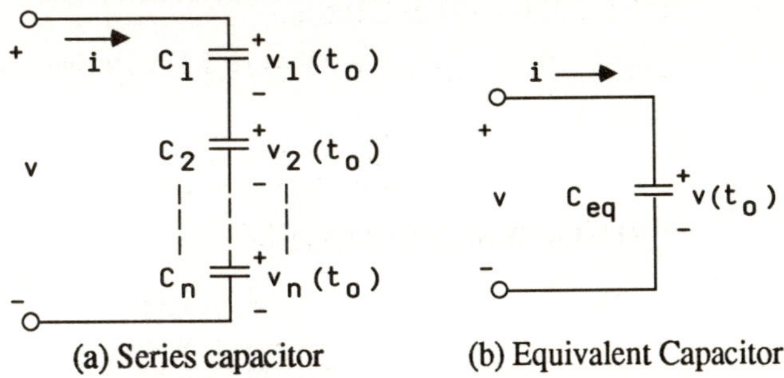

(a) Series capacitor   (b) Equivalent Capacitor

Figure 5.9: Capacitors connected in series

$$\frac{1}{C_{eq}} = \frac{1}{C_1} + \frac{1}{C_2} + \cdots + \frac{1}{C_n}$$

and

$$v_{(t_o)} = v_i(t_o) + v_2(t_o) + \cdots + v_n(t_o)$$

$$\frac{1}{C_{eq}} = \frac{1}{C_1} + \frac{1}{C_2} + \cdots + \frac{1}{C_n}$$
$$v_{(t_o)} = v_i(t_o) + v_2(t_o) + \cdots v_n(t_o)$$

(a) Series inductor carrying an
initial current $i(t_o)$

(b) Equivalent inductor

Figure 5.10: Inductors connected in series

Inductors connected in series (Figure 5.10(a)) can be reduced to a single equivalent inductor (Figure 5.10(b)). In the series connection, the inductors are required to carry the same current, but the voltages across each inductor can be summed to find the voltage across the series connection by using KVL. The voltage across the series connection in Figure 5.10 is

$$v = v_1 + v_2 + v_3 = (L_1 + L_2 + L_3)\frac{di}{dt} \tag{5.22}$$

The equivalent inductance for n inductors in series is

$$L_{eq} = L_1 + L_2 + L_3 + \cdots + L_n \tag{5.23}$$

5.5 Parallel Elements and Current Division

Circuit elements are connected in parallel when the same voltage is common to each of them. The circuit in Figure 5.11 is an illustration of resistors connected in parallel.

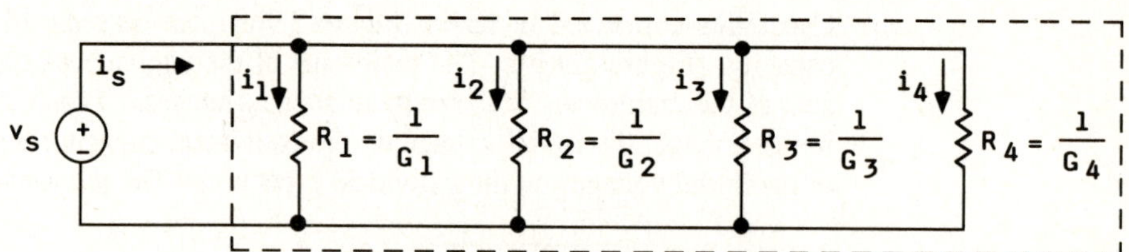

Figure 5.11: Resistors in parallel

From KCL, we have

$$i_s = i_1 + i_2 + i_3 + i_4 \tag{5.24}$$

and from Ohm's law, we have

$$i_1 = v_s/R_1 = v_s G_1 \tag{5.25}$$
$$i_2 = v_s/R_2 = v_s G_2$$
$$i_3 = v_s/R_3 = v_s G_3$$
$$i_4 = v_s/R_4 = v_s G_4$$

Substituting Equation (5.25) into Equation (5.24) results in

$$i_s = v_s(1/R_1 + 1/R_2 + 1/R_3 + 1/R_4) \tag{5.26}$$

from which we obtain

$$i_s/v_s = 1/R_{eq} = 1/R_1 + 1/R_2 + 1/R_3 + 1/R_4 \tag{5.27}$$

Equation (5.27) shows that the four resistors in the circuit of Figure 5.11 can be replaced by a single equivalent resistor, as shown in Figure 5.12.

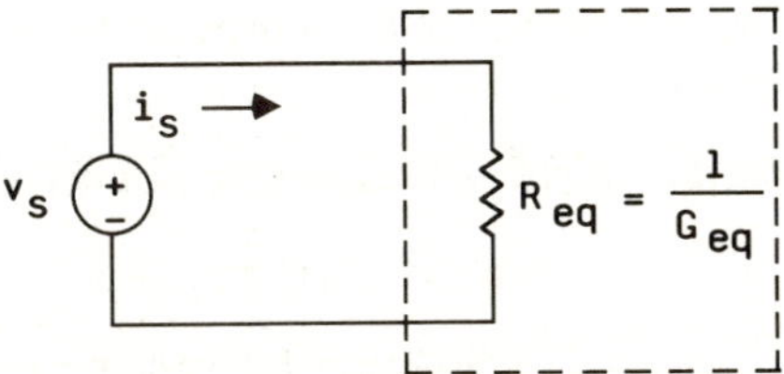

Figure 5.12: Equivalent circuit
of the circuit of Figure 5.11

For N resistors connected in parallel, Equation (5.27) becomes

$$1/R_{eq} = \sum_{i=1}^{N} 1/R_i = 1/R_1 + 1/R_2 + \cdots + 1/R_N \tag{5.28}$$

$$R_{eq} = 1/(1/R_1 + 1/R_2 + \cdots + 1/R_N) \tag{5.29}$$

In dealing with resistors connected in parallel, it may be convenient to use the concept of conductance. Then Equation (5.28) becomes

$$G_{eq} = \sum_{i=1}^{N} G_i = G_1 + G_2 + G_3 + \dots + G_N \tag{5.30}$$

From Equation (5.26),

$$v_s = i_s/(1/R_1 + 1/R_2 + 1/R_3 + 1/R_4) \tag{5.31}$$

Substituting Equation (5.31) into Equation (5.25) yields

$$i_1 = i_s \frac{\left(\dfrac{1}{R_1}\right)}{\left(\dfrac{1}{R_1} + \dfrac{1}{R_2} + \dfrac{1}{R_3} + \dfrac{1}{R_4}\right)} \tag{5.32}$$

and

$$i_2 = i_s \frac{\left(\dfrac{1}{R_2}\right)}{\left(\dfrac{1}{R_1} + \dfrac{1}{R_2} + \dfrac{1}{R_3} + \dfrac{1}{R_4}\right)}$$

In terms of conductance,

$$i_1 = \frac{i_s G_1}{(G_1 + G_2 + G_3 + G_4)} \tag{5.33}$$

and

$$i_2 = \frac{i_s G_2}{(G_1 + G_2 + G_3 + G_4)}$$

Thus, the current of the source, i_s, divides between conductances in direct proportion to their values.

For the case of two resistors in parallel, Equations (5.32) reduce to

$$i_1 = \frac{i_s(\frac{1}{R_1})}{\left[\frac{1}{R_1} + \frac{1}{R_2}\right]} = \frac{R_2 i_s}{(R_1 + R_2)} \tag{5.34}$$

and

$$i_2 = \frac{R_1 i_s}{(R_1 + R_2)}$$

Thus, the current divides in inverse proportion to the resistance. That is, larger current flows through the smaller resistance.

Capacitors connected in parallel (Figure 5.13) can be replaced by a single capacitor that is the sum of the capacitances of the individual capacitors. Capacitors connected in parallel must carry the same voltage. If there is an initial voltage across the original parallel capacitors, this initial voltage appears across the equivalent capacitance C_{eq}.

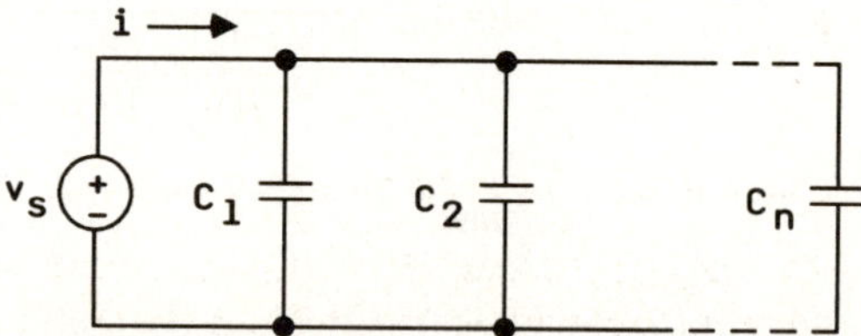

Figure 5.13: Capacitors connected in parallel

Inductors in parallel (Figure 5.14) have the same terminal voltage. The current for three individual inductors in parallel is

$$i = (\frac{1}{L_1} + \frac{1}{L_2} + \frac{1}{L_3})\int_{t_o}^{t} v d\tau + i_1(t_o) + i_2(t_o) + i_3(t_o) \tag{5.35}$$

In terms of a single inductor, Equation (5.35) becomes

$$i = L_{eq} \int_{t_o}^{t} v d\tau + i(t_o) \tag{5.36}$$

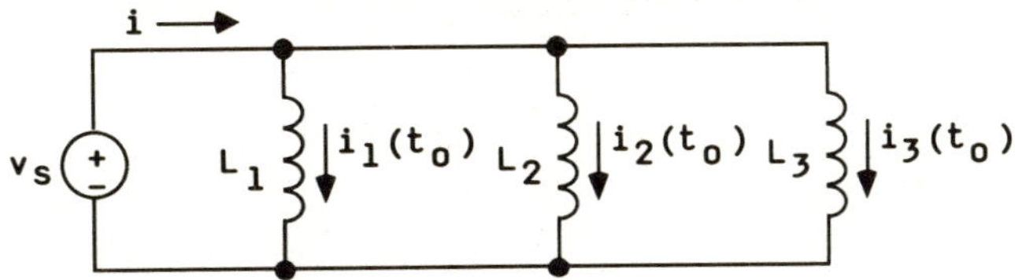

Figure 5.14: Inductors in parallel

Comparing Equation (5.35) with Equation (5.36), we obtain

$$\frac{1}{L_{eq}} = \frac{1}{L_1} + \frac{1}{L_2} + \frac{1}{L_3} \tag{5.37}$$

and

$$i(t_o) = i_1(t_o) + i_2(t_o) + i_3(t_o)$$

5.6 Dependent Voltage and Current Sources

An ideal dependent, or controlled, voltage source is a source in which the voltage across its terminal is determined by either a voltage or a current existing at some other location in the circuit. The circuit symbol for a dependent voltage source is shown in Figure 5.15. The voltage source v_s is controlled by a voltage or a current somewhere else in the circuit.

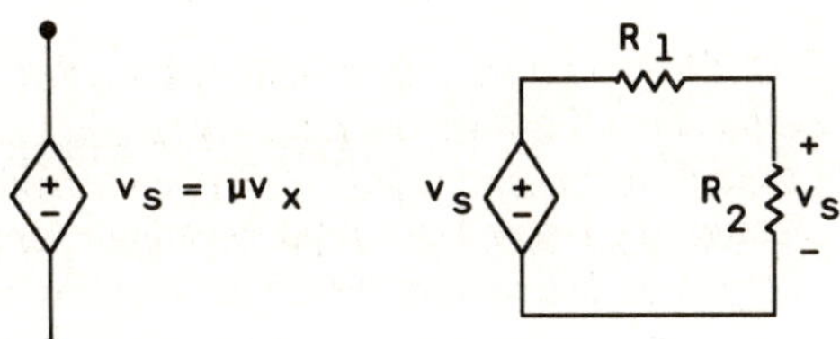

Figure 5.15: Dependent voltage source

Let v_x or i_x represent the controlling variable. Then

$$v_s = \mu v_x \tag{5.38}$$

or

$$v_s = \gamma i_x,$$

where μ and γ are constants. μ is unitless, but γ has the units of volts/ampere, or ohms.

An ideal dependent, or controlled, current source is a source in which the terminal current is determined by either a voltage or a current existing at some other location in the circuit. We can have either a voltage-controlled current source or a current-controlled current source. The circuit symbol for a dependent current source is shown in Figure 5.16.

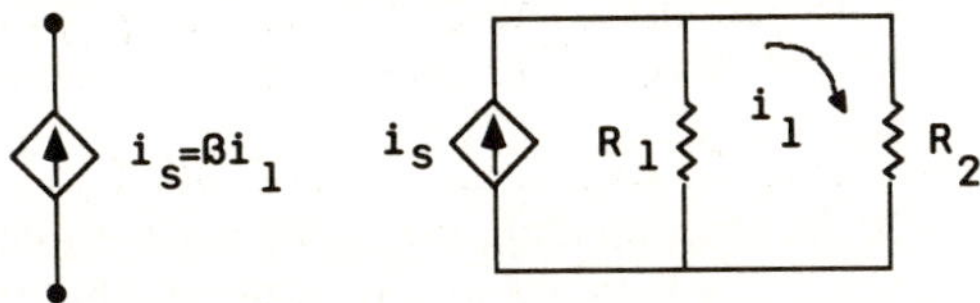

Figure 5.16: Dependent current source

The current source i_s is controlled by either a voltage v_x or a current i_x somewhere else in the circuit, so that

$$i_s = \alpha v_x \tag{5.39}$$

or

$$i_s = \beta i_x$$

where α and β represent constants. β is dimensionless, and α has the units of amperes/volt, or siemens.

It is important to note that knowing the terminal voltage of a dependent or independent voltage source is not sufficient to state what current the source may be carrying. A similar statement obtains for the dependent or independent current source.

5.7 Operational Amplifiers[1],[2]

The operational amplifier (op amp) is basically a voltage-controlled voltage source. The dependent voltage source appears at the output terminals of the op amp, and the voltage on which it depends is applied to the input terminals.

The op amp symbol is shown in Figure 5.17.

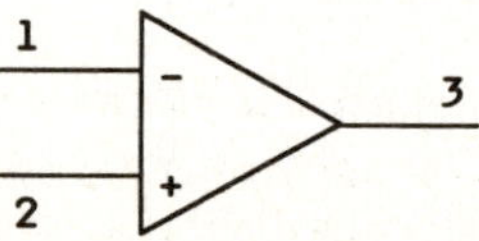

Figure 5.17: Circuit symbol for the op amp

Terminals 1 and 2 are input terminals, and terminal 3 is the output terminal. The op amp is an active and, therefore, requires dc power to operate. Most integrated circuit op amps require two dc power supplies. Two terminals are brought out of the op amp package and are connected to a positive voltage and a negative voltage as shown in Figure 5.18.

It is important to note that no terminal of the op amp is physically connected to ground. The reference grounding point is just the common terminal of the two power supplies. The presence of op amp power supplies are assumed, but not normally shown.

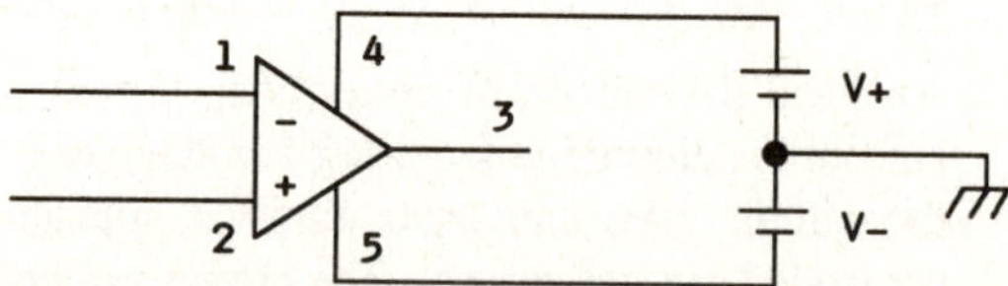

Figure 5.18: The op amp with dc power supplies

The op amp senses the difference $v_2 - v_1$ between the voltage signals applied at its two input terminals, and multiplies this quantity by a gain factor A, causing the resulting output $A(v_2-v_1)$ to appear at the output terminal. The voltage v_1 is the voltage between terminal 1 and ground, and v_2 is the voltage between terminal 2 and ground.

In the ideal op amp, the signal current into terminal 1 and the signal current terminal into 2 are both zero. This suggests that the input impedance of the ideal op amp is infinite. The output terminal, terminal 3, acts as the output terminal of an ideal voltage source. The voltage between terminal 3 and ground will always be equal to $A(v_2-v_1)$ independent of the current drawn from terminal 3 into a load. That is, the ideal output impedance is zero.

The equivalent circuit of the ideal op amp is shown in Figure 5.19.

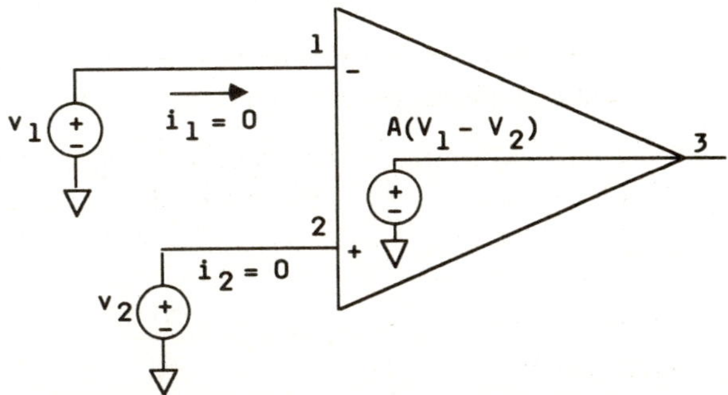

Figure 5.19: Equivalent circuit of the ideal op amp.

The sign of the output is the same as that of v_2 and opposite to that of v_1. Thus, terminal 1 is called the **inverting terminal** ("-" sign), while terminal 2 is called the **noninverting terminal** ("+" sign).

The op amp responds only to the difference signal (v_2-v_1) and ignores any signal common to both inputs. That is, $v_1=1V$ and $v_2=1V$ produces an ideal voltage of $A(v_2-v_1) = 0$ at terminal 3.

The ideal op amp should have a gain A whose value is infinite. In addition, A should remain constant over all frequencies. This infinite op amp gain suggests that the voltage v_1 approaches v_2, and gain A approaches infinity, but the output voltage at terminal 3 is a finite value. Therefore, the voltage between the op amp input terminals should be negligibly small. That is,

$$v_2-v_1 = \frac{v_o}{A} \approx 0 \qquad (5.40)$$

and, therefore,

$$v_1 \approx v_2 \qquad (5.41)$$

This means that whatever voltage is at terminal 2 will appear at terminal 1 because of the infinite gain A. Thus, a "virtual short" exists between the two input terminals. It should be noted, however, that this does not mean that terminals 1 and 2 should be shorted together while analyzing a circuit.

In sum, the ideal op amp has two important properties that the student needs to know: (1) the currents into both input terminals are zero, and (2) the voltage between the input terminals is zero. As an example, consider the inverting op amp of Figure 5.20. In this circuit,

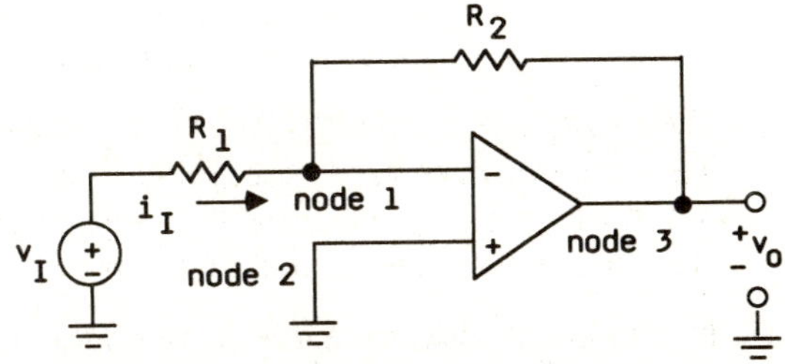

Figure 5.20: Inverting op amp

resistor R_2 is connected from the output terminal of the op amp, terminal 3, back to the inverting or negative input terminal, terminal 1. R_2 closes the loop around the op amp. In addition, terminal 2 has been grounded and resistor R_1 has been added between terminal 1 and the input signal source, v_I. The output is taken between terminal 3 and ground and is not dependent on the value of the current supplied to a load impedance connected between terminal 3 and ground.

Since there is no voltage across the input terminals of the op amp, the voltage at node 1 is zero. Applying KCL at the inverting node (node 1) of the op amp gives

$$\frac{(v_I - 0)}{R_1} = \frac{(0 - v_3)}{R_2} \tag{5.42}$$

or

$$v_o = v_3 = \frac{-R_2 v_I}{R_1}$$

The input current is

$$i_I = i_o = \frac{v_I}{R_1} \tag{5.43}$$

The circuit of Figure 5.20 is called an inverter because the polarity of v_3 is opposite to that of the input voltage v_I. An equivalent circuit of the inverter is shown in Figure 5.21.

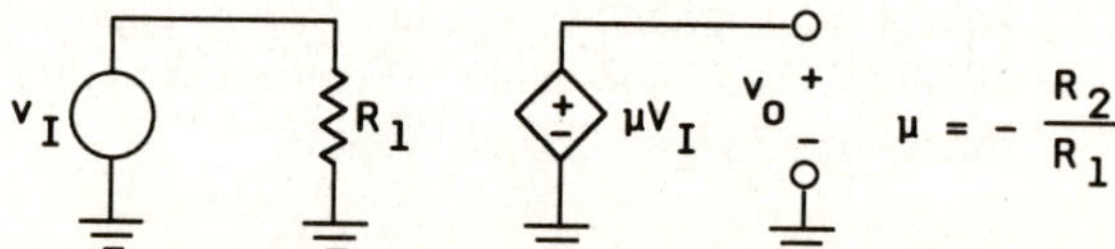

Figure 5.21: Equivalent circuit of the inverter

The circuit of Figure 5.21 is a voltage controlled dependent voltage source.

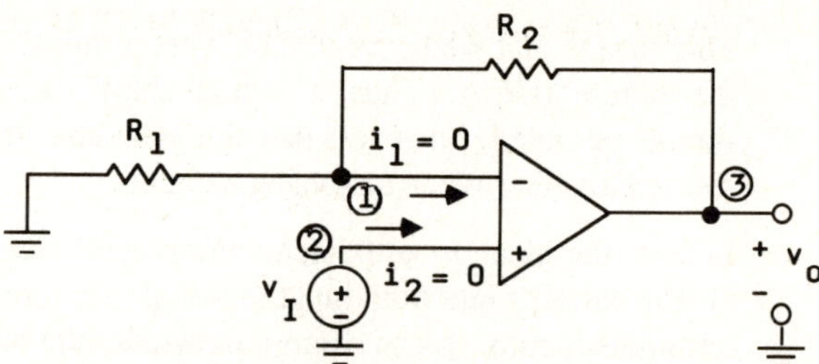

Figure 5.22: Noninverting op amp

The noninverting op amp is shown in Figure 5.22. Since there is no voltage across the input terminals of the op amp, the voltage of node 1 is v_I. Applying KCL at node 1 results in

$$\frac{(v_I - 0)}{R_1} = \frac{(v_o - v_I)}{R_2}$$

or

$$v_o = \mu v_I \qquad\qquad (5.44)$$

where

$$\mu = 1 + \frac{R_2}{R_1}$$

A voltage-controlled source equivalent circuit for the noninverting op amp is shown in Figure 5.23.

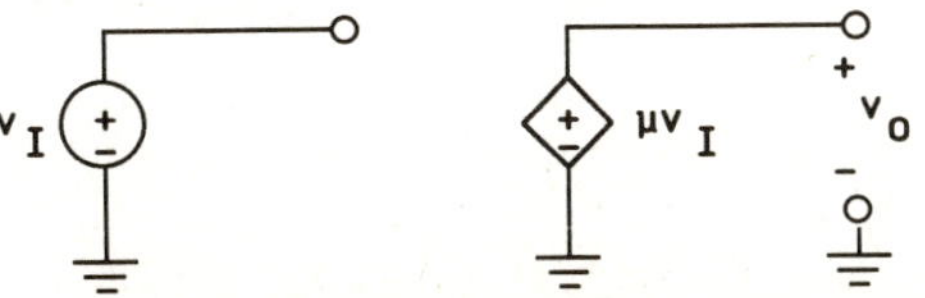

Figure 5.23: Equivalent circuit
for the noninverting op amp

5.8 State Variables[3],[4]

The state variables used for network analysis are the capacitor voltages and the inductor currents. They replace the loop currents and the node voltages introduced in previous sections. Their use allows all other voltages and currents in the network to be found.

The advantage of the state variable is that its form is especially suited for computer simulation. In addition, use of the state variables is a popular technique for describing control systems.

A series R-L-C circuit, shown in Figure 5.24, is driven by a voltage v_s.

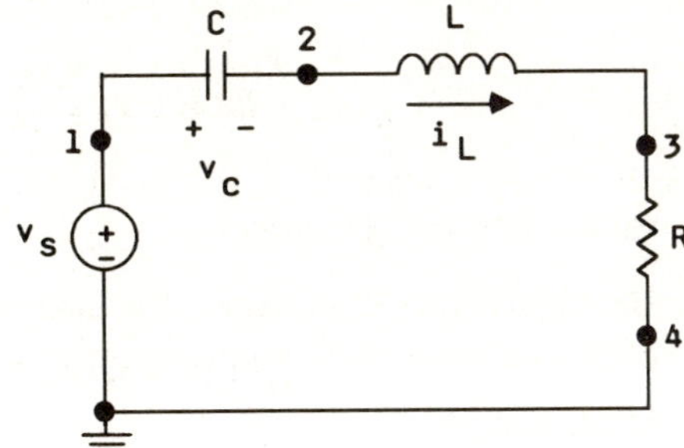

Figure 5.24: Series R-L-C circuit

The state variables are the capacitor voltages v_c and the inductor current i_L. At node 2, we write, from Kirchhoff's current law,

$$C\frac{dv_C}{dt} = i_L \qquad\qquad (5.45)$$

From Kirchhoff's voltage law around the only loop, we get

$$L\frac{di_L}{dt} = v_s - i_L R - v_C \tag{5.46}$$

When these equations are written in the form

$$\frac{dv_C}{dt} = 0v_C + \frac{1}{C}i_L \tag{5.47}$$

$$\frac{di_L}{dt} = \frac{-1}{L}v_C - \frac{R}{L}i_L + v_s \tag{5.48}$$

the equations are said to be in state form. The generalization of these equations is constructed using x as the general state variable and y as the general input:

$$\frac{dx_1}{dt} = a_{11}x_1 + a_{12}x_2 + \cdots + a_{1n}x_n + y_1$$

$$\frac{dx_2}{dt} = a_{21}x_1 + a_{22}x_2 + \cdots + a_{2n}x_n + y_2 \tag{5.49}$$

$$\cdot$$

$$\cdot$$

$$\cdot$$

$$\frac{dx_n}{dt} = a_{n1}x_1 + a_{n2}x_2 + \cdots + a_{nn}x_n + y_n$$

The equations are written in a form especially suited for computer simulation.[4]

5.9 Summary

In this chapter, we have discussed three fundamental laws: Ohm's law and Kirchhoff's current and voltage laws. These laws were used to analyze simple voltage and current divider circuits. We also introduced both the ideal operational amplifier as a dependent voltage source and the concept of the state variable.

LAB 5L1

TITLE:	Passive Resistor Networks
OBJECTIVE:	The objective is to introduce the fundamental concepts of KVL, KCL, power, and worst case analysis. The circuits are single-loop and three-node networks.
EQUIPMENT:	Power Supply

EQUIPMENT continued:

Power Supply
  voltage 0-10 volts
  current 500 mA
Two multimeters capable of measuring volts, ohms, and amps. Determine and record the accuracy of the meters.

RECOMMENDED: DIGITAL METERS

Variety of Resistors
(all resistors are 1/8W)
10Ω (**Note:** Resistor substitution boxes may
100Ω be used, but they are cumbersome)
150Ω
300Ω
650Ω
Check and record the tolerance of the resistors.

Breadboard

COLOR CODING: Most larger resistors have their resistance values and tolerances stamped on their bodies. For those resistors that are too small to use this method, a color code is used. Four color bands are painted on the resistor body to specify the resistance and tolerance.

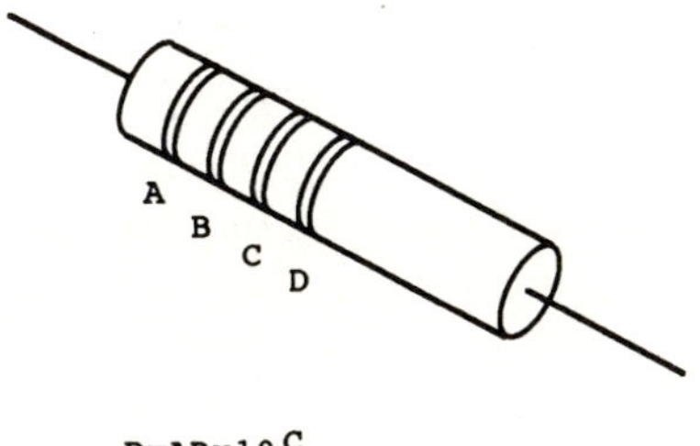

COLOR CODE

Black	=	0	Green	=	5
Brown	=	1	Blue	=	6
Red	=	2	Violet	=	7
Orange	=	3	Grey	=	8
Yellow	=	4	White	=	9

D = Tolerance digit

No Band = $\pm\,20\%$
Silver = $\pm\,10\%$
Gold = $\pm\,5\%$

PRELABORATORY:

1. In the circuit of Figure 5L1.1

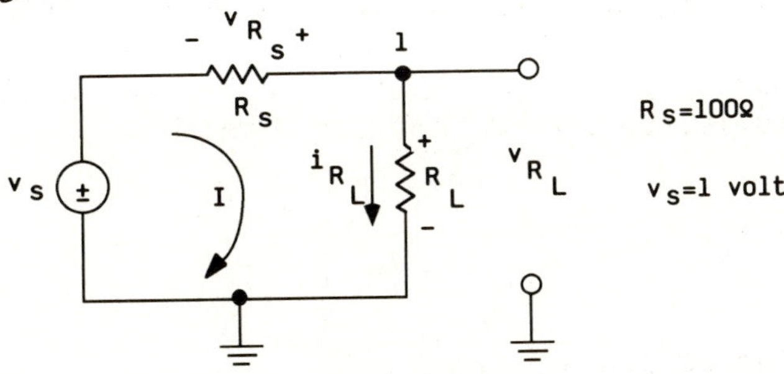

Figure 5L1.1

a. Write the KCL equation that describes the current flowing out of the source at node 1.

b. Write the KVL equation around the loop.

c. Reverse all the current and voltage designations in the circuit, and repeat parts a and b. (Do not change the polarity of the independent voltage source.) Compare the resulting equations.

d. Let R_s=100 Ω and R_L=1 kΩ. Calculate the power delivered by the source and the power dissipated in each resistor.

2. Analyze the circuit shown in Figure 5L1.2

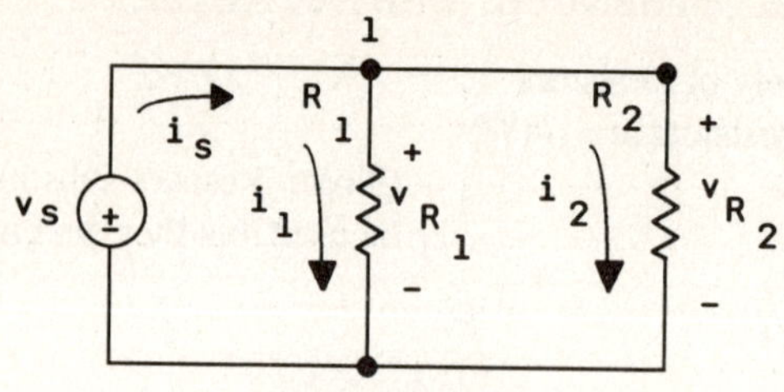

Figure 5L1.2

a. Determine v_{R_1} and v_{R_2} as a function of i_s. That is,

$$v_{R_1} = f_1(i_S)$$

and

$$v_{R_2} = f_2(i_S)$$

b. Sum the currents flowing out of node 1.

c. Determine i_s as a function of v_s, R_1, and R_2. That is,

$$i_s = f_3(v_S)$$

d. Determine i_1 and i_2 as a function of i_s. That is,

$$i_1 = f_4(i_S)$$

and

$$i_2 = f_5(i_S)$$

e. What is the maximum and minimum value of i_S if R_1 and R_2 each have a tolerance of $\pm$ 10% and v_S is within $\pm1\%$ of its nominal value? Let $R_1 = R_2 = 10\ k\Omega$ and $v_s = 1$ V.

3. Consider the circuit shown in Figure 5L1.3.

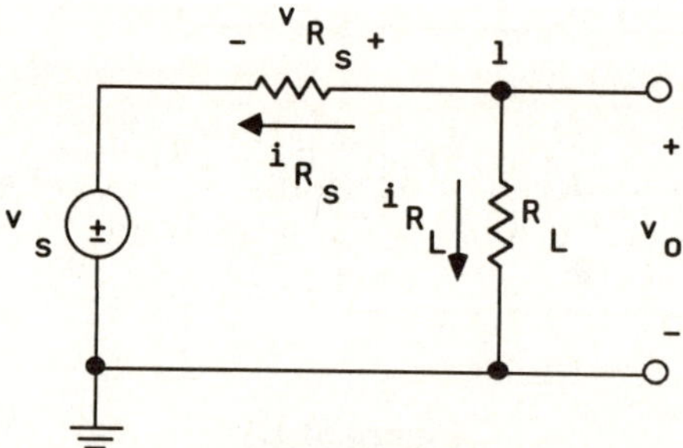

Figure 5L1.3

a. Calculate the values needed to fill in Table 5L1.1. (Let $v_s = 1$ V and $R_s = 100\ \Omega$.)

Table 5L1.1

R_L	P_{V_s} (watts)	i_{R_s} (A)	i_{R_L} (A)	v_{R_L} (v)	P_{R_s} (w)	P_{R_L} (w)
$0\,\Omega$						
$10\,\Omega$						
$50\,\Omega$						
$100\,\Omega$						
$200\,\Omega$						
$300\,\Omega$						
∞						

b.   Plot i_{R_L} versus v_{R_L} on the axes shown in Figure 5L1.4.

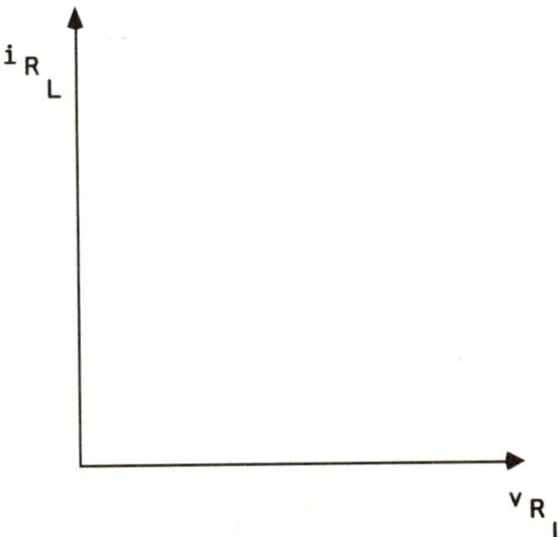

Figure 5L1.4

c.   What would the current through R_L be if $R_L = 0\,\Omega$? What would the voltage drop across R_L be if $R_L \to \infty$?

d.   Plot the power dissipated in R_L versus R_L. What value of R_L results in this resistor dissipating its maximum power ($R_s = 100\,\Omega$)? Compare the value of R_s to the value of R_L for the condition of maximum power dissipation.

4.   For the circuit shown in Figure 5L1.5,

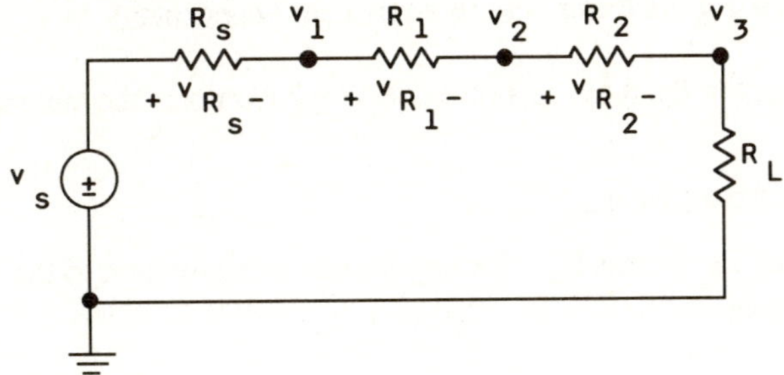

Figure 5L1.5

a.   Compare the value of the currents in each resistor.

b.   Determine the node voltages v_1, v_2, and v_3 as a function of v_S and the resistors.

c.   Express the resistor voltage drops in terms of v_S and/or the node voltages.

d.   What happens to the equations in parts a and b when $R_1 = R_2 = 0$?

e.   If each resistor has a $\pm 10\%$ tolerance and v_S is set to within 1% of its nominal values, what are the maximum and minimum values that can be expected for v_3? Use the total derivative technique explained in Section 2.3.3 of Chapter 2.

Laboratory Procedure:

Build the circuit in Figure 5L1.6. A voltmeter and an ammeter will be necessary.

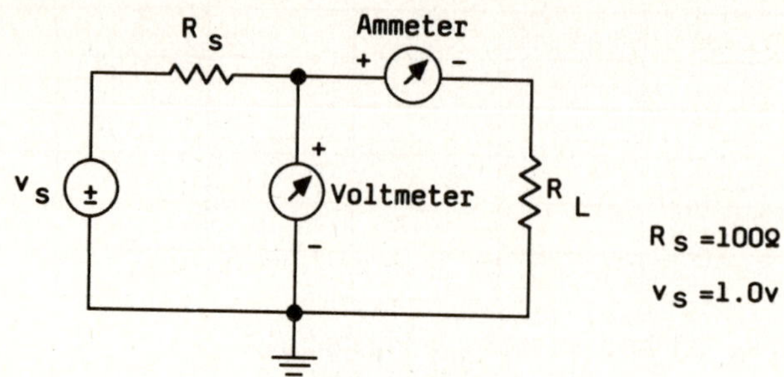

Figure 5L1.6

Use the passive sign convention to indicate the direction for positive resistor voltages and positive resistor currents. The passive sign convention states that whenever the reference direction for the current in an element is in the direction of the voltage drop across the element, use a positive sign in the expression that relates the voltage to the current. Otherwise use a negative sign.

a. Fill in the Table 5L1.2. (Notice that you are varying the value of R_L.)

Table 5L1.2

R_L (ohms)	i_{R_s} mA	i_{R_L} mA	v_{R_s} V	v_{R_L} V	Calculate from i x v	
					P_{R_L} W	P_{R_s} W
0						
10						
50						
100						
200						
300						
∞						

b. Repeat the measurements of part a, but now use only one meter at a time. (Remove the unused meter from the circuit by replacing the meter with its equivalent impedance.)

Are there any differences in the three sets of data? Why? Compare the table with the theoretical results of the prelab.

c. Plot the following data from part a:

1. i_{R_L} versus v_{R_L} and P_{R_L} versus R_L. Give a physical interpretation to the x- and y-intercepts in the i_{R_L} versus v_{R_L} curve.

2. Plot P_{R_L} versus R_L and P_{R_s} versus R_L.

    a. What value of R_L results in the maximum power dissipation in R_L?

    b. What is the power dissipation in R_s when the power dissipation in R_L is maximized?

d. Build the circuit shown in Figure 5L1.7.

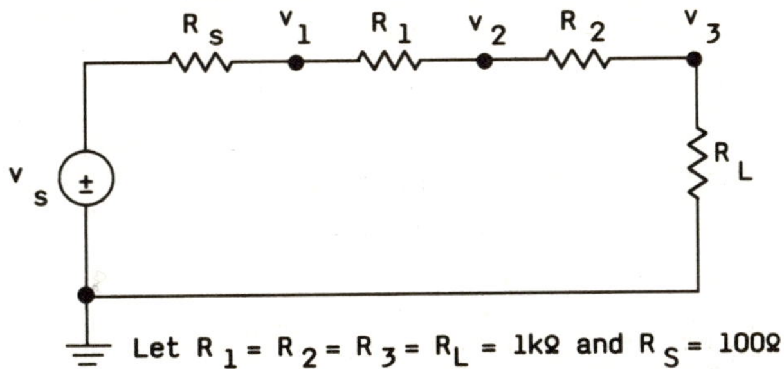

Let $R_1 = R_2 = R_3 = R_L$ = 1kΩ and R_S = 100Ω

v_1, v_2, v_3 are node voltage measured
with respect to ground.

Figure 5L1.7

Define the currents and voltages as you assign them to each resistor. (Use the passive sign convention.)

a.   Given your polarity convention, measure and record the following values:

$v_{R_s}, v_{R_1}, v_{R_L}, i_{R_s}, i_{R_1}, i_{R_2}, v_1, v_2, v_3$. Show analytically that the value you measured is within an

acceptable tolerance based on the tolerance of the resistors, the accuracy of the power supply setting, and the accuracy of the meter.

DISCUSSION:

CONCLUSIONS:

LAB 5L2

TITLE: Dependent Sources

OBJECTIVE: This laboratory will use operational amplifiers to demonstrate dependent voltage sources. The analysis will require a good working knowledge of KVL, KCL, and Ohm's law.

EQUIPMENT: Three Power Supplies
         12 V, -12 V, and 1 V @ 500 mA.
       One multimeter (Determine and record the accuracy of the meter)
       Variety of Resistors:
         1/8 W and ±10% tolerance (or better)
       Operational Amplifier:
         Any general purpose op amp
           unity-gain-compensated potentiometers: ranges 0-100 Ω, 0-10 kΩ, and 0-100 kΩ

PRELABORATORY:

1.   Given the circuit in Figure 5L2.1;

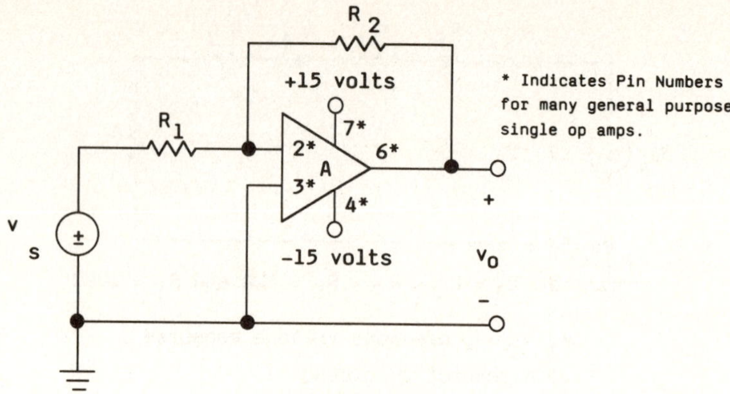

Figure 5L2.1

a. Redraw the circuit using the equivalent circuit shown in Figure 5L2.2 for the nonideal operation amplifier.

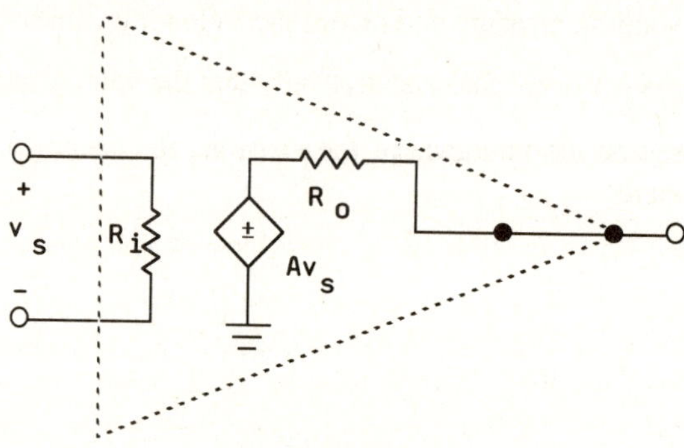

Figure 5L2.2

b. Using the op-amp model shown in Figure 5L2.2 solve for v_{out} in Figure 5L2.1.

c. Using your answer in part b, assume the following:

> Let R_{in} approach infinity
> Let R_o approach zero
> Let A approach infinity

Given these three conditions, what is the resulting answer to part b? (You should get $v_o = -\dfrac{R_2}{R_1} v_s$)

d. Use the result of part b, and answer the following questions.
(Let $R_1 = 1 \text{ k}\Omega$, $R_2 = 2 \text{ k}\Omega$, $R_{in} = 10^6 \ \Omega$, $R_o = 10 \ \Omega$)

1. What is the value of v_o / v_s for the following values of A: $A = 10^2$, 10^3, ... , 10^6?

2. How do these values of v_o / v_s compare to the value obtained by using the equation in part c?

Laboratory Procedure:

1. Build the circuit in Figure 5L2.3.

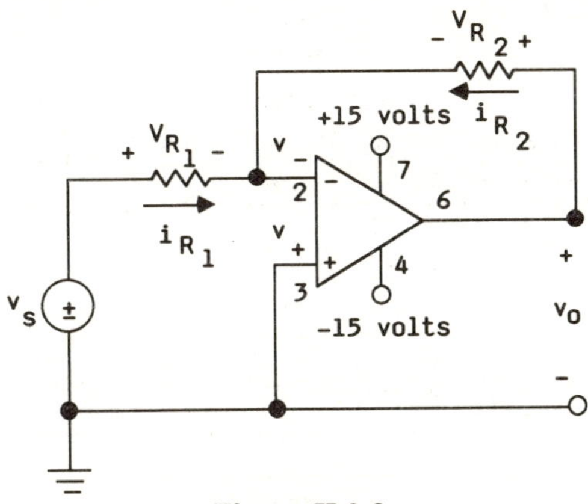

Figure 5L2.3

Let $R_1 = 1$ kΩ, $R_2 = 2$ kΩ, and $v_s = 1$ V. Measure and record the following currents and voltages:

$$i_{R_1}, v_{R_1}, v_{R_2}, v_o.$$

From these measurements, calculate the current flowing into the inverting input.

2. Build the circuit in Figure 5L2.4:

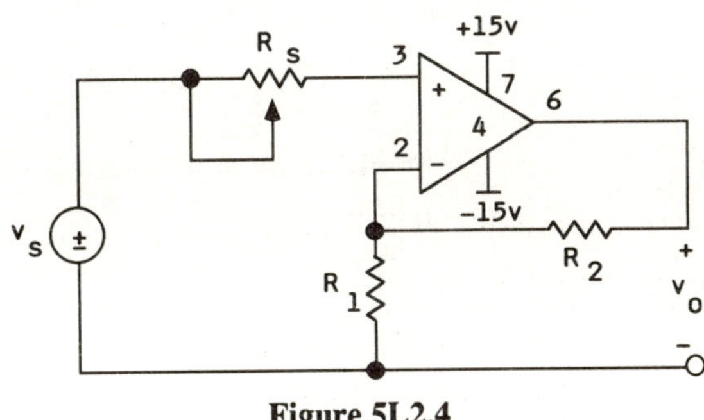

Figure 5L2.4

Let $v_s = 1$ V, $R_1 = 1$ kΩ, $R_2 = 10$ kΩ. R_s is a potentiometer that ranges from 0 to 10 kΩ.

a. Analyze the circuit using the equivalent op amp circuit given in the prelaboratory. What effect does varying R_s have on the voltage v_o? Why?

b. Analyze the circuit using the nonideal operational amplifier model from the prelab. Obtain values for A, R_i (or R_{in}) and R_o from your op amp specification sheet.
Measure and record v_s and v_o.
Vary R_s over its full range. What effect does this have on the output? Compare this result to your calculations.

c. Interchange R_s and R_1.
Vary the potentiometer. Observe and record.

d. Replace R_2 with a potentiometer and let $R_1 = 1$ kΩ. Complete Table 5L2.1.

Table 5L2.1

	Potentiometer Value	Measured v_o (V)	Calculated v_o (V)	Observations
R_s=potentiometer	100 Ω			
R_1=1 kΩ	500 Ω			
R_2=10 kΩ	1000 Ω			
	10,000 Ω			
R_1=Potentiometer	1 kΩ			
R_2=1 kΩ	2 kΩ			
R_s=10 kΩ	5 kΩ			
	10 kΩ			
R_2=potentiometer	1 kΩ			
R_1=1 kΩ	2 kΩ			
R_s=1 kΩ	5 kΩ			
	10 kΩ			

3. Build the circuit shown in Figure 5L2.5.

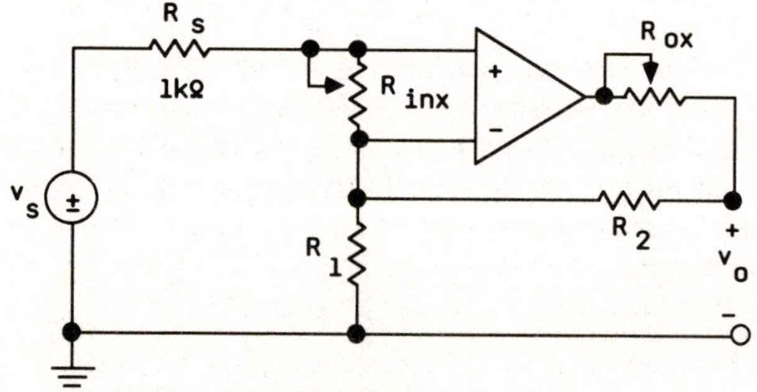

Figure 5L2.5

Let R_{inx} be a potentiometer capable of 0-100 kΩ resistance full scale.
Let R_{ox} be a potentiometer capable of 0-100 Ω resistance full scale.

a. Analyze the circuit using the nonideal model from the prelab, but assume R_{in} is infinite and R_o is zero. Therefore, the open loop gain A is the only internal op amp parameter that will enter into your calculations.

b. Find the relationship between v_o and v_S (v_o=f(v_S)) where f is a function of A, R_s, R_{inx}, R_1, R_2, and R_{ox}.

c. Complete Table 5L2.2.

Table 5L2.2

R_{inx}	R_{ox}	v_o (measured)	v_o (calculated)
open circuit	0		
100 kΩ	0		
50 kΩ	0		
10 kΩ	0		
1 kΩ	0		
open circuit	10		
open circuit	100		
open circuit	1 kΩ		

DISCUSSION: Include an error analysis to verify that the results from the prelab agree with the experimental results.

CONCLUSIONS:

LAB 5L3

TITLE: State Variable Analysis and Computer Simulation[3],[4]

OBJECTIVE: To use the state variable technique to analyze a circuit containing inductors, capacitors, and resistors. Then, using a numerical integrator, to convert first-order differential equations into first-order difference equations and solve the equations numerically with a computer or a programmable calculator. If the loading effects of the instrumentation are taken into account, the results of the experimental portion of the laboratory will measure the accuracy of your simulation. Remember, your simulation is only as good as your model.

EQUIPMENT: Function generator: 1 kHz sine wave, 1-V amplitude.
Pulse generator: needed to simulate an impulse and a step function. A pulse of appropriately long duration can yield the step response. A pulse of large amplitude and narrow width can simulate an impulse.
Components:

3	1-kΩ resistors (1/8 W or greater)
1	0.0022-μF capacitor
1	0.22-μH inductor

Breadboard
Oscilloscope

PRELABORATORY:

For the circuit in Figure 5L3.1;

R	=	1 kΩ
R_L	=	1 kΩ
C	=	0.0022 μF
L	=	0.22 μH

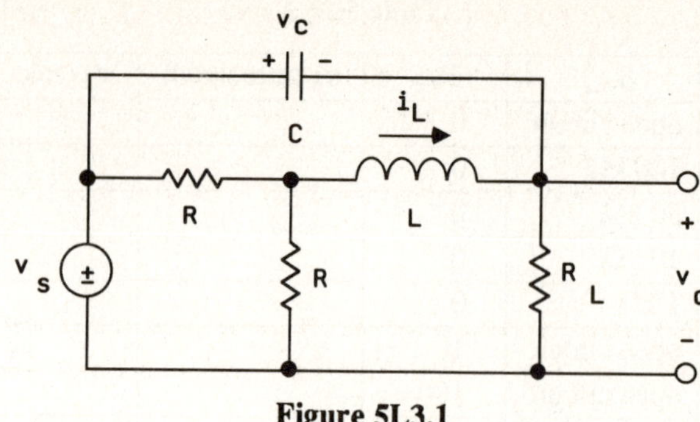

Figure 5L3.1

1. Write the state equations describing this system. Assume that all initial conditions are zero. (Remember to use inductor current and capacitor voltage as your state variables.)

2. Substitute the following approximation for a derivative into the equations written in part 1.

$$\frac{dx}{dt} \equiv \lim_{T \to 0} \frac{x(nT+T) - x(nT)}{T}$$

$$
\begin{array}{lll}
\text{where} \quad x(nT) & \equiv & \text{the value of the variable x at time } nT \\
x(nT+T) & \equiv & \text{the value of the variable x at time } nT+T \\
T & \equiv & \text{the time between samples of variable x} \\
X & \equiv & \text{the capacitor voltage or inductor current.}
\end{array}
$$

3. Solve the first-order difference equations of part 2 using a programmable calculator or a higher level language on a computer. This means writing your own numerical integration routine. Solve these equations with the following inputs: a unit amplitude step, a unit area impulse, and a 1-V amplitude sine wave (1 kHz). Also, do each of these with three different time steps: 1 ns, 1 μs and 1 ms. (This entails nine simulations.)

Laboratory Procedure:

1. Build the circuit given in your prelab.

2. Use a pulse generator to simulate an impulse. Observe your simulated impulse and the circuit's impulse response on a dual-trace oscilloscope. Increase the pulse amplitude and decrease the pulse width, until no measurable change in output occurs. Remember to maintain unity area for your impulse. Sketch and compare this response to the numerical solution in the prelab.

3. Change the input to simulate a step function. (A train of pulses will work if the pulses are wide enough and there is enough time between pulses so that all previous responses decay to steady state.)

4. Sketch the input and response of your circuit for part 3 and compare the results with those of your step simulation.

5. Use a function generator and apply a 1 kHz, 1-V amplitude sinusoidal input to your circuit. Again observe and sketch the input and output of your circuit on a dual trace oscilloscope. (Remember, for a linear circuit, when a sinusoidal voltage/current is applied, only the amplitude and phase angle of the output will differ from the input. The frequency remains the same.)

6. For the situation described in step 3, use a simulation program, such as MICROCAP or SPICE, and input the data for your circuit to the simulation program. Get a transient response plot and include it in your report.

DISCUSSION:

1. Discuss the effect of different values of T on the match between your circuit response and the response of your simulation.

2. Describe the change in circuit response as you vary the impulse.

3. Assume that all components except the capacitor are at their nominal values. If the capacitor is actually 20% below its given (nominal) value, how will the response of part 2 be affected? Let $T=10^{-5}$ seconds for your simulation. Also, determine the effect of this variation on the SPICE simulation. Use the .SENS statement.

CONCLUSIONS: Summarize the various factors that contribute to differences in response between the actual circuit and the simulations. One example might be the parasitic shunt capacitance across an inductor.

LAB 5L4

TITLE: Network Analysis of a Multiple-Loop, Resistive System

OBJECTIVE: To analyze a multiple-loop network using loop and node equations. The results should verify KVL and KCL. The equation "bookkeeping" is done with matrix notation.

EQUIPMENT: Power Supply:
Multimeter: two if available
Resistors:

2   10 kΩ
2   1k Ω
2   5.11 kΩ
1   10 Ω

Note and record the tolerance and accuracy of all
equipment and components.

PRELABORATORY:

1. Analyze the circuit shown in Figure 5L4.1.

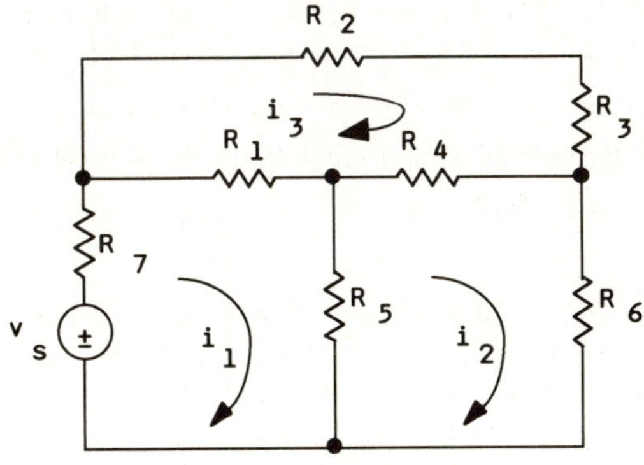

Figure 5L4.1 (a)

a. Use mesh (loop) analysis to find three independent equations describing this circuit.

b. Place your equations from part a into the matrix form

$$Ai_i = \begin{bmatrix} a_{11} & a_{12} & a_{13} \\ a_{21} & a_{22} & a_{23} \\ a_{31} & a_{32} & a_{33} \end{bmatrix} \begin{bmatrix} i_1 \\ i_2 \\ i_3 \end{bmatrix} = \begin{bmatrix} V_s \\ 0 \\ 0 \end{bmatrix}$$

c. Find the inverse of the matrix A, and premultiply by A^{-1}. (Remember, A^{-1} is the transpose of the cofactors divided by the determinant.) Keep all terms in symbolic form. Mathematically,

$$A^{-1} Ai_i = \begin{bmatrix} I_1 \\ I_2 \\ I_3 \end{bmatrix} = A^{-1} \begin{bmatrix} v_s \\ 0 \\ 0 \end{bmatrix}$$

d. Solve for i_1, i_2, and i_3. Use the resistor values and power supply setting specified in the laboratory.

e. Use the loop currents to find the node voltages.

2. Analyze the circuit of Figure 5L4.1 again, but this time, a little differently, as follows.

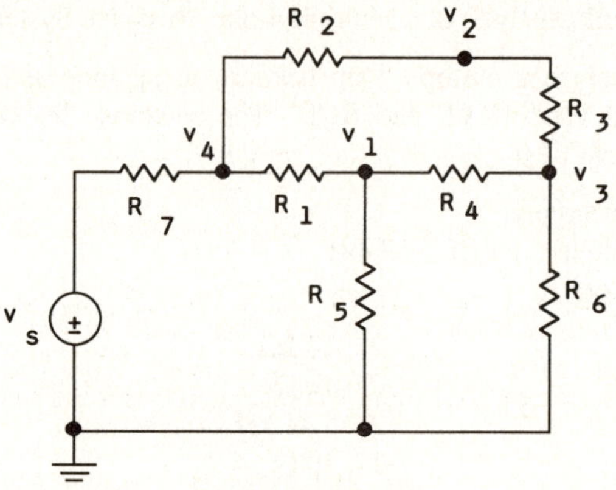

Figure 5L4.1 (b)

a. Analyze the circuit using node analysis.

b. Put your equations into the matrix form

$$Av = \begin{bmatrix} a_{11} & a_{12} & a_{13} & a_{14} \\ a_{21} & a_{22} & a_{23} & a_{24} \\ a_{31} & a_{32} & a_{33} & a_{34} \\ a_{41} & a_{42} & a_{43} & a_{44} \end{bmatrix} \begin{bmatrix} v_1 \\ v_2 \\ v_3 \\ v_4 \end{bmatrix} = \begin{bmatrix} v_s \\ 0 \\ 0 \\ 0 \end{bmatrix}$$

c. Find the inverse of the matrix A, and premultiply by A^{-1}. Keep all terms in symbolic form. Mathematically,

$$A^{-1} AV = V = A^{-1} \begin{bmatrix} v_s \\ 0 \\ 0 \\ 0 \end{bmatrix}$$

d. Solve for v_1, v_2, v_3, and v_4. Use the resistor values and voltage source setting specified.

e. Find the node currents from v_1, v_2, and v_3.

Laboratory Procedure:

1. Let the circuit of Figure 5L4.1 have the following characteristics:

$$v_s = 1 \text{ V}$$
$$R_1 = R_2 = 10 \text{ k}\Omega$$
$$R_3 = R_4 = 1 \text{ k}\Omega$$
$$R_5 = R_6 = 5.11 \text{ k}\Omega$$
$$R_7 = 10 \text{ }\Omega$$

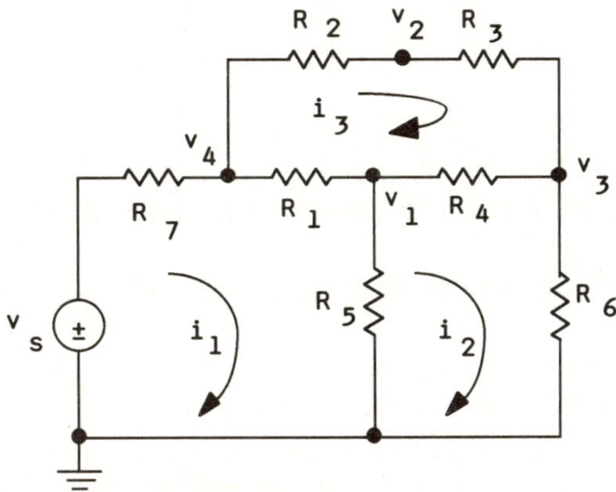

Figure 5L4.1 (c)

2. Use a voltmeter and ammeter to measure all branch currents and voltages, loop currents, and node voltages.

Record all the specified measurements on the following chart:

Branch Voltages	Branch Currents	Loop Currents
$v_{R_1} = \underline{\quad} \pm \underline{\quad}$	$i_{R_1} = \underline{\quad} \pm \underline{\quad}$	$i_1 = \underline{\quad} \pm \underline{\quad}$
$v_{R_2} = \underline{\quad} \pm \underline{\quad}$	$i_{R_2} = \underline{\quad} \pm \underline{\quad}$	$i_2 = \underline{\quad} \pm \underline{\quad}$
$v_{R_3} = \underline{\quad} \pm \underline{\quad}$	$i_{R_3} = \underline{\quad} \pm \underline{\quad}$	$i_3 = \underline{\quad} \pm \underline{\quad}$
$v_{R_4} = \underline{\quad} \pm \underline{\quad}$	$i_{R_4} = \underline{\quad} \pm \underline{\quad}$	
$v_{R_5} = \underline{\quad} \pm \underline{\quad}$	$i_{R_5} = \underline{\quad} \pm \underline{\quad}$	Node voltages
$v_{R_6} = \underline{\quad} \pm \underline{\quad}$	$i_{R_6} = \underline{\quad} \pm \underline{\quad}$	$v_1 = \underline{\quad} \pm \underline{\quad}$
$v_{R_7} = \underline{\quad} \pm \underline{\quad}$	$i_{R_7} = \underline{\quad} \pm \underline{\quad}$	$v_2 = \underline{\quad} \pm \underline{\quad}$
		$v_3 = \underline{\quad} \pm \underline{\quad}$
		$v_4 = \underline{\quad} \pm \underline{\quad}$

3. Given 10%-tolerance resistors, determine the maximum and minimum values for the node voltages and loop currents. (Assume a 1% accuracy for V_s).

DISCUSSION: Compare the maximum and minimum values for all voltages and currents obtained in the prelab with the values obtained in the experimental section of the lab. Do your data agree with your calculations? Explain.

CONCLUSIONS: Did the experiment verify KVL, KCL, and Ohm's law? Comment.

LAB 5L5

TITLE: Analysis and Design Using an RC Network

OBJECTIVE: To apply Kirchhoff's and Ohm's laws to the analysis and design of a network to meet a given set of specifications.

EQUIPMENT:

Pulse generator
A selection of resistors and capacitors (see Laboratory Procedure)
Oscilloscope (dual trace)
Breadboard

BACKGROUND:

The lab requires the generation of pulses. These can be developed from a pulse generator or a function generator has square waves and an offset control.

A pulse can be modeled or represented by two step functions.

A mathematical expression for a step function, usually denoted by u(t), can be written as

$$u(t)=1, \quad t \geq 0$$
$$u(t)=0, \quad t < 0$$

Graphically, u(t) is shown in Figure 5L5.1.

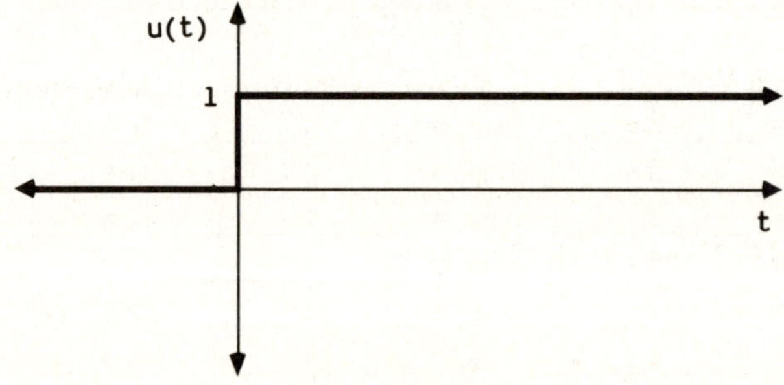

Figure 5L5.1

If we want to delay a step function by τ seconds, we simply write

$$u(t-\tau)$$

For any function u(x) to be nonzero, the argument of u, namely x, must be greater than or equal to zero. Thus, for u(t - τ) to equal 1, t - $\tau \geq 0$. the graph of u(t - τ) is shown in Figure 5L5.2.

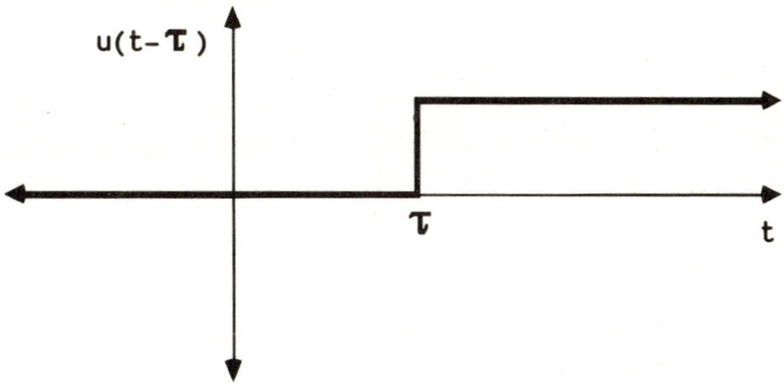

Figure 5L5.2

Also, -u(t) is defined by

$$-u(t) = -1, \qquad t \geq 0$$
$$-u(t) = \ \ 0, \qquad t < 0$$

Figure 5L5.3 shows the graph of -u(t).

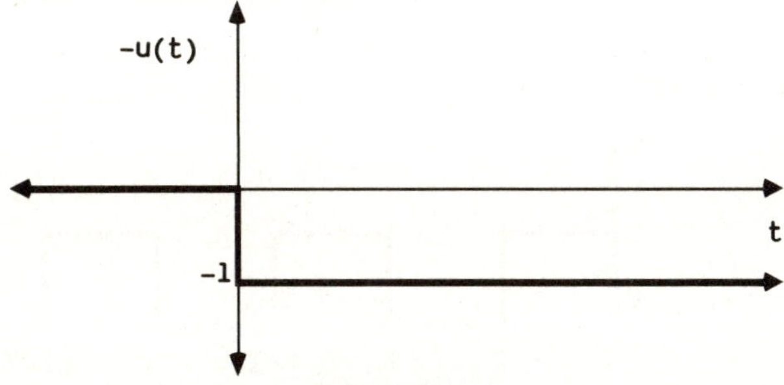

Figure 5L5.3

Applying the preceding concepts allows us mathematically to describe a pulse function as two unit step functions:

$$p(t) = u(t) - u(t-\tau)$$

Figure 5L5.4 shows the curve p.

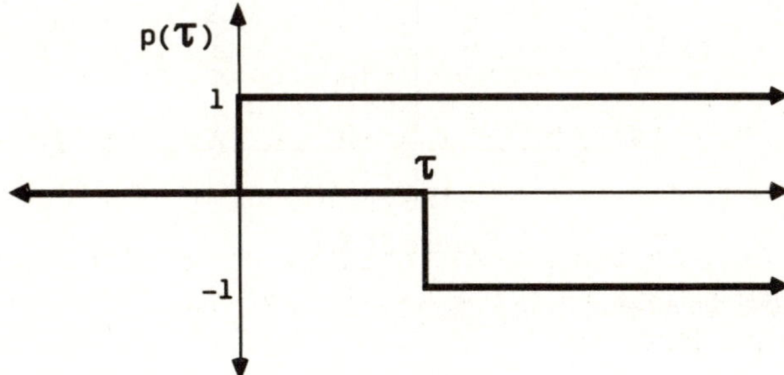

Figure 5L5.4

Adding these two waveforms yields,

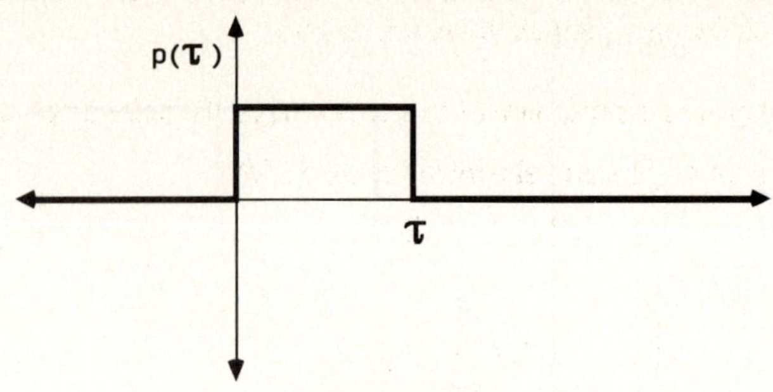

Figure 5L5.5

where τ is referred to as a pulse width.

These concepts can be applied to generate the series of pulses shown in Figure 5L5.6, where the pulse train is given by

$$\text{pulse train} = u(t) - u(t-\tau) + u(t-T) - u(t-(T+\tau)) + u(t-2T) - u(t-(2T+\tau))$$

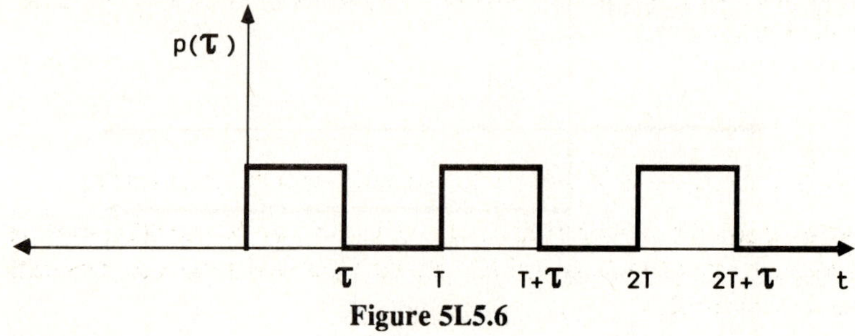

Figure 5L5.6

PRELABORATORY:

1. For the circuit of Figure 5L5.7,

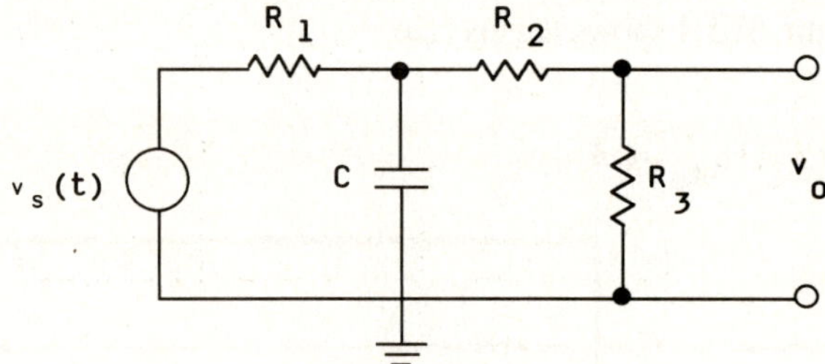

Figure 5L5.7

a. Let v(t) be a pulse waveform, as in Figure 5L5.8.

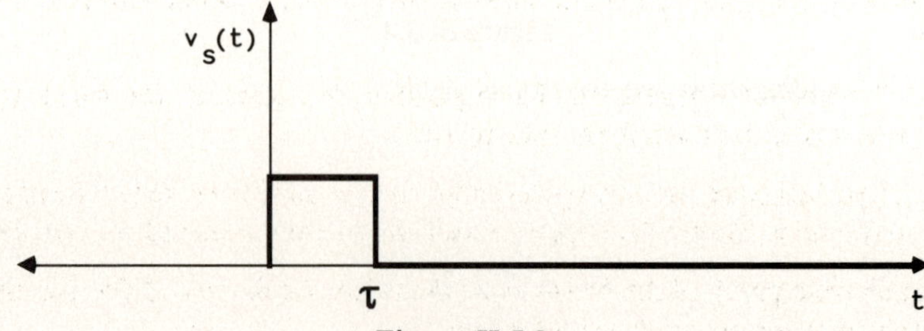

Figure 5L5.8

With $R_1=R_2=R_3=1k\Omega$ and $C=1\mu f$, determine a pulse width and amplitude to place 3 volts on the capacitor when the pulse amplitude returns to zero.

Calculate and sketch the response of this circuit when the pulse amplitude returns to zero.

   b.   Let v(t) be a train of pulses as shown in Figure 5L5.9.

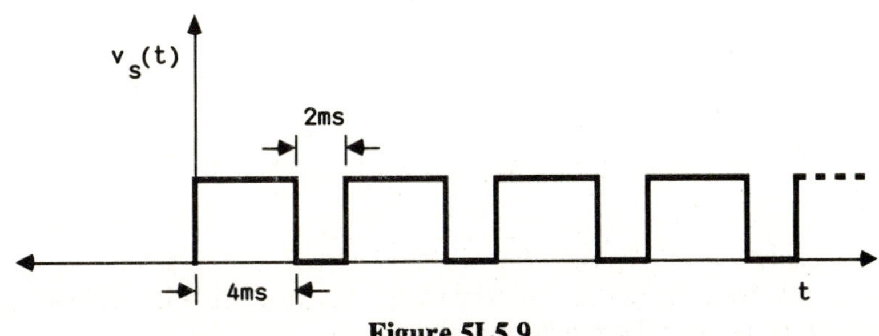

Figure 5L5.9

Determine values for the resistors and capacitors that will result in a peak-to-peak variation of 0.5 V for v_o after 4 cycles of $v_s(t)$. After a large number of cycles of $v_s(t)$, v_o should meet the following specifications:

$$v_{o_{max}} \leq 5 \text{ V}$$
$$v_{o_{min}} \geq 4.5 \text{ V}$$

After large number of cycles, the output waveform reaches steady state.

   2.   Do a worst-case analysis of your circuit design to ensure that the variation in v_o is met after 4 cycles. You should consider the tolerances of all R's and the C, as well as the accuracy with which $v_s(t)$ can be set and v_o can be measured.

Laboratory Procedure:

   1.   Build the circuit of Figure 5L5.10.

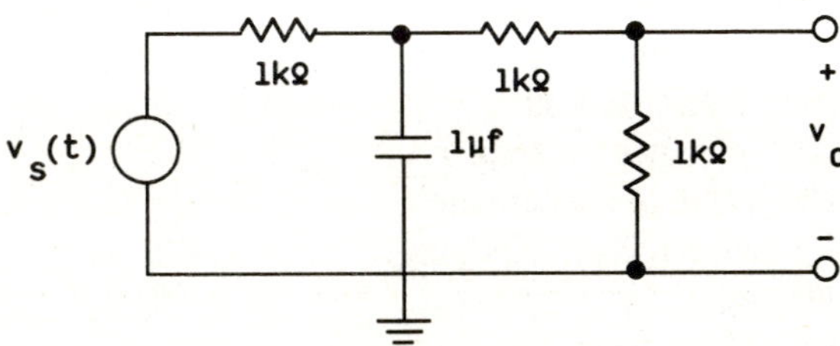

Figure 5L5.10

   a.   Use a pulse generator to obtain a train of pulses that allows you to verify the width and amplitude settings you determined in part a of the prelab. Remember, the time between pulses must be long enough to allow the charge on the capacitor to return to zero before the next pulse arrives.

      1.   Why would the generation of a single pulse be inadequate for this measurement (assuming you don't have a storage oscilloscope)?

      2.   Use a dual-trace oscilloscope in chop mode or an alternative oscilloscope system to display $v_s(t)$ and $v_o(t)$ simultaneously. Sketch these results. Record your oscilloscope settings.

   2.   Build the circuit from part b of the prelab using the values for R's and C that you determined. Display v(t) and $v_o(t)$ in chop mode and sketch the results.

    a. Does your design meet the specifications?

    b. Sketch the response $v_o(t)$ of your circuit for each change described below. Apply each change individually by always returning your circuit to its original values before the next change.

        1. Double the capacitor value

        2. Use half the capacitor value.

        3. Double all resistor values.

        4. Divide each resistor value by two.

Describe the effect each change has on the response and relate this to the equations describing the system.

3. Use SPICE to analyze your system further. Run SPICE (or a similar program) to verify all experimental results from part II of the laboratory. Use the .SENS statement of SPICE to do a sensitivity analysis of v_{out} for all component variations.

DISCUSSION:

CONCLUSIONS:

LAB 5L6

TITLE: Resistivity and a Two-Loop Network

OBJECTIVE: Two concepts are discussed in this laboratory: analysis of a simple two-loop network using KVL, KCL, and Ohm's law, and the concept of resistivity.

EQUIPMENT:

Power supply
   voltage 0-10 V @ 500 mA.
Two multimeters capable of measuring volts, ohms, and amperes
Resistors
   $100\,\Omega$, $1\,k\Omega$ 1/8 watt
   $\pm 10\%$ tolerance (or better).
1-meter length of copper wire
   wire between 20 and 28 gauge
Breadboard
Beaker and hot plate
Ice

PRELABORATORY:

1. Given the circuit of Figure 5L6.1,

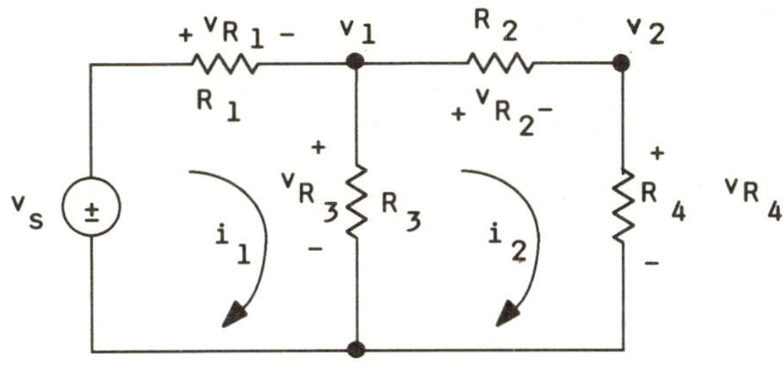

Figure 5L6.1

a. Write the Mesh equations for the circuit, and solve for i_1 and i_2. Use these values to find v_1 and v_2. (Let $v_s = 1$ V; $R_1 = R_2 = 100\ \Omega$, and $R_3 = R_4 = 1\ k\Omega$).

b. Write the node equations for the circuit, and solve for v_1 and v_2. Use v_1 and v_2 to find i_1 and i_2. (Assume the same values as in part A.)

c. Replace R_2, R_3 and R_4 with an equivalent resistance R_1 that "sees" the same load at node 1. Solve for v_1. (Use the voltage divider theory.)

d. Assume the circuit resistors have a $\pm 10\%$ tolerance and v_s is set to 1 V $\pm 1\%$. What are the maximum and minimum values of v_1 and v_2?

2. Resistivity

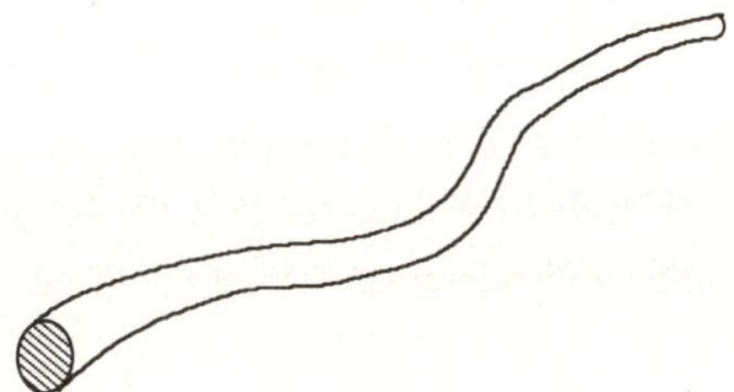

Figure 5L6.2

a. Given a 1-meter length of copper wire (Figure 5L6.2), without mechanically measuring the cross sectional area of the wire, how can you use the resistivity of copper to determine the cross sectional area? (The resistivity of copper wire, ρ_{cu}, is $1.673\text{x}10^{-8}\ \Omega$-m at $20°$C. Copper's temperature coefficient of resistivity is $4.3\text{x}10^{-11}\ \Omega$-m/$°$C for temperatures between 0 and $100°$C.)

b. If the resistance of a 1-meter length of copper is $0.131\ \Omega$, what is the wire gauge? (Assume that the temperature is $20°$C.)

c. Given a 1-meter length of copper wire with a diameter of $0.0005"$, how much would the resistance of the wire change per $°$C? If the wire were replaced by a platinum wire, what would be the change in resistance per $°$C? Which wire would be better for measuring temperature? Why?

LABORATORY PROCEDURE:

1. Build the circuit shown in Figure 5L6.3.

Measure the following quantities and put the data into tabular form: resistor voltages (v_{R_1}, v_{R_2}, v_{R_3}, and v_{R_4}), loop currents (i_1 and i_2), node voltages (v_1 and v_2), and component currents (i_{R_1}, i_{R_2}, i_{R_3}, i_{R_4} and i_{v_s}). Also, determine the power delivered by v_s and the power dissipated in each resistor. Does the total power delivered equal the total power dissipated? Compare these results to those of the prelab. Are your experimental results within worst-case maximum and minimum values?

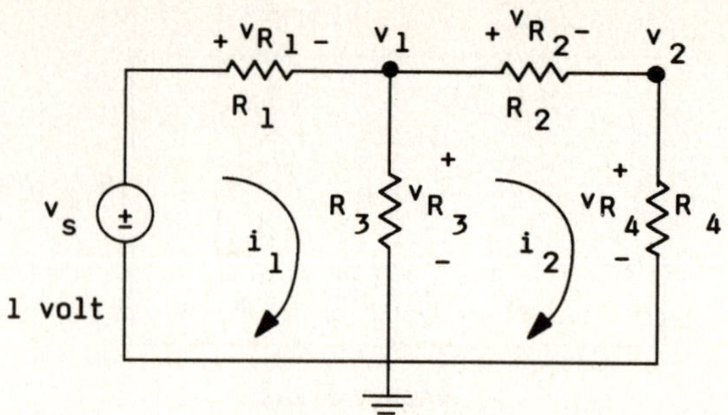

Figure 5L6.3

2. Each student or lab group will receive a 1-meter length of copper wire of an unknown gauge.

    Coil the wire and place it into a cup of distilled ice water. What is the gauge of the wire? (What should you do about your meter's lead resistance?) Predict what the resistance will be at 100°C. Place the coiled copper wire in a cup of boiling water and record the resistance. Compare the theoretical results for a temperature of 100°C to your experimental data.

DISCUSSION:

CONCLUSIONS:

REFERENCES

[1] C. J. Savant, Jr., M. S. Roden, and G. L. Carpenter, *Electronic Circuit Design: An Engineering Approach* (Menlo Park, CA: Benjamin/Cummings, 1987), pp. 324-38.

[2] A. S. Sedra and K. C. Smith, *Microelectronic Circuits,* 2d ed. (New York: Holt, Rinehart and Winston, 1987), p. 87-113.

[3] M. E. Van Valkenburg, *Network Analysis,* 3d ed. (Englewood Cliffs, NJ: Prentice-Hall, 1974), pp. 83-4.

[4] A. W. Al-Khafaji and J. R. Tooley, *Numerical Methods in Engineering Practice* (New York: Holt, Rinehart and Winston, 1986), pp. 412-40.

Chapter 6

LINEAR NETWORK THEOREMS

6.1 Introduction

The analysis of circuits can be shortened considerably by the use of certain network theorems. As an illustration, if we are interested in what happens at a particular element in a circuit, it may be possible to replace the remainder of the circuit by an equivalent, but simpler, circuit through the use of one of the network theorems. The new circuit is usually much easier to solve.

In this chapter, we discuss a number of network theorems and illustrate their use in laboratory experiments.

6.2 Thevenin and Norton Equivalents[1],[2],[3]

The use of either Thevenin or Norton's theorem enables us to replace an entire circuit, seen at a pair of terminals, by an equivalent circuit made up of a single impedance and a source. Therefore, we may determine the voltage or current of a single element of a relatively complex circuit by replacing the rest of the circuit with an equivalent impedance and source and analyzing the resulting circuit. Thevenin and Norton equivalent circuits can be used to represent any circuit made up of <u>linear</u> elements. An element is said to be linear if multiplying a variable x (current or voltage) by a constant K results in the multiplication of variable y (voltage or current) by the same constant K.

Figure 6.1(a) represents any circuit made up of sources, both independent and dependent, and linear elements. We identify the terminal pairs as 1 and 2. In Figure 6.1(b), we show the Thevenin equivalent to the circuit of Figure 6.1 (a).

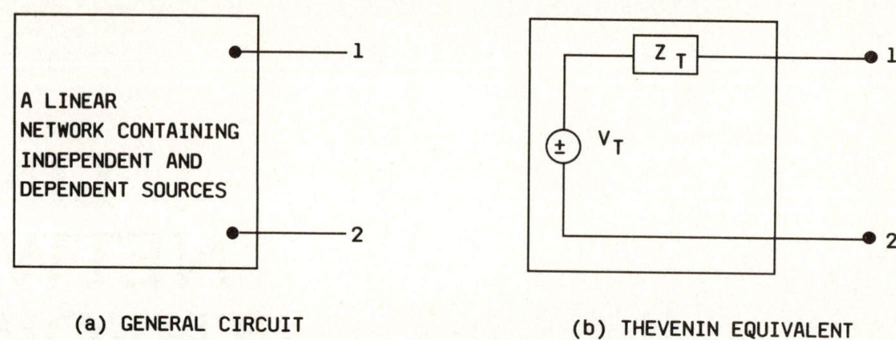

Figure 6.1: Thevenin equivalent circuit.

The original interconnection of sources and impedances can be replaced by an independent voltage source v_T, in series with an impedance Z_T. This series combination is equivalent to the original circuit because, if we connect the same load across terminals 1 and 2 of each circuit, we get the same voltage and currents at the terminals of the load. The equivalence will hold for all possible values of load impedance.

In order to represent the original circuit by its Thevenin equivalent, we have to be able to determine the Thevenin voltage v_T and the Thevenin impedance Z_T. To calculate v_T, it is only necessary to calculate the open-circuit voltage in the original circuit. Thus, $v_T = v_{OC}$. Now, if we place a short circuit across terminals 1 and 2 of the Thevenin equivalent circuit, the short-circuit current from 1 to 2 will be

$$i_{SC} = \frac{v_T}{Z_T} \tag{6.1}$$

This current i_{sc} and the one that exists in a short circuit placed across terminals 1 and 2 of the original network are identical. So from Equation (6.1), we can write

$$Z_{TH} = \frac{v_T}{i_{SC}} \tag{6.2}$$

Therefore, the Thevenin impedance is the ratio of the open-circuit voltage to the short-circuit current. If the network under investigation contains independent sources, we can deactivate all such sources and then calculate the impedance seen looking into the network at the designated terminal pair. A voltage source is deactivated by replacing it with a short circuit. A current source is deactivated by replacing it with an open circuit. If the network contains dependent sources, an alternative procedure for finding Z_T is to deactivate all independent sources. By applying either a test voltage source or a test current source to the Thevenin terminals 1 and 2, the Thevenin impedance will equal the ratio of the voltage across the test source to the current delivered by the test source.

The Norton equivalent circuit consists of an independent current source in parallel with the Norton equivalent resistance. This circuit can be derived from the Thevenin equivalent circuit simply by making a source transformation. Thus, the Norton current equals the short-circuit current at the terminals of interest and the Norton impedance is identical to the Thevenin impedance (see Figure 6.2).

A source transformation is a technique, Figure 6.2, that allows us to replace a voltage source in series with an impedance by a current source in parallel with the same impedance.

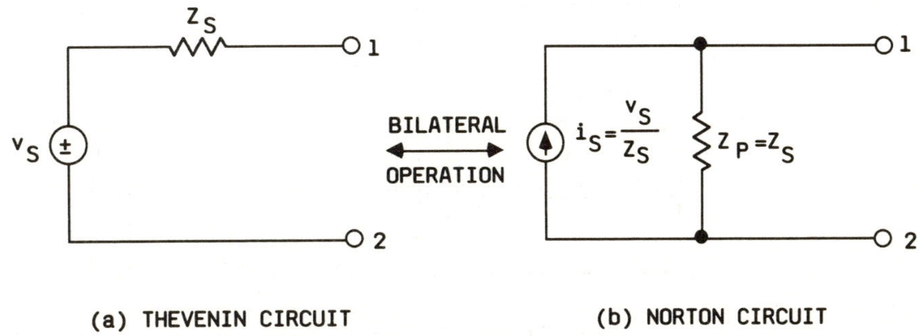

Figure 6.2: Source transformation

The two configurations shown in the figure are equivalent with respect to the terminals 1 and 2 provided that

$$i_S = \frac{v_s}{Z_S} \tag{6.3}$$

and

$$Z_S = Z_p \tag{6.4}$$

We can use the technique of source transformations[4] to derive a Thevenin or Norton equivalent circuit. This technique is particularly useful when a network contains only independent sources. The presence of dependent sources complicates the procedure in that retention of the identity of the controlling voltage and currents is required, which usually limits circuit reduction using source transformation.

Example: Use of source, Thevenin, and Norton transformations
to find the current I of voltage source v_{S2}.

Source Transformation[5],[6]

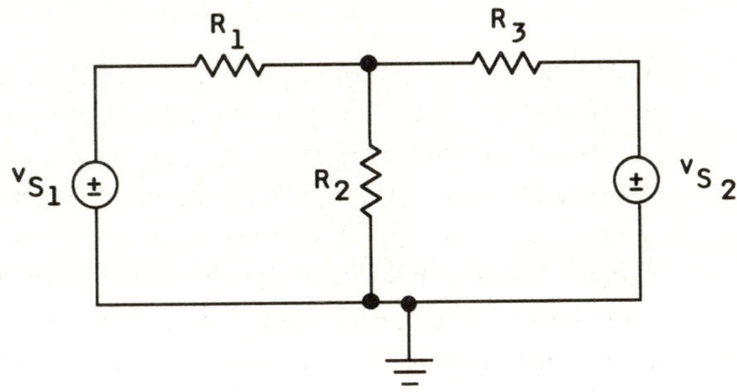

Figure 6.3(a): Original circuit

Step 1: Transform v_{S_1} from a voltage source, Figure 6.3(a), to a current source,
Figure 6.3(b)

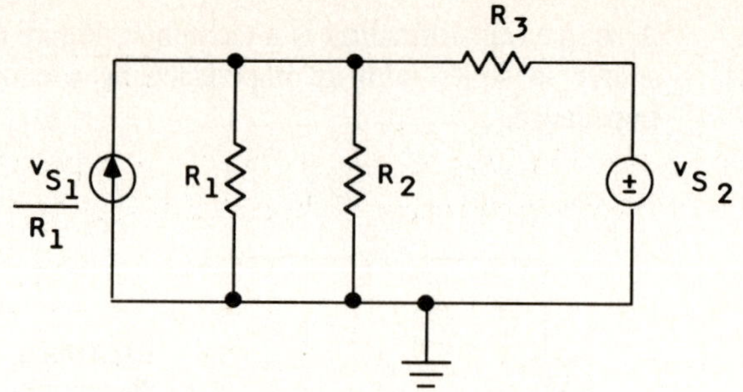

Figure 6.3(b): Step 1

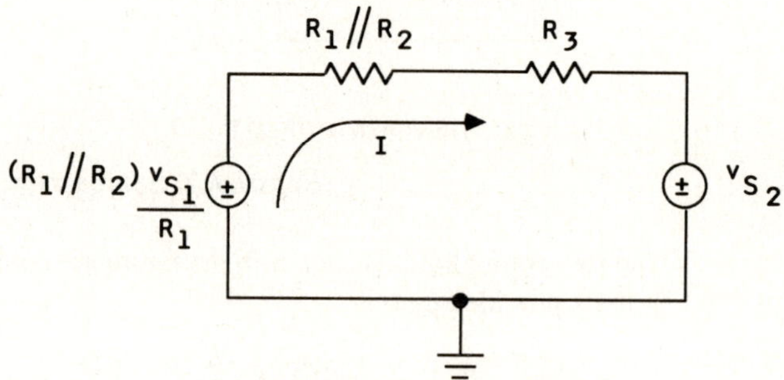

Figure 6.3(c): Step 2

Step 2: Combine R_1 and R_2 and do a second source transformation, Figure 6.3(c).

Step 3: Find the current I associated with v_{S_2}:

$$I = \frac{-(R_1//R_2)\dfrac{v_{S_1}}{R_1} + v_{S_2}}{R_1//R_2 + R_3} \tag{6.5}$$

Thevenin Equivalent

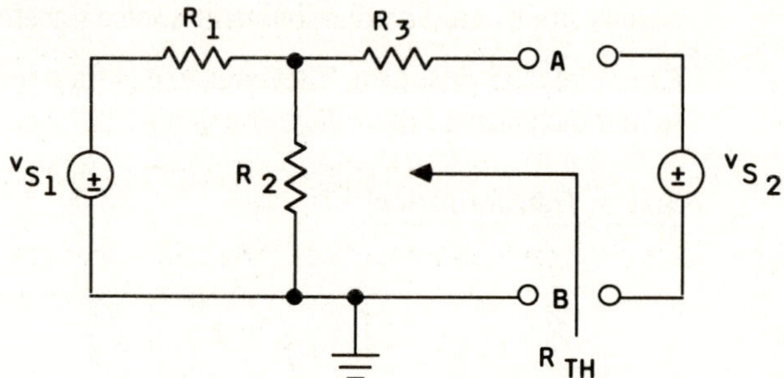

Figure 6.4: Thevenin equivalent circuit

R_T is found by shorting v_{S_1} and calculating the resistance looking into nodes A-B:

$$R_T = R_1//R_2 + R_3 \tag{6.6}$$

v_T is the open circuit voltage of nodes A-B. So

$$v_T = v_{OC} \tag{6.7}$$

Because there is no current in resistor R_3, the voltage across resistor R_2 is also v_T. We can then use voltage division to find the voltage across R_2.

$$v_T = v_{OC} = v_{S_1} \frac{R_2}{R_1 + R_2} \tag{6.8}$$

Therefore, v_{S_1}, R_1, R_2 and R_3 can be replaced by v_T and R_T to yield Figure 6.5, the Thevenin equivalent of the circuit of Figure 6.4.

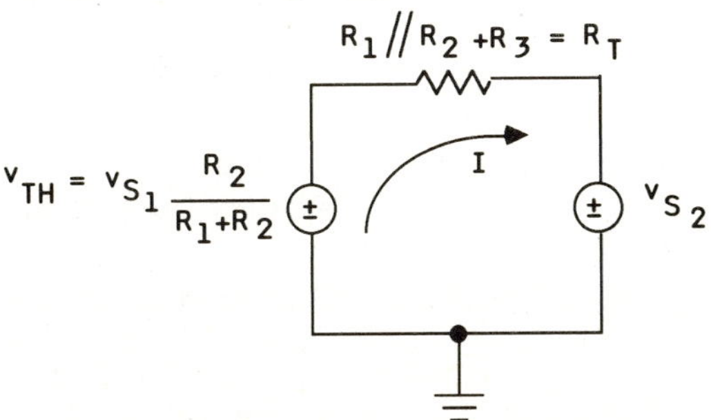

Figure 6.5: Thevenin equivalent of Figure 6.4.

The current I associated with v_{S_2} is

$$I = \frac{-v_{S_1} \dfrac{R_2}{R_1 + R_2} + v_{S_2}}{R_1 /\!/ R_2 + R_3} \tag{6.9}$$

The Norton equivalent, then, is the circuit of Figure 6.6

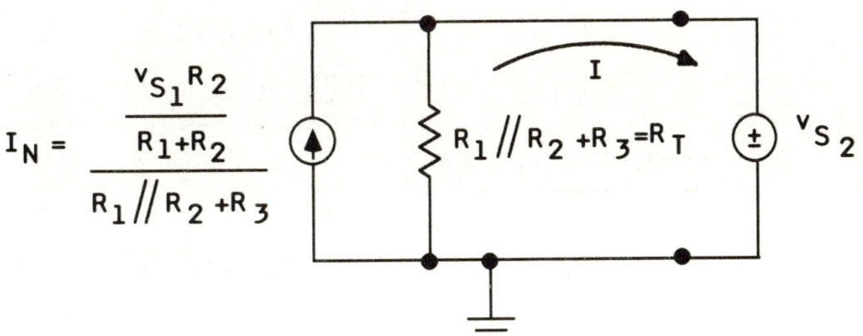

Figure 6.6: Norton equivalent of Figure 6.4

Notice that the solution for the current I is the same in all three cases.

6.3 Superposition[7],[8],[9]

The superposition theorem states that, whenever a <u>linear</u> system is driven by more than one independent source of energy, the total <u>response</u> can be found by finding the response to each independent source separately and summing the individual responses. The principle is applicable to any linear system.

Example: Use of superposition
to find the current of v_{S_2}.

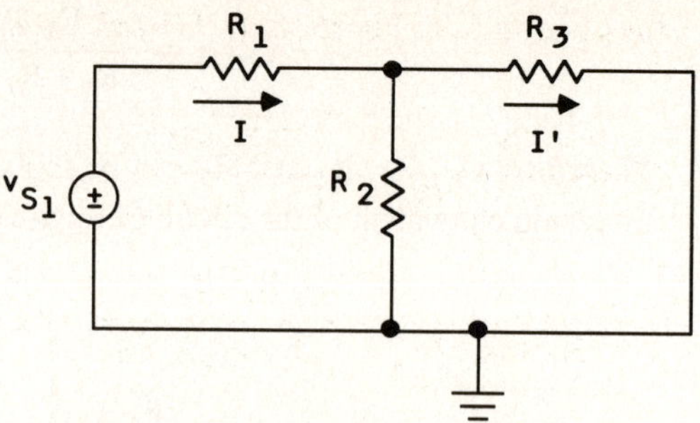

Figure 6.7(a): Superposition with $v_{S_2} = 0$

Step 1: From Figure 6.3 we short circuit source v_{S_2} and apply source v_{S_1}, Figure 6.7(a). The resistance v_{S_1} sees is $R_1 + R_2//R_3$; therefore,

$$I = \frac{v_{S_1}}{R_1 + R_2//R_3} \tag{6.10}$$

and

$$I' = \left[\frac{v_{S_1}}{R_1 + R_2//R_3} \right] \frac{R_2}{R_2 + R_3} \tag{6.11}$$

Step 2: We short source v_{S_1}, apply source v_{S_2}, and solve for the current in R_3:

$$I'' = \frac{v_{S_2}}{R_3 + R_1//R_2} \tag{6.12}$$

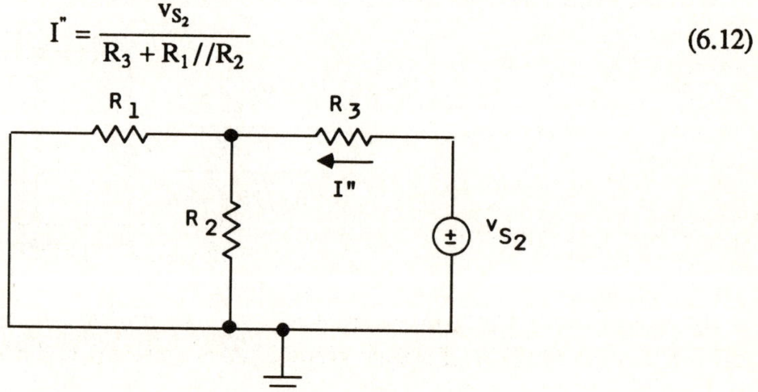

Figure 6.7(b): Superposition with $v_{S_1} = 0$

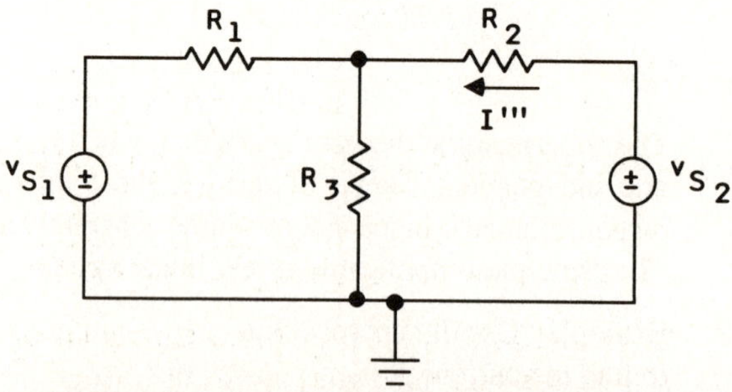

Figure 6.7(c): Superposition with both v_{S_1} and v_{S_2} applied from Figures 6.7(a), 6.7(b) and 6.7(c),

$$I''' = I'' - I' \tag{6.13}$$

Substituting from Equations (6.11) and (6.12) yields

$$I''' = \frac{v_{S_2}}{R_3 + R_1//R_2} - \frac{v_{S_1}\left(\dfrac{R_2}{R_2 + R_3}\right)}{R_1 + R_2//R_3}$$

$$I''' = \frac{v_{S_2}}{R_3 + R_1//R_2} - \frac{v_{S_1}\,R_2\,(R_1 + R_2)}{R_1//R_2 + R_3} = -\frac{v_{S_1}\dfrac{R_2}{R_1 + R_2} - v_{S_2}}{R_1//R_2 + R_3} \tag{6.14}$$

A comparison of Equations (6.9) and (6.14) shows that the result for the current of v_{S_2} is the same as was obtained using the Thevenin and Norton equivalents.

6.4 *Maximum Power Transfer Theorem*[10],[11]

When a source with constant voltage v supplies a load of impedance Z_L through a fixed value of series impedance Z_S, maximum power is received by the load if its impedance is the conjugate of Z_S (Figure 6.8). For example, if Z_S were a pure resistance, Z_L would be an equal resistance in order to receive maximum power. If Z_S were somewhat inductive, Z_L would need to be somewhat capacitive. So for maximum power transfer,

$$Z_L = Z_s^*, \; R_L = R_s, \; X_L = -X_s \tag{6.15}$$

where Z_s^* denotes the conjugate of Z_s.

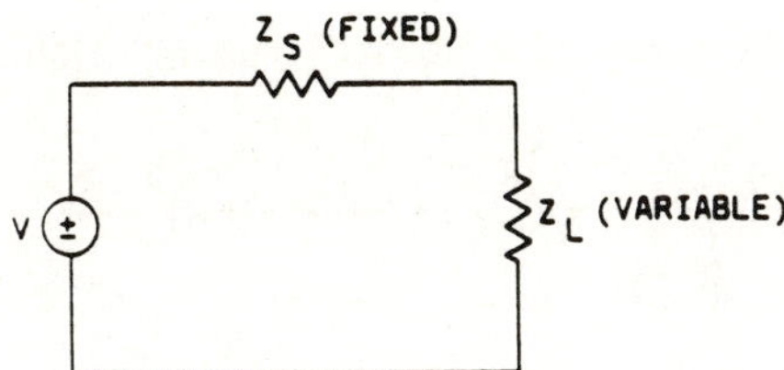

Figure 6.8: Maximum power transfer

The power dissipated in Z_L is a function of the three circuit parameters v, Z_S, and Z_L. Therefore, the average power delivered to the load is

$$P = |I|^2 R_L \tag{6.16}$$

where

$$I = \frac{v}{(R_S + R_{L)} + j(X_S + X_{L)}} \tag{6.17}$$

Substituting I into Equation (6.16), we can write

$$P = \frac{|v|^2 R_L}{(R_S + R_L)^2 + (X_S + X_L)^2} \tag{6.18}$$

To maximize P, we need to find the values of R_L and X_L, for which $\partial P/\partial R_L$ and $\partial P/\partial X_L$ are both zero. This results in

$$X_L = -X_S \tag{6.19}$$

and $\quad R_L = \sqrt{R_S^2 + (X_L + X_X)^2} \tag{6.20}$

Combining the two results shows that both derivatives are zero when $Z_L = Z_S^*$, and the maximum average power delivered to the load is

$$P_{max} = \frac{|v^2|R_L}{4R_L^2} = \frac{1}{4}\frac{|v|^2}{R_L} \tag{6.21}$$

6.5 Tellegan's Theorem[12],[13],[14]

Tellegan's theorem, which depends on KVL and KCL, is a general network theorem that is valid for any lumped network that contains any elements, i.e., linear or nonlinear, passive or active, time varying or time invariant.

Tellegan's theorem states that for an arbitrary lumped network with b branches and n nodes, and with each branch assigned arbitrarily a branch voltage, v_k, satisfying KVL and a branch current I_k satisfying KCL, for $k = 1, 2, 3, \ldots, b$, if the voltages and currents are measured with respect to arbitrary selected associated reference directions, then

$$\sum_{k=1}^{b} v_k(t)i_k(t) = 0 \tag{6.20}$$

Since $v_k(t)i_k(t)$ is the power delivered at time t by the network to branch k, the theorem may be interpreted to mean that as far as lumped circuits are concerned, KVL and KCL imply conservation of energy and that the sum of the power delivered by the independent sources to the network is equal to the sum of the power absorbed by all of the other branches of the network.

As an example consider Figure 6.9. Here, we

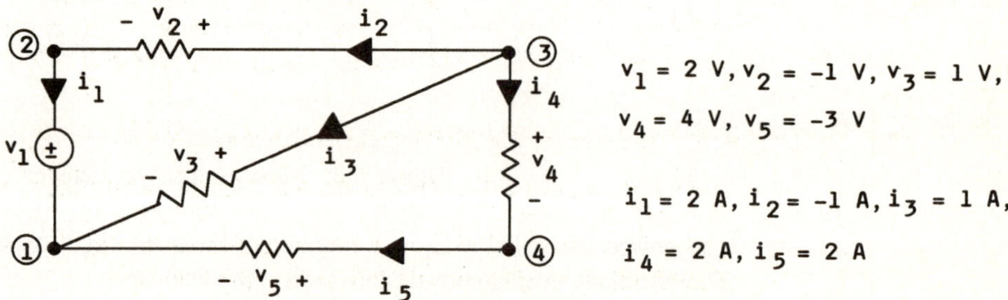

Figure 6.9: Example of Tellegan's theorem

arbitrarily assign branch voltages and branch currents subject only to satisfying KVL and KCL for all the loops and nodes. KVL is satisfied since

$$v_1 + v_2 = v_3 = v_4 + v_5 \tag{6.21}$$

$$2 + (-1) = 1 = 4 + (-3)$$

$$1 = 1 = 1$$

and KCL is satisfied since

$$i_1 = i_2, \; i_4 = i_5, \; i_1 + i_3 + i_5 = 0 \tag{6.22}$$

$$1 = 1, \; 2 = 2, \; 1 + (-3) + 2 = 0$$

Tellegan's theorem states that

$$\sum_{k=1}^{5} v_k i_k = v_1 i_1 + v_2 i_2 + v_3 i_3 + v_4 i_4 + v_5 i_5$$

$$= (2)(1) + (-1)(1) + (1)(-3) + (4)(2) + (-3)(2)$$

$$\text{So } \sum_{k=1}^{5} v_k i_k = 2 - 1 - 3 + 8 - 6 = 0 \tag{6.23}$$

6.6 Miller's Theorem[15]

Consider the circuits of Figure 6.10. The broken lines

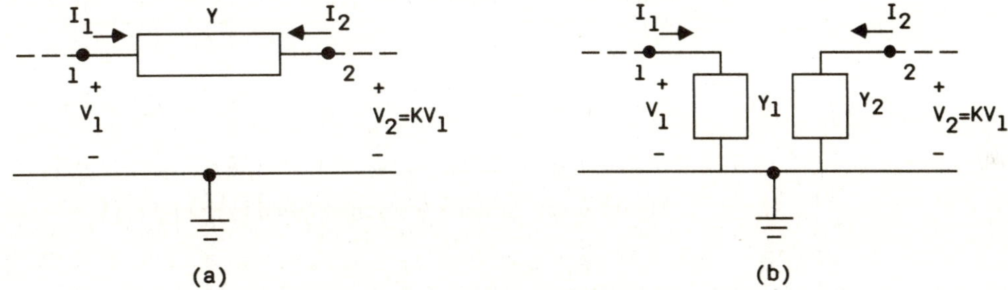

Figure 6.10: Miller's theorem

indicate that the circuits may be part of a network. Miller's theorem states that the admittance Y of Figure 6.14 may be replaced by two admittances Y_1, between node 1 and ground and y_2 between node 2 and ground, as shown in Figures 6.14(b).

The voltage gain from node 1 to node 2 is denoted by $K \equiv \dfrac{V_2}{V_1}$. Miller's theorem is dependent on the condition that K can be determined by independent means.

It is easy to show that for the currents I_1 and I_2 of Figure 6.14(a) to be equal to the corresponding currents of Figure 6.14(b), the conditions

$$Y_1 = Y(1 - K) \tag{6.24}$$

and

$$Y_2 = Y(1 - \frac{1}{K}) \tag{6.25}$$

must apply. Thus, for the circuits of the two figures to be equivalent, Equations (6.24) and 6.25) must be satisfied. It is important to note that Miller's theorem applies only when the conditions that existed in the network when K was determined are not changed.

As an example, consider the circuit of Figure 6.11.

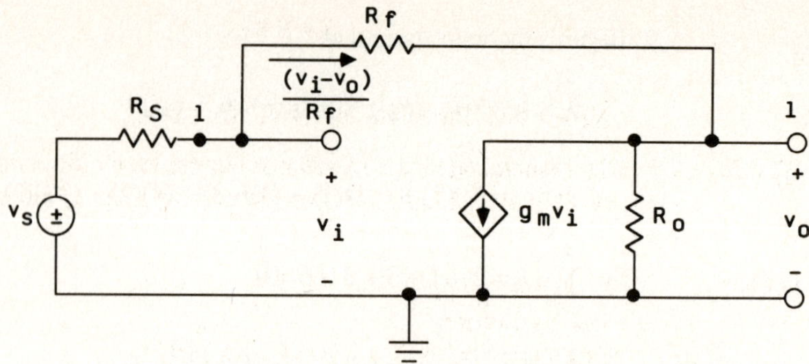

Figure 6.11: Equivalent circuit for a feedback amplifier

Miller's theorem can be used to simplify this circuit to that shown in Figure 6.12, with

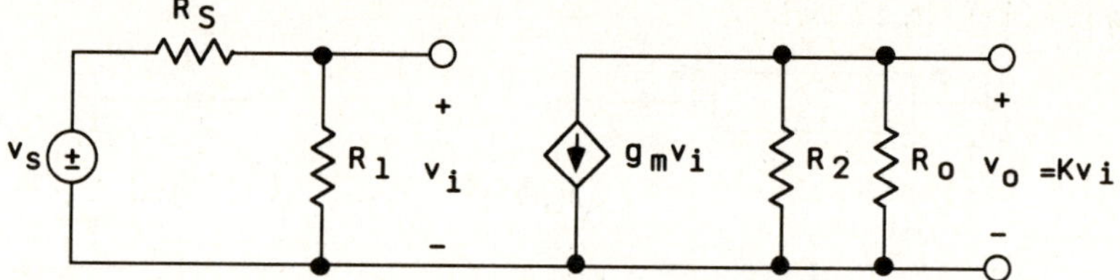

Figure 6.12: Miller's theorem applied to Figure 6.11

$$R_1 = \frac{R_f}{1-K} \text{ and } R_2 = \frac{R_f}{1 - \dfrac{1}{K}}.$$ The term $K = \dfrac{v_o}{v_i}$ and can be found by writing a node

equation at node 2 of Figure 6.11:

$$\frac{v_i - v_o}{R_f} = g_m v_i + \frac{v_o}{R_o} \tag{6.26}$$

Solving for the voltage gain yields

$$K = \frac{v_o}{v_i} = \frac{-g_m + \dfrac{1}{R_f}}{\left[\dfrac{1}{R_o} + \dfrac{1}{R_f}\right]} \tag{6.27}$$

6.7 Summary

The theorems introduced in this chapter are all quite general. They apply equally to power systems, communication systems, low-frequency systems, and high-frequency systems. Superposition permits sources to be treated separately and so is used when there are several sources. Source transformation permits substitution of a constant-current source for a constant-voltage source or vice versa. Thevenin's and Norton's theorems allow us to substitute a simple circuit for the part of a network that remains constant while we analyze another part of the network.

Tellegan's theorem and maximum power transfer, although not true network theorems, will prove useful in the solution of circuit problems. Miller's theorem or the Miller equivalent circuit is useful for determining the input impedance and the forward transmission, or gain, of an amplifier.

LAB 6L1

TITLE: Norton and Thevenin Equivalent Circuits[16]

OBJECTIVE: To demonstrate the equivalence between a multiple resistive network and its Norton or Thevenin circuit. The concept of load lines will also be used.

EQUIPMENT: One power supply, 5 V
One multimeter
Resistors: 4 x 1 kΩ, 1 x 10 kΩ, 1 x 100 Ω.
Breadboard

PRELABORATORY:

1. In parts a, b, and c below, start with the original circuit before doing each modification. Given the circuit in Figure 6L1.1;

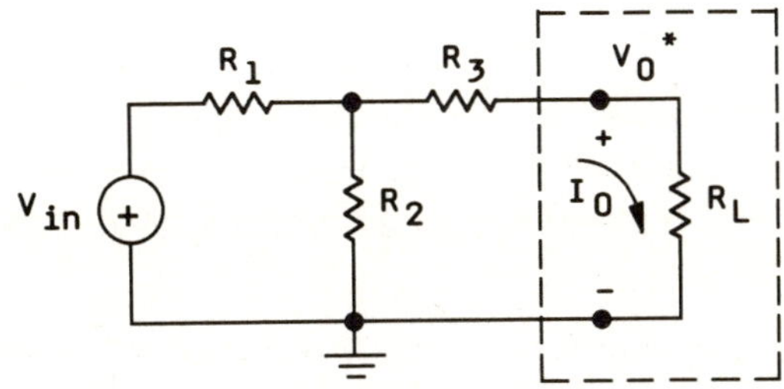

Figure 6L1.1: Test circuit for Thevenin and Norton circuits

  a. Disconnect the load resistor R_L and calculate the voltage at node V_o relative to ground. What is this voltage called?

  b. Short circuit the V_o node (connect V_o to ground) and calculate the current flowing in this short circuit. What is this current called?

  c. Short circuit V_{in} (disconnect V_{in} and connect a short circuit between the open end of R_1 and ground). Remove the load resistor R_L and determine the resistance of the network "looking into" this port. What is this resistance called?

  d. Draw the Norton and Thevenin equivalent circuits. Show the values of V_T, I_T, and Z_T (R_T in this case).

Laboratory Procedure:

1. Build the circuit shown in Figure 6L1.1, and repeated below in Figure 6L1.2. Let $R_1 = R_2 = R_3 = 1$ kΩ.

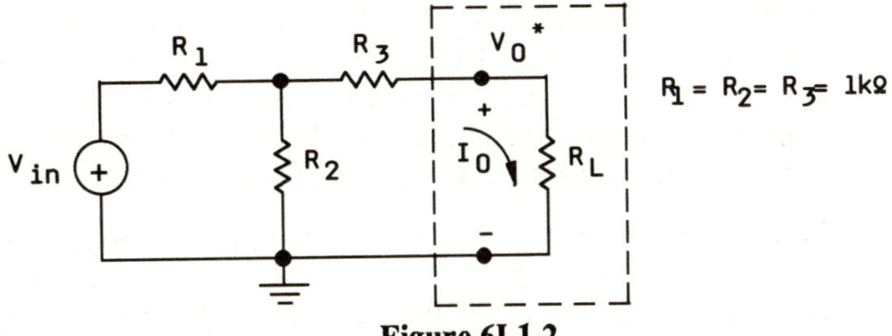

Figure 6L1.2

a. Given the polarity and direction for V_o and I_o, record and plot I_o versus V_o for the five different values of R_L shown below. Also, plot the lines representing the five values of R_L -- in other words, V_L/i_L.

$$R_L = \quad \text{infinite (open circuit)}$$
$$10 \text{ k}\Omega$$
$$1 \text{ k}\Omega$$
$$100 \text{ }\Omega$$
$$0 \text{ (short circuit)}$$

b. Determine the Thevenin equivalent circuit from the data of part a.

2. Build the Thevenin circuit found in part 1b, and measure I_o and V_o for this network using the same load resistors as in part 1a. Plot and compare these data with those of part 1a.

3. From your plots, determine the values of I_o and V_o for a 500 Ω load. Verify this experimentally for both the original circuit and the equivalent circuit.

4. What would the maximum variation in the plots be if all resistors had a 10% tolerance and the power supply was set to within 1% of the nominal 5 V value?

5. Based on your error analysis, was the Thevenin circuit equivalent to the original network?

DISCUSSION:

CONCLUSIONS:

LAB 6L2

TITLE: Thevenin's, Norton's, Tellegan's, Superposition, and Maximum Transfer Theorems[16],[17],[18]

OBJECTIVE: The purpose of this laboratory is to introduce the student to Thevenin's, Norton's, Tellegan's, superposition, and the maximum power transfer theorems. The laboratory will use only resistive circuits.

EQUIPMENT: Multimeter
Two single-voltage power supplies
Components: see lab for specific resistor values
Breadboard

PRELABORATORY:

1. Given the circuit of Figure 6L2.1;

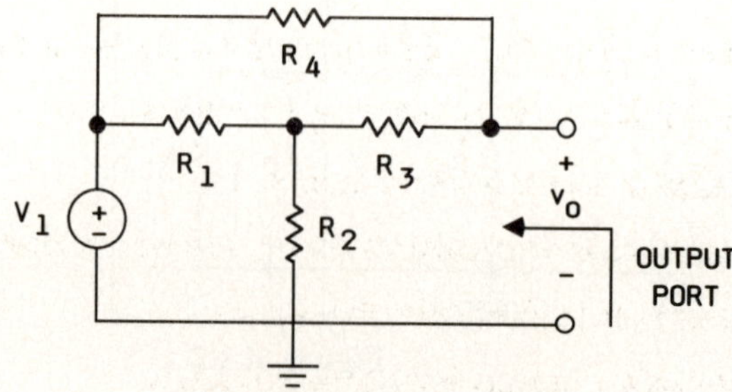

Figure 6L2.1: Test circuit

a. Find the short-circuit current for this network in terms of V_1 and the resistors. Remember, the short-circuit current is the current that flows through a zero-resistance wire placed across the output.

b. Find the equivalent resistance "looking into" the circuit from the output port (expressed in terms of R's). Remember to replace V_1 with its equivalent resistance. In other words, short V_1 to ground.

c. Find the open-circuit voltage for this network, expressed in terms of R's and V_1.

d. Draw the Norton and Thevenin equivalent circuits for the network.

2. Connect a load resistor R_L across the output of the circuit in part 1.

a. Express the power delivered to the load in terms of the Thevenin voltage and resistance. Then express these values in terms of the R's and V_1 in the circuit.

b. Take the derivative of the expression found for the power in part a with respect to R_L and equate this to zero. Solve for R_L, and prove that this represents the maximum power delivered to R_L.

c. What is the relationship between the Thevenin resistance and R_L in part b?

3. Find the Thevenin and Norton equivalent circuits for the circuit shown in Figure 6L2.2.

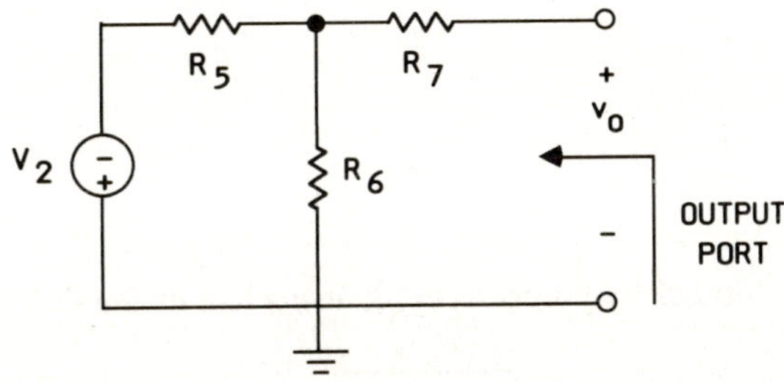

Figure 6L2.2

4. Connect the output ports of the circuits from Figures 6L2.1 and 6L2.2 to form the circuit shown in Figure 6L2.3(a).

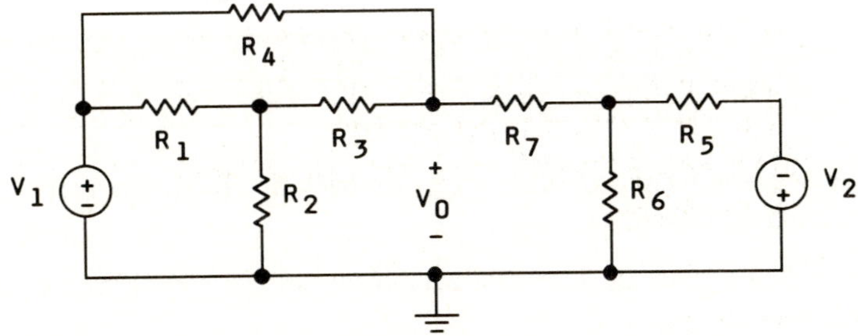

Figure 6L2.3(a)

a. Redraw the circuit using the Thevenin equivalent circuits for each subcircuit.

b. Find the voltage at node V_0 for the following conditions:

1. V_1 is replaced by its equivalent resistance.

2. V_2 is replaced by its equivalent resistance. (Be sure to reconnect V_1.)

3. Both V_1 and V_2 are connected to the circuit.

c. Explain how the results from part b verify superposition.

    d. Find the load resistor value R_L that results in the maximum power dissipated in the load. (Hint: find the Thevenin resistance of the overall circuit.)

5. Using the circuit from part 4,

    a. Find the power, in watts, dissipated in each resistor.

    b. Find the power delivered to the circuit by the two voltage sources.

    c. What can you say about the total power delivered and the total power dissipated in a resistive circuit?

Laboratory Procedure:

1. Build the three circuits shown in Figures 6L2.1, 6L2.2, and 6L2.3. For all cases let:

$$R_1 = R_2 = 20 \text{ k}\Omega \qquad\qquad V_1 = 10 \text{ V}$$
$$R_3 = 10 \text{ k}\Omega$$
$$R_4 = 1 \text{ k}\Omega \qquad\qquad V_2 = -5 \text{ V}$$
$$R_5 = R_6 = 20 \text{ k}\Omega$$
$$R_7 = 10 \text{ k}\Omega$$

Assume that positive current flow is out of V_o.

    a. Plot I_o as a function of V_o for each circuit, using five different load resistors.

Load Resistor	Circuit 1		Circuit 2		Circuit 3	
	v_o	i_o	v_o	i_o	v_o	i_o
Open Circuit						
10 kΩ						
1 kΩ						
100 Ω						
Short Circuit						

    b. Find the negative reciprocal of the slope for each line plotted in part a. How does this relate to the Thevenin resistance?

    c. Use your data to find the Thevenin equivalent voltage and Norton equivalent current for each circuit.

2. Build the circuit shown in Figure 6L2.3(b).

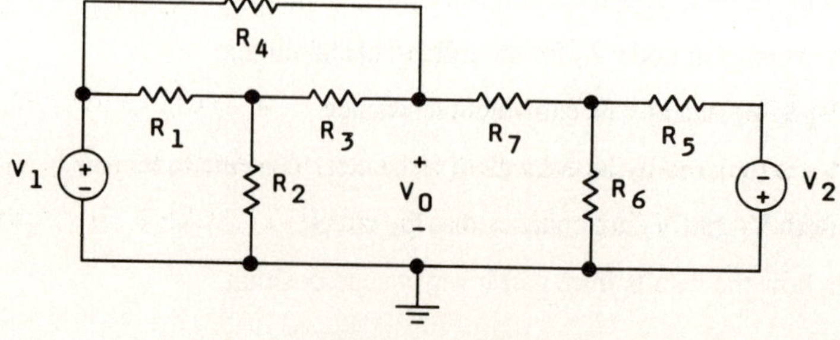

Figure 6L2.3(b)

Consider this circuit to be modeled by the "black box" of Figure 6L2.4.

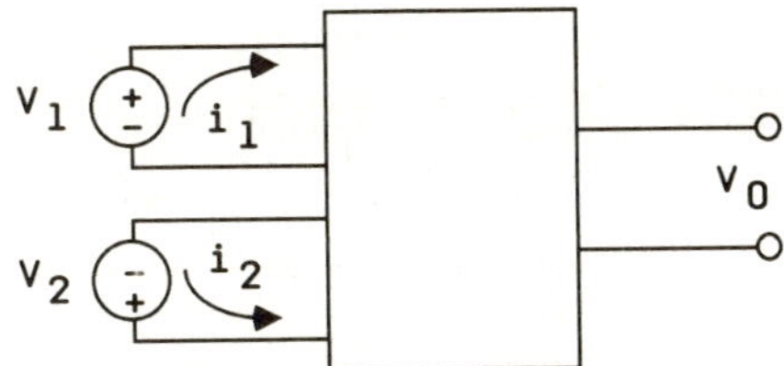

Figure 6L2.4

Since the circuit of Figure 6L2.4 can be modeled by its Thevenin equivalent shown in Figure 6L2.5,

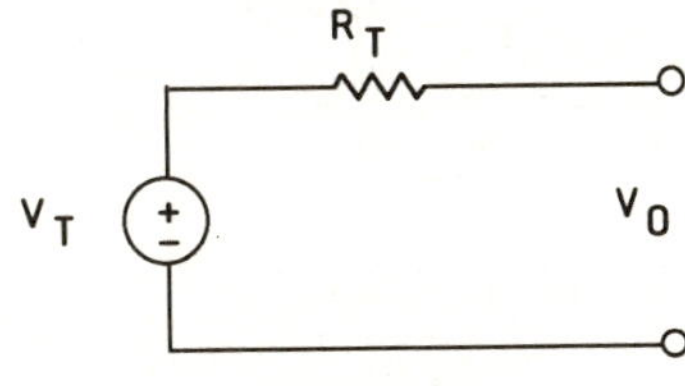

Figure 6L2.5

the total power dissipated with a short circuit output would be:

$$P_{dissipated} = (I_{shortcircuit})^2 R_T \qquad (6L2.1)$$

We can use this idea to determine the network's Thevenin equivalent resistance.

    a.   Measure and record the total power delivered to the circuit when V_o is short circuited.

    b.   Measure the short-circuit current and record it.

    c.   Determine the Thevenin resistance for this network.

    d.   Verify that this is the equivalent resistance of the network.

        1.   Measure V_{oc} and record it.

        2.   Write these values (except R_L) on the circuit model.

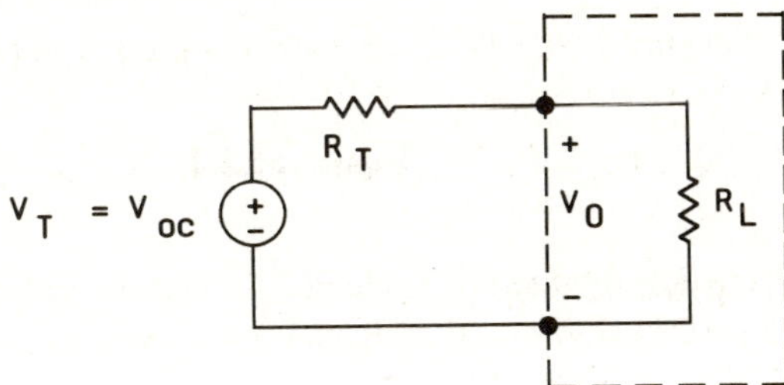

Figure 6L2.6

        3.   Build the circuit of Figure 6L2.6, and plot V_o versus I_o for five different load resistors.

3.   Build the circuit in Figure 6L2.2 and repeat part 1 for this network, finding I_{sc}, V_{oc}, and R_T.

4.   Connect the output ports of Figure 6L2.1 and Figure 6L2.2 together to form the circuit shown in Figure 6L2.3. Then

   a. Redraw the circuit using the Thevenin equivalent circuit for each subcircuit.

   b. Find the voltage at node V_o for the following conditions:

      1. V_1 is replaced by its equivalent resistance.

      2. V_2 is replaced by its equivalent resistance.

      3. V_1 and V_2 are in the circuit.

   c. Add the values of V_o found in parts 1 and 2, and verify that they equal V_o in part 3. This is the expected result due to what theorem?

5. Tellegan's theorem

     1. Build the two circuits of Figure 6L2.7. (Note: The circuits have identical topologies but different component values. (They could also be nonresistive elements or nonlinear components.))

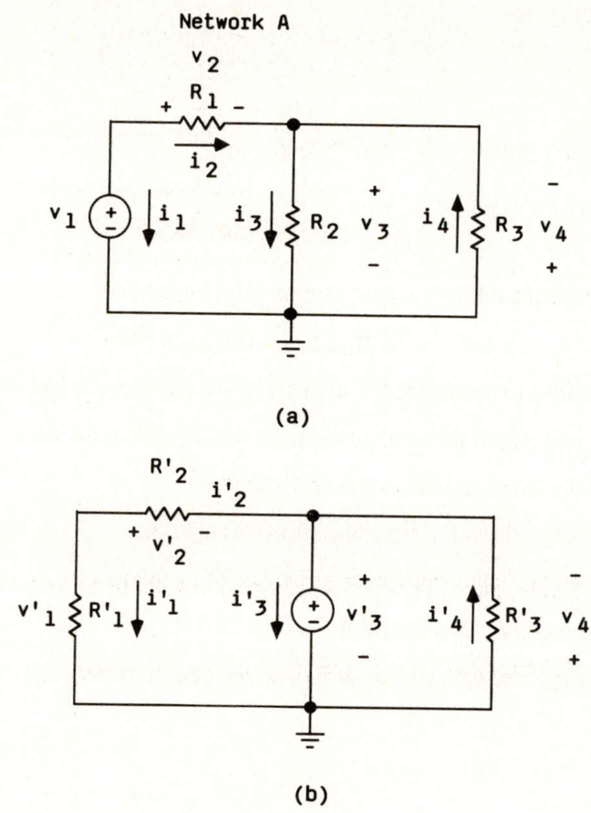

Figure 6L2.7

     2. Measure and record the data required in Table 6L2.1. Use the following resistors values: $R_1 = 10\,k\Omega$, $R_2 = R_3 = 20\,k\Omega$, $V_1 = 10\,V$, $R_1' = R_2' = 10\,k\Omega$, $V_3' = -5\,V$, $R_3' = 20\,k\Omega$.

Table 6L2.1

	Branch			
Item	1	2	3	4
v_k				
i_k				
v_k'				
i_k'				
$v_k i_k$				
$v_k' i_k'$				
$v_k i_k'$				
$v_k' i_k$				

Note: $v_k \equiv$ voltage across branch k
$i_k \equiv$ current through branch k

3. Use your data to find the following summations:

   a. $\displaystyle\sum_{k=1}^{4} v_k i_k$

   b. $\displaystyle\sum_{k=1}^{4} v_k' i_k'$

   c. $\displaystyle\sum_{k=1}^{4} v_k i_k'$

   d. $\displaystyle\sum_{k=1}^{4} v_k' i_k$

DISCUSSION: Compare your experimental results with those predicted by theory.

CONCLUSION: Summarize your findings.

LAB 6L3

TITLE: Analysis and Design of an Analog-to-Digital Converter Using an R-2R Ladder Network

OBJECTIVE: To use the principle of superposition and the Thevenin equivalent circuit to simplify the analysis of a practical circuit used in digital-to-analog converters.

EQUIPMENT: Power Supply, 8 V
Breadboard
1/8 W resistors:
   $10 - 1\,k\Omega$
   $10 - 2\,k\Omega$
or two groups of resistors that are between $1\,k\Omega$ and $100\,k\Omega$ and that are in a ratio of 2 to 1
Multimeter
2 SPDT (single-pole, double-throw) switches

PRELABORATORY:

1. Analyze the (3-Bit DAC) circuit of Figure 6L3.1, in which

   S_1, S_2 and S_3 are SPDT switches and all independent voltage sources V are equal to 8 V.

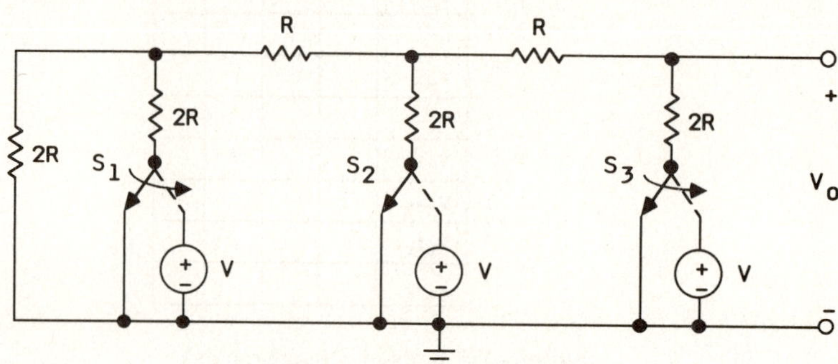

Figure 6L3.1: 3 Bit DAC

For parts a, b, and c, use the Thevenin equivalent circuit to simplify the analysis.

   a. Find V_o when S_3 connects to V and S_1 and S_2 are connected to ground.

   b. Find V_o when S_2 connects to V and S_1 and S_3 are connected to ground.

   c. Find V_o when S_1 connects to V and S_2 and S_3 are connected to ground.

   d. What are the values for V_o for the eight possible combinations of S_1, S_2, and S_3 given in Table 6L3.1? (Let V = 8 V.) For $S_x = 0$, the switch x connects the resistor to ground. For $S_x = 1$, the switch x connects the resistor to v.

Table 6L3.1

S_3	S_2	S_1	V_o (volts)
0	0	0	
0	0	1	
0	1	0	
0	1	1	
1	0	0	
1	0	1	
1	1	0	
1	1	1	

2. Do an error analysis on the R-2R circuit. Determine the change in V_o for the following situations:

   a. All resistors are above their nominal (designated) value by 10%.

   b. All resistors are below their nominal value by 10%.

   c. The resistor values are high or low by 10% of the nominal value, so as to yield the worst case error in V_o.

3. Show the circuit for an 8-bit DAC using the R-2R ladder approach.

Laboratory Procedure:

1. Build the circuit in Figure 6L3.1, and verify the circuit performance by completing table 6L3.2. Be sure to record the tolerance and accuracy of the resistors and voltmeter, respectively.

Table 6L3.2

S_3	S_2	S_1	V_o (volts)
0	0	0	±
0	0	1	±
0	1	0	±
0	1	1	±
1	0	0	±
1	0	1	±
1	1	0	±
1	1	1	±

2. Build and verify the performance of an 8-bit DAC.

   a. What is the smallest voltage change possible in V_o with a single switch change? What is the largest?

   b. Record V_o for the following switch positions:

S_8	S_7	S_6	S_5	S_4	S_3	S_2	S_1	V_o
0	0	0	0	0	0	0	0	
0	0	0	0	0	0	0	1	
0	0	0	0	0	0	1	0	
0	0	0	0	0	1	0	0	
0	0	0	0	1	0	0	0	
0	0	0	1	0	0	0	0	
0	0	1	0	0	0	0	0	
0	1	0	0	0	0	0	0	
1	0	0	0	0	0	0	0	

Also, verify superposition for this network for the following two conditions.

S_8	S_7	S_6	S_5	S_4	S_3	S_2	S_1	V_o
1	0	1	0	1	0	1	0	
0	1	0	1	0	1	0	1	

DISCUSSION:

1. Use error analysis to compare the experimental and calculated values for V_o.

2. What is the maximum power dissipation in the 8-bit DAC? Minimum power dissipation?

3. Explain how this circuit is used to translate a digital bit pattern into an analog voltage.

CONCLUSIONS:

LAB 6L4

TITLE: Application of Network Theorems to a Design Problem[19],[20]

OBJECTIVE: This laboratory gives the student practice using network theorems that will make circuit analysis easier. The laboratory will also show the student the essential elements of real design, starting with specifications and a circuit topology.

EQUIPMENT AND COMPONENTS:

Voltmeter
Three power supplies, ±15 V and 10 V
Function generator
Operational amplifier
Resistors,   1/8 W ±1%

BACKGROUND: The design in this lab contains the essential elements of a subcircuit used in the Hewlett-Packard 5890A's thermal conductivity detector (TCD). Essentially, v_s is an ac-coupled 5-Hz square wave input signal with an amplitude proportional to the thermal conductivity difference between two gases. How this signal is generated is not relevant to the lab, but for additional information, see References [21] and [22].

The design specifications are as follows:

1. The voltage gain $\dfrac{v_o}{v_s}$=300 @ 5 Hz.

2. The voltage gain $\dfrac{v_{ref}}{v_o} = -\dfrac{1}{2}$; where $v_{ref} = 10$ V.

3. $I_{ref} < 1$ mA.

4. $I_o < 10$ mA.

5. $V_{omax} = 0$.

6. $V_{omin} = -10$ V.

7. V_s peak-to-peak max = 15 mV. $v_{s_{min}} = \pm 1$ μV.

8. The 5-Hz square wave input is shown in Figure 6L4.1(a), and the output signal is shown in Figure 6L4.1(b).

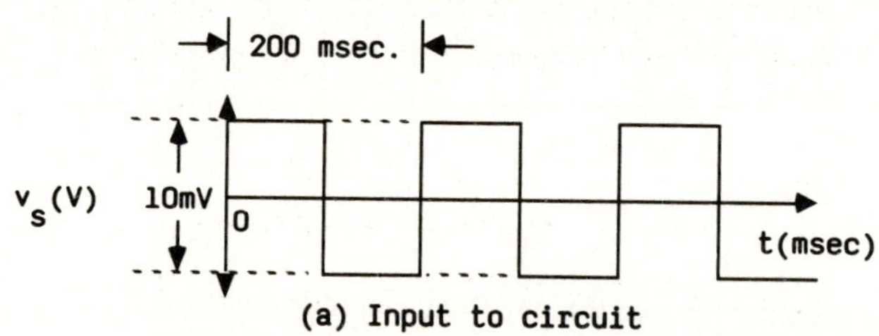

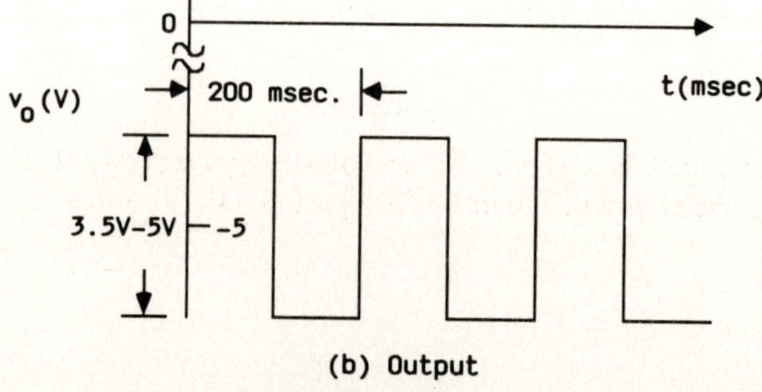

Figure 6L4.1

The input signal v_s needs to be amplified before it is digitized (converted to a digital signal by an analog-to-digital-converter (ADC)), and the continuously integrating ADC used in this design accepts negative voltages. The input signal is amplified by a gain of 300 and is biased in the middle of its dynamic range (0 V to -10 V). The essential elements contained in this circuit application are given in Figure 6L4.2.

PRELABORATORY:

1. For the circuit in Figure 6L4.2,

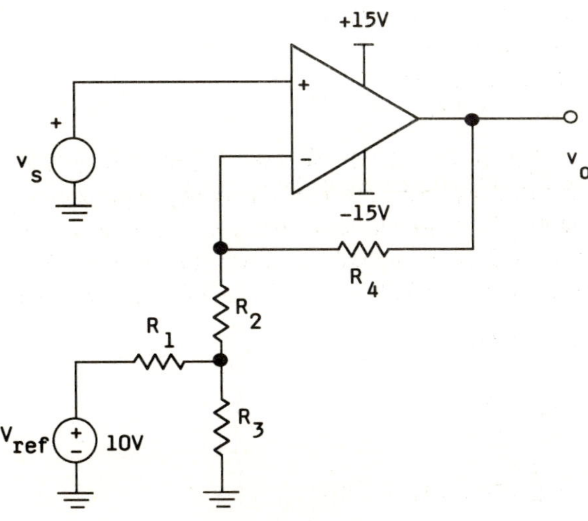

Figure 6L4.2

a. Find K_1 and K_2 given in the equation

$$V_o = K_1 V_S + K_2 V_{ref} \qquad\qquad (6L4.1)$$

with both V_S and V_{ref} connected to the circuit.

b. Now use superposition, and again find K_1 and K_2. In other words, find V_o with V_S connected and V_{ref} replaced by its equivalent resistance. Then find V_o with V_{ref} connected and V_S replaced by its equivalent resistance.

Thevenin Equivalent

2. Use the Thevenin equivalent concept to evaluate the network composed of V_{ref}, R_1, R_2, and R_3. This is one of the circuits that results when we use superposition on the circuit of Figure 6L4.2.

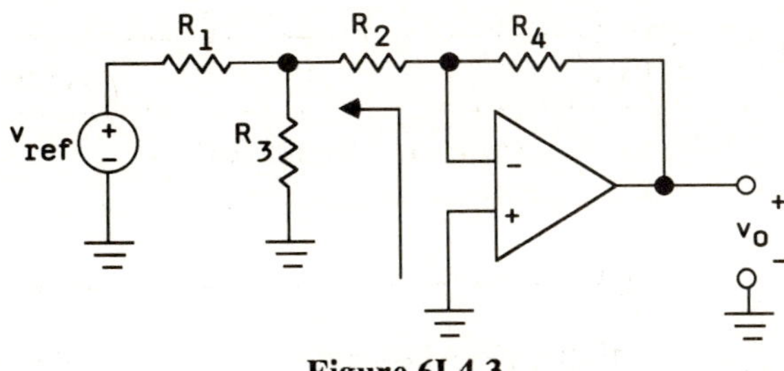

Figure 6L4.3

a. Looking into the circuit of Figure 6L4.3, as indicated by the arrow, find the Thevenin equivalent network. The resulting circuit will look like Figure 6L4.4.

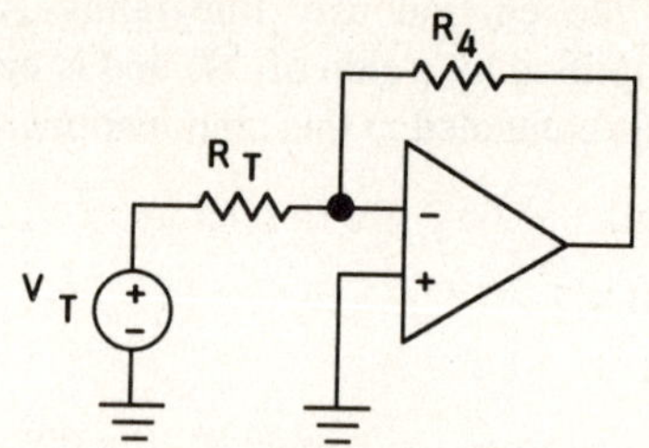

Figure 6L4.4

b. Determine V_T and R_T for the circuit in Figure 6L4.3 and represented in Figure 6L4.4.

c. Find the transfer function K_2 for the circuit in Figure 6L4.4. Use the circuit components that represent V_T and R_T, and compare K_2 to the K_2 found in part 1.

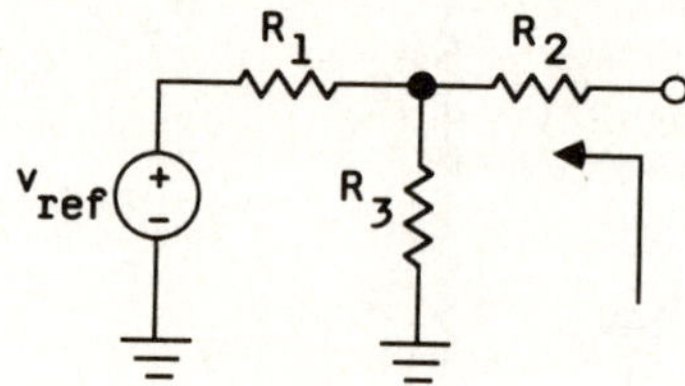

Figure 6L4.5

3. Find the Norton equivalent of the circuit of Figure 6L4.5. In the Norton equivalent, I_{SC} is the current that flows out of the output with the output shorted to ground. Since R_2 is connected to a virtual ground, I_{SC} flows through R_4. Therefore,

$$V_o = -I_{SC} R_4 \qquad\qquad (6L4.2)$$

Use I_{SC} from the Norton equivalent of Figure 6L4.5 in this equation. Compare this result with the previous value of K_2 from part 1.

Design Problem

4. Given the circuit of Figure 6L4.2, the transfer function derived in part 1 of this prelab, and the specifications from the "background" section, determine values for R_1, R_2, R_3, and R_4.

5. Verify your design using a circuit simulation program (e.g., SPICE). Use the circuit in Figure 6L4.6 to model the operational amplifier.

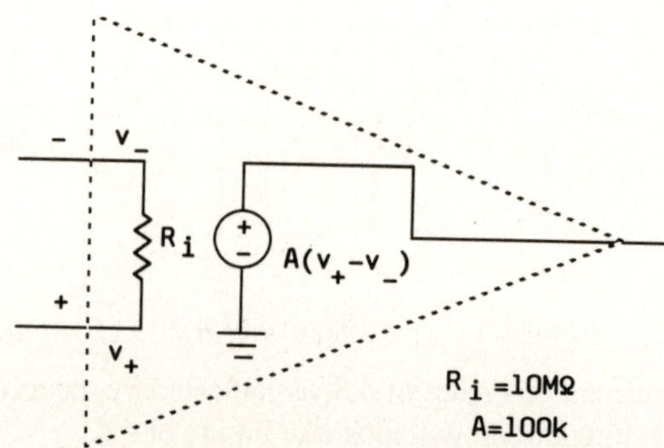

Figure 6L4.6: Operational amplifier model

Laboratory Procedure:

1. Build the circuit in Figure 6L4.2 using the resistor values determined by your design.

    a. Let v_S equal a 25-mV (peak-to-peak) square wave with a 200-msec period.

    b. Use an oscilloscope to observe and sketch the output v_o and input v_S. Be sure to use the chop mode for easier viewing.

    c. Determining the dynamic range.

        1. Find the largest peak-to-peak value v_S can have and still allow the output v_o to stay within the design specifications.

        2. Find the smallest allowed amplitude for v_S with the constraint that the peak-to-peak output values are still detectable at two times the noise level.

        3. What is the dynamic range $\dfrac{v_{S\ max}}{v_{S\ min}}$?

2. Replace V_{ref}, R_1, R_2, and R_3 in your circuit design with the Thevenin equivalent circuit you determined in the prelab. The new circuit is shown in Figure 6L4.7. Then redo parts a, b, and c from part 1 of the laboratory procedure. Is there any observable difference in performance of the circuit?

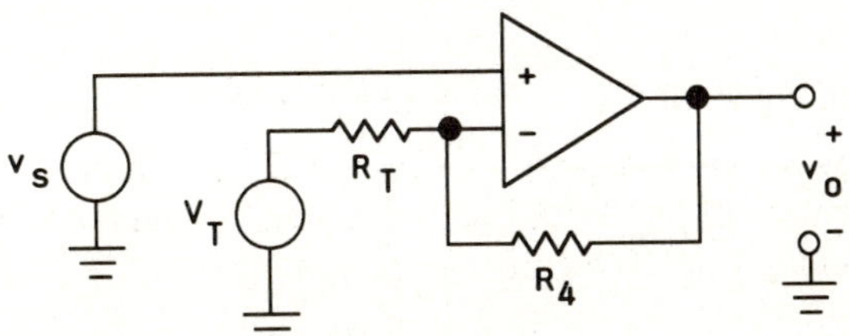

Figure 6L4.7

3. Replace V_{ref}, R_1, R_2, and R_3 in your circuit design with the Norton equivalent circuit you determined in the prelab. The new circuit is shown in Figure 6L4.8. Redo parts 1, 2, and 3 from Part 1 of the laboratory. Is there any observable difference in the performance?

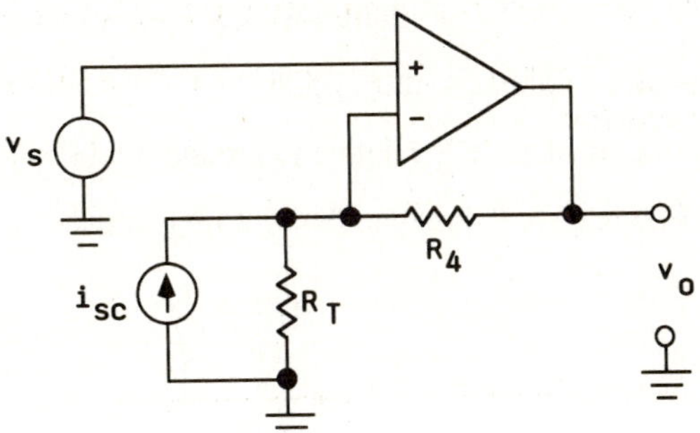

Figure 6L4.8

Note: i_{SC} can be simulated with a 20 V source and
a resistor that results in $20\dfrac{V}{R} = i_{SC}$.

4. For the circuit of Figure 6L4.2, experimentally determine the change in v_o for a 10% change in R_4. Verify this change using the .SENS statement in SPICE.

DISCUSSION: Include a sensitivity analysis to verify your experimental results in part 4. Also, compare your simulation results with the lab results.

CONCLUSIONS:

LAB 6L5

TITLE: Network Theorems

OBJECTIVE: This laboratory gives the student several circuits upon which to apply various theorems. The student must verify the theorems, where possible, and explain the cases where a particular theorem is not applicable.

EQUIPMENT AND COMPONENTS:
   DVM

   Powersupplies, ±15 V and 0-5 V   (adjustable)

   Resistors   (1/8 W, 1% tolerance)

   Op-amp (general purpose)

   diode (general purpose)

PRELABORATORY:

1.   Consider the circuit of Figure 6L5.1.

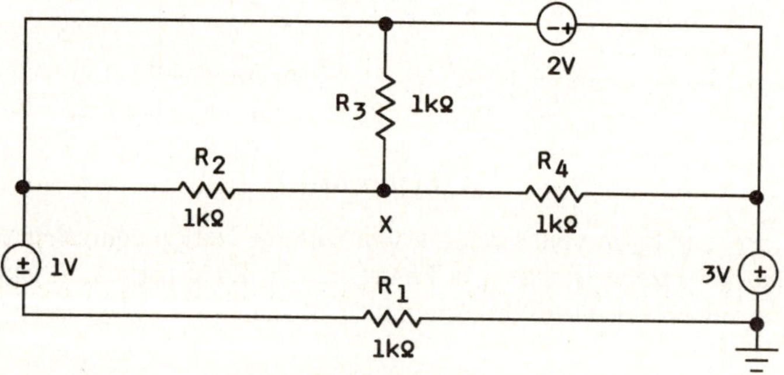

Figure 6L5.1

   a.   Show that superposition applies to this circuit.

   b.   Looking into the circuit, at node X relative to ground, find the Thevenin equivalent circuit.

2.   Consider the circuit of Figure 6L5.2. Assume the op amp is ideal.

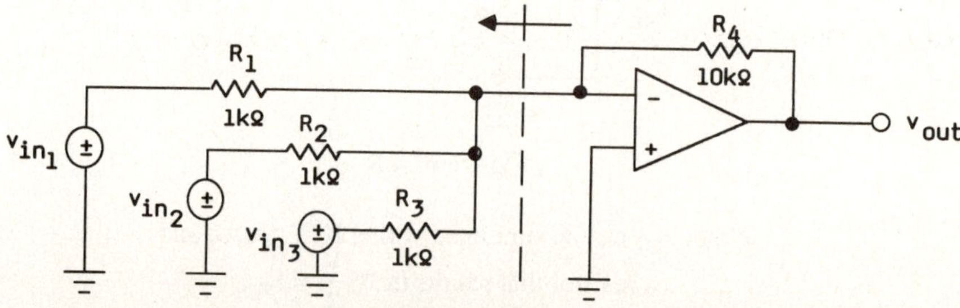

Figure 6L5.2

   a.   Find the Thevenin equivalent as one looks back into the source as shown by the arrow.

   b.   Let $V_{in_1} = 1$ V, $V_{in_2} = 2$ V, and $V_{in_3} = 3$ V. Does superposition apply to this circuit? If so, use superposition and verify that you get the same result as in part a.

3.   Consider the circuit of Figure 6L5.3.

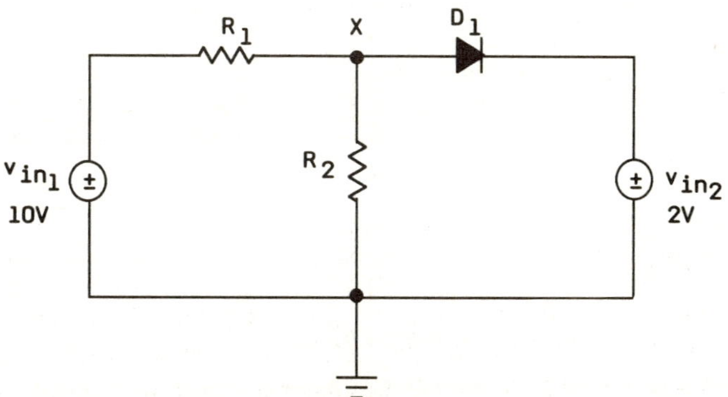

Figure 6L5.3

a.   Use superposition to find the voltage at node X.

b.   Find the voltage at node X with both sources applied. Do you get the same result as in part a? Why?

4.   Consider the circuit of Figure 6L5.4. Assume the op amp is ideal.

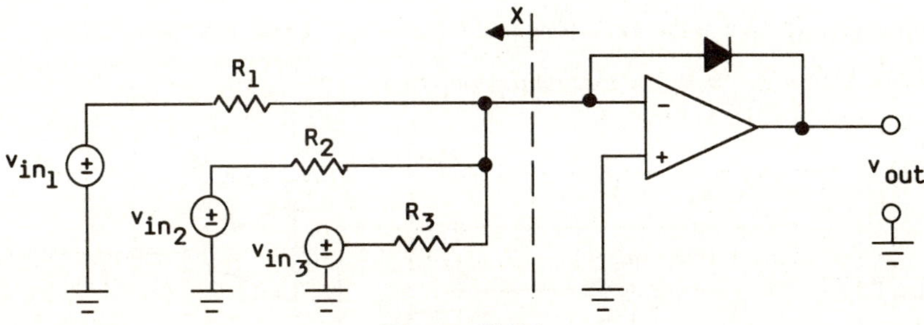

Figure 6L5.4

a.   If $R_1 = R_2 = R_3 = 1\ k\Omega$ and $V_{in_1} = V_{in_2} = V_{in_3} = 1$ V can superposition be applied to find V_{out}? If so, apply it. Can the resistors and input voltages be replaced by a Thevenin equivalent circuit? If so, show it.

b.   If the current-voltage relationship of the diode is given by $I_D = I_s e^{\frac{V_D}{V_T}}$, where I_D and V_D are the diode current and voltage, respectively, find a relationship between V_D and the Thevenin equivalent input voltage.

c.   Is this a linear circuit? Prove it!

Laboratory Procedure:

1.   Resistor network. Build in the circuit of Figure 6L5.1 and shown in Figure 6L5.5.

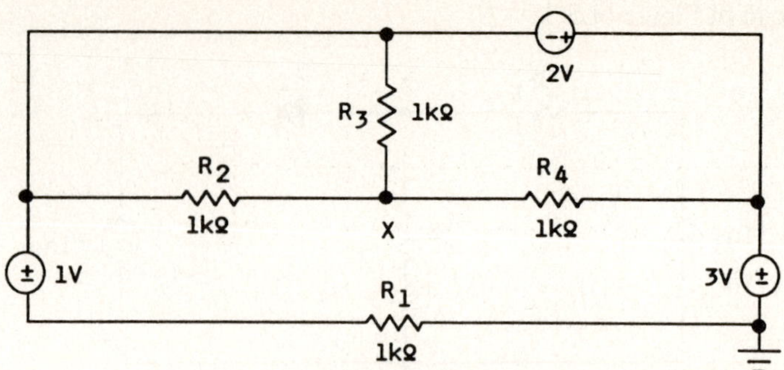

Figure 6L5.5

a. Measure the voltage at node X and confirm the result from your prelab. Apply the voltages one at a time and confirm superposition.

b. Build the Thevenin equivalent circuit and show that the output with no load is equal to the voltage at node X.

2. Summing amplifier. Build the circuit of 6L5.2 and shown in Figure 6L5.6. Be sure to connect ±15 volts to the V_{CC_+} and V_{CC_-} connections of the op amp.

a. Let $V_{in_1}=V_{in_2}=V_{in_3}=0.2$. Apply each input supply separately and record the respective V_{out}. Then apply all three at the same time and record V_{out}. Does superposition apply? Why?

b. Now let $V_{in_1} = V_{in_2} = V_{in_3} = 1V$ and repeat part a.

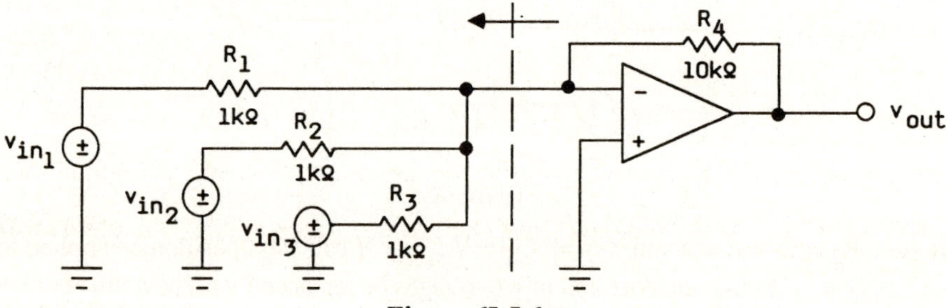

Figure 6L5.6

3. Diode Circuit. Build the circuit of Figure 6L5.3 and repeat in 6L5.7.

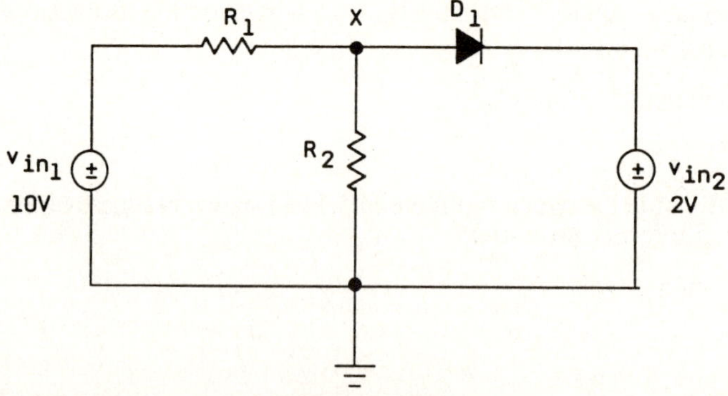

Figure 6L5.7

Apply V_{in_1} and V_{in_2} simultaneously and find the voltage at node X. Then apply them separately. Is this a linear circuit? Why?

4. Diode-op amp circuit. Build the circuit of Figure 6L5.4 and repeated in Figure 6L5.8.

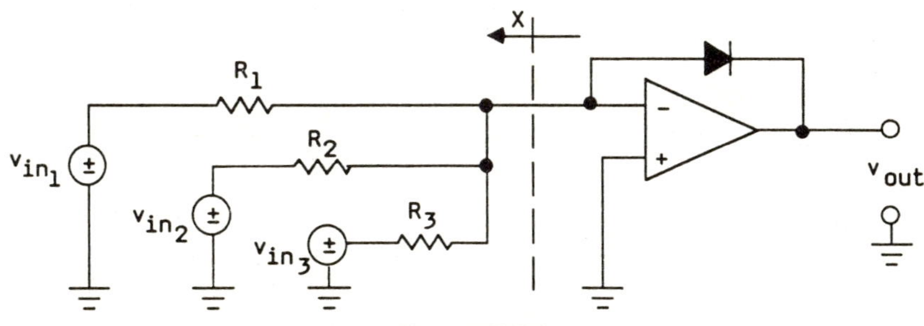

Figure 6L5.8

a. Use the component values given in the prelab and see whether superposition applies to find V_{out}.

b. Replace the circuit to the left of the X with its Thevenin equivalent and solve for V_{out}. Is this a linear circuit? Why?

c. Take the circuit of part b and vary V_{TH} from 0- to 1-V in 0.1-V increments. Measure and record V_{out}. Plot V_{out} versus V_{TH} on semilog paper and V_{TH} on the log axes. What does this plot tell you about the relationship between V_{out} and V_{TH}. Is it linear? Why?

DISCUSSION:

CONCLUSIONS:

REFERENCES

[1]   J. W. Nilsson, *Electric Circuits,* 2d ed. (Reading, MA: Addison-Wesley, 1986), pp. 117-30.

[2]   W. H. Hayt, Jr., and J. E. Kemmerly, *Engineering Circuit Analysis,* 4th ed. (New York: McGraw-Hill, 1986), pp. 84-91.

[3]   D. E. Johnson, J. L. Hilburn, and J. R. Johnson, *Basic Circuit Analysis,* 3d ed. (Englewood Cliffs, NJ: Prentice-Hall, 1986), pp. 122-9.

[4]   Nilsson, pp. 114-17.

[5]   Hayt and Kemmerly, pp. 78-83.

[6]   Johnson, Hilburn and Johnson, pp. 130-5.

[7]   Nilsson, pp. 133-7.

[8]   Hayt and Kemmerly, pp. 74-8.

[9]   Johnson, Hilburn and Johnson, pp. 116-21.

[10]   Nilsson, pp. 130-3, and pp. 396-402.

[11]   Johnson, pp. 136-7.

[12]   Johnson, p. 33.

[13]   M. E. Van Valkenburg, *Network Analysis,* 3d ed. (Englewood Cliffs, NJ: Prentice-Hall, 1974), pp. 437-41.

[14]   Desoer and Kuh, *Basic Circuit Theory* (New York: McGraw-Hill, 1969), pp. 392-402.

[15]   A. S. Sedra and K. C. Smith, *Microelectronic Circuits,* 2d ed. (New York: Holt, Rinehart and Winston, 1987), pp. 52-6.

[16]   L. S. Bobrow, *Elementary Linear Circuit Analysis* (New York: Holt, Rinehart and Winston, 1987), pp. 121-50, and pp. 251-60.

[17]   Van Valkenburg, pp. 420-37.

[18]   Desoer and Kuh, pp. 2-10.

[19]   V. H. Grinich and H. G. Jackson, *Introduction to Integrated Circuits* (New York: McGraw-Hill, 1975), pp. 528-33.

[20]   A. S. Sedra and K. C. Smith, pp. 88-113.

[21]   J. Peters, D. Kolloff, D. Messaros, G. Law, and L. Plotczyk, *Design Improvements of the 5880A Modulated Flow Thermal Conductivity Detector - Operating Considerations*, Technical Paper 96, Pub. 43-5953-1638, Copyright 7-82, Hewlett-Packard Co.

[22]   D. W. Messaros, C. E. Law, D. Kolloff, D. Gerhardt, and R. Freeman, *A New Low Volume, Single Filament-Single Column Thermal Conductivity Detector for Use with Open Tubular Columns*, Technical Paper 104, Pub. No. 43-5953-1772, Copyright 6-84, Hewlett-Packard Co.

Chapter 7

FREQUENCY AND TRANSIENT RESPONSE

7.1 Introduction

In this chapter, we study the frequency and transient characteristics or response of networks. We shall determine what happens to the amplitude and phase of the steady-state currents and voltages as the frequency of the source is varied. Circuits that transmit signals at some frequencies much better than other frequencies will be examined for their use as filters. The ability of a circuit to be frequency-selective has applications in the communications and control fields.

We show that the frequency response of a circuit is related to other circuit responses. That means that we can measure the frequency response of a circuit, devise a model for the circuit, and use the model to predict the response of the circuit to other types of inputs.

We introduce the topic of frequency response by studying the Bode plot, pole-zero plot, resonance, bandwidth and quality factor, and filters. We conclude the chapter with a discussion of the Fourier series and the Laplace and Fourier transforms. These techniques allow us to find the steady-state response of a circuit using a Fourier series when the signal sources are periodic, and Fourier and Laplace transforms for nonperiodic sources. The Laplace and Fourier transforms also allow us to transform back and forth between the frequency and time domains. A transformation is performed to create a domain where the mathematical procedures to describe the circuit conditions are easier to carry out. We introduce the concepts of the step response and impulse response in the context of our discussion of the Laplace transform.

7.2 *Poles and Zeros*[1],[2],[3],[4]

In general, in the frequency domain, the network function takes the form

$$\frac{V_{out}(s)}{V_{in}(s)} = H(s) = \frac{N(s)}{D(s)} = \frac{a_n^n + a_{n-1}s^{n-1} + \cdots + a_1 s + a_o}{b_m s^{2m} + b_{m-1}s^{m-1} + \cdots + b_1 s + b_o} \qquad (7.1)$$

The constants a and b are real and the integers m and n are positive. The ratio N(s)/D(s) is called a **proper ratio function** if m>n and an **improper ratio function** if m<n.

The rational network function of Equation (7.1) can be expressed as the ratio of two factored polynominals. That is, Equation (7.1) can be written as

$$H(s) = \frac{K(s + z_1)(s + z_2) + \cdots + (s + z_{n)}}{(s + p_1)(s + p_2) + \cdots + (s + p_m)} \qquad (7.2)$$

where K is the constant $\dfrac{a_n}{b_m}$.

The numbers z_1, z_2, ..., z_n are called **zeros** of the network function because these are the values of s for which the network function becomes zero. The numbers p_1, p_2, ..., p_m are the values of s for which the network function becomes infinite and are called **poles**. The values of the poles and zeros, together with the values of the factors a_n and b_m, uniquely determine the network function.

We can evaluate the network function of Equation (7.2) at some frequency $s=j\omega_o$ such that Equation (7.2) becomes

$$H(s=j\omega_o) = \frac{K(j\omega_o + z_1)(j\omega_o + z_2) \cdots (j\omega_o + z_m)}{(j\omega_o + p_1)(j\omega_o + p_2) \cdots (j\omega_o + p_n)} \qquad (7.3)$$

Each term of the form $(j\omega_o + z_m)$ and $(j\omega_o = p_n)$ is a complex number and may be represented by a vector having a magnitude and phase.

7.3 *Amplitude and Phase Response: Bode Diagrams*[5],[6],[7]

The Bode diagram is a quick method for getting an idea of the amplitude and phase variation of a network transfer function as a function of the frequency of the network. Consider the transfer function of Equation (7.3). The first step in making Bode diagrams is to put the expression for $H(j\omega)$, in the standard form

$$H(j\omega) = \frac{Kz_1 z_2 z_3 \cdots z_n (1+\frac{j\omega}{z_1})(1+\frac{j\omega}{z_2})(1+\frac{j\omega}{z_3}) \cdots (1+\frac{j\omega}{z_n})}{(p_1 p_2 p_3 \cdots p_m)j\omega(1+\frac{j\omega}{p_1})(1+\frac{j\omega}{p_2})(1+\frac{j\omega}{p_3}) \cdots (1+\frac{j\omega}{p_m})} \qquad (7.4)$$

Let the constant

$$\frac{Kz_1 z_2 z_3 z_4 \cdots z_n}{p_1 p_2 p_3 p_4 \cdots p_m} = K_o$$

Now write Equation (7.4) in polar form:

$$H(j\omega) = \frac{K_o \mid 1+\frac{j\omega}{z_1} \mid \psi_1 \mid 1+\frac{j\omega}{z_2} \mid \mid \psi_2 \mid 1+\frac{j\omega}{z_3} \mid \psi_3 + \cdots + \mid 1+j\omega z_n \mid \psi_n}{\mid \omega \; 90° \mid 1+\frac{j\omega}{p_1} \mid \beta_1 \mid 1+\frac{j\omega}{p_2} \mid \beta_2 \mid 1+\frac{j\omega}{p_3} \mid \beta_3 + \cdots + \mid 1+\frac{j\omega}{p_m} \mid \beta_m} \qquad (7.5)$$

Rearranging Equation (7.5), we obtain

$$H(j\omega) = \frac{K_o}{\omega} \frac{\left|1+\dfrac{j\omega}{z_1}\right| \left|1+\dfrac{j\omega}{z_2}\right| + \cdots \left|1+\dfrac{j\omega}{z_n}\right|}{\left|1+\dfrac{j\omega}{p_1}\right| \left|1+\dfrac{j\omega}{p_2}\right| + \cdots + \left|1+\dfrac{j\omega}{p_m}\right|} \underline{\psi_1 + \cdots + \psi_n - 90° - \beta_1 - \beta_m} \qquad (7.6)$$

Then the magnitude of the transfer function is

$$|H(j\omega)| = \frac{K_o}{\omega} \frac{\left|1+\dfrac{j\omega}{z_1}\right| \left|1+\dfrac{j\omega}{z_2}\right| \left|1+\dfrac{j\omega}{z_3}\right| + \cdots + \left|1+\dfrac{j\omega}{z_n}\right|}{\left|1+\dfrac{j\omega}{p_1}\right| \left|1+\dfrac{j\omega}{p_2}\right| \left|1+\dfrac{j\omega}{p_3}\right| + \cdots + \left|1+\dfrac{j\omega}{p_m}\right|} \qquad (7.7)$$

and the phase of the transfer function is

$$\theta(\omega) = \psi_1 + \psi_2 + \cdots + \psi_n - 90° - \beta_1 - \beta_2 - \cdots - \beta_m \qquad (7.8)$$

The phase angles ψ and β are defined by

$$\psi_n = \tan^{-1} \frac{\omega}{z_n} \qquad (7.9)$$

and

$$\beta_m = \tan^{-1} \frac{\omega}{p_m} \qquad (7.10)$$

The Bode diagrams then consist of plotting amplitude, Equation (7.7), and phase, Equation (7.8) as functions of frequency. The amplitude calculation can be simplified by expressing Equation (7.7) in terms of decibels, i.e.,

$$A_{dB} = 20\log_{10} |H(j\omega)| = 20\log_{10} \frac{K_o}{\omega} \frac{\left|1+\dfrac{j\omega}{z_1}\right| \left|1+\dfrac{j\omega}{z_2}\right| \left|1+\dfrac{j\omega}{z_3}\right| + \cdots + \left|1+\dfrac{j\omega}{z_n}\right|}{\left|1+\dfrac{j\omega}{p_1}\right| \left|1+\dfrac{j\omega}{p_2}\right| \left|1+\dfrac{j\omega}{p_3}\right| + \cdots + \left|1+\dfrac{j\omega}{p_m}\right|} \qquad (7.11)$$

$$A_{dB} = 20\log_{10} K_o + 20\log_{10} \left|1+\frac{j\omega}{z_1}\right| + 20\log_{10} \left|1+\frac{j\omega}{z_2}\right| \qquad (7.12)$$

$$+ 20\log_{10} \left|1+\frac{j\omega}{z_3}\right| + \cdots + 20\log_{10} \left|1+\frac{j\omega}{z_n}\right| - 20\log_{10}\omega - 20\log_{10} \left|1+\frac{j\omega}{p_1}\right|$$

$$- 20\log_{10} \left|1+\frac{j\omega}{p_2}\right| - 20\log_{10} \left|1+\frac{j\omega}{p_3}\right| + \cdots - 20\log_{10} \left|1+\frac{j\omega}{p_m}\right|$$

In performing the magnitude plot of Equation (7.12), we plot each term of the equation separately and then combine the separate plots graphically. The individual terms can be approximated in all cases by straight lines on semilog graph paper.

The plot of $20\log_{10}K_o$ is a horizontal straight line, since K_o is not a function of frequency. The plot of $20\log_{10} |1+j\omega/z_n|$ is approximated by two straight lines. For small values of ω, the magnitude of $|1+j\omega/z_1| \rightarrow 1$, and

$$20\log_{10} |1+j\omega/z_n| \rightarrow 0 \text{ as } \omega \rightarrow 0 \qquad (7.13)$$

For large values of ω, the magnitude of $|1+j\omega/z_n| \rightarrow \omega/z_n$; therefore,

$$20\log_{10} |1+j\omega/z_n| \rightarrow 20\log_{10}(\omega/z_n) \text{ as } \omega \rightarrow \infty \qquad (7.14)$$

On semilog paper, $20\log_{10}(\omega/z_n)$ is a straight line with a slope of 20 dB/decade where a decade is a 10-to-1 change in frequency. The frequency is plotted on the log scale, while the amplitude in dB is plotted on the linear scale. The straight-line plot of Equation (7.14) intersects the 0-dB axis at $\omega = z_n$. This value of ω is called the **corner frequency**. Equations (7.13) and (7.14) are plotted in Figure 7.1.

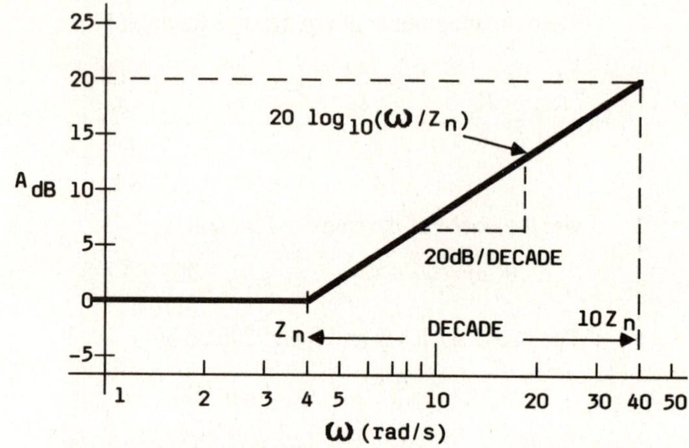

Figure 7.1: Straight-line approximation of the amplitude plot of a first- order zero

The plot of $-20\log_{10}\omega$ is a straight line having a slope of -20 dB/decade that intersects the 0-dB axis at $\omega=1$.

The plot of $-20\log10 \left| 1 + j\dfrac{\omega}{p_m} \right|$ is approximated by two straight lines intersecting the 0-dB axis at $\omega = p_m$. For large values of ω, the straight line $-20\log_{10}(\omega/p_m)$ has a slope of -20 dB/decade.

The straight-line plots for first-order poles and zeros can be made more accurate by correcting the amplitude values of the corner frequency, one-half the corner frequency, and twice the corner frequency. At the corner frequency, the actual value in db is

$$A_{db_{corner}} = \pm 20\log_{10} | 1 + j | \approx \pm 3 \text{ dB} \tag{7.15}$$

The actual value of the amplitude in db at one-half of the corner frequency is

$$A_{db}\frac{corner}{2} = \pm 20\log_{10} \left| +\frac{j}{2} \right| \approx \pm 1 \text{ dB} \tag{7.16}$$

At twice the corner frequency, the actual value is

$$A_{dB2xcorner} = \pm 20\log_{10} | 1 + j2 | = \pm 7 \text{ dB} \tag{7.17}$$

In Equations (7.15), (7.16), and (7.17), the plus sign refers to the zero and the minus sign to the pole.

Phase-angle plots can also be made using straight-line approximations. The phase angle associated with the constant K_o is zero, and the phase angle associated with a first-order zero or pole at the origin is a constant $\pm 90°$. For first-order poles or zeros not at the origin, the straight-line approximations are those for which, for frequencies less than $1/10\ f_{corner}$, the phase angle is assumed to be zero. For frequencies greater than $10 \times f_{corner}$, the phase angle is assumed to be $\pm 90°$. Between $1/10\ f_C$ and $10 \times f_C$, the phase-angle plot is a straight line that goes through $0°$ at $1/10f_C$, $\pm 45°$ at f_C, and $\pm 90°$ at $10f_C$. Again, the plus sign applies to zeros and the minus sign to poles. Figure 7.2 is a plot of the phase-angle for a zero and pole.

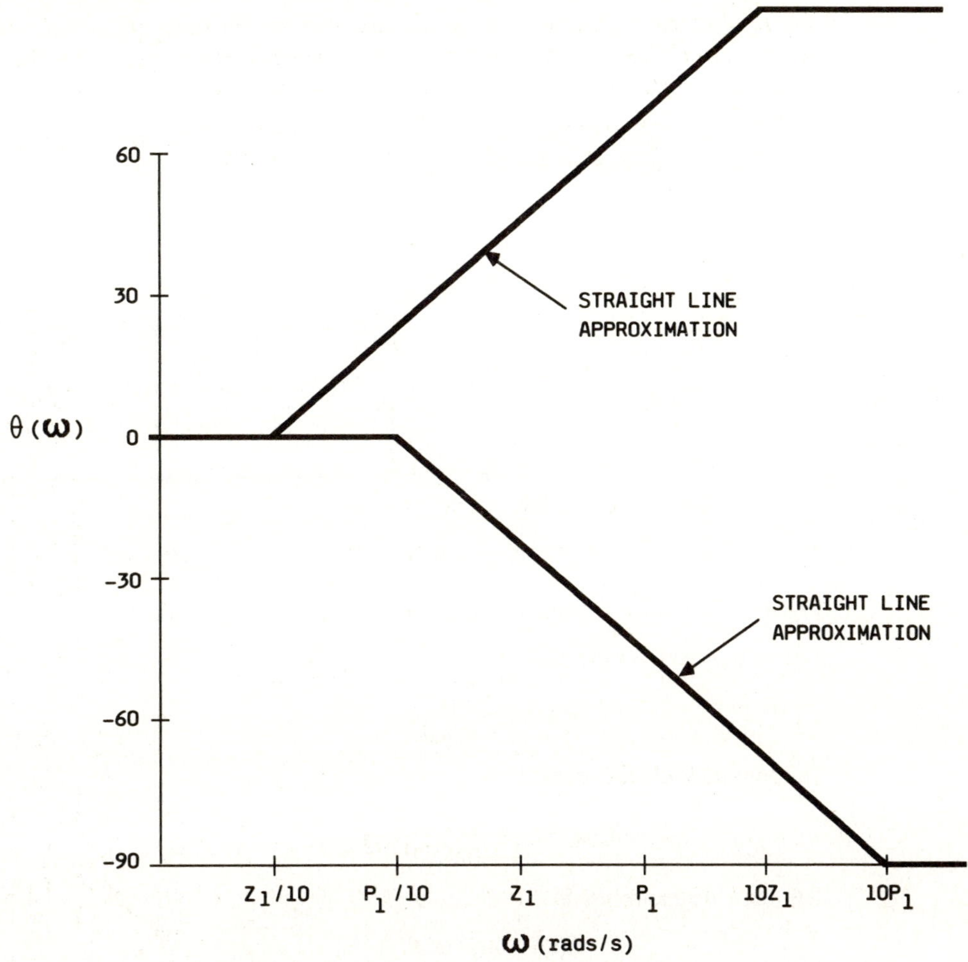

Figure 7.2: Plot of the phase angle for a zero and pole

The actual phase angle can be calculated by solving

$$\psi_n = \tan^{-1}(\omega/z_n) \tag{7.18}$$

for a zero and

$$-\beta_n = -\tan^{-1}(\omega/p_n) \tag{7.19}$$

for a pole.

When complex poles or zeros occur in the expression for H(s), additional methods must be applied so that amplitude and phase-angle plots can be drawn. There are many excellent discussions and examples of the use of Bode diagrams for complex poles and zeros.[6],[7],[8]

7.4 Filters[8]

A filter network is designed to pass signals with a specific frequency range and reject signals outside the frequency range. Filters that pass low frequencies but reject high frequencies are called **low-pass filters**; filters that pass high frequencies but block low frequencies are called **high-pass filters**. Filters that pass some particular frequency band but reject frequencies outside this band are called **band-pass filters**. Filters that reject a particular band of frequencies but pass all other frequencies are called **band- reject filters**. Filters that are constructed of only passive elements, such as resistors, capacitors, and inductors are called **passive filters**.

The ideal and typical characteristics for a low-pass filter are shown in Figure 7.3. Also shown is a simple passive low-pass filter.

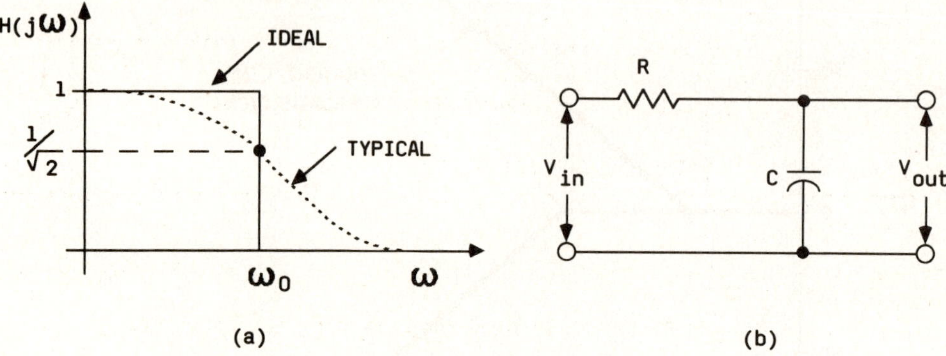

Figure 7.3: Frequency characteristics and low-pass filter circuit

The voltage gain for the network is

$$H(j\omega) = \frac{1}{1 + j\omega RC} \tag{7.20}$$

The amplitude characteristic is

$$M(\omega) = \frac{1}{[1 + (\omega RC)^2]^{1/2}} \tag{7.21}$$

The phase characteristic is

$$\phi(\omega) = \tan^{-1}\omega RC \tag{7.22}$$

At the half-power, or break, frequency, the amplitude is

$$M(\omega = 1/\tau) = \frac{1}{\sqrt{2}} \tag{7.23}$$

The term "half-power frequency" derives from the fact that at this frequency the power is one-half its maximum value.

A Bode plot for the low-pass filter is shown in Figure (7.4).

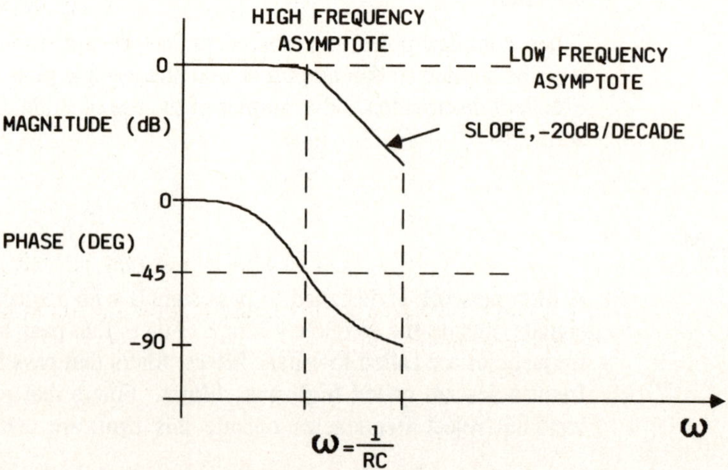

Figure 7.4: Bode plot of low-pass filter

Note that the magnitude plot is flat for low frequencies, but then rolls off at high frequencies at 20 dB/decade.

The passive high-pass filter and characteristics are shown in Figure 7.5.

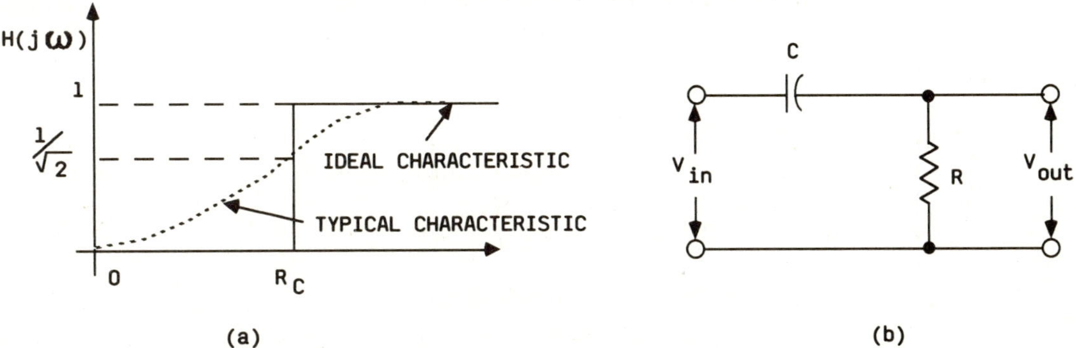

Figure 7.5: Frequency characteristics and high-pass filter circuit

The voltage gain of the circuit is

$$H(j\omega) = \frac{j\omega RC}{1 + j\omega RC} \qquad (7.24)$$

The magnitude characteristic is

$$M(\omega) = \frac{\omega RC}{[1 + (\omega RC)^2]^{1/2}} \qquad (7.25)$$

The phase is

$$\pi(\omega) = \frac{\pi}{2} - \tan^{-1}\omega \qquad (7.26)$$

The Bode plot is shown in Figure 7.6.

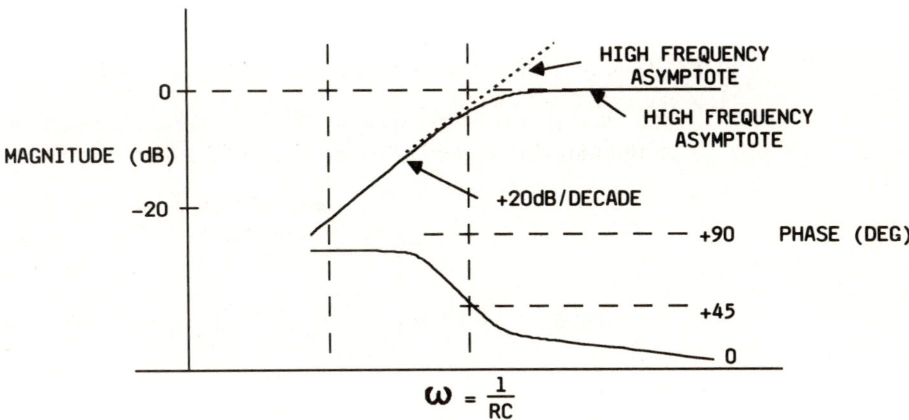

Figure 7.6: Bode plot for high-pass filter

It should be clear from the figure that the high-pass filter attenuates low frequencies, but passes high frequencies.

7.5 Fourier Series[9],[10],[11]

In this section, we consider periodic input or forcing functions that satisfy certain mathematical restrictions and that can be generated in the laboratory. These functions may be represented as the sum of an infinite number of sine and cosine functions that are harmonically related. The response of the linear network may then be found by superimposing these sine and cosine functions.

The period function $f(t)$ must satisfy the following conditions:

1. $f(t)$ must be single valued.

2. $f(t)$ must have a finite number of discontinuities in the periodic interval.

3. $f(t)$ must have a finite number of maxima and minima in the periodic interval.

4. the integral $\int_{t_o}^{t_o+T} |f(t)|\, dt$ must exist.

If these conditions are fulfilled, then $f(t)$ can be expressed as an infinite series, namely,

$$f(t) = a_o + \sum_{n=1}^{\infty}(a_n \cos n\omega_o t + b_n \sin n\omega_o t) \tag{7.27}$$

where

$$\omega_o = \frac{2\pi}{T} = 2\pi f_o \tag{7.28}$$

The coefficients of the Fourier series are given by

$$a_o = \frac{1}{T}\int_o^T f(t)dt \tag{7.29}$$

$$a_n = \frac{2}{T}\int_o^T f(t)\cos n\omega_o t\, dt \tag{7.30}$$

$$b_n = \frac{2}{T}\int_o^T f(t)\sin n\omega_o t\, dt \tag{7.31}$$

The Fourier series of Equation (7.27) can be expressed in exponential form by substituting the exponential forms of the sine and cosine. The result is

$$f(t) = \sum_{n=-\infty}^{\infty} C_n e^{jn\omega_o t} \tag{7.32}$$

where $\qquad C_n = \frac{1}{T}\int_{-T/2}^{+T/2} f(t)e^{-jn\omega_o t}dt \tag{7.33}$

for $\qquad \omega_o = 2\pi/T$

The amplitude of the component of the exponential Fourier series at $\omega = n\omega_o$ is $|C_n|$. As an example, consider the train of pulses of Figure 7.27.

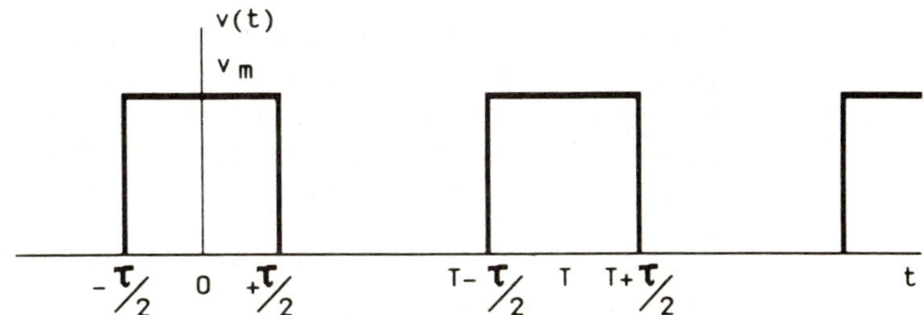

Figure 7.7: Train of pulses with period T

For this train,

$$C_n = \frac{1}{T} \int_{-\tau/2}^{-\tau/2} V_m e^{-jn\omega_o \tau} dt \tag{7.34}$$

$$= \frac{jV_m}{n\omega_o T} (e^{-jn\omega_o \tau/2} - e^{+jn\omega_o \tau/2}) \tag{7.35}$$

$$= \frac{2V_m}{n\omega_o T} \sin n\omega_o \tau/2 \tag{7.36}$$

$$= \frac{V_m \tau}{T} \frac{\sin(n\omega_o \tau/2)}{n\omega_o \tau/2} \tag{7.37}$$

The exponential series of v(t) is then

$$v(t) = \sum_{n=-\infty}^{\infty} (\frac{V_m T}{\tau}) \frac{\sin(n\omega_o \tau/2)}{n\omega_o \tau/2} e^{jn\omega_o t} \tag{7.38}$$

7.6 Fourier Transform[12],[13],[14]

The Fourier transform is an operation that converts a function in time into a function in frequency, i.e., jω. Given a function in time, there is one and only one Fourier transform that corresponds to it, and given a function in ω, there is one and only one function of time that corresponds to it. The Fourier transform allows us to deal with nonperiodic functions and negative-time functions, as well as positive-time functions. Therefore, the Fourier transform is suited for problems that are described by events that start at t=-∞. A transformation is performed to create a domain in which the mathematical procedures that describe the circuit conditions are easier to carry out. It turns out that we can obtain the Fourier transform of virtually any arbitrarily chosen function f(t) for any voltage or current that can be produced in the laboratory.

Our objective in linear circuit analysis for the Fourier transform requires the evaluation of the Fourier transform of the forcing function, the multiplication of the transform by the system function to obtain the transform of the response function, and, finally, using the inverse transform to obtain the response function. In this process, complicated integrodifferential equations are replaced by simple linear polynominal equations that can be solved using algebraic techniques. The Fourier transform pair is

$$F(\omega) = \int\limits_{-\infty}^{+\infty} e^{-j\omega t} \, f(t)dt \tag{7.39}$$

and

$$f(t) = \frac{1}{2\pi} \int\limits_{-\infty}^{+\infty} e^{j\omega t} F(\omega)d\omega \tag{7.40}$$

The function $F(\omega)$ is the Fourier transform of $f(t)$, and $f(t)$ is the inverse Fourier transform of $F(\omega)$.

$F(\omega)$ is a complex quantity that can be described in either rectangular or polar form. From

$$F(\omega) = \int\limits_{-\infty}^{+\infty} f(t) \, (\cos \omega t - j\sin \omega t)dt \tag{7.41}$$

$$= \int\limits_{-\infty}^{+\infty} f(t) \cos \omega t \, dt - j \int\limits_{-\infty}^{+\infty} f(t)\sin \omega t \, dt \tag{7.42}$$

we can write

$$A(\omega) = \int\limits_{-\infty}^{+\infty} f(t) \cos \omega t \, dt \tag{7.43}$$

and

$$B(\omega) = - \int\limits_{-\infty}^{+\infty} f(t) \sin \omega t \, dt \tag{7.44}$$

Finally,

$$F(\omega) = A(\omega) + jB(\omega) = \mid F(\omega) \mid \, e^{j\theta(\omega)} \tag{7.45}$$

We can conclude that

1. The real part of $F(\omega)$ is an even function of ω.

2. The imaginary part of $F(\omega)$ is an odd function of ω.

3. The magnitude of $F(\omega)$ is an even function ω.

4. The phase angle of $F(\omega)$ is an odd function of ω.

5. Replacing ω by $-\omega$ yields the conjugate of $F(\omega)$, i.e., $F(-\omega) = F^*(\omega)$.

The theorem that relates the energy associated with a time domain function of finite energy content to the Fourier transform of the function is called **Parseval's theorem**, written

$$\int\limits_{-\infty}^{+\infty} f^2(t)dt = \frac{1}{2\pi} \int\limits_{-\infty}^{+\infty} \mid F(\omega) \mid^2 d\omega \tag{7.46}$$

For example, if $f(t)$ represents either the voltage across or the current through a 1-Ω resistor, then the energy associated with $f(t)$ can be calculated either by integration of the square of $f(t)$ over all times or by integrating $1/2\pi$ times the square of the Fourier transform of $f(t)$ over all frequencies.

Table 7.1 gives a summary of the transform pairs of important elementary functions. Also listed in the table are operational transforms. Note, that the Fourier transform is written as

$$F\{f(t)\} = F(\omega) \tag{7.47}$$

Table 7.1 Fourier Transform Pairs
(Reprinted with permission of Macmillan Publishing Company from *Basic Engineering Circuit Analysis*, Second Edition, by J. D. Irwin. © 1987 by Macmillan Publishing Co.)

Elementary Functions

$f(t)$	$\mathcal{F}\{f(t)\} = F(\omega)$
1. $\delta(t)$	1
2. $\delta(t - t_o)$	$e^{-j\omega t_o}$
3. $e^{j\omega_o t}$	$2\pi\delta(\omega - \omega_o)$
4. A	$2\pi A\delta(\omega)$
5. $\cos \omega_o t$	$\pi[\delta(\omega + \omega_o) + \delta(\omega - \omega_o)]$
6. $\sin \omega_o t$	$j\pi[\delta(\omega + \omega_o) - \delta(\omega - \omega_o)]$
7. $\operatorname{sgn}(t)$	$\dfrac{2}{j\omega}$
8. $\mu(t)$	$\pi\delta(\omega) + \dfrac{1}{j\omega}$
9. $e^{-at}\mu(t)$	$\dfrac{1}{a + j\omega}$
10. $e^{-a|t|}$	$\dfrac{2a}{\omega^2 + \omega^2}$
11. $e^{-at}\cos \omega_o t \cdot \mu(t)$	$\dfrac{a + j\omega}{(a + j\omega)^2 + \omega^2}$
12. $e^{-at}\sin \omega_o t \cdot \mu(t)$	$\dfrac{\omega_o}{(\omega + j\omega)^2 + \omega^2}$
13. $\mu(t + \tfrac{1}{2}T) - \mu(t - \tfrac{1}{2}T)$	$T\dfrac{\sin \dfrac{\omega T}{w}}{\dfrac{\omega T}{2}}$

Operational Transforms

$f(t)$	$\mathcal{F}\{f(t)\} = F(\omega)$
14. $Kf(t)$	$Kf(\omega)$
15. $f_1(t) - f_2(t) + f_3(t)$	$F_1(\omega) - F_2(\omega) + F_3(\omega)$
16. $\dfrac{d^n f(t)}{dt^n}$	$(j\omega)^n F(\omega)$
17. $\displaystyle\int_{-\infty}^{t} f(x)\,dx$	$F(\omega)/j\omega$
18. $f(at)$	$\dfrac{1}{a}F\left(\dfrac{\omega}{a}\right),\ a>0$
19. $f(t-a)$	$e^{-j\omega a}\,F(\omega)$
20. $e^{j\omega_o t}f(t)$	$F(\omega - \omega_o)$
21. $f(t)\cos\omega_o t$	$\dfrac{1}{2}F(\omega - \omega_o) + \dfrac{1}{2}F(\omega + \omega_o)$
22. $\displaystyle\int_{-\infty}^{+\infty} x(\lambda)h(t-\lambda)\,d\lambda$	$X(\omega)H(\omega)$
23. $f_1(t)f_2(t)$	$\dfrac{1}{2\pi}\displaystyle\int_{-\infty}^{+\infty} F_1(u)F_2(\omega-\mu)\,du$
24. $t^n f(t)$	$(j)^n \dfrac{d^n F(\omega)}{d\omega^n}$

Transform pair number 22 of the table is of significance because it informs us that the transform of the response function $Y(\omega)$ is the product of the input transform $X(\omega)$ and the system function $H(\omega)$, (see Figure 7.8). In the time domain, the result is

$$\text{is} \qquad y(t) = \int_{-\infty}^{+\infty} x(\lambda)h(t-\lambda)\,d\lambda \qquad\qquad (7.48)$$

$$\text{or} \qquad y(t) = x(t) * h(t). \qquad\qquad (7.49)$$

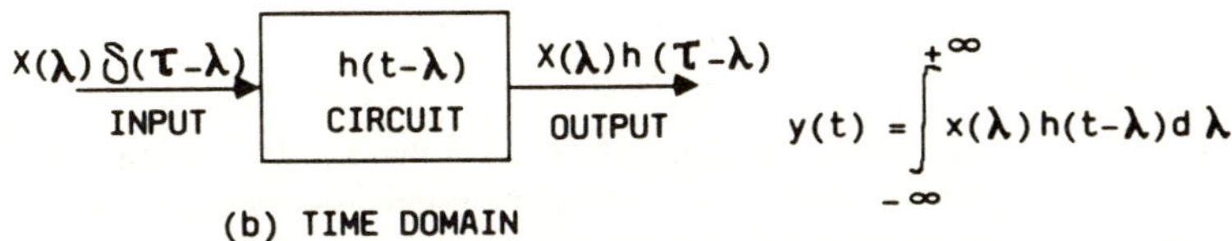

$$y(t) = \int_{-\infty}^{+\infty} x(\lambda)h(t-\lambda)d\lambda$$

Figure 7.8: Illustration of convolution integral

Equation (7.49) states that as the output is equal to the input convolved with the unit impulse response of the circuit. That is, $h(t-\lambda)$ is the response function resulting from a unit impulse.

As an example of convolution theory, consider the circuit of Figure 7.9. The forcing function is given as

$$i_g(t) = 5e^{-3t}u(t) \tag{7.50}$$

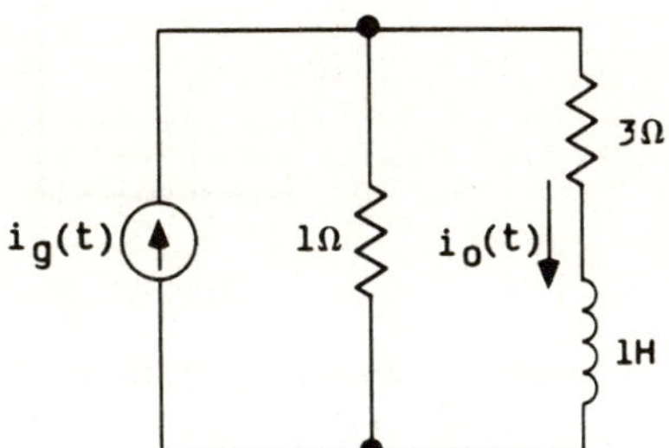

Figure 7.9: Example circuit for convolution theory

The transfer function of the circuit is the ratio of I_o/I_g, so

$$H(\omega) = \frac{I_o}{I_g} = \frac{1}{4 + j\omega} \tag{7.51}$$

From Fourier transform pair number 9 of Table 7.1, the Fourier transform of the driving source is

$$I_g(\omega) = F\{5e^{-3t}u(t)\} = \frac{5}{3 + j\omega} \tag{7.52}$$

The Fourier transform of $i_o(t)$ is

$$I_o(\omega) = I_g(\omega)H(\omega) = \frac{5}{3 + j\omega} \frac{1}{4 + j\omega} \tag{7.53}$$

Using partial fractions, we obtain

$$i_o(t) = F^{-1}\left\{\frac{5}{3+j\omega} - \frac{5}{4+j\omega}\right\} \qquad (7.54)$$

$$i_o(t) = 5e^{-3t}\,u(t) - 5e^{-4t}\,u(t) \qquad (7.55)$$

$$i_o(t) = 5(e^{-3t} - e^{-4t})u(t) \qquad (7.56)$$

7.7 *Step Function*[15],[16],[17]

A forcing function, such as a voltage or current, whose value changes abruptly or whose derivative is discontinuous is called a **singular function**. There are two singular functions that are important in circuit analysis: the unit impulse function and the unit step function. In this section, we discuss the unit step, deferring until a later section a discussion of the unit impulse.

The unit step function, $u(t)$, is <u>dimensionless</u> and is defined mathematically by

$$\begin{aligned}u(t) &= 0, \quad t<0\\ u(t) &= 1, \quad t>0\end{aligned} \qquad (7.57)$$

That is, function is equal to zero for negative values of t, equal to one for positive values of t, and is discontinuous at t=0. The unit step function is shown in Figure 7.10.

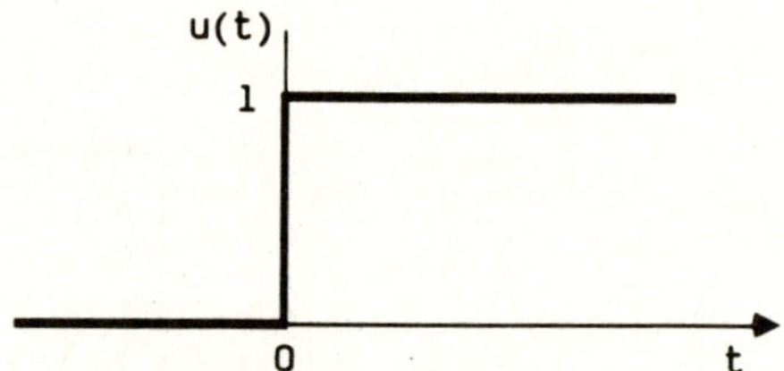

Figure 7.10: The unit step function

The unit step is dimensionless; therefore, a voltage step of V volts and a current step of I amperes are written as Vu(t) and Iu(t), respectively.

The step function of Equation (7.57) can be generalized by replacing t by t - t_o. Then we obtain

$$\begin{aligned}u(t - t_o) &= 0, \quad t<t_o\\ u(t - t_o) &= 1, \quad t>t_o\end{aligned} \qquad (7.58)$$

A graph of this function is shown in Figure 7.11. In effect,

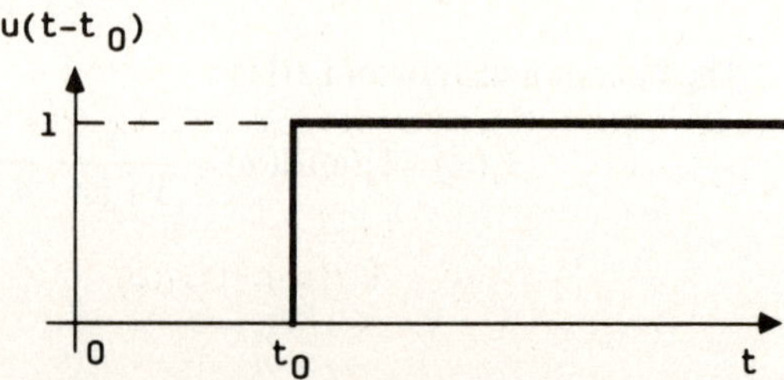

Figure 7.11: The unit step function u(t - t_o)

The function u(t-t_o) is the function u(t) delayed by t_o seconds.

7.8 Impulse Function[18],[19]

Figure 7.12 shows a pulse defined by

$$f(t-\tau) = \frac{1}{a}, \qquad \frac{-a}{2} < t - \tau < \frac{a}{2} \tag{7.59}$$

$$= 0, \quad \text{elsewhere}$$

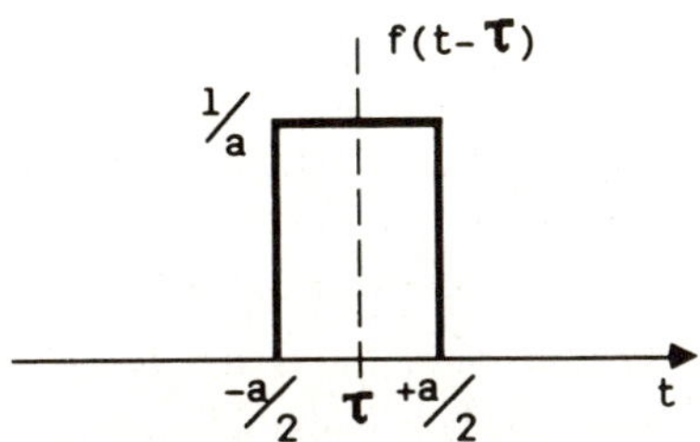

Figure 7.12: Finite pulse

The area under the pulse is 1 independently of the value of a. If $a \to 0$, the base of the pulse shrinks towards zero and its height increases towards ∞, maintaining the constant area (or strength) of 1. The graph at $a = \infty$ is shown in Figure 7.13 and is called the **impulse function** or **unit impulse function**. It is symbolized by $\delta(t - \tau)$ or $\delta(t)$. Like the step function, the impulse function is a singular function, i.e., its value changes abruptly or its derivative is discontinuous.

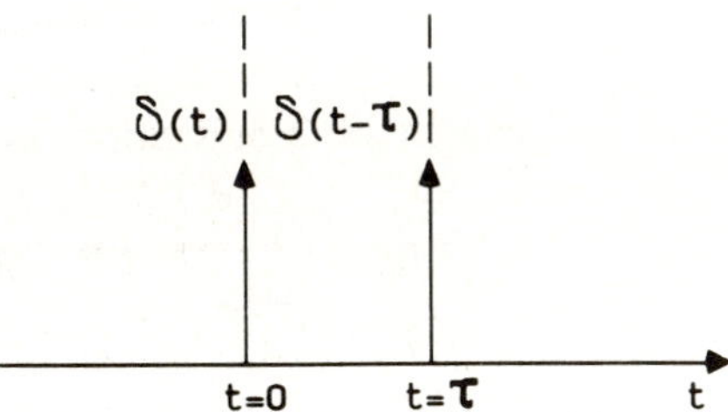

Figure 7.13: Graphical symbol for the
impulse functions $\delta(t)$ and $\delta t - \tau)$

If the unit impulse is multiplied by a constant, the area under the impulse will be equal to the constant multiplying factor. The area, again, is called the strength of the impulse. As an example, the impulse $10\delta(t-5)$ has a strength of 10 at time t=5. Mathematically, the impulse is defined by

$$\delta(t - \tau) = 0, \quad t \neq \tau \tag{7.60}$$

and

$$\int_{-\infty}^{+\infty} \delta(t - \tau)dt = 1 \tag{7.61}$$

An important property of the impulse function is its sifting or sampling property, defined by

$$\int_{-\infty}^{+\infty} f(t)\delta(t)dt = f(0) \tag{7.62}$$

or $\qquad\qquad \displaystyle\int_{-\infty}^{+\infty} f(t)\delta(t - \tau)dt = f(\tau) \tag{7.63}$

Examples of impulse functions are $\int_{-\infty}^{+\infty} e^{-5t} \delta(t - \frac{1}{5})dt = 0.367$ and $\int_{-\infty}^{+\infty} \sin(4\pi t + \frac{\pi}{2})\delta(t) = 1$. They say, respectively, that the strength of the impulse $e^{-5t}\delta(t - \frac{1}{5})$ is .367 and the strength of the impulse $\sin(4\pi t + \frac{\pi}{2})\delta(t)=1$. The sampling or sifting property of the impulse will be useful in determining the Laplace transform of the impulse function.

7.9 Laplace Transform[20],[21],[22]

Like the Fourier transform, the Laplace transform is a systematic treatment for converting between the time and frequency domains. The Laplace transform can be applied to a much wider variety of excitation functions than can the Fourier method. An example is the impulse function. The Laplace transform transforms a set of linear constant-coefficient integrodifferential equations into a set of linear polynomial equations. The Laplace transform also has the facility to deal with initial energy storage conditions, i.e., the initial values of the current and voltage variables.

The Laplace transform of a function is defined by

$$L\left\{f(t)\right\} = \int_{0}^{\infty} f(t)e^{-st}dt \tag{7.64}$$

This equation can also be written as

$$F(s) = L\left\{f(t)\right\} \tag{7.65}$$

The inverse Laplace transform is written as

$$f(t) = \frac{1}{2\pi j} \int_{\sigma_o - j\infty}^{\sigma_o + j\infty} e^{st}F(s)ds \tag{7.66}$$

or

$$f(t) = L^{-1}\left\{F(s)\right\} \tag{7.67}$$

Table 7.2 presents Laplace transform pairs.

Table 7.2: Table of Laplace Transform Pairs

$f(t) = L^{-1}\{F(s)\}$	$F(s) = L\{f(t)\}$
$\delta(t)$ (impulse)	1
$u(t)$ (step)	$\dfrac{1}{s}$
$tu(t)$ (ramp)	$\dfrac{1}{s^2}$
$\dfrac{t^{n-1}}{(n-1)!}u(t), n=1,2,\ldots$	$\dfrac{1}{s^n}$
$e^{-\alpha t}u(t)$ (exponential)	$\dfrac{1}{s + \alpha}$
$te^{-\alpha t}u(t)$ (damped ramp)	$\dfrac{1}{(s + \alpha)^2}$

Table 7.2: Table of Laplace Transform Pairs

$f(t) = L^{-1}\{F(s)\}$	$F(s) = L\{f(t)\}$
$\dfrac{t^{n-1}}{(n-1)!}e^{-\alpha t}u(t)$, n=1,2,...	$\dfrac{1}{(s+\alpha)^n}$
$\dfrac{1}{\beta-\alpha}(e^{-\alpha t}-e^{-\beta t})u(t)$	$\dfrac{1}{(s+\alpha)(s+\beta)}$
$\sin \omega t u(t)$ (sine)	$\dfrac{\omega}{s^2+\omega^2}$
$\cos \omega t u(t)$ (cosine)	$\dfrac{s}{s^2+\omega^2}$
$\sin(\omega t + \theta)u(t)$	$\dfrac{s\sin\theta + \omega\cos\theta}{s^2+\omega^2}$
$\cos(\omega t + \theta)u(t)$	$\dfrac{s\cos\theta - \omega\cos\theta}{s^2+\omega^2}$
$e^{-\alpha t}\sin \omega t\, u(t)$ (damped sine)	$\dfrac{\omega}{(s+\alpha)^2+\omega^2}$
$e^{-\alpha t}\cos \omega t\, u(t)$ (damped cosine)	$\dfrac{s+\alpha}{(s+\alpha)^2+\omega^2}$
$f_1(t) \pm f_2(t)$ (addition)	$F_1(s)\pm F_2(s)$
$Kf(t)$ (scalar multiplication)	$KF(s)$
$\dfrac{df(t)}{dt}$	$sF(s) - f(0^-)$
$\dfrac{d^2f(t)}{dt^2}$ (time)	$s^2F(s)-sf(0^-)-\dfrac{df(0^-)}{dt}$
$\dfrac{d^nf(t)}{dt^n}$ (differentiation)	$s^nF(s)-s^{n-1}f(0^-)-s^{n-2}\dfrac{df(0^-)}{dt}-s^{n-3}\dfrac{d^2f(0^-)}{dt}-...-\dfrac{d^{n-1}f(0^-)}{dt}$
$\displaystyle\int_{0^-}^{t} f(t)dt$ (time)	$\dfrac{1}{s}F(s)$
$\displaystyle\int_{-\infty}^{t} f(t)dt$ (integration)	$\dfrac{1}{s}F(s) + \dfrac{1}{s}\displaystyle\int_{-\infty}^{0^-} f(t)dt$
$f_1(t)*f_2(t)$ (convolution)	$F_1(s)F_2(s)$
$f(t-a)u(t-a)$ (time shift) $a\geq 0$	$e^{-as}F(s)$
$f(t)e^{-at}$ (frequency shift)	$F(s+\omega)$
$f(t) = L^{-1}\{F(s)\}$	$F(s) = L\{F(t)\}$
$-tf(t)$ (frequency)	$\dfrac{dF(s)}{ds}$
$t^nf(t)$ (differentiation)	$(-1)^n\dfrac{d^nF(s)}{ds^n}$
$\dfrac{f(t)}{t}$ (frequency) (integration)	$\displaystyle\int_{s}^{\infty} F(s)ds$
$f(at)$, $a \geq 0$ (scaling)	$\dfrac{1}{a}F(\dfrac{s}{a})$
$f(0^+)$ (initial value)	$\lim_{s\to\infty} s\,F(s)$
$f(\infty)$ (final value)	$\lim_{s\to 0} sF(s)$, all poles of $sF(s)$ in left half-plane
$f(t)=f(t+nT)$ n=1,2, ... (time periodicity)	$\dfrac{1}{1-e^{-Ts}}F_1(s)$ where $F_1(s) = \displaystyle\int_{0^-}^{T} f(t)e^{-st}dt$

Table 7.2: Table of Laplace Transform Pairs

$f(t) = L^{-1}\{F(s)\}$	$F(s) = L\{f(t)\}$
$Ke^{-\alpha t}u(t)$	$\dfrac{K}{s+\alpha}$
$Kte^{-\alpha t}u(t)$	$\dfrac{K}{(s+\alpha)^2}$
$2\|K\|e^{-\alpha t}\cos(\beta t+\theta)u(t)$	$\dfrac{K}{s+\alpha-j\beta}+\dfrac{K^*}{s+\alpha+j\beta}$ where K is a complex quantity
$2t\|K\|e^{-\alpha t}\cos(\beta t+\theta)u(t)$	$\dfrac{K}{(s+\alpha-j\beta)^2}+\dfrac{K^*}{(s+\alpha+j\beta)^2}$ where K is a complex quantity

Example

Given the circuit of Figure 7.14 with no energy stored in the circuit at the time the switch is closed, find i(t) for $t \geq 0$ for R = 15 Ω, L = 1 H and C = 0.020 F with $V_{dc} = 12$ V.

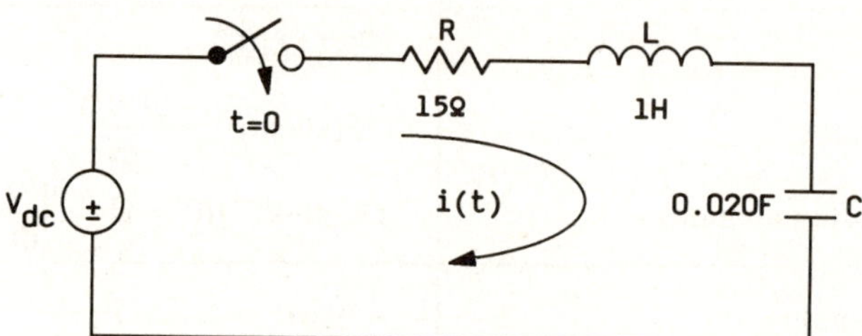

Figure 7.14: Example of the use of Laplace transforms

First, we write the integrodifferential equation for $t \geq 0$:

$$V_{dc} = L\frac{di(t)}{dt} + \frac{1}{C}\int_0^t i(t)dt + Ri(t) \tag{7.68}$$

Next, we transform the equation to the s domain using the appropriate pairs from Table 7.2. These are the step, time differentiation, time integration, and scalar multiplication pairs. We obtain

$$\frac{1}{s}V_{dc} = RI(s) + LsI(s) + \frac{1}{C}\frac{I(s)}{s} \tag{7.69}$$

(Remember, the initial current and voltage conditions are 0.) The third step is to solve Equation (7.69) for I(s) to obtain

$$I(s) = \frac{\dfrac{V_{dc}}{L}}{s^2 + s\dfrac{R}{L} + \dfrac{1}{LC}} \tag{7.70}$$

Substituting in the values of the components yields

$$I(s) = \frac{15}{s^2 + 15s + 50}$$

$$I(s) = \frac{15}{(s+10)(s+5)} \tag{7.71}$$

Table 7.3: Summary of S-Domain Equivalent Circuits

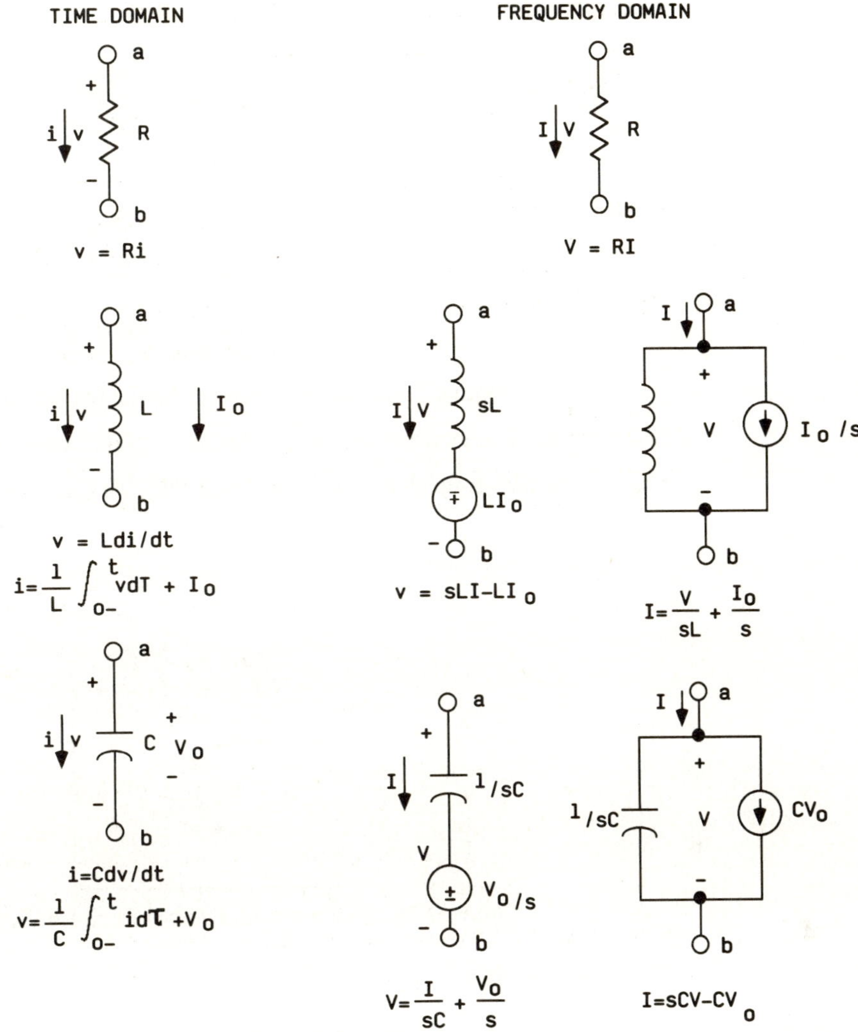

Next, we use partial fraction expansion:

$$I(s) = \frac{15}{(s+10)(s+5)} = \frac{A}{s+10} + \frac{B}{s+5} \tag{7.72}$$

$$.spA = \frac{15}{s+5}\bigg|_{s=-10} = \frac{15}{-5} = -3 \tag{7.73}$$

$$B = \frac{15}{s+10}\bigg|_{s=-5} = \frac{15}{5} = 3 \tag{7.74}$$

$$I(s) = \frac{-3}{s+10} + \frac{3}{s+5} \tag{7.75}$$

Finally, we take the inverse Laplace transform of Equation (7.75) to obtain

$$i(t) = (3e^{-5t} - 3e^{-10t})u(t) \tag{7.76}$$

By developing s-domain equivalent circuits for the circuit elements, we can eliminate the step of writing the integrodifferential equations that describe the circuit. Table 7.3 is a summary of the s-domain equivalent circuits.

Example

Given the RL circuit of Figure 7.15 with R=10 Ω, L = 5 H, and a voltage source of 5u(t),

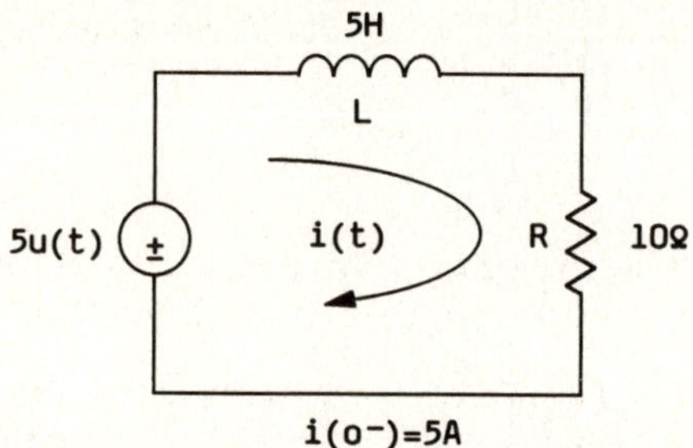

Figure 7.15: Example illustrating the Laplace transform

Solve for the circuit current i(t). The initial value of the current is 5 A.

For the s-domain equivalent circuit of Figure 7.16, we can write the s-domain equation as

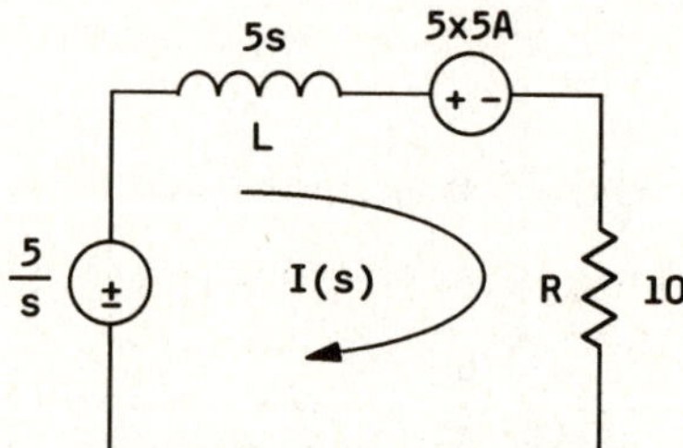

Figure 7.16: S-domain equivalent circuit for Figure 7.15

The equation is then

$$\frac{5}{s} = 5s(I(s)-5) + 10I(s) \tag{7.77}$$

Solving for I(s) yields

$$5(s + 2)I(s) = \frac{5}{s} + 25 \tag{7.78}$$

$$I(s) = \frac{1}{s(s + 2)} + \frac{5}{s + 2}$$

Using partial fraction expansion, we obtain

$$I(s) = \frac{1}{s(s + 2)} + \frac{5}{s + 2} = \frac{A}{s} + \frac{B}{s + 2} \tag{7.79}$$

$$A = \frac{1}{s + 2}\bigg|_{s=0} = \frac{1}{2} = 0.5 \tag{7.80}$$

$$B = -\frac{1}{2} + 5 = 4.5 \tag{7.81}$$

Then

$$I(s) = \frac{0.5}{s} + \frac{4.5}{s + 2} \tag{7.82}$$

So from Table 7.2,

$$i(t) = 0.5u(t) + 4.5e^{-2t}u(t) \qquad (7.83)$$

or

$$i(t) = (0.5 + 4.5e^{-2t})u(t) \qquad (7.84)$$

7.10 Summary

In this chapter, we have studied methods for determining the frequency response of circuits. Among these methods are Bode and pole-zero plots.

Use of the Fourier series allows us to find the response of a circuit to a periodic input. Laplace and Fourier transforms are systematic methods for finding the steady state response of a circuit for nonperiodic input sources. The Laplace transform transforms a set of linear, constant-coefficient differential equations into a set of linear polynominal equations. The initial values of the current and voltage variables are automatically introduced into the polynominal equations. The Fourier transform is useful for problems in communications theory and signal processing. The impulse and step functions are introduced in relation to the Laplace transform.

LAB 7L1

TITLE: Frequency and Transient Responses of First-Order Linear Systems

OBJECTIVE: This laboratory introduces the frequency and transient (time domain) responses of first-order systems. A system is first order when the highest order derivative appearing in the equation is the first derivative. The transient response is the output of the system due to various input signals, such as the impulse and step. The output is expressed in the time domain, as are the inputs. Next, the steady-state response of the system due to sinusoidal inputs is examined, with the output expressed in terms of phase and magnitude relative to the input. Also, the transient and frequency responses of the first-order systems presented are evaluated using SPICE.

EQUIPMENT AND COMPONENTS:
Function generator
Pulse generator
Oscilloscope
2 - 1/8 W resistors
    1 MΩ
    10 kΩ
2 - capacitor
    1 μF
    0.1 μF

PRELABORATORY:

1. Consider the circuit of Figure 7L1.1. Use the Laplace Transform to determine the circuit's response for each of the three input signals.

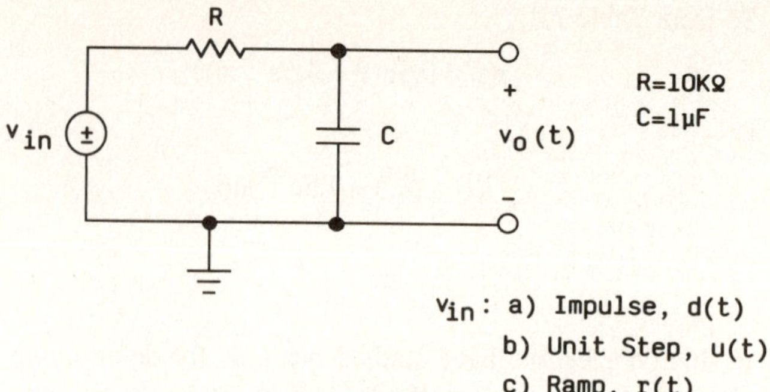

Figure 7L1.1

   a. Discuss how a train of pulses could be used to approximate the unit step input and appear on a nonstorage type of oscilloscope. (Include the height, width, and duty cycle that the pulse train would need to give a reasonable step response.)

   b. Discuss the simulation of the impulse using the same ideas from part a. Given the function generator available in your laboratory, how large should R be to make this approach reasonable?

2. Consider the circuit of Figure 7L1.1 again.

   a. If the input signal v_{in} is a sinusoidal, use Laplace transforms to calculate the time domain response.

   b. Are the frequencies of the input and output signals the same? Why?

   c. If the magnitude of the sinusoidal input signal v_{in} is 1 V, find the magnitude and phase angle of the output as the frequency varies from 10 Hz to 20 KHz. Evaluate the magnitude and phase angle at 3 frequencies per decade, and express the magnitude in dB's ($V_{odB} = 20\log | \frac{v_o}{v_{in}} |$). Plot the magnitude versus the frequency on semilog paper. (Frequency goes in the log axes.) Also, plot the phase angle versus the frequency.

   d. What is the rate of change in the magnitude if the frequency is above 500 Hz?

   e. What is the maximum phase angle? Minimum?

3. Repeat parts 1 and 2 using the circuit of Figure 7L1.2.

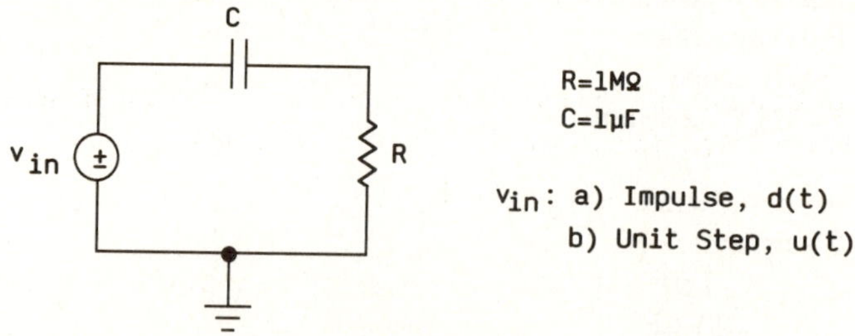

Figure 7L1.2

Laboratory Procedure:

1. Impulse and step.

   a. Build the circuit of Figure 7L1.3.

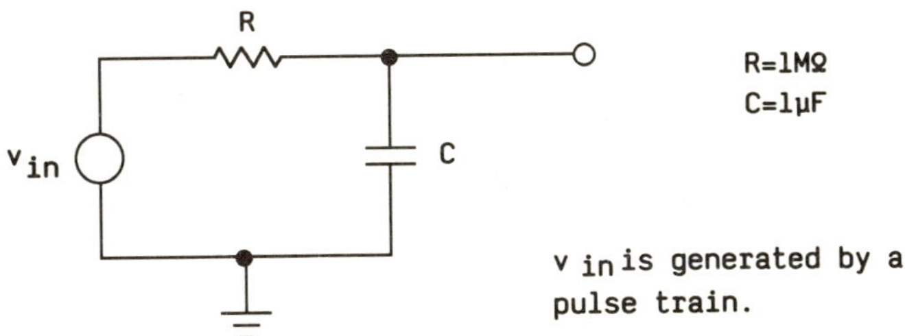

Figure 7L1.3

Adjust the width and height of the pulse so that a unit area is maintained, i.e., let height = 1 V and width = 1 S, height = 2 V and width = 0.5 S, etc. Continue this process until the output v_o doesn't change in appearance. How does the final output compare to the theoretical plot for the unit impulse?

1. Set the pulse amplitude to 1 V and increase the pulse width while displaying v_{in} and v_o on the oscilloscope. How wide does the pulse need to be so at least four time constants ($4RC=4\tau$) of the response can be seen during the pulse width? The repetition rate may need to be adjusted. Be sure that the time between pulses is long enough so that no charge is left on the capacitor at the beginning of the next pulse. Plot this response and compare it to the theoretical result. Graphically determine the derivative of the output plot and compare it to the impulse response obtained in the previous section. Change resistor R to 1 kΩ and repeat.

2. For both step responses of part 1 (R=1 MΩ and R=1 kΩ) extrapolate the initial slope with a straight line until it intersects 1 V, as shown in Figure 7L1.4.

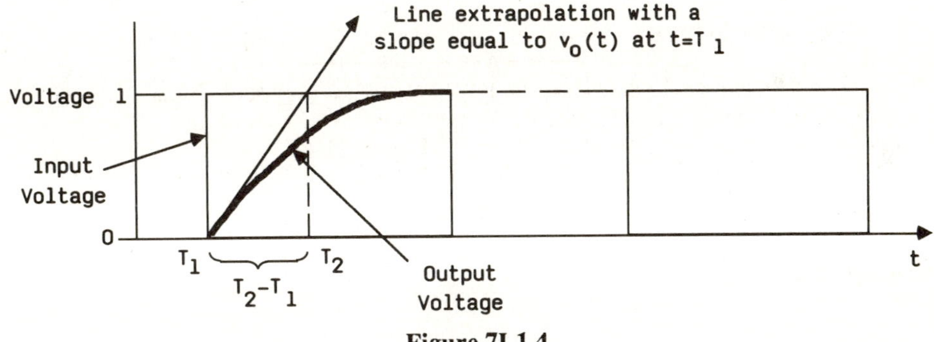

Figure 7L1.4

Then drop a vertical line from the point where the straight-line extrapolation intersects the input voltage. Compare this time difference to one RC time constant.

b. Repeat part a with the circuit of Figure 7L1.5.

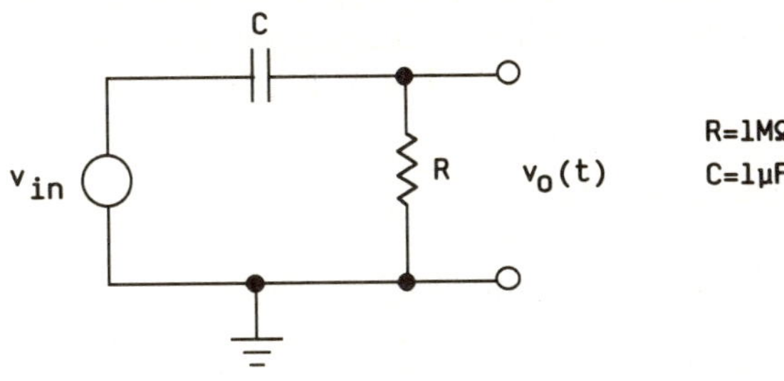

Figure 7L1.5

2. Sinusoidal response.

   a. Build the circuit of Figure 7L1.6.

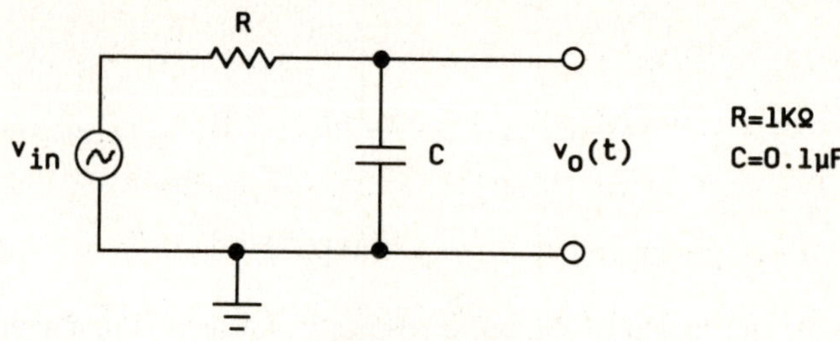

Figure 7L1.6

Let v_{in} be sinusoidal with a 1-V amplitude, and vary the frequency from 10 Hz to 10 KHz collecting three magnitude and phase angle points per decade as done in the prelab. Plot the data on the same graph as in the prelab. Compare the rate of change in the magnitude with respect to frequency found in the laboratory with the prelab results.

   b. Repeat part a with the circuit of Figure 7L1.7.

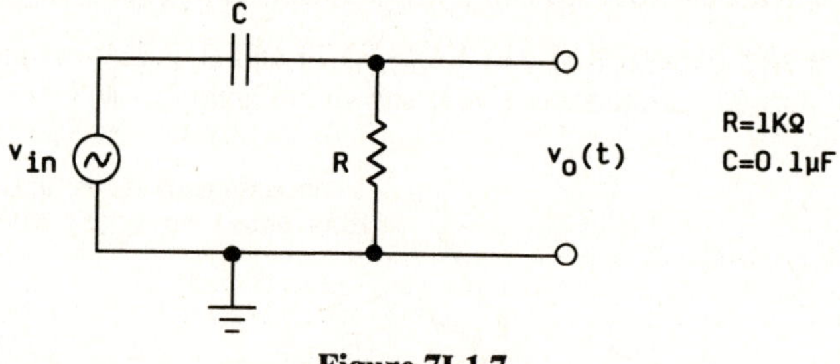

Figure 7L1.7

DISCUSSION:

CONCLUSIONS:

LAB 7L2

TITLE: First-Order Design Problem

OBJECTIVE: This laboratory presents several circuit requirements situations that can be solved with an RC circuit. Your objective is to determine whether a frequency analysis or a transient analysis is more appropriate and to use that analysis to calculate the needed values for R and C. This is done in the prelab and verified using SPICE and the laboratory procedure.

EQUIPMENT AND COMPONENTS:
    Function generator
    Resistors (determined by your design)
    Capacitors (nonpolar, determined by your design)
    Oscilloscope
    Low-frequency spectrum analyzer

PRELABORATORY:

1. Design a circuit consisting of a resistor R and capacitor C that will have a maximum output voltage when the input voltage has maximum slope. The only restriction on the input is that the signal must be periodic and must have a maximum amplitude of 10 V.

    a. Design your circuit so that it will work on sine waves up to 10 kHz.

    b. Test your circuit using SPICE with the following inputs:

        1. A sine wave with a 1-V amplitude and a 1-kHz frequency.

        2. A 5-V amplitude square wave and a 1-millisecond period. The pulse width is 0.5 ms.

        3. A triangular wave with a 1-V amplitude and a 0.1-millisecond period. The pulse width is 0.05 ms.

2. Design a circuit consisting of a resistor R and a capacitor C that will reduce all harmonic content in a square wave so that the maximum amplitude above the third harmonic is no more than one-tenth the amplitude of the fundamental frequency. Also, the amplitude of the output's fundamental frequency is to be unchanged from the amplitude of the input.

    a. Determine the resistor and capacitor values.

    b. Test your design using SPICE and the following inputs:

        1. a square wave with a 1-V amplitude and a 1-millisecond period. Pulse width is 0.5 ms.

        2. a triangular wave with a 1-V amplitude and a 1-millisecond period.

Laboratory Procedure:

1. Slope detector.

    a. Build your design for part 1 of the prelaboratory.

    b. Verify the circuit using the input signals defined in part 1 of the prelab.

    c. Try a 100 kHz sine wave as an input. Does the circuit still show where the maximum slope is?

2. Harmonic frequency filter.

    a. Build the circuit you designed in part 2 of the prelab.

    b. Use a spectrum analyzer to verify that your circuit works for the inputs defined in part 2b. Sketch the harmonics of the input and output waveforms.

DISCUSSION:

CONCLUSIONS:

LAB 7L3

TITLE: Frequency and Transient Responses of the RLC Circuit: A Second-Order System

OBJECTIVE: This laboratory contains a second-order circuit that is analyzed for its transient response to a unit step input and its frequency response to sinusoidal inputs. First, the resistor R, inductor L, and capacitor C are first treated as ideal components and analyzed in the prelaboratory as such. Then various parasitics are included to yield a more accurate model and simulation. Finally, these effects are confirmed in the laboratory.

EQUIPMENT AND COMPONENTS:

Universal impedance bridge
or a capacitance and inductance bridge
or other equipment capable of giving
values of L, C, and R at some frequency
Capacitor (nonelectrolytic)
range of values: 0.01 μF to 0.1 μF
Inductor
range of values: 10 mH to 1 H
Resistors
range of values: 1 kΩ to 10 kΩ
Function generator
Oscilloscope

PRELABORATORY:

1. Consider the circuit of Figure 7L3.1.

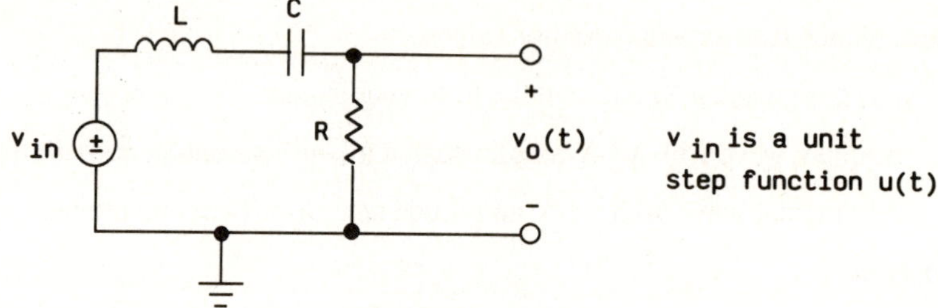

Figure 7L3.1

a. Using Laplace transforms, determine the transfer function $v_o(S)/v_{in}(S)$.

b. The general form of the denominator of a second-order system is

$$D(s) = s^2 + (2\zeta\omega_n)s + \omega_n^2$$

For the circuit of Figure 7L3.1, identify the following:

   1. $\zeta \equiv$ the damping ratio.

   2. $\omega_n \equiv$ the undamped natural frequency.

c. Consider three cases for ζ: $\zeta=0.5$, $\zeta=1$; and $\zeta=2$. For $\omega_n=10$ krad/sec, determine values for R, L, and C for each case. Then sketch the step response. Be sure that the values you use are available in your lab.

d. Knowing the mathematical relationship between the step response and the impulse response, determine the impulse response for the three cases of part 1c.

2. In the circuit of Figure 7L3.1 let $v_{in}(t)$ be a sinusoidal input with a magnitude of 1 V.

a. Consider again the three cases for ζ: $\zeta=0.5$; $\zeta=1$; and $\zeta=2$. For $\omega_n=10$ krad/s, plot the magnitude and phase angle of the transfer function $V_o(j\omega)/V_{in}(j\omega)$ from 100 rad/krad/s to 1 Mrad/krad/s for each case. Do not use any approximations for the graphs, and plot at least three points per decade. Your plots should be done on semilog paper, with frequency on the log axes and magnitude (in dB) and phase angle (in degrees) on the linear axes.

3. Consider again the circuit of Figure 7L3.1.

a. Use the component values that yield the three cases mentioned in part 1, and do a SPICE analysis for these cases to show the transient step response.

b. Use the same component values, and do a frequency analysis using SPICE from 100 rad/sec to 1 Mrad/sec for each of the three cases. Assume a sinusoidal input with magnitude 1 and phase 0°.

Laboratory Procedure:

1. LRC circuit (assumed ideal).

a. Build the circuit shown in Figure 7L3.2 three times, once for each value of ζ.

b. Let v_{in} be a train of pulses with amplitude 1 V, pulse width 5 millisecond and duty cycle 50%. Explain the reason for the circuit's generating a small output. Plot the step response from your oscilloscope for all three cases.

c. Let v_{in} be a sinusoidal waveform with a 1-V amplitude (2-V peak to peak). Vary the frequency from 100 Hz to 100 kHz and record the output voltage's magnitude and phase angle relative to the input waveform. Be sure to put your scope in the chop mode to make these phase measurements.

d. Compare your experimental results with those of your prelab.

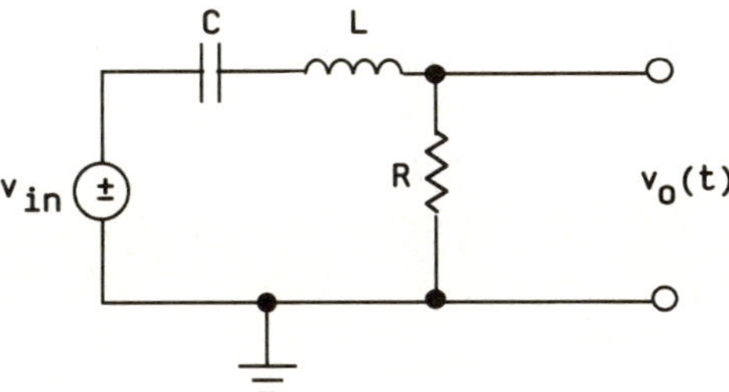

Figure 7L3.2

2. Nonideal case (for all three circuits).

a. Measure your capacitors using a capacitance bridge or universal bridge.

1. If the dissipation factor D is given by equation 7L3.1, determine the values of the capacitor and the equivalent parallel resistance R_p and series resistance R_s.

$$D = \frac{1}{\omega R_p\, C_p} = \omega R_s C_s \qquad\qquad (7L3.1)$$

The two models are shown in Figure 7L3.3.

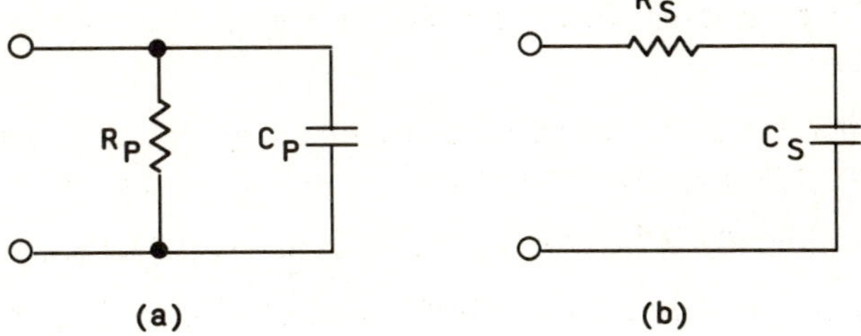

(a)   (b)

Figure 7L3.3

b. Measure your inductor using an inductance bridge or a universal bridge.

1. Determine and record the value of the inductor and the effective series resistance R_s given in the model of Figure 7L3.4.

2. Determine the value of the internal parasitic inductance that is modeled as a lumped inductor L. The parasitic elements to be measured are very small and become significant only at very high frequencies. In measuring these parasitics, the sweep frequency must be increased up to 10 Mhz or higher. If the universal bridge or vector impedance meter cannot go up to these frequencies, the parasitics may not be able to be determined.

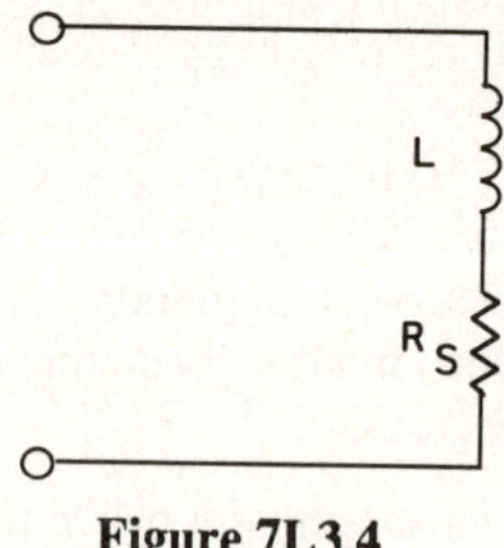

Figure 7L3.4

c. Measure the value of the resistor using a universal bridge or vector impedance meter. Determine the value of the associated parasitics, L and C_p.

d. The selectivity or sharpness of the peak in a resonant current is called the **quality factor Q**, defined as the ratio of the resonant frequency to the bandwidth:

$$Q = \frac{\omega_o}{\beta} \tag{7L3.2}$$

For a parallel RLC resonant circuit,

$$Q = \omega_o\, RC = \frac{R}{\omega_o L} = R\sqrt{\frac{C}{L}} \tag{7L3.3}$$

For a series RLC resonant circuit,

$$Q = \frac{\omega_o\, L}{R} = \frac{1}{\omega_o\, C\, R} = \frac{1}{R}\sqrt{\frac{L}{C}} \tag{7L3.4}$$

Determine the Q of the RLC circuit.

3. SPICE simulation using more accurate models.

a. Repeat the SPICE simulation of the prelab for all three cases using the models found in part 2 of the lab.

b. Where possible, do a comparison of the theoretical results obtained in the prelaboratory with the results obtained by measurement, and discuss any differences.

DISCUSSION:

CONCLUSIONS:

LAB 7L4

TITLE:	Transient Response for Op Amp-Based Filters (Slew Rate Limiting)
OBJECTIVE:	This laboratory investigates the frequency response of simple filter designs employing operational amplifiers and explores the limitations of their response due to the limited slew rate of the op amp. The laboratory includes a theoretical analysis of the source of slew rate limitation in an op amp, as well as an examination of the practical effects of this limitation as seen in actual circuits.

EQUIPMENT AND COMPONENTS:

        Oscilloscope
        Function generator (square and sine waves)
        Power Supplies (± 15 V @ 100 mA)
        Op amps (2)
            a high-slewing-rate op amp
            $\approx 10\text{V}/\mu\text{s}$, such as the OP-01 (PMI)
            a low-slewing-rate op amp $\approx .1\text{V}/\mu\text{s}$,
            such as the OP-07 (PMI)

        Resistors $\Big\}$ as per design
        Capacitors

PRELABORATORY:

1. Analyze the circuits of Figure 7L4.1 to determine their response to the given inputs. (Sketch their response.) Assume that the op amps are ideal.

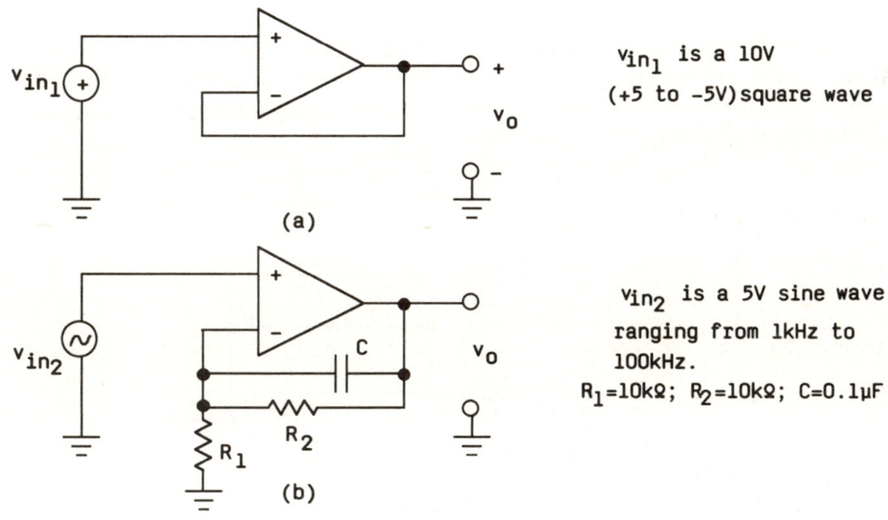

Figure 7L4.1

2. Define the term **slew rate**, and from the data sheets find the slew rate for the op amps given in this laboratory.

    a. What is the maximum-frequency sine wave that could be applied as the input to the circuit of Figure 7L4.1(a) and still not be limited by the slew rate? Assume that the sine wave has a 10 V amplitude.

    Apply two 10 kΩ resistors as shown in Figure 7L4.2 and reduce the sine wave amplitude to 5 V. Is there any change in the maximum frequency? Why?

    Now determine the maximum frequency with a 5-V sine wave applied to the circuit of Figure 7L4.1(a). Is there any change in the maximum frequency? Why? Do this analysis for both op amps.

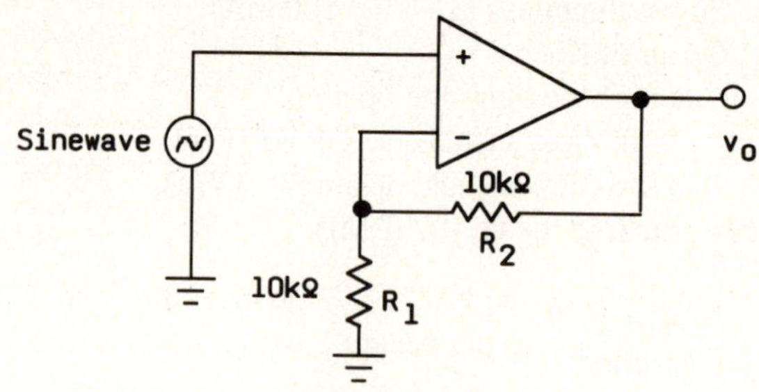

Figure 7L4.2

   b.   What is the maximum-frequency 5-V sine wave that can be applied as an input to the circuit of Figure 7L4.1(b) and not be limited by the slew rate? Determine a new value for R_1 that could double the maximum frequency where slew-rate limitation starts to occur. What is the dc gain for the circuit? Do this analysis for both op amps.

3. Consider the low-pass filter of Figure 7L4.3. Design this circuit so that it has a dc gain of 10 V/V and a pole at 10 kHz. Using both op amps, determine the maximum amplitude of the input sine wave that would allow the frequency response of the circuit to be obtained from 100 Hz to 100 kHz with no distortion due to slew rate.

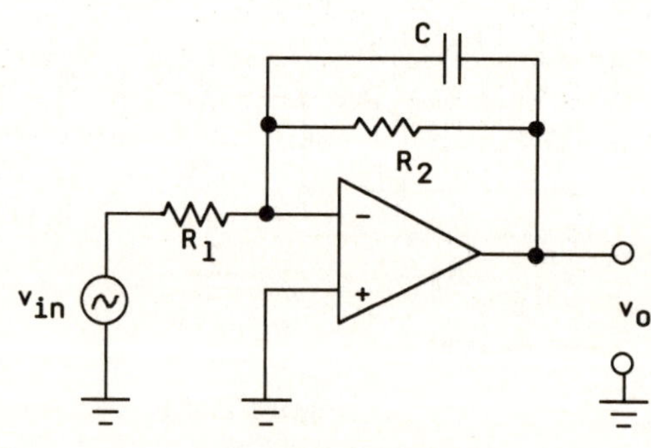

Figure 7L4.3

Laboratory Procedure:

1. Slew Rate.

   a.   Build the circuit of Figure 7L4.4. Let v_{in} be a square wave of 10 V amplitude (+10 V to -10 V). Using the op amp with the

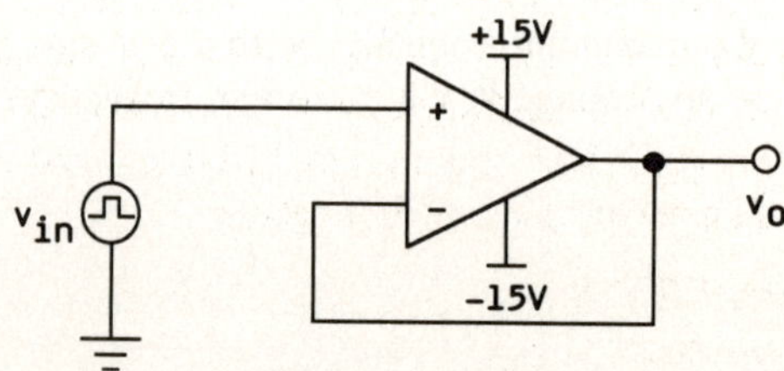

Figure 7L4.4

lower slew rate and the delayed sweep capability of the oscilloscope, sketch the leading and trailing edge of the output. Determine the slew rate of your op amp, and compare it to the value given on the data sheet. Repeat this procedure for the faster op amp.

b. Build the circuit of Figure 7L4.2. Let v_{in} be a sine wave of 5-V amplitude ranging in frequency from 100 Hz to 100 kHz. Plot the theoretical magnitude response over this range and the actual observed response for both op amps. At what frequency does the output magnitude deviate from the theoretical? How does this compare to the results of part 3b in the prelaboratory?

2. Low-pass active filter design.

a. Build the circuit designed in part 4 of the prelaboratory and shown in Figure 7L4.5.

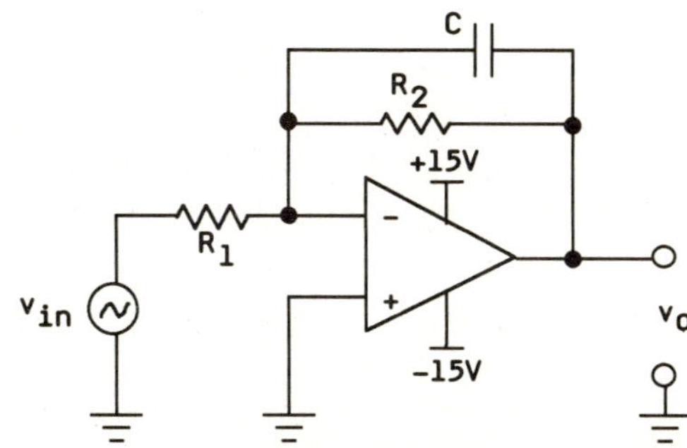

Figure 7L4.5

b. Plot the frequency response of the circuit against that of the ideal circuit, and compare the two. Do this for both op amps. Is the amplitude of the op amp circuit greater or less than that predicted by the theoretical calculation? If you used the slew rate determined in part 1 of the laboratory, rather than the worst case value given on the data sheets, how do your results compare in terms of where distortion would start to take place?

DISCUSSION:

CONCLUSIONS:

LAB 7L5

TITLE: The Function Generator

OBJECTIVE: Because one of the most important tools used to evaluate an actual circuit's frequency response is the function generator, the chapter would be incomplete if this device were not discussed. One way to address the topic is to design a simple function generator and, in the process, examine a few subcircuits of the generator in terms of time and frequency domain analysis. The design starts with a general-purpose timer chip (555) to generate a square wave. Then a low-pass-filter and an integrator are used to produce a sine wave and a triangular wave from the square wave. The interesting point here is that the low-pass filter and the integrator are the same circuit looked at in two different ways: from the time domain, and frequency domain perspective, respectively.

EQUIPMENT AND COMPONENTS:
Oscilloscope
Function generator (square and sine waves)
Power Supply (±15 V @ 100 mA)
Resistors (±1%, 1/8 W, and values determined by students)

Capacitors (nonpolar, with values determined by students)
Op amp
555 timer

PRELABORATORY:

1. Learn about the 555 timer chip from any analog data book (e.g. from National, Raytheon, or Signetics Corporations). Find the circuit configuration for free-running operation (astable mode), shown in Figure 7L5.1. Note that since R_A is set equal to zero, the square waves will have a 50% duty cycle.

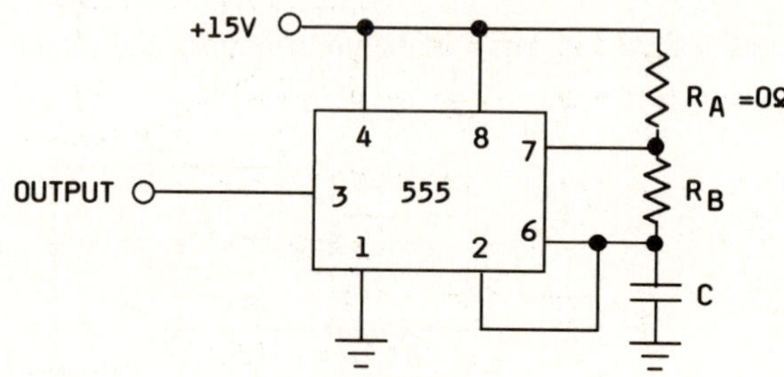

Figure 7L5.1

Sketch the output for R_B ranging from 1 kΩ to 1 MΩ with C equal to 0.01 μF.

2. Consider the circuit of Figure 7L5.2.

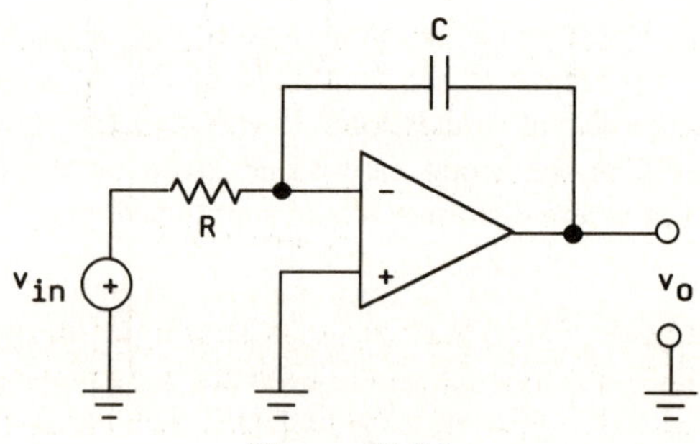

Figure 7L5.2

   a. Determine the transfer function $V_o(j\omega)/V_{in}(j\omega)$ for this circuit. Is the circuit a low-pass or high-pass filter? Explain why. If a dc input were applied, what would happen at the output?

   b. Analyze the circuit in the time domain with a 1-V amplitude sine wave (1 kHz), a unit step, and a 1-V amplitude square wave as inputs. Sketch the output for each input. From the time domain perspective, a unit step is a dc input after t=0. What appears at the output as time goes to infinity for a unit step input?

3. The circuit of Figure 7L5.3, with dc bias currents and offset voltages of the op amp, illustrates some dc problems with op amps.

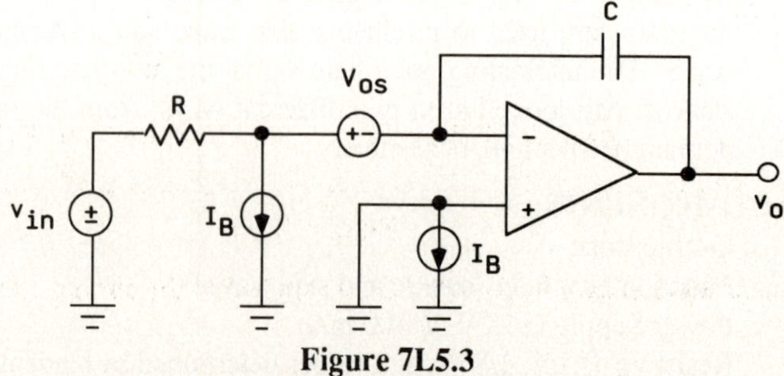

Figure 7L5.3

Set v_{in} equal to zero and let C=1 μF and R=1 kΩ. What is the value of v_o after 1 millisecond? 10 milliseconds? 1 second? 10 seconds? What is the maximum value that the output of an actual op amp circuit will reach?

Now suppose R′ is added to the circuit, as shown in Figure 7L5.4.

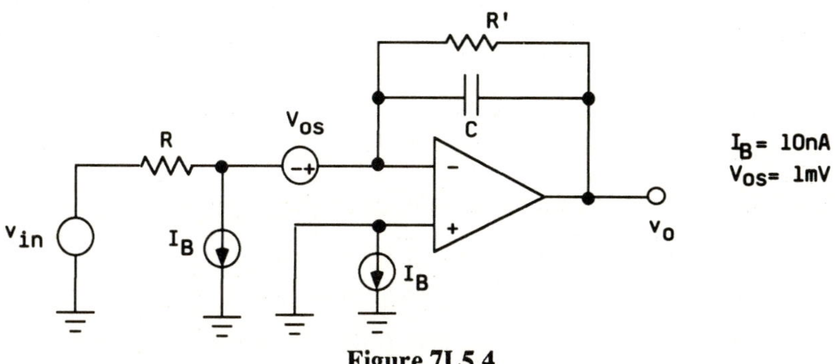

Figure 7L5.4

Determine the output for $v_{in} = 0$, with R′ = R = 10 kΩ, and C=1 μF.

4. Consider a square wave with amplitude 1 V and period 1 ms as shown in Figure 7L5.5.

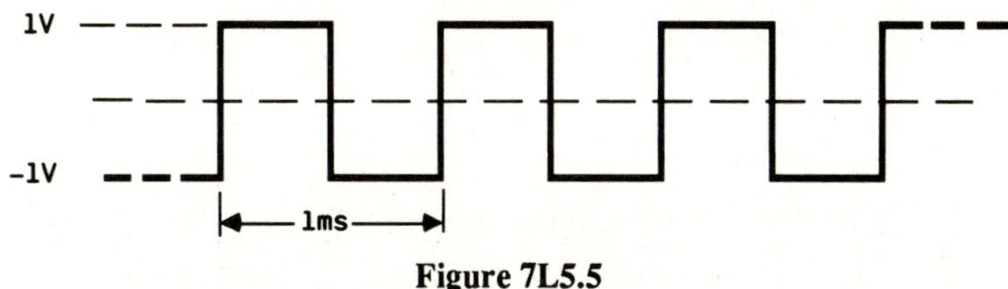

Figure 7L5.5

Do a Fourier series analysis of the square wave and determine the magnitude of the various sine wave components, starting with the fundamental to the third harmonic. Determine how many low-pass filters, as shown in Figure 7L5.6, need to be cascaded together so that the third harmonic contributes less than 10% distortion to the fundamental sine wave.

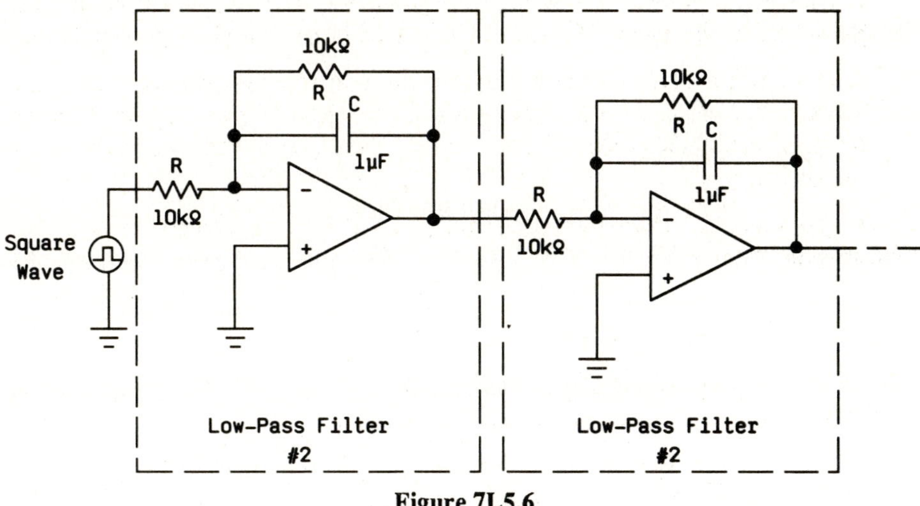

Figure 7L5.6

5. Suppose the square wave shown in Figure 7L5.5 is applied to the circuit shown in Figure 7L5.7.

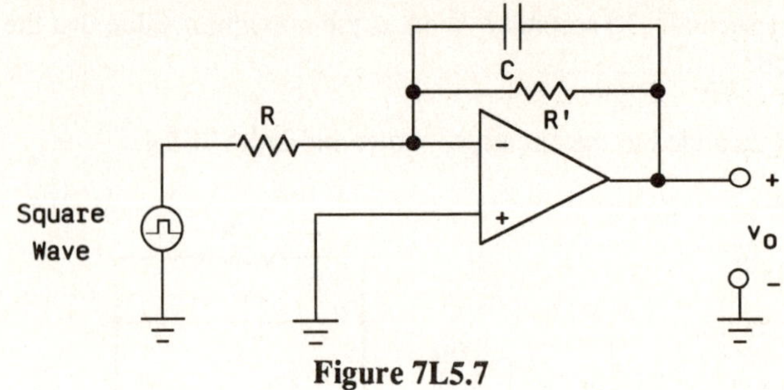

Figure 7L5.7

Let R=10 kΩ and R$'$=100 kΩ. Determine the value of C required to produce a triangular wave at the output.

Laboratory Procedure:

1. Design a function generator that can produce 1-kHz sine waves, square waves, and triangular waves with periods of 1 millisecond. The basic design can follow the block diagram of Figure 7L5.8.

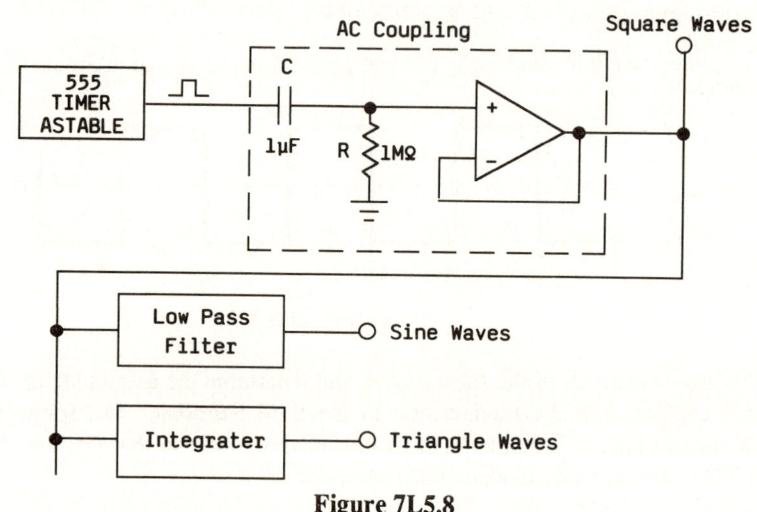

Figure 7L5.8

DISCUSSION:

CONCLUSIONS:

REFERENCES

[1]   J. W. Nilsson, *Electric Circuits*, 2d ed. (Reading, MA: Addison-Wesley, 1986), pp. 584-6.

[2]   D. E. Johnson, J. L. Hilburn and J. R. Johnson, *Basic Electric Circuit Analysis*, 3d ed. (Englewood Cliffs, NJ: Prentice-Hall, 1986), pp. 476-9.

[3]   J. D. Irwin, *Basic Engineering Circuit Analysis*, 2d ed. (New York: Macmillan Publishing Company, 1987), pp. 496-8.

[4]   W. H. Hayt, Jr., and J. E. Kemmerly, *Engineering Circuit Analysis*, 4th ed. (New York: McGraw-Hill, 1986), pp. 349-57.

[5]   Hayt and Kemmerly, pp. 403-15.

[6]   Nilsson, pp. 666-87.

[7] Irwin, pp. 499-511.

[8] Irwin, pp. 512-20.

[9] Hayt and Kemmerly, Chapter 17.

[10] Nilsson, Chapter 8.

[11] Irwin, pp. 637-58.

[12] Irwin, pp. 658-73.

[13] Nilsson, Chapter 19.

[14] Hayt and Kemmerly, Chapter 18.

[15] Johnson, Hilburn and Johnson, pp. 222-32.

[16] Irwin, pp. 263-7.

[17] Nilsson, pp. 557-60

[18] Hayt and Kemmerly, pp. 536-40.

[19] Nilsson, pp. 560-5, pp. 617-29

[20] Nilsson, Chapters 15 and 16.

[21] Irwin, Chapter 17.

[22] Hayt and Kemmerly, Chapter 19.

Chapter 8

AC STEADY STATE POWER

8.1 Introduction

In previous chapters, we have been concerned with the voltage and current response of an electrical network. In this chapter, we shall be concerned with the power that is supplied or absorbed by electrical or electronic elements.

In most instances, electrical and electronic systems have peak power or maximum instantaneous power ratings associated with them. For example, transistor power amplifiers produce a distorted output and speakers give a distorted sound when the peak power exceeds a certain rated value. The power relationships that we shall be considering are the product of periodic currents and voltages, usually sinusoidal.

A very important measure of power is the average power, or the average rate at which energy is absorbed by an element. Average power is independent of time. A quantity called the effective value is a mathematical measure of the effectiveness of waves, including sinusoidal waves, in delivering power.

In this chapter, we summarize the results of dealing with power in ac circuits. We consider such topics as instantaneous power, average power, superposition, effective or rms values, and the power factor.

8.2 Instantaneous Power[1],[2]

Using the passive sign convention, let the steady-state voltage and current sinusoidal signals associated with a circuit respectively be

$$\cos (\omega t + \theta_v - \theta_i)$$

$$v(t) = V_m \cos (\omega t + \theta_v - \theta_i) \tag{8.1}$$

and

$$i(t) = I_m \cos (\omega t) \tag{8.2}$$

where θ_v is the voltage phase angle and θ_i is the current phase angle. Then the instantaneous power can be written as

$$p(t) = i(t)v(t) = V_m I_m \cos (\omega t + \theta_v - \theta_i) \cos (\omega t) \tag{8.3}$$

Using the trigonometric identity

$$\cos \theta_1 \cos \theta_2 = \frac{1}{2} \left[\cos (\theta_1 - \theta_2) + \cos(\theta_1 + \theta_2) \right] \tag{8.4}$$

we can write the instantaneous power as

$$p = \frac{V_m I_m}{2} \cos (\theta_v - \theta_i) + \frac{V_m I_m}{2} \cos (2\omega t + \theta_v \, \theta_i) \tag{8.5}$$

Then, using the trigonometric identity

$$\cos(\theta_1 + \theta_2) = \cos\theta_1 \cos\theta_2 - \sin\theta_1 \sin\theta_2 \tag{8.6}$$

we can expand the second term on the right of equation (8.5) to obtain

$$\frac{V_m I_m}{2} \cos (\theta_v - \theta_i) \cos 2\omega t$$

$$-\frac{V_m I_m}{2} \sin (\theta_v - \theta_i) \sin 2\omega t \tag{8.7}$$

The following points can be made concerning Equation (8.7):

1. The average value of the power is given by the first term on the right, with the last two terms individually integrating to zero over one period.

2. The frequency of the instantaneous power is twice the frequency of the voltage or current.

3. If the circuit under consideration is resistive, $\theta_v = \theta_i$ and Equation (8.7) reduces to

$$p = \frac{V_m I_m}{2} + \frac{V_m I_m}{2} \cos 2\omega t \tag{8.8}$$

The power expressed by this equation is called the **instantaneous real power** and can never be negative. In the case of a resistive network, the electrical energy is transformed into thermal energy.

4. If the circuit under consideration is purely inductive, the current will lag the voltage by 90°; that is, $\theta_v - \theta_i$ will equal +90°, and equation (8.7) reduces to

$$p = -\frac{V_m I_m}{2} \sin 2 \, \omega t \tag{8.9}$$

Here the average power is zero, and in this circuit the instantaneous power oscillates between the circuit and the source. When p is positive, energy is stored in the magnetic fields of the inductor; when p is negative, energy is extracted from the fields.

5. If the circuit under consideration is purely capacitive, the current will lead the voltage by 90° and $\theta_v - \theta_i$ will equal -90°. Then Equation (8.7) becomes

$$p = \frac{V_m I_m}{2} \sin 2\omega t \tag{8.10}$$

Here again, the average power is zero and the power oscillates between the source and the electric field of the capacitor.

The term **reactive power** denotes the power associated with capacitive and inductive circuits. These elements are then termed **reactive elements** in steady-state sinusoidal analysis.

8.3 Average Power[3],[4]

The average power of a periodic wave can be derived by integrating the power function over a complete period and dividing the result by the period. Thus from Equation (8.5) the average power for a sinusoidal function is

$$p = \frac{1}{nT} \int_{t_0}^{t_0 + nT} p(t) dt$$

$$= \frac{1}{nT} \int_{t_0}^{t_0 + T} \frac{V_m I_m}{2} \left[\cos(\theta_v - \theta_i) + \cos(2\omega t + \theta_v + \theta_i) \, dt \right] \tag{8.11}$$

where t_0 is arbitrary, $T = 2\pi/\omega$ is the period of the voltage or current, and n is a positive integer.

Inspection of Equation (8.11) shows that it reduces to

$$p = \frac{1}{2} V_m I_m \cos(\theta_v - \theta_i) \tag{8.12}$$

For a purely resistive circuit,

$$p = \frac{1}{2} V_m I_m$$

and for a purely reactive circuit,

$$p = \frac{1}{2} V_m I_m \cos(90°)$$

$$= 0$$

These two results verify points 3, 4, and 5 of Section 8.2.

8.4 Superposition and Power[5],[6]

Network analysis techniques can be used to find the voltage and current that are necessary to determine the power of a circuit. But in general, we cannot apply superposition to instantaneous power.

If current is of the form

$$i(t) = I_1 \cos(\omega_1 t + \theta_1) + I_2 \cos(\omega_2 t + \theta_2) \tag{8.13}$$

with $\omega_1 \neq \omega_2$, the average power absorbed by a resistor R is given by

$$p = \frac{1}{T}\int_0^T \left[I_1 \cos(\omega_1 t + \theta_1) + I_2 \cos(\omega_2 t + \theta_2) \right]^2 R\, dt \tag{8.14}$$

Equation (8.14) can be rewritten as

$$p = \frac{1}{T}\int_0^T \left[I_1^2 \cos^2(\omega_1 t + \theta_1) + I_2^2 \cos^2(\omega_2 t + \theta_2) \right.$$
$$\left. + 2 I_1 I_2 \cos(\omega_1 t + \theta_1) \cos(\omega_2 t + \theta_2) \right] R\, dt$$

Using the identity of Equation (8.4) and integrating, we obtain

$$p = \frac{I_1^2}{2} R + \frac{I_2^2}{2} R + \frac{1}{T}\int_0^T \left\{ I_1 I_2 \cos\left[(\omega_1 + \omega_2)t + \theta_1 + \theta_2 \right] \right.$$
$$\left. + I_1 I_2 \cos\left[(\omega_1 - \omega_2)t + \theta_1 - \theta_2 \right] \right\} R\, dt \tag{8.15}$$

The last term in Equation (8.15) will be zero if $\omega_1 - \omega_2 \neq 0$. Therefore, superposition will hold if the terms are of different frequencies, and Equation (8.15) reduces to

$$P = \frac{I_1^2 R}{2} + \frac{I_2^2 R}{2} \tag{8.16}$$

Thus, the average power due to the sum of components is the sum of the average power due to each component acting alone, provided that the frequencies of the components are not equal or that only one of the components is dc.

8.5 *Effective or RMS Power*[7],[8],[9],[10]

We can define the effective value of a periodic waveform, current, or voltage as a constant that is equal to the dc current or voltage that would deliver the same average power to a resistance R. Therefore, if I_{rms} is the effective value of a current i, then

$$P = R I_{rms}^2 = \frac{1}{T}\int_0^T R i^2\, dt$$

and

$$I_{rms} = \sqrt{\frac{1}{T}\int_0^T i^2\, dt} \tag{8.17}$$

The effective voltage can be written as

$$V_{rms} = \sqrt{\frac{1}{T}\int_0^T v^2\, dt} \tag{8.18}$$

Thus, both I_{rms} and V_{rms}, we take the square root of the average, or mean, value of the square of the current or voltage. Another name for the effective value is the root mean square value, abbreviated rms.

The rms or effective value for a sinusoidal current (or voltage) of the form $i = I_m \cos(\omega t + \theta)$ can be found from Equation (8.17) and is

$$I_{rms} = \frac{I_m}{\sqrt{2}} \tag{8.19}$$

Equation (8.19) says that a sinusoidal current that has an amplitude I_m delivers the same average power to a resistor R as does a dc current equal to $I_m / \sqrt{2}$.

The average power delivered by a sinusoidal current or voltage, given by Equation (8.12), can be rewritten in terms of the effective current and voltage in the form

$$P = V_{rms} I_{rms} \cos(\theta_v - \theta_i) \tag{8.20}$$

The factor $\sqrt{2}$ occurs in Equation (8.19) because the periodic function is sinusoidal. The factor by which the maximum value must be divided to obtain the effective value depends on the mathematical form of the periodic function being considered.

For the case of a current or voltage made up of sinusoids of <u>different</u> frequencies, the effective value is

$$I_{rms} = \sqrt{I_{dc}^2 + I_{1rms}^2 + I_{2rms}^2 + \ldots + \pm I_{Nrms}^2} \tag{8.21}$$

A similar result obtains for the effective voltage.

8.6 Power Factor[11],[12],[13]

Consider again Equation (8.20). In this equation, the product of the effective voltage and the effective current is referred to as the **apparent power**. Clearly the average power can never be greater than the apparent power. The unit of apparent power is the volt-ampere, and the unit of average power is the watt.

The ratio of the average power to the apparent power is defined as the **power factor** and is written as

$$PF = \frac{P}{V_{rms} I_{rms}} = \cos(\theta_v - \theta_i) \tag{8.22}$$

In this equation,

$$\cos(\theta_v - \theta_i) = \cos\theta_z \tag{8.23}$$

where θ_z is the phase angle of the load impedance and is called the **power factor angle**.

The power factor angle can take two extreme positions, the one corresponding to the case of a purely resistive load where $\theta_z = 0$ and $PF = 1$, and the other corresponding to the case of a purely reactive load where $\theta_z = \pm 90°$ and $PF = 0$.

A load for which $-90° < \theta_z < 0$ describes an RC combination, whereas a load for which $0 < \theta_z < 90°$ is an equivalent RL combination. But, because $\cos\theta_z = \cos(-\theta_z)$ and the PF for an RC load is equal to that of an RL load, the PF is characterized as leading or lagging by the phase of the current with respect to that of the voltage. This definition implies that an RC load has a leading PF and an RL load has a lagging PF.

8.7 Summary

Power calculations are an important part of electrical circuit and electronic device characterization. In this chapter, we have introduced the basic concepts used to make these calculations. We have discussed the differences between real and reactive power and introduced the term **apparent power**. Also, we have examined the commonly used concept of the effective or rms value of a periodic waveform.

LAB 8L1

TITLE: Instantaneous and Average Power in Resistive Loads

OBJECTIVE: In this laboratory exercise, the student will calculate the average and instantaneous power dissipated by a resistive load. The use of superposition will be illustrated. The student will verify his or her results by using dc voltmeters and ammeters to measure average voltage and by using ac measurements to calculate average power. If the student has access to a power meter, then he or she can independently verify the voltage-current calculations of average power.

EQUIPMENT and MATERIALS:
Function generator
   triangular, square, and sine waves with a dc offset
1   1 kΩ, 1/8 W resistor
Oscilloscope
Breadboard
Power meter (optional)

PRELABORATORY:

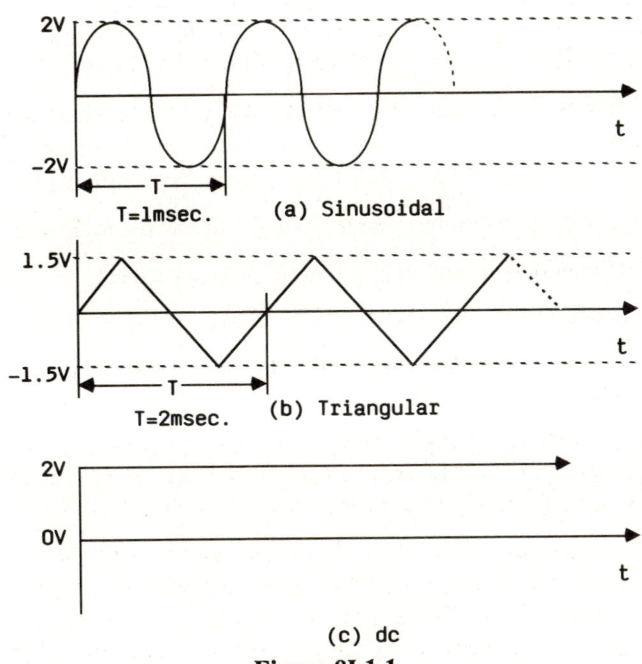

Figure 8L1.1

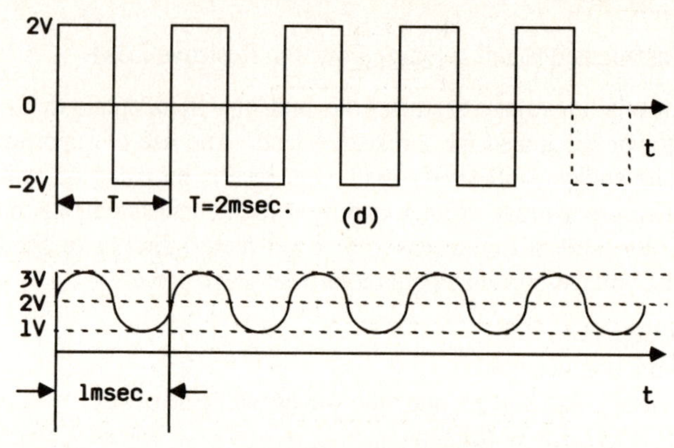

(e) Sinusoidal
Figure 8L1.1(continued)

1. For each waveform shown in Figure 8L1.1, determine the following:

    a.   A mathematical expression that describes each waveform over one period.

    b.   The instantaneous voltage at times, t = 0.3 msec. and t=1.4 msec.

    c.   The average voltage.

    d.   The RMS voltage.

2. If the voltages of part 1 are connected across a 1-kΩ resistor, what value of instantaneous current flows at times t = 0.3 msec., and t = 1.4 msec. Find the average and rms currents flowing in the 1-kΩ resistor.

3. Use the voltages and currents determined in parts 1 and 2 to find the following:

    a.   The instantaneous power at times, t = 0.3 msec. and t = 1.4 msec.

    b.   The average power.

4. Since the voltage in Figure 8L1.1(e) is the sum of two waveforms -- a dc value and a sine wave -- can superposition be used to determine power? If so, use it to find the average power of each component of the waveform, combine the two components, and verify the result of part 3.

Laboratory Procedure:

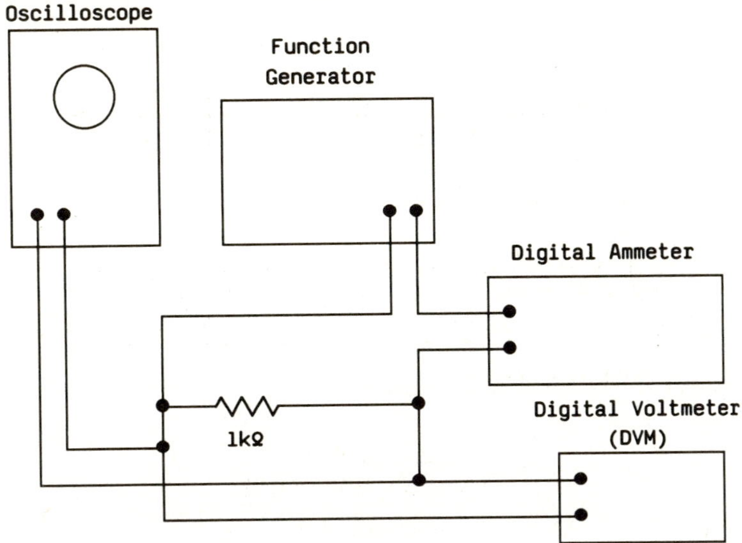

Figure 8L1.2

Set up the system shown in Figure 8L1.2, and use the function generator to produce the five waveforms given in the prelaboratory.

1. Set the voltmeter and ammeter to ac. Measure and record the rms current and voltage for each waveform. You should study the instruction manual to determine how the meter responds to nonsinusoidal waveforms.

2. Set the voltmeter and ammeter to dc. Measure and record the average voltage and current for each waveform.

3. Complete Table 8L1.1.

Table 8L1.1

	Prelaboratory Calculations						Laboratory Measurements			
	Power		Voltage		Current		Power (optional)	Voltage	Current	IxV
Waveform	Average	Instant	Average	rms	Average	rms	Average	Average	Average	
a										
b										
c										
d										
e										

DISCUSSION: Explain any discrepancy between the prelaboratory calculations and the measured laboratory results. Include the accuracy and tolerance of the equipment and the components used.

CONCLUSIONS:

LAB 8L2

TITLE: Application of Superposition to Power and Maximum Power Transfer for Resistive Networks[19],[20],[21]

OBJECTIVE: This laboratory exercise illustrates the restrictions on the use of superposition for power calculations in linear resistive networks. Also, the concepts of maximum power transfer and impedance matching are demonstrated through the use of an 8-Ω load and transformer.

EQUIPMENT and COMPONENTS:
>2 function generators
>4 1/8-W, 1-kΩ resistors
>Wattmeter
>Digital multimeter (DMM)
>Breadboard
>Audio transformer
>8-Ω 1/2-W resistor

PRELABORATORY:

1. Consider the circuit of Figure 8L2.1. For each waveform,

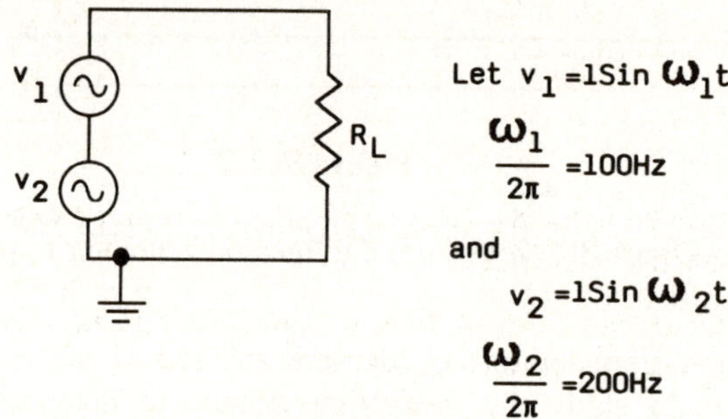

Figure 8L2.1

   a. Derive the expression for the average power dissipated in R_L.

   b. Verify that for the given values of v_1 and v_2, superposition applies in calculating the individual average powers due to v_1 and v_2 and that they equal the total average power when both are applied simultaneously.

   c. Show this is not the case for $\omega_1 = \omega_2$.

2. Consider the circuit of Figure 8L2.2.

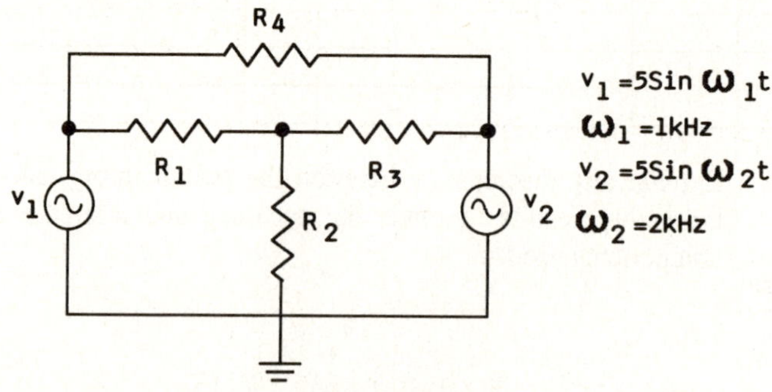

Figure 8L2.2

   a. Determine the voltage and current in each resistor due to each voltage source applied separately. Use the resistor values of Figure 8L2.3 ($f_1 = 1$ kHz and $f_2 = 2$ kHz).

   b. Calculate the average power dissipated in resistor R_4 due to the individual application of each voltage source.

   c. Apply both voltage sources simultaneously, and show that the total average power equals the sum of the individual average powers.

   d. Show that superposition does not apply if $\omega_1 = \omega_2$.

3. Consider the circuit of Figure 8L2.2 again, but with the slight modification shown in Figure 8L2.3, wherein that voltage source $v_2(t)$ has been replaced by a load resistor R_L.

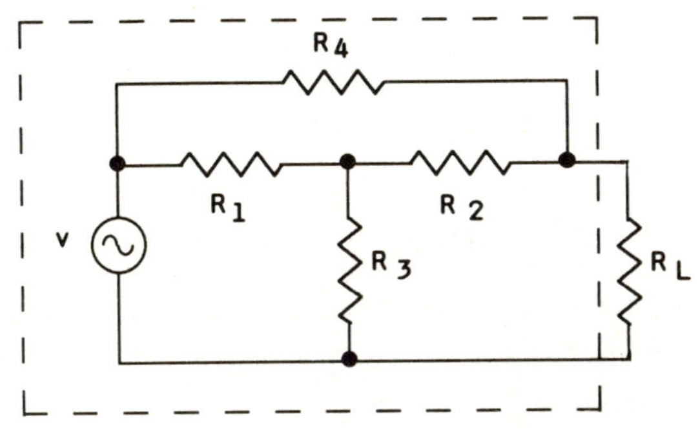

Figure 8L2.3

   a. Find the Thevenin equivalent circuit for the network contained within the dashed box. Show that $V_T = 0.8\ V_{rms}$ and $R_T = 600\ \Omega$.

   b. Let R_L vary from $100\ \Omega$ to $1\ k\Omega$ in $100\text{-}\Omega$ increments. Show that the maximum power dissipated in R_L occurs when $R_L = 600\ \Omega$.

4. Let the actual load impedance of the circuit shown in Figure 8L2.4 be $8\ \Omega$, the resistance of some loudspeakers. A transformer can be used to make the $8\text{-}\Omega$ load resistor "look like" a $600\text{-}\Omega$ load to the network, thereby impedance matching the load to the circuit. This allows the maximum power available to drive the speaker.

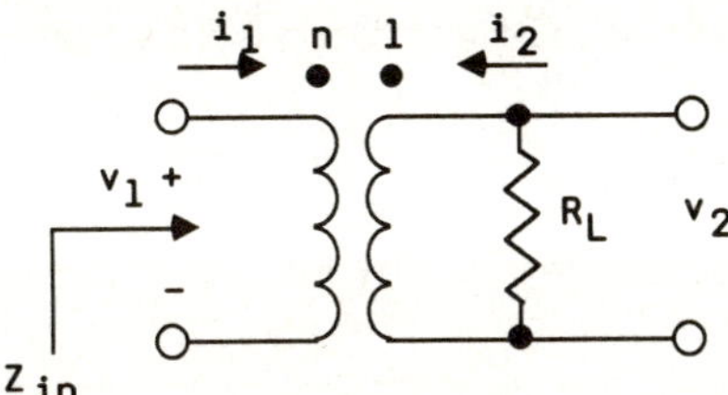

Figure 8L2.4

The equations describing the two-port circuit are

$$v_1 = nv_2 \qquad\qquad (8L2.1)$$
$$i_2 = -ni_1 \qquad\qquad (8L2.2)$$

or, in matrix form,

$$\begin{bmatrix} v_1 \\ i_2 \end{bmatrix} = \begin{bmatrix} 0 & -n \\ n & 0 \end{bmatrix} \begin{bmatrix} i_1 \\ v_2 \end{bmatrix}$$

   a. Show that $Z_{in} = R_{in} = \dfrac{v_1}{i_1} = n^2 R_L$.

   b. What must the turn ratio n be to make the $8\text{-}\Omega$ load receive the maximum power possible?

Laboratory Procedure:

1. Build the circuit of Figure 8L2.2 repeated below in Figure 8L2.5, where the voltage sources are function generators. Let all resistors have a resistance of $1\ k\Omega$.

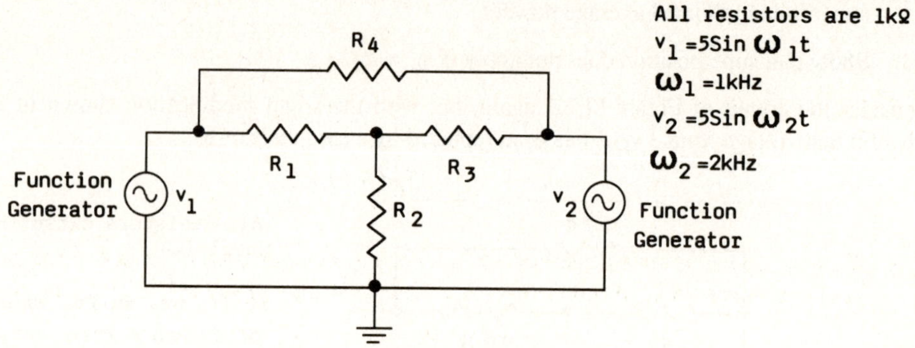

Figure 8L2.5

a. Using a multimeter, measure the rms voltage and current in each resistor. Calculate and record the average power dissipated in each resistor.

b. Verify the calculations of Part a using a wattmeter. (Optional)

c. Repeat the procedure of Parts a and b for each voltage source applied individually, with the source that is removed replaced by its equivalent impedance ($R_T = 0$ for a voltage source).

d. Do the data substantiate the application of superposition for power determinations?

e. Repeat Parts a, b, c, and d with $\omega_1 = \omega_2 = 1$ kHz.

2. Build the circuit of Figure 8L2.3 from the prelab but without resistor R_L, as shown in Figure 8L2.6.

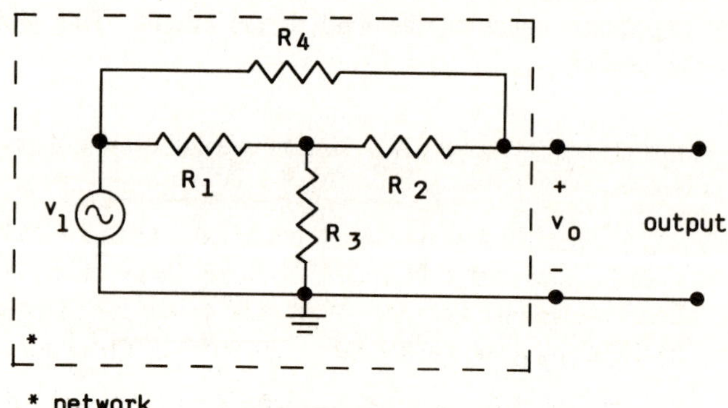

* network

Figure 8L2.6

a. Find the open-circuit voltage (v_o with no load) and the short-circuit current (the current flowing in the output when the output is short circuited). Then calculate the Thevenin equivalent impedance.

b. Place the load resistor R_L across the output, as in Figure 8L2.3. Measure and record the average power dissipated in R_L. Use a wattmeter, if available, to verify your calculation. Increase R_L in increments of 100 Ω. Repeat the procedure for R_L up to and including $R_L = 1$ kΩ. Plot the power dissipated in R_L as a function of R_L. What value of R_L causes maximum power dissipation? How does this value of R_L compare with R_T, and does that result agree with your prelab results?

c. Keep the circuit built here for Part 3 of the lab.

3. Impedance Matching

a. Connect an 8-Ω load across the output of the network of Part 2, as shown in Figure 8L2.7. Note the volume of the audible 1-kHz tone.

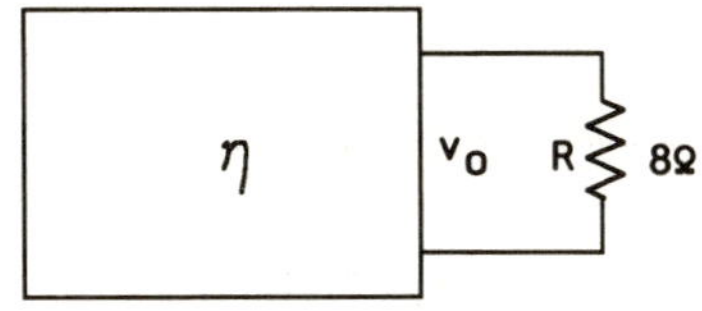

Figure 8L2.7

b.   Connect an audio transformer with a turn ratio as close as possible to the value calculated in the prelab between the load and the network. Figure 8L2.8 shows the resulting circuit.

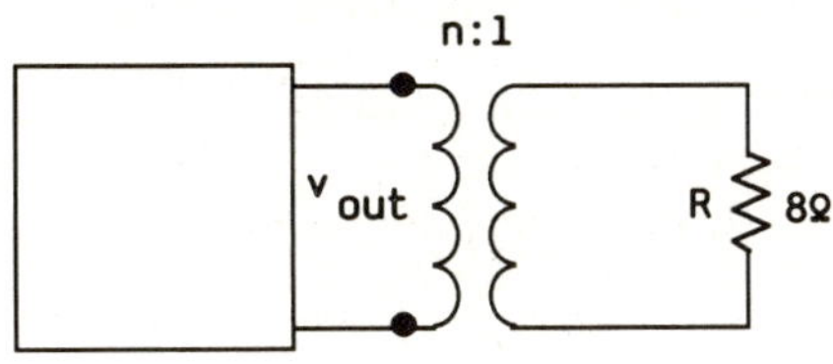

Figure 8L2.8

c.   What is the audible effect created by connecting the transformer into the circuit?

DISCUSSION:

CONCLUSIONS:

LAB 8L3

TITLE:	Introduction to Apparent, Average, and Complex Power in Complex Impedance Circuits[19],[20],[21]
OBJECTIVE:	Using digital multimeters and a wattmeter, the student will measure the apparent and average power of a simple RLC circuit in this laboratory exercise. By means of these measurements, the student will then determine the power factor and phase angle between voltage and current for each component.

The student will characterize a "black box" constructed of unknown R's, L's, and C's in terms of apparent and average power and will determine the equivalent circuit of the black box.

EQUIPMENT and COMPONENTS:
Digital multimeter (DMM)
Wattmeter
Function generator
Resistors: 2 - 1 KΩ, 1/8 W
Capacitors: 2 - 4,700 pF.
Inductor: 6.8 μH
Black box

PRELABORATORY:

1. Define the following power-related terms, and describe how you could measure them using an ac voltmeter, an ac ammeter, and a wattmeter:

   a. Instantaneous power

   b. Average power

   c. Apparent power

   d. Complex power

   e. Power factor

2. What does a wattmeter measure? Explain your answer in terms of average and apparent power.

3. For the circuit of Figure 8L3.1 ($\omega = 2\pi \times 1$ kHz),

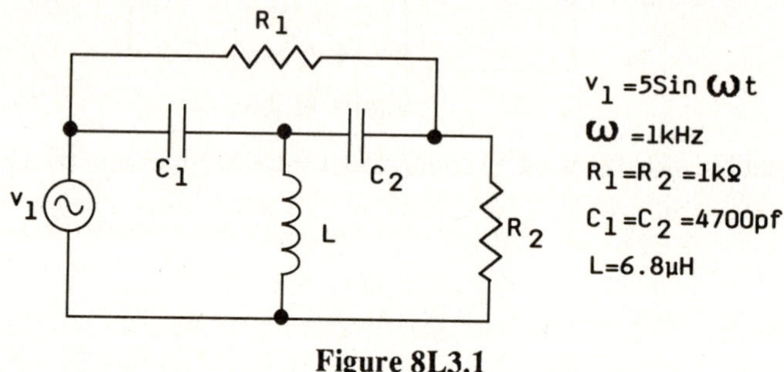

Figure 8L3.1

use phasor notation and solve the following problems:

   a. Determine the voltage and current in each component.

   b. Calculate the apparent power in each component and the apparent power delivered by v_i.

   c. Calculate the average power in each component and the average power delivered by v_i.

   d. Calculate the power factor for each component, given by

$$\text{p.f.} = \cos\theta = \frac{\text{average power}}{\text{apparent power}}$$

Laboratory Procedure:

1. Determination of average, apparent, and complex power for a complex impedance circuit.

   a. Build the circuit of Figure 8L3.1.

   b. Use an ac voltmeter and ammeter to measure and record the rms value of current and voltage in each component. (Include the voltage source.)

   c. Use a wattmeter to measure and record the average power dissipated in each device. (Include the voltage source.)

   d. Calculate the power factor for each device.

2. Determination of the real and imaginary components of a complex impedance using apparent and average power measurements.

   a. Build the circuit of Figure 8L3.2.

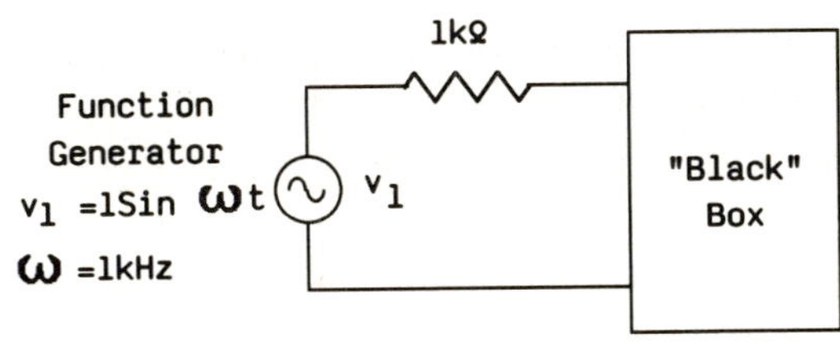

Figure 8L3.2

b. Measure and record the average power dissipated in the black box by using a wattmeter.

c. Measure and record the apparent power using an ac ammeter and ac voltmeter.

d. For the black box, calculate the phase angle between the voltage and the current. From this information, determine an equivalent circuit that could replace the black box.

e. Construct the substitution circuit, and test it by repeating the measurements made in Parts b and c.

DISCUSSION:

CONCLUSIONS:

LAB 8L4

TITLE: Superposition and Maximum Power Transfer for Complex Impedance Circuits[19], [20], [21]

OBJECTIVE: This laboratory exercise demonstrates the limited use of superposition to find the total average power in a complex impedance circuit. Complex impedance matching for maximal power transfer and an alternative method of measuring a circuit's phase angle without the use of a wattmeter are also presented.

EQUIPMENT and COMPONENTS:
Function generator
Wattmeter
Oscilloscope (dual trace)
Power supply
Resistors: 1 kHΩ, 3x10 kΩ, 100 Ω
Capacitors: 2 - 4,700 pF.
Inductors: 1,000 µH, 100 µH

PRELABORATORY:

1. Measuring average power and power factor without a wattmeter. Consider the circuit of Figure 8L4.1.

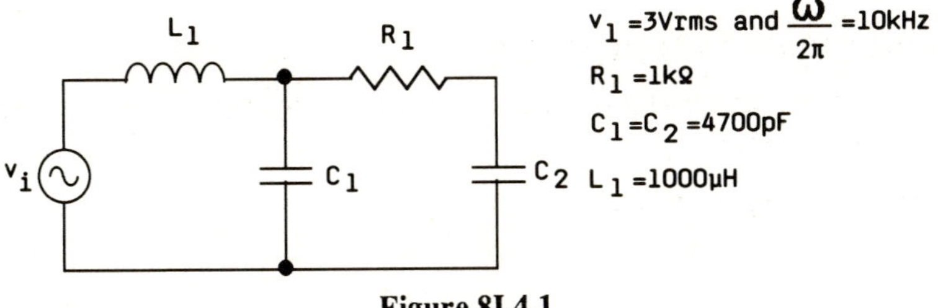

Figure 8L4.1

a. Using phasor notation, find the voltage and current in each component. Also, find the apparent and average power for each device.

b. Place a 1-Ω resistor in series with each component to sense current. R_1 serves this purpose for capacitor C_2.

c. Recalculate the voltage drops across each component, and determine the apparent and average power dissipated in each component, including the 1-Ω sensing resistors.

d. Discuss how these sensing resistors, along with a dual-trace oscilloscope, could be used to measure the phase angle between the voltage and the current in a given component. Remember, an ac voltmeter and ammeter can show apparent power directly.

e. Given a black box containing R's, L's, and C's, how could you use the method of Part d to characterize the complex impedance of the black box and determine the power factor?

f. What percent error was introduced in comparing the results of Parts a and c?

2. Superposition of power for complex impedance circuits. Consider the circuit of Figure 8L4.2.

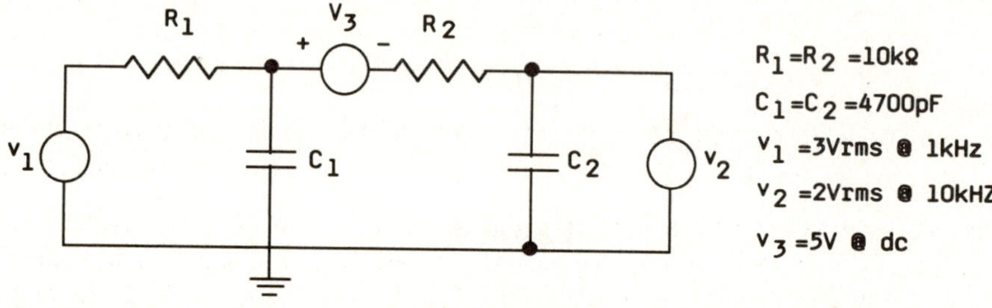

Figure 8L4.2

a. Calculate the current and voltage in each circuit element resulting from V_1, V_2, and V_3 applied separately. Remember to replace a voltage source that has been removed with its equivalent impedance. Calculate the apparent and average power for each component.

b. Repeat the exercise with the three voltage sources applied simultaneously. Does superposition apply to power for this circuit? Why?

3. Impedance matching. Consider the circuit of Figure 8L4.3, where N_1 represents a model for a filtered voltage source, and N_2 is a typical industrial load, e.g., a motor.

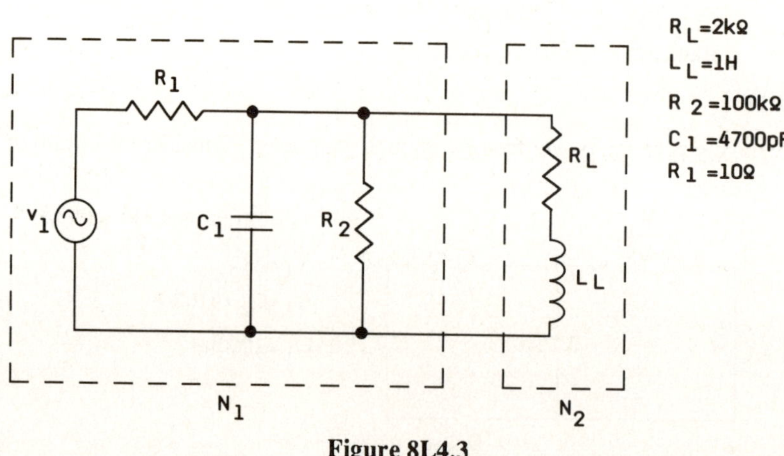

Figure 8L4.3

a. Determine the complex impedance of the networks N_1 and N_2. Assuming that the load could not be modified, how could capacitor C_1 be changed to improve the maximum power transfer? Would it make sense to have the source resistance equal to the load resistance? How much power would then be wasted in R_1 ll R_2?

Laboratory Procedure:

1. Compare resistive current sensing with wattmeter measurements.

    a. Build the circuit of Figure 8L4.1, repeated in Figure 8L4.4, where the voltage source is a function generator.

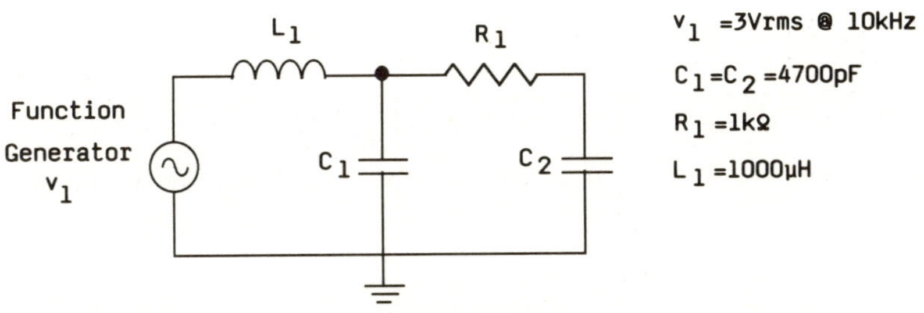

Figure 8L4.4

    b. Use a wattmeter and an ac voltmeter and ac ammeter to measure the apparent and average power in each component. Is the power factor what you expected for the inductor and capacitor? Why?

    c. Connect two 1-Ω current-sensing resistors in series with L_1 and C_1, respectively. Remeasure and record the current-voltage phase angle using a dual-trace oscilloscope. One channel is connected to the current-sensing resistor. The other channel is connected to the device being tested in series with the current-sensing resistor. Compare the phase angle between the two waveforms with the angle indicated by the ratio of average to apparent power. (R_1 can be used as the sensing resistor for C_2.) Note that an oscilloscope used to measure a branch voltage (e.g., V_{R_1}) must be in differential mode.

    d. Repeat Part 1c using 100-Ω sensing resistors.

2. Superposition of power for complex impedances.

    a. Build the circuit of Figure 8L4.2 and repeated in Figure 8L4.5, where the voltage sources are function generators.

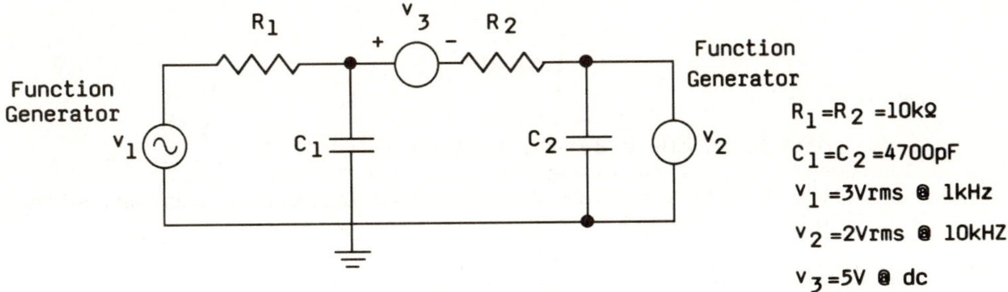

Figure 8L4.5

b. Measure and record the average power in R_1 and R_2 with all three voltage sources in the circuit.

c. Measure and record the average power in R_1 and R_2 with each voltage source applied separately. Be sure to replace the sources that are removed with their equivalent resistances.

d. Does superposition of average powers apply?

e. Repeat Parts a, b, c, and d with the frequency of v_1 and v_2 set equal to 1 kHz.

3. Complex impedance matching.

a. Build the circuit of Figure 8L4.6.

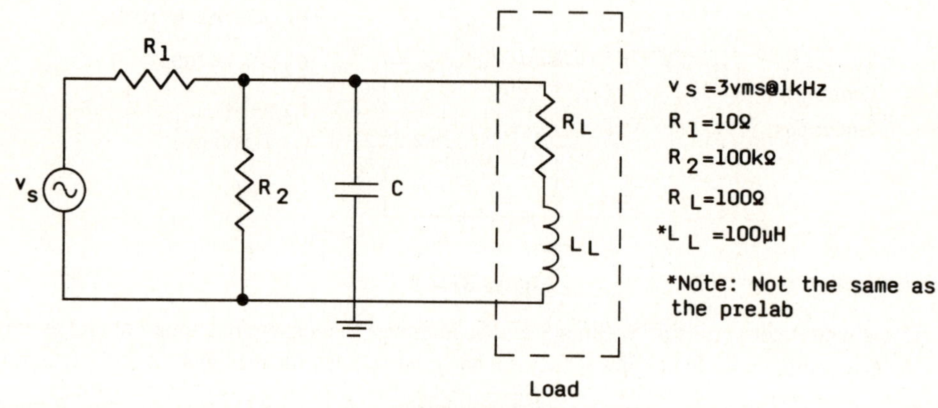

Figure 8L4.6

b. Remove capacitor c from the circuit, and measure the apparent and average power delivered to the load and the average power dissipated in R_1 and R_2.

c. Determine a value of C such that $X_L = -X_{eq}$. Use this value (or one as close as possible), and repeat the procedure from part B with C in the circuit.

d. Change $R_1 \parallel R_2$ so that it matches R_L, and repeat Part c. Is the maximum possible average power transferred to the load. How much power is wasted in R_1 and R_2?

DISCUSSION:

CONCLUSIONS:

LAB 8L5

TITLE Average and Effective Power, in Decibels[19], [20], [21]

OBJECTIVE: This laboratory exercise discusses three fundamental concepts: average power, effective power, and the decibel.

EQUIPMENT and COMPONENTS:
Function generator
  triangular, square, and sine waves with a dc offset
Oscilloscope
  AC voltmeter
1 - 5 kΩ resistor, 1/8 W
1 - 1 kΩ resistor, 1/8 W
3 - 10 kΩ resistors, 1/8 W
1 - µF capacitor

PRELABORATORY:

1. Find the average, rms, and peak-to-peak values of the periodic voltage waveforms shown in Figure 8L5.1.

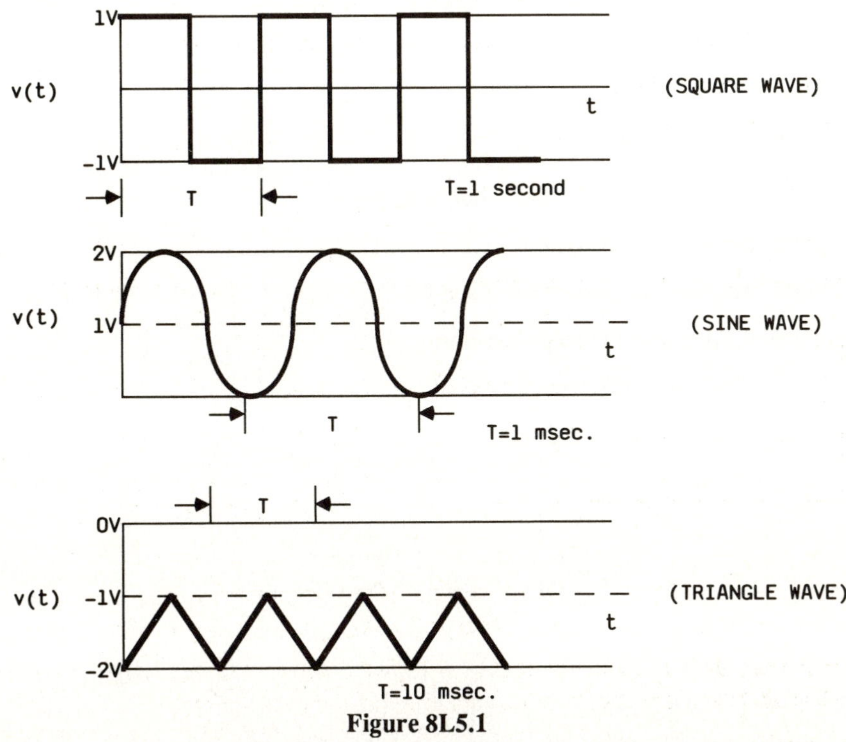

Figure 8L5.1

2. If the waveforms in the figure represent voltages across a 10-Ω resistor, then what would be the rms power dissipated in the resistor? Would this be the same as the average power? Why?

3. From question 1, what is the rms voltage for the square wave if the frequency is doubled? Does the same result occur for the triangular wave? Why?

4. For the circuit of Figure 8L5.2, can the output voltage (v) be expressed in units of volts? Why? Can the output voltage be expressed in dBs? Why?

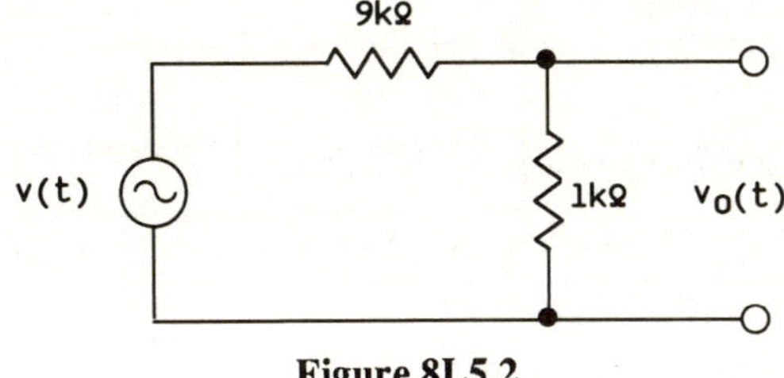

Figure 8L5.2

5. In the circuit of Figure 8L5.3,

$$
\begin{aligned}
v(t) &= 10\sin \omega t \\
\omega &= 1000 \text{ rad/sec} \\
R_1 &= 7 \text{ k}\Omega \\
R_2 &= 3 \text{ k}\Omega
\end{aligned}
$$

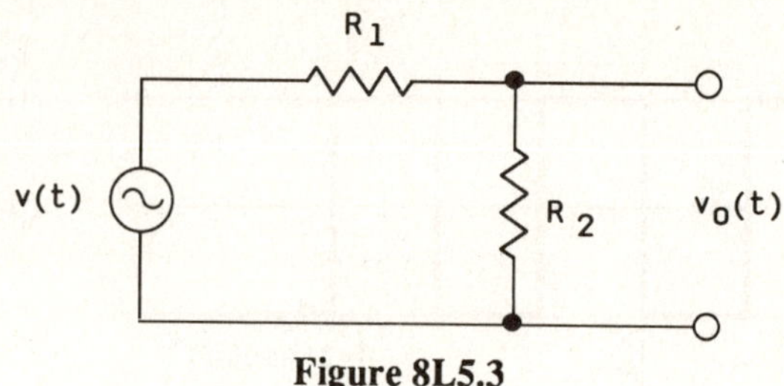

Figure 8L5.3

   a.   Find the rms power in the output.

   b.   Find the rms power dissipated in R_1 and R_2.

   c.   Find the rms power in the input waveform.

   d.   Show that Tellegan's theorem is satisfied.

   e.   Express the output power in dB.

6.   For the circuit of Figure 8L5.4, in which $R_1 = R_2 = 200\ \Omega$,

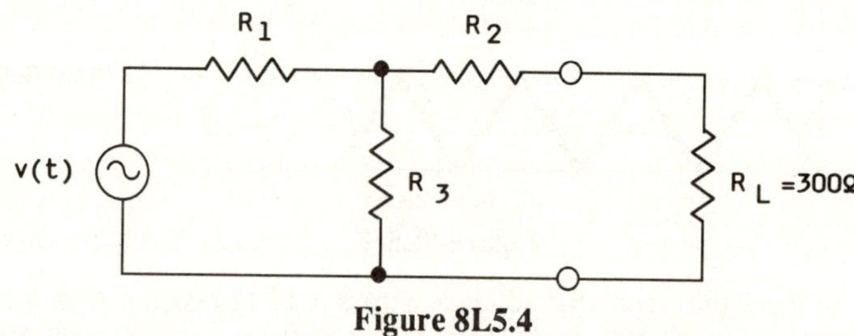

Figure 8L5.4

Find the value of R_3 that results in maximum power dissipation in the output load resistor R_L.

Laboratory Procedure:

1.   Set up the circuit shown in Figure 8L5.5, and use the circuit to verify your rms calculations in Question 1 of the prelab.

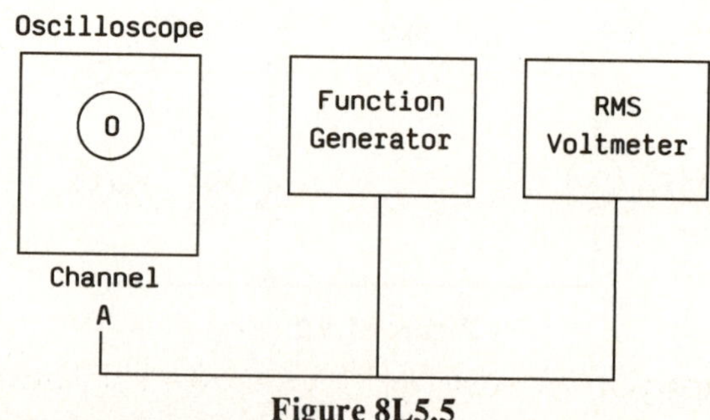

Figure 8L5.5

2.   Construct the circuit shown in Figure 8L5.6.

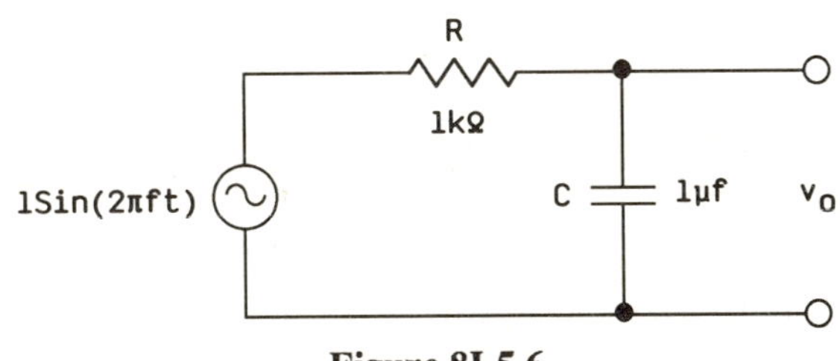

Figure 8L5.6

   a. Measure the output peak-to-peak voltage, with the frequency varying between 10 Hz and 1200 kHz. Record at least 10 data points.

   b. Calculate the decibel value of the data collected in Part a. Plot voltage in decibels versus frequency in Hz on semilog paper.

   c. What is the slope of the curve?

3. For the circuit of Figure 8L5.6, use your multimeter to measure the decibel values calculated in Part 2.

   a. Comparing your decibel results in Part 3 to those of 2, what is the constant decibel offset?

   b. What is the reference voltage used by your meter in measuring decibels?

4. Construct the circuits of Figure 8L5.7.

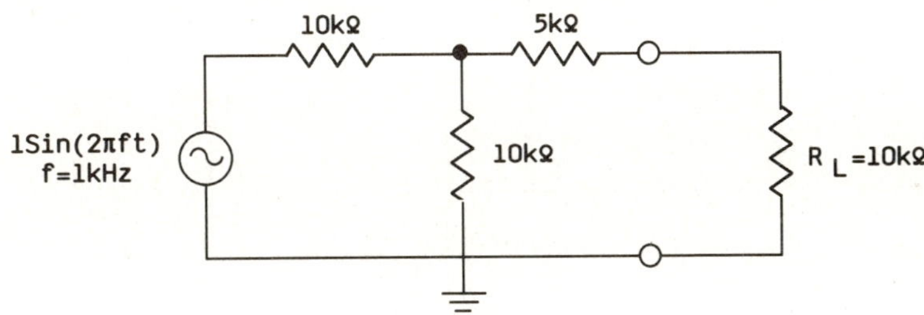

Figure 8L5.7

   a. Measure the rms voltages and currents throughout the circuit.

   b. Replace the network with its Thevenin equivalent and repeat the measurements of Part a.

   c. Calculate the average power delivered by the source in Parts a and b. Use Tellegan's theorem to show that the circuits are equivalent.

DISCUSSION:

CONCLUSIONS:

REFERENCES

[1] J. D. Irwin, *Basic Engineering Circuit Analysis,* 2d ed. (New York: Macmillan, 1987), pp. 410-2.

[2] J. W. Nilsson, *Electric Circuits,* 2d ed. (Reading, MA: Addison-Wesley, 1986), pp. 373-8.

[3] Irwin, 1987, pp. 412-8.

[4] W. H. Hayt, Jr., and J. E. Kemmerly, *Engineering Circuit Analysis,* 4th ed. (New York: McGraw-Hill, 1986), pp. 286-93.

[5] Irwin, 1987, pp. 415-7.

[6] D. E. Johnson, J. L. Hilburn, and J. R. Johnson, *Basic Electric Circuit Analysis,* 3d ed. (Englewood Cliffs, NJ: Prentice-Hall, 1986), pp. 371-5.

[7] Irwin, 1987, pp. 423-7.

[8] Nilsson, 1986, pp. 379-82.

[9] Hayt and Kemmerly, pp. 294-7.

[10] Johnson, Hilburn, and Johnson, 1986, pp. 376-8.

[11] Irwin, 1987, pp. 427-32.

[12] Hayt and Kemmerly, 1986, pp. 297-301.

[13] Johnson, Hilburn, and Johnson, 1986, pp. 378-82.

Chapter 9

CHARACTERIZATION OF NONLINEAR NETWORKS

Networks Using Diodes

9.1 Introduction

The diode is the simplest and most fundamental nonlinear electronic element. In addition to being an important electronic device in its own right, the diode forms the basis for nearly all other solid-state devices, including the integrated circuit.

We begin this chapter by discussing the current-voltage characteristics of the diode through examination of the diode equation. Next, we discuss diode parameters, including reverse saturation current, breakdown voltage, dynamic resistance, junction capacitance, and temperature effects. Simple circuit models to simulate the diode's characteristics at low and high frequencies are also presented. Finally, we show applications of diodes to typical circuits, such as rectifier circuits, voltage regulators, clippers and clampers, power supplies, and logic circuits. Included in some of these diode circuits is the ideal operational amplifier. We analyze all of the circuits presented through the use of the diode models we develop.

9.2 The Diode Equation and Terminal Characteristics of Real Junction Diodes

The current-voltage characteristic curve of a silicon junction diode is illustrated in Figure 9.1.

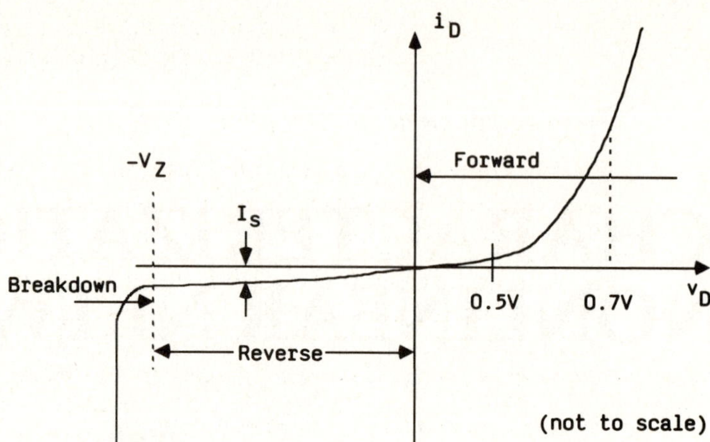

Figure 9.1: The diode current-voltage relationship

The curve can be divided into three regions:

1. The region in which $v_D > 0$, called the forward bias region,

2. The region in which $v_D < 0$, called the reverse-bias region, and

3. The region in which $v_D < -V_z$, called the breakdown region.

Forward Region
____

It can be demonstrated through the use of solid-state physics[1],[2],[3] that the forward current i_D can be mathematically related to the temperature T and the applied bias v_D in the following manner:

$$i_D = I_s \exp\left(\frac{v_D}{nV_T} - 1\right) \tag{9.1}$$

where

$$
\begin{array}{lll}
I_S & = & \text{reverse saturation current} \\
V_T & = & \text{thermal voltage} \\
n & = & \text{constant between 1 and 2 depending} \\
& & \text{on the material, the physical} \\
& & \text{structure, and the operating} \\
& & \text{region of the diode.}
\end{array}
$$

I_S is directly proportional to the cross-sectional area of the diode. So if we double the junction area of the diode, the result is a doubling of the value of I_S and thus a doubling of the value of i_D for a given forward voltage v_D. The saturation current is also a strong function of temperature. Experimentally, for a silicon diode, the reverse current doubles for every 10°C rise in temperature.[4]

The thermal voltage, V_T is

$$V_T = kT/q$$

where k =   Boltzmann's constant ($1.38 \times 10^{-23}\ \dfrac{\text{joules}}{\text{kelvin}}$)

         T =   absolute temperature in kelvins

         q =   charge on the electron (1.6×10^{-19} coulombs).

At room temperature ($20°C$), $V_T \approx 25$ mV.

Diodes made using the standard integrated-circuit process usually display n=1. Discrete silicon diodes generally display values of n=2 at low current levels, but n=1 at high current levels. Discrete germanium diodes display values of n=1.[5]

At a forward current of about 1 mA, the diode voltage temperature coefficient for silicon is approximately -2mV/C. This means that for an increase in temperature of 1°C, the forward voltage v_D decreases by 2 mV. Also, for a fixed forward bias of 700 mV at room temperature, the silicon diode forward current i_D doubles every 12°C.[6]

For $i \gg I_S$, Equation (9.1) can be written as

$$i_D \simeq I_S e^{v_D/nV_T} \tag{9.2}$$

If we now rewrite equation (9.2) so that voltage is expressed as a function of current, we obtain

$$v_D = nV_T \ln \frac{i_D}{I_S} \tag{9.3}$$

It can be shown from Equation (9.3) that for a tenfold change in current, the diode voltage drop changes by $2.3nV_T$, which is a 60-mV change when n=1 and a 120-mV change when n=2.[7]

The relationship shown in Equation (9.3) suggests that i_D and v_D should be plotted on semilog paper using the linear abscissa for voltage and the logarithmic ordinate for current. The result is a straight line with a slope of $2.3nV_T$ per ten units of current. The simple experiment in Laboratory 9.2 shows how to determine the value of n.

For silicon diodes in the forward region, the current i_D is negligibly small for values of v_D smaller than about 0.5 V. This value is called the **turn-on, cut-in,** or **threshold voltage.** For fully conducting silicon diodes, the cut-in voltage lies in the range of 0.6 V to 0.8 V. For germanium diodes, the cut-in voltage is approximately 0.3 V, and for gallium arsenide it is about 1.2 V.

The resistance of a diode at a particular operating point (I_D, V_D) is called the dc or **static resistance** of the diode and is given by

$$R_{dc} = \frac{V_D}{I_D} \tag{9.4}$$

Figure 9.2 shows that the dc resistance of a diode is independent of the shape of the characteristic; this is not true for an ac source. In the case of an ac source, the varying input will move the operating point up and down the characteristic curve and define a specific change in current and voltage. Figure 9.2 shows a part of the forward characteristic of the diode. The instantaneous signal i_D varies about the quiescent or dc point.

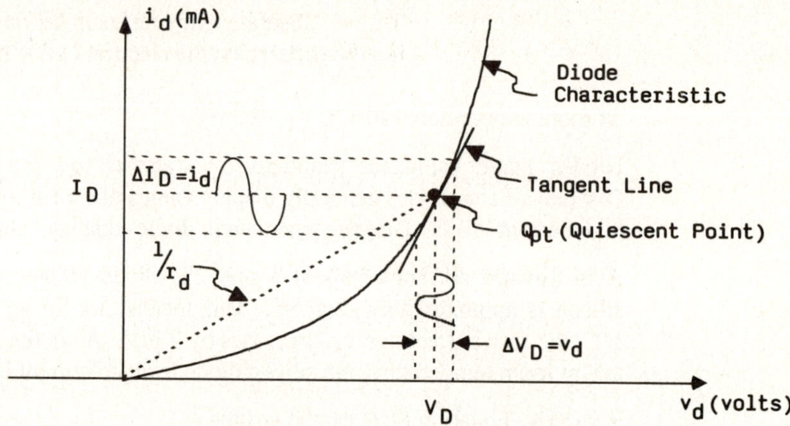

Figure 9.2: Dynamic diode resistance

A tangent to the curve through the dc operating point defines the dynamic resistance for this region of the diode characteristic. The dynamic resistance is then given by

$$r_d = \frac{di_D}{dv_D} \tag{9.5}$$

The diode small-signal resistance, or incremental resistance, is found from Equations (9.1) and (9.5). Substituting Equation (9.1) into (9.5) gives

$$\frac{di_D}{dv_D} = \frac{I_S e^{v_D/nV_T}}{nV_T}$$

and

$$\frac{1}{r_d} = \frac{i_D + I_s}{nV_T}$$

But

$i_D \gg I_S,$ and $I_D \gg i_d,$ where i_d is the ac component of $i_D.$

Then

$$r_d = \frac{nV_T}{i_D} \approx \frac{nV_T}{I_D} \tag{9.6}$$

Although r_d changes when the diode current changes, r_d is fixed for a specific operating condition, and the approximation in Equation (9.6) will be used to denote the diode forward resistance. For n=1 and for room temperature, this resistance takes on the value

$$r_d = \frac{25 \text{ mV}}{I_D \text{ (mA)}} \ \Omega \tag{9.7}$$

Reverse Region

When the diode voltage v_D becomes negative, we enter the reverse-bias region of operation. From Equation (9.1), when v_D is negative and satisfies the condition

$$\frac{-v_D}{nV_T} \gg 1 \tag{9.8}$$

the exponential term becomes small compared to 1, and Equation (9.1) becomes

$$I = -I_S \tag{9.9}$$

Thus, the current in the reverse direction is constant and equal to $-I_S$.

Real diodes show reverse currents that are quite small, but not nearly as small as I_S. For the small-signal diode, I_S is on the order of 10^{-14} to 10^{-15} A, but measurements show that the value of the saturation current is about 1nA-1pA. The difference in these two values is due to surface leakage effects, which are proportional to the surface area[8] of the diode.

The Breakdown Region

The breakdown region is entered when the magnitude of $-v_D$ exceeds a critical value called the **breakdown voltage**. This voltage is specified as v_Z in Figure 9.1.

In the breakdown region, the reverse current $-i_D$ increases rapidly while the associated change in voltage is small. This condition of breakdown is not normally destructive of the diode if the power dissipated by the device is limited to a safe value within the power rating of the device. The reverse current of the diode is limited by the circuit in which the diode is included. The diode current-voltage characteristic in the reverse region is shown in Figure 9.3.

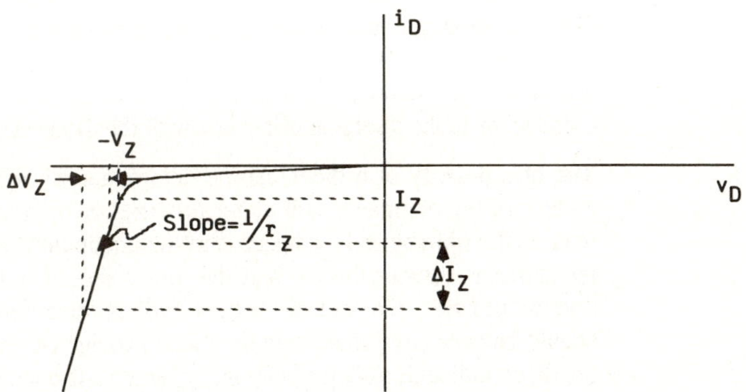

Figure 9.3: Diode i_D–v_D characteristic in the breakdown region

The fact that the characteristic in breakdown is almost vertical suggests that the diode can be used in voltage regulation.

A voltage regulator is a circuit that provides a constant dc voltage between its output terminals. The voltage should remain constant despite changes to the load current or changes in the input dc power supply. Zener diodes, breakdown diodes, and reference diodes are manufactured to operate in the breakdown region and to be constant and almost independent of the current through the diode. We shall discuss voltage regulator circuits in Section 9.4.4.

From Figure 9.3, the value of r_Z is equal to the inverse of the slope of the i_D–v_D curve at the operating point. When the diode is an effective regulator, r_Z is small (typically, a few tens of ohms), but at the knee of the characteristic, where the diode is not an effective regulator, r_Z reaches a high value.

The temperature coefficient of reference diodes is given as a percentage change in reference voltage per Celsius degree change in diode temperature. The coefficient is normally in the range $\pm 0.1\%/^\circ$C. If the breakdown mechanism involves avalanche breakdown, the temperature coefficient is positive; if it involves Zener breakdown, the temperature coefficient is negative.

There are two different mechanisms for diode breakdown: Zener breakdown, responsible for breakdown that occurs at low voltages (2-5 V), and avalanche breakdown, responsible for breakdown that occurs at voltages above 7 V.[9] Avalanche breakdown results in diodes with sharper knees; thus, its use leads to the manufacture of better voltage regulators.

<u>Diode Notation</u>

The notation most frequently used for semiconductor diodes is shown in Figure 9.4.

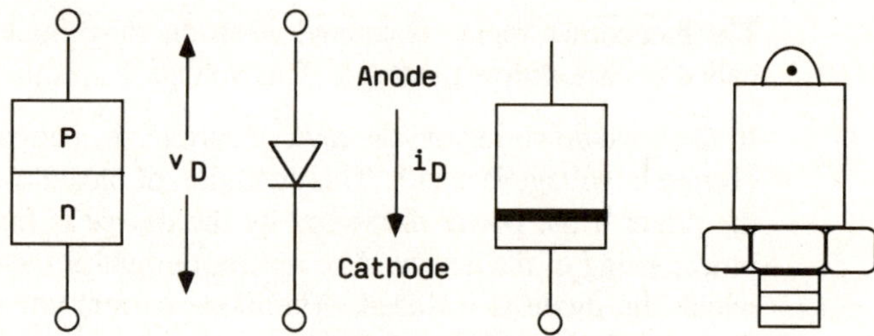

Figure 9.4: Diode symbols

For most diodes, markings such as a dot or band appear at the cathode end. The anode is the higher or more positive terminal, the cathode the lower or more negative terminal. Conventional (not electron) forward current flows from the anode to the cathode, or in the direction of the arrow of the diode symbol.

The bias polarity of a diode can be determined by using an ohmmeter. The internal battery of the ohmmeter will either forward- or reverse-bias the diode. If the positive lead of the ohmmeter is connected to the anode and the negative lead to the cathode, the diode is forward biased and the meter should indicate low resistance. With the reverse polarity, the internal battery will reverse-bias the diode and the resistance should be very large since only the diode leakage current flows. A small resistance for the diode indicates a short circuit and a large resistance indicates an open circuit.

Another common technique for determining the polarity of a diode requires the use of a power supply capable of displaying voltage and current simultaneously. Connect the diode across the supply, and slowly increase the voltage until current begins to flow. For a silicon diode this is about 0.6 V. If you get to 1 V and no current flow, the diode is reverse biased.

9.3 The Diode as a Circuit Element

An equivalent circuit is a group of elements chosen to represent the actual terminal characteristics of a device or system. Once the model is selected, the device symbol is removed from the circuit and the model is substituted without grossly affecting circuit behavior. There is no one correct equivalent for a given system, nor is there any model that is absolutely accurate. All models are just approximations, and which model is used is governed by the accuracy required and the time available to devote to the solution of the system.

One approach is to analyze the circuit using the simplest model and repeat the analysis with the next more complex model. If the change in the result is less than a certain predetermined percentage, no further iteration is necessary.

Linear DC Models

Figure 9.5 shows five piecewise linear diode models.[10] Linear equations are obtained from linear circuit models and are much easier to solve than non-linear equations. The choice of which model to use is a matter of engineering judgment.

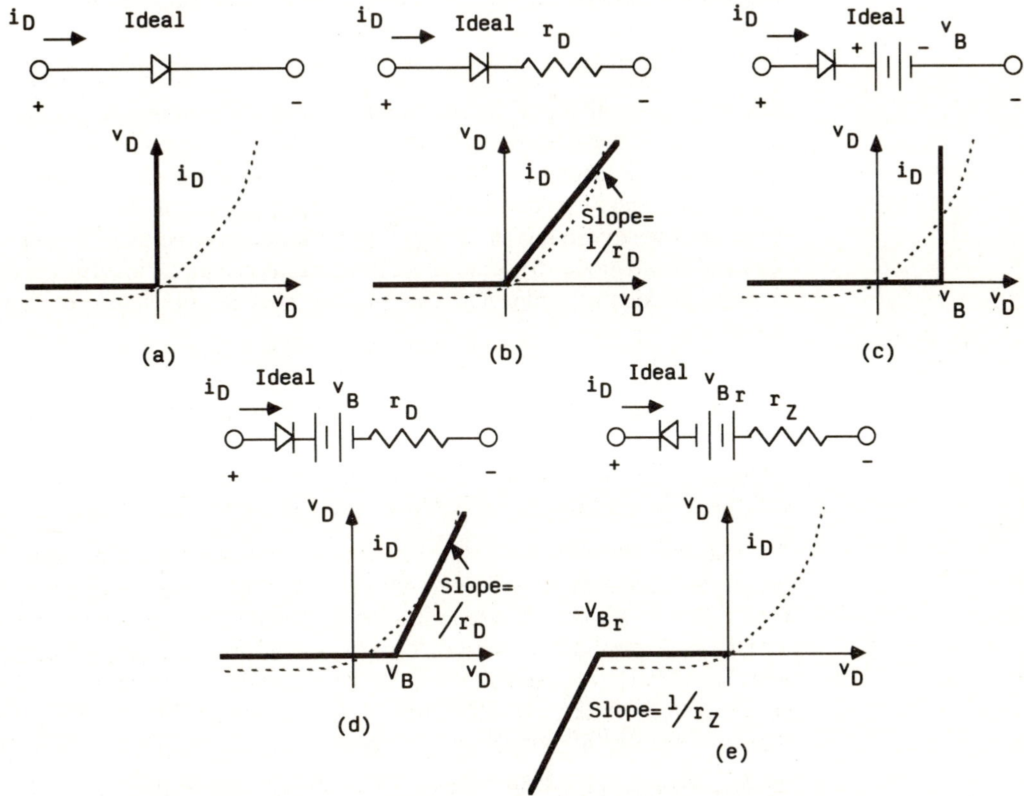

Figure 9.5: Five piecewise linear DC diode models (Reprinted by permission from *Electronic Circuit Analysis and Design*, Second Edition, by William H. Hayt, Jr. and Gerold W. Neudeck. Copyright © 1983 by Houghton Mifflin Company.)

Figure 9.5(a) shows the simplest, or ideal, diode model. Here, the diode is assumed to act as a simple switch so that when the switch is opened no current flows and when the switch is closed a short circuit is obtained. If the voltages involved in the circuit are much larger than the diode drop, the ideal model becomes a good model for the real diode. The second dc linear model, in Figure 9.5(b), contains a resistor r_D that allows the forward voltage to take on nonzero values of voltage. The value of r_D is a function of the circuit conditions. The model of Figure 9.5(c) assumes that the conducting diode has a constant voltage drop across it that is independent of the current. For silicon diodes, this voltage is 0.7 V. For germanium diodes, the voltage is approximately 0.3 V, and for gallium arsenide, 1.2 V.

Figure 9.5(d) shows a model that contains an ideal diode plus two linear elements. The circuit leads to the linear equation

$$v_D = V_B + i_D r_D \quad \text{for } v_D > V_B \tag{9.10}$$

The circuit of Figure 9.5(e) models the reverse-bias region of the diode, shown in Figure 9.3. The battery included in the circuit has a voltage that is equal to V_{BR} for the diode, and $1/r_Z$ is the slope of the curve during breakdown. Both the diode and the battery have a polarity in the breakdown model that is opposite that in the forward models.

Small Signal Models

In the case of the small-signal model, a diode is biased to operate at a point on the forward i-v characteristic and a small ac signal is superimposed on the dc bias. Here, the diode is modeled by a resistance equal to the inverse of the slope of the tangent to the i-v characteristic at the bias point. This resistance is given by Equation (9.6).

The ac diode small-signal models are more complex than the dc models because with ac the diode operation depends upon frequency. A simple ac model for a reverse-biased diode is shown in Figure 9.6(a). The capacitor C_j represents the junction capacitance, and r_Z represents the reverse-bias resistance of the diode before the diode enters the breakdown region. Figure 9.6(b) shows the ac equivalent circuit model for a forward-biased diode. This model includes two capacitors, the diffusion capacitors C_D and the junction capacitor C_j. The quantity r_d is the previously explained dynamic resistance.

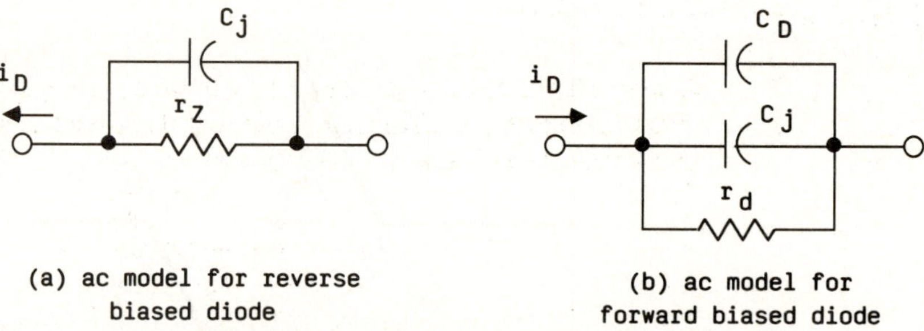

Figure 9.6: Small-signal AC diode model

The depletion or junction capacitance depends on the dc operating point of the diodes and the diode voltage and is given by

$$C_j = \frac{C_{jo}}{(1 - v_D/V_B)^N} \quad (v_D \leq 0) \tag{9.11}$$

The capacitance C_{jo} is the depletion capacitance when the external voltage across the diode is zero. The voltage V_B is the built-in, threshold, or turn-on voltage (.7 V for Si), and N is a constant that depends on the junction itself. Typical values of N range from 1/2 for an abrupt, or step, junction to 1/3 for a linear junction. The parameter v_D is the reverse voltage across the diode.

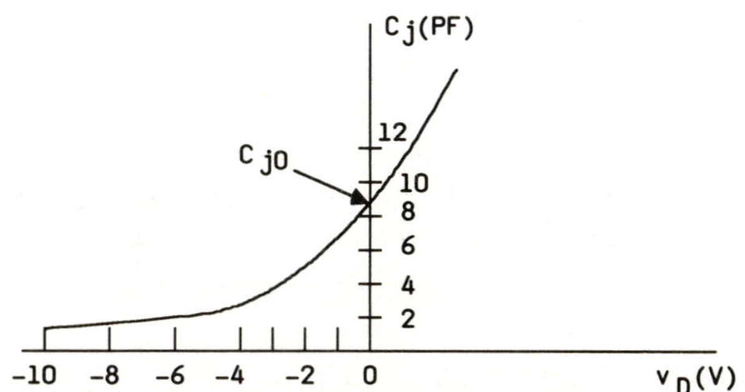

Figure 9.7: Depletion capacitance
versus the diode voltage

Figure 9.7 shows the graph of a typical diode junction capacitance plotted against applied reverse voltage. When the junction is forward biased, there is an additional component of capacitance called the **diffusion capacitance** that is much larger than the junction capacitance. The diffusion capacitance is related to charge storage that results when a pn junction is forward-biased and minority charges are injected across the junction. The diffusion capacitance C_D is proportional to the forward current i_D; that is,

$$C_D = K\, i_D \tag{9.12}$$

where K is a constant at any given temperature.[11]

9.4 Diode Applications

9.4.1 Load Line or Graphical Solution

A simple diode circuit consisting of a voltage source v_S, a resistor R, and a diode D is given in Figure 9.8. We shall solve for the circuit current i_D by using the diode characteristic and Kirchhoff's voltage law.

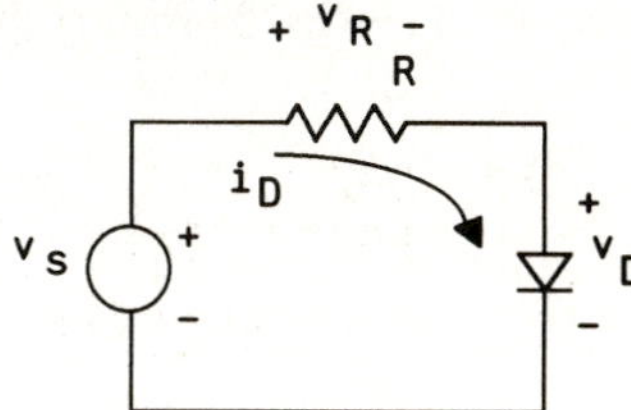

Figure 9.8: Simple diode circuit

Writing the voltage equation around the loop yields

$$v_D = v_S - i_D R \tag{9.13}$$

We can plot this equation, known as a **load line**, on the same axis as a plot of the diode characteristic, Figure 9.9. The intersection of these two plots is called the **operating point** of the circuit and yields the circuit current, i_D, and the voltage across the diode, v_D. Note that we have assumed the diode to be ideal.

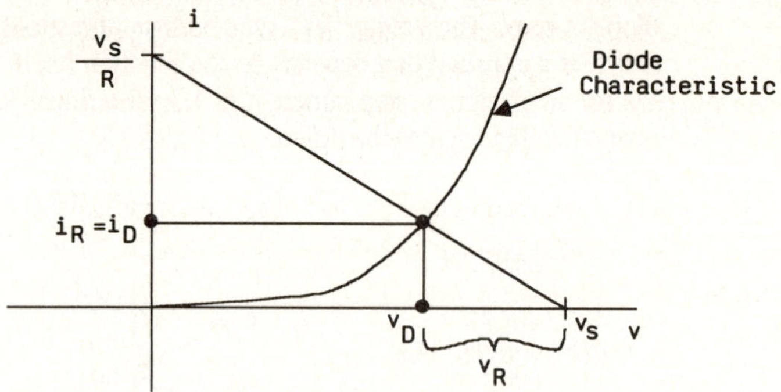

Figure 9.9: Graphical analysis of the circuit of Figure 9.8

Approximate Solution

A forward-conducting diode has a voltage drop that lies in a narrow range, from about 0.6 to 0.8 V. As an approximation, we may assume that a diode operating at its rated current will have a voltage drop of 0.7 V. Then, using Kirchhoff's loop equation, we can determine the current of the circuit of Figure 9.8 to be

$$\frac{v_S - 0.7\,\text{V}}{R} \tag{9.14}$$

9.4.2 Current Steering

Consider the circuit of Figure 9.10.

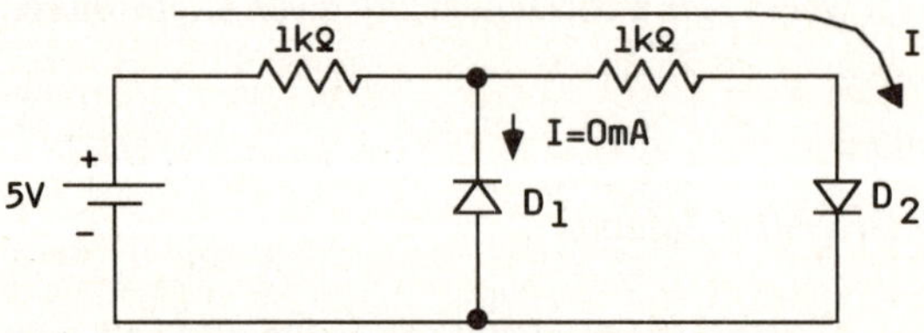

Figure 9.10: Diode circuit with only D_2 conducting

Assuming ideal diodes, by tracing the current path, it can be seen that D_1 is off but D_2 is on.

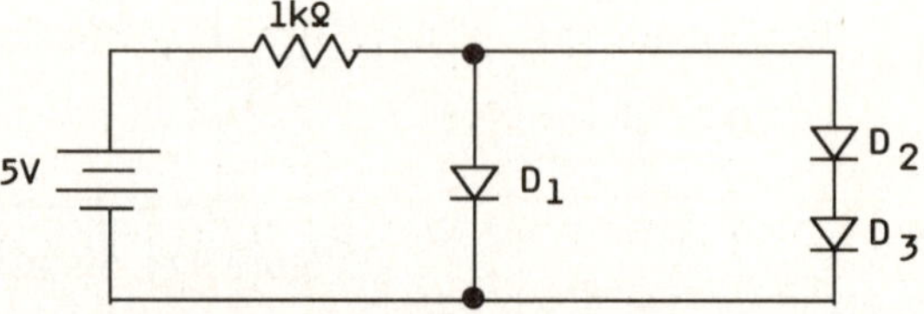

Figure 9.11: Only D_1 conducts

A similar analysis of Figure 9.11 shows that if each of the diodes is assumed to be ideal, then all three diodes are assumed to be conducting. However, for real silicon diodes, a threshold voltage of 0.6 to 0.8 V must be reached before the diode starts to conduct. Thus, there is 0.7 V across diode D_1. Now, since the series diode combination D_2 and D_3 is in parallel with D_1, the total voltage across the sum of D_2

and D_3 is also only about 0.7 V, but a minimum of 1.2 V is necessary to turn on the diode combination. Thus, only D_1 is conducting, and D_2 and D_3 each drops only 0.35 V. The circuit current is calculated to be

$$\frac{5\,V - 0.7\,V}{1\,k\Omega} = 4.3\,mA$$

and flows through the $1\,k\Omega$ resistor and diode D_1. Again, this result is correct, assuming a particular model for the diode.

9.4.3 Diode Rectifiers

<u>Half–wave Rectifier</u>

Figure 9.12 displays a simple half-wave rectifier.

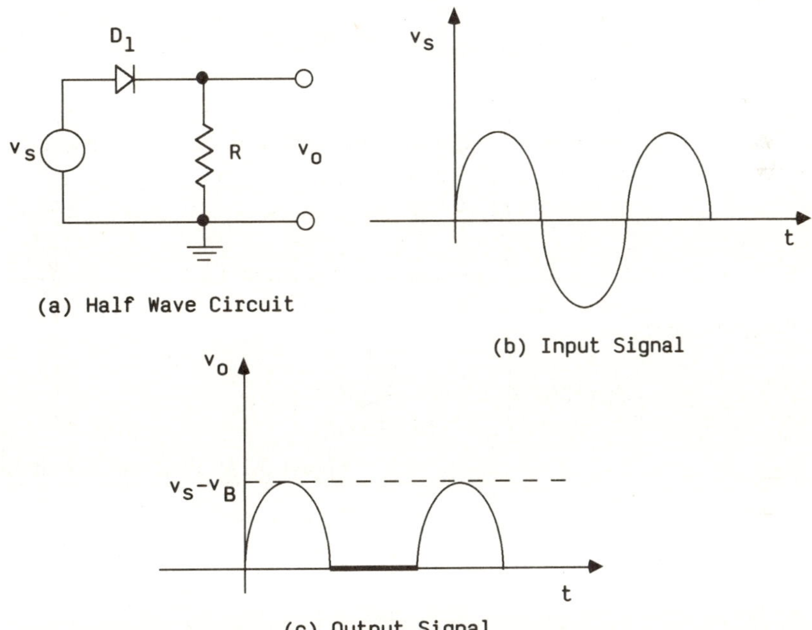

Figure 9.12: Half-wave rectifier with input and output waveforms

As the input voltage goes positive with respect to ground, the diode is forward biased and conducts. The output voltage v_o is 0.7 V below v_s due to the diode voltage drop. But when v_s goes negative with respect to ground, the diode turns off, no current flows, and $v_o = 0$ V. Because v_o never goes negative, the output current is always in the same direction. Accordingly, we obtain direct current or pulsating direct current.

<u>The Peak Rectifier[12]</u>

The peak rectifier makes use of the interaction between a diode and a capacitor to produce a dc output voltage that is almost equal to the peak of the input.

Figure 9.13(a) shows the basic peak rectifier circuit and the sinusoidal input signal is shown in Figure 9.13(b). Initially, the capacitor is assumed to be discharged. Then, as the input voltage increases above zero, the diode conducts and the capacitor charges up to the instantaneous value of the input voltage. If the diode is ideal, the voltage across the capacitor will be equal to the input voltage. At $t=t_1$, the capacitor voltage will be equal to the peak of the input, V_m. At $t>t_1$, the input decreases and reverse biases the diode. The capacitor voltage thus remains constant at the peak value V_m, unless the value of a later positive peak of the input signal is greater. Therefore, the circuit of Figure 9.13(a) detects the positive peak of the input signal and provides an output dc voltage equal to the value of the peak voltage.

In Figure 9.14, we examine the effect of adding a load resistor across the resistor of the circuit of Figure 9.13(a). Here, the diode is assumed to be nonideal.

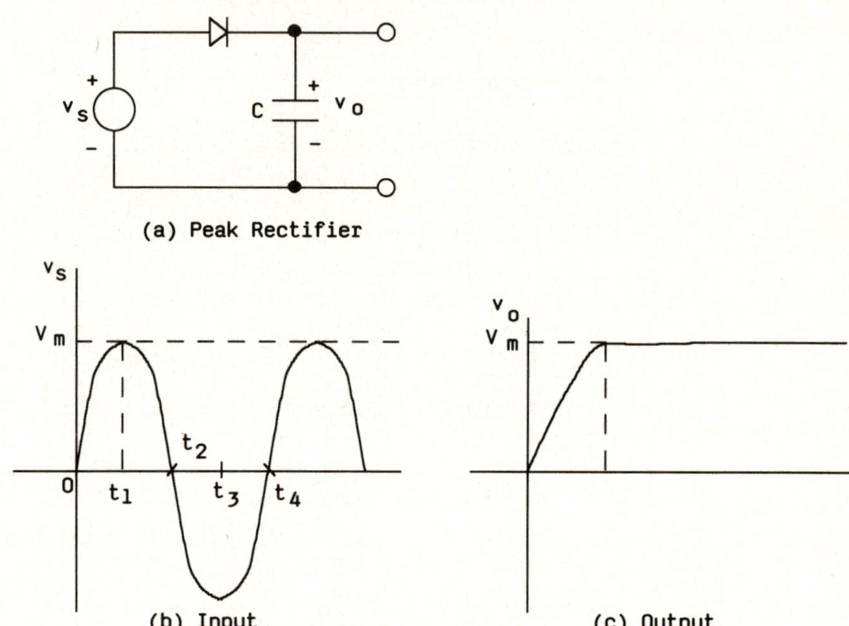

Figure 9.13: Peak rectifier circuit

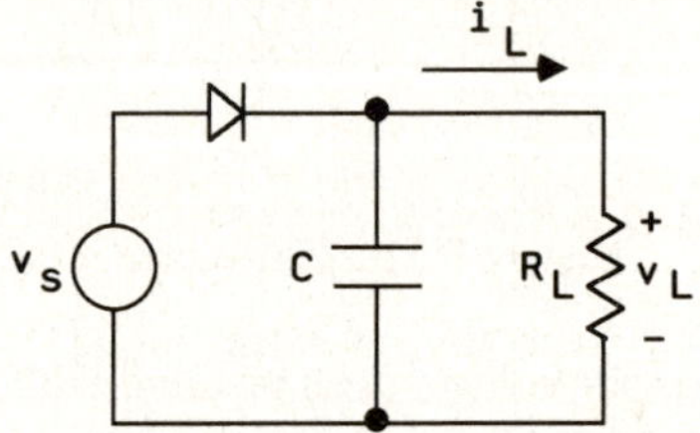

Figure 9.14: Peak rectifier circuit with nonideal diode and load resistor

The capacitor charges to the peak value of v_S equal to V_m less the diode drop. If R_L is very large, i_L is small and the capacitor does not discharge as fast as the decrease in v_S after its peak. The capacitor can only discharge through R_L because the diode is off when the source voltage is less than the load voltage. The discharge has a time constant of $R_L C$, so during discharge the load voltage is

$$v_L = (V_m - V_B)e^{-t/R_L C} \qquad (9.15)$$

where V_B is the diode turn-on voltage and V_m is the peak voltage of the input sine wave. The capacitor discharge continues until the source voltage v_s equals $v_L + V_B$ at time T_1. Figure 9.15 shows the entire waveform of the half-wave peak rectifier.

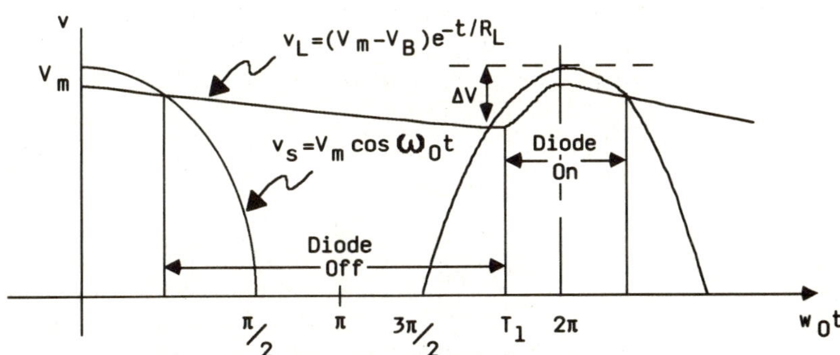

Figure 9.15: Waveform of the half-wave peak rectifier

Equating v_s - V_B with v_L gives

$$V_m \cos \omega_o T_1 - V_B = (V_m - V_B)e^{-T_1/R_L C} \qquad (9.16)$$

The discharge time T_1 may then be found from Equation (9.15) by iteration. The charge lost during discharge will be restored to the capacitor during that part of the cycle when the source voltage is rising above the minimum load voltage. The amount that the load voltage is reduced while the diode is off is called the **ripple voltage**.

When the ripple voltage is less than a tenth of V_m, the load current

$$I_L = \frac{V_m - V_B}{R_L}$$

is assumed constant, and the discharge time T_1 is assumed equal to the period of the charging cycle. Then, in the steady state, the charge out of the capacitor during discharge, $Q_D = I_L T$, must be equal to the charge added during recharge, $Q_C = C\Delta V$. We can solve for the ripple voltage by equating the two expressions, thus,

$$\Delta V = \frac{I_L T}{C} = \frac{(V_p - V_B)}{R_L C}T \qquad (9.17)$$

Actually, Equation (9.17) predicts a ripple voltage slightly larger than the actual value, but the error is on the conservative side because a small ripple is desired. A more detailed calculation of ripple voltage and diode current is found in Savant, Roden, and Carpenter.[13]

Full–Wave Rectifier[14]

A full-wave rectifier transfers input energy to the output during both halves of the input cycle and provides increased average current per cycle over that of the half-wave rectifier.

The average of a periodic function is defined as the integral of the function over one period divided by the period and is equal to the first term in a Fourier series expansion of the function. The result is a full-wave rectifier that produces twice the average current of the half-wave rectifier.

The circuit of Figure 9.16 is a full-wave rectifier using a center-tapped transformer. The input voltages V_{s_1} and V_{s_2} are shown in Figure 9.17 (a) and (b), and the output voltage v_o is shown in Figure 9.17(c).

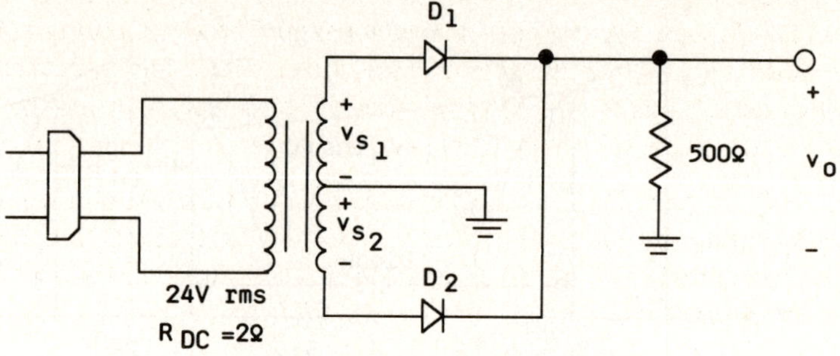

Figure 9.16: Full-wave rectifier

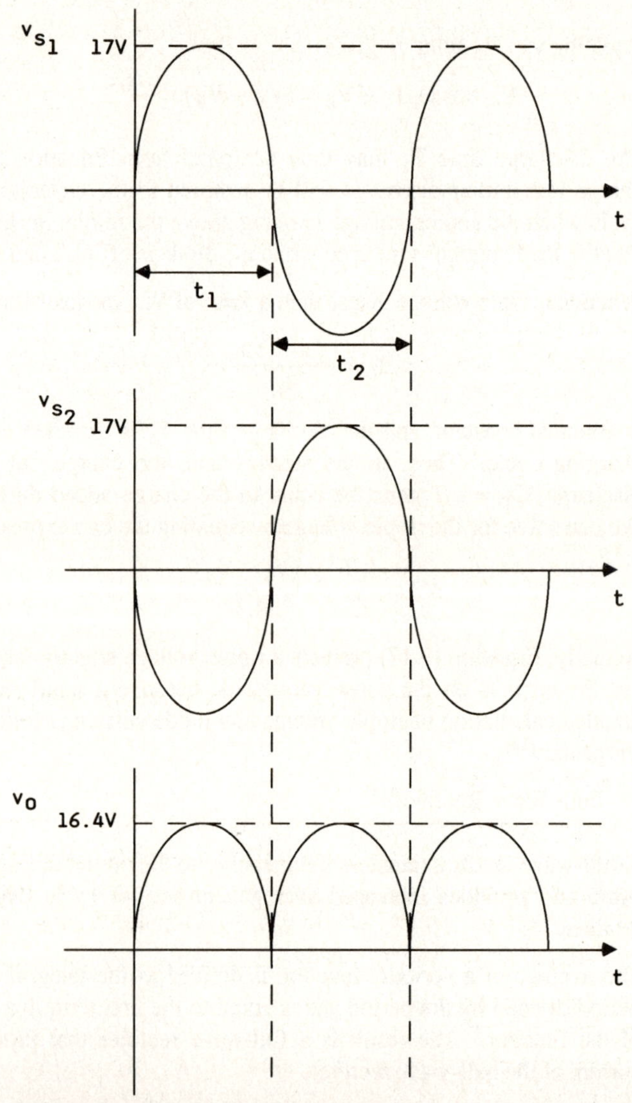

Figure 9.17: Transformer secondary voltages (a and b) and output voltage (c) of full-wave rectifier

The phase of the ac waveform at the topmost winding of a transformer is always 180°
out of phase with the bottommost winding. Therefore, during t_1 D_1 is conducting, and
during t_2 D_2 is conducting. For this center-tapped full-wave rectifier circuit, only half
of the total secondary voltage is used for each half-cycle and the time between voltage
pulses is now 1/120 sec rather than 1/60 sec.

Full-wave rectification is also possible without the use of a center-tapped transformer
by using a full-wave bridge rectifier or operational amplifiers. Both of these
procedures are discussed in Reference [14]. Full-wave peak rectifiers are demonstrated
in Lab 9L5.[15]

9.4.4 Voltage Regulators

The diode $i_D - v_D$ characteristic in breakdown is almost constant and independent of
the current through the diode. This characteristic enables the diode to be used in
voltage regulation. A voltage regulator is a circuit that provides a constant dc voltage
between its output terminals. This output voltage must remain as constant as possible,
in spite of changes in the load current drawn from the regulator or the dc power supply
that feeds the regulator circuit. Diodes called **breakdown diodes** or **Zener diodes** are
manufactured to operate specifically in the breakdown region.

The circuit symbol of the Zener or breakdown diode is shown in Figure 9.18. In normal
application, current flows into the cathode, and the cathode is positive with respect to
the anode. Therefore, I and V in the figure have positive values. The equivalent circuit
for the breakdown diode is illustrated in Figure 9.5(e) and consists of an ideal diode in
series with a voltage source, V_Z, and a small resistance, r_Z. Applications of voltage
regulators are discussed in Savant, Roden, and Carpenter.[16]

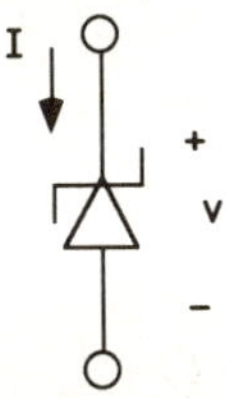

Figure 9.18: Circuit symbol for the breakdown diode

9.4.5 Clipping and Clamping Circuits

Another common circuit application of diodes is wave shaping, in which portions of
waveforms may be clipped or clamped to specific dc levels by the diode circuits in
question. Clipping circuits generally consist of a diode, resistor, and voltage source,
while clamping circuits are similar to clipping circuits, but include a capacitor in
addition to the other components.

Clipping Circuits

Clipping circuits are used to eliminate a part of a waveform that lies above or below
some reference level. The rectifier circuits discussed earlier use clipping action at the
zero level. If a battery is added in series with the diode, a rectification circuit will clip
everything above or below the battery voltage, depending upon the orientation of the
diode.

Four examples of clipping circuits are shown in Figure 9.19.

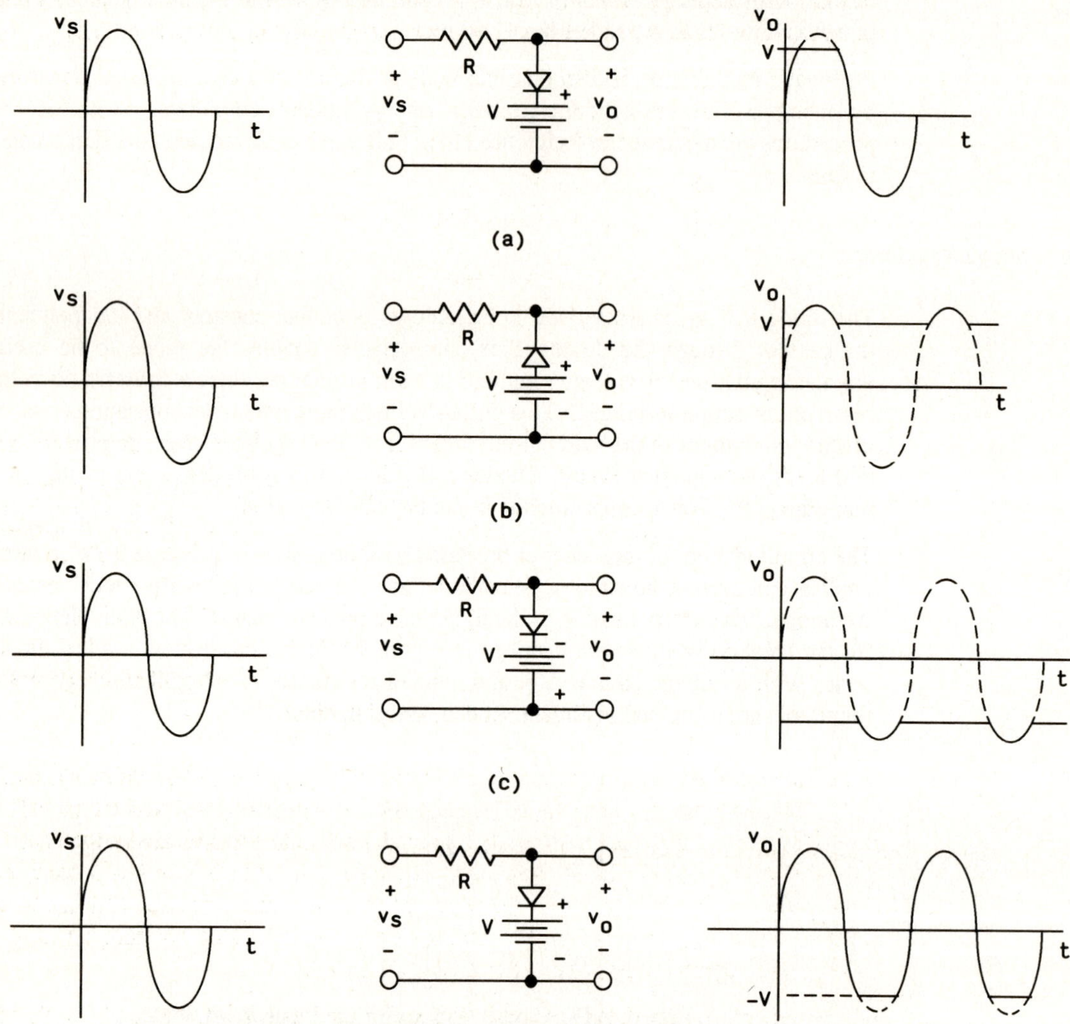

Figure 9.19: Ideal clipping circuits (Reprinted by permission from *Electronic Circuit Design: An Engineering Approach*, by C. J. Savant, Jr., M. S. Roden, and G. L. Carpenter. Copyright © 1987 by Benjamin/Cummings.)

In each case, the diode is assumed to be ideal.[17] If the diode is not ideal, the parameters V_B and r_D must be included in the circuit evaluation. The voltage V_B must be overcome before the diode will conduct, and when the diode does conduct, a forward resistance r_D is included.   The effect of V_B is to make the clipping level $V_B + V$ instead of V. The effect of the resistance is to change the flat clipping action to one that is proportional effect to the input voltage through the voltage divider.

Clamping Circuits

The function of a clamping circuit, shown in Figure 9.20, is to fix one edge of a periodic waveform to some reference voltage.

The RC time constant must be long compared to the period of v_S, so that the capacitor will not discharge appreciably during one cycle. If the capacitor is initially uncharged, then when the positive extreme of v_S exceeds V_R, the diode will be forward biased and will allow the capacitor to charge to the difference between the positive peak of v_S and the reference voltage V_R. Here V_C is positive and the resistor R is not necessary.

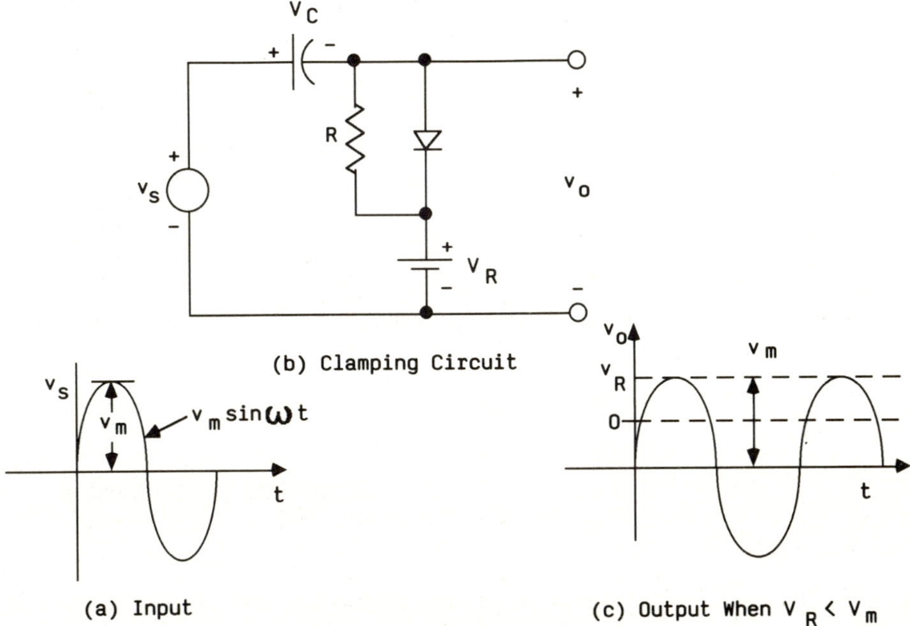

Figure 9.20: Clamping circuit

If the diode is ideal, then after the initial charge of V_C, $v_o = v_S - V_C$, and the positive peak of V_o appears to be "clamped" to V_R. For a real diode, the peak of v_o will be clipped because the diode must conduct enough charge to replace the charge that has leaked off through R during the remainder of the period. The resistor R is needed only if the positive peak of v_S is less than V_R. In this case, the capacitor charges to the negative value $V_m - V_R$ through the resistor, and the diode does not conduct until v_o exceeds v_R. Again, the positive extreme of V_o is at V_R.

The circuit of Figure 9.21 is a doubler circuit. In this circuit, C_1, and D_1 form a clamping circuit while D_2 acts as a rectifier and C_2 as a filter capacitor. When D_1 conducts, C_1 charges to V_m with the polarity shown. The waveform across D_1 is a sinewave level shifted up by V_m as shown in Figure 9.21(c). The peak value of this sine wave is now $2V_m$. D_2 allows current to charge C_2 from the $C_1 - D_1$ combination, but does not allow C_2 to discharge back towards the input; as a result, C_2 charges to $2 V_S$. This circuit functions as a voltage doubler because the output voltage is twice what would be obtained from a normal rectifier circuit. Using similar principles voltage triplers and quadruplers can also be analyzed. These types of wave shaping circuits are further illustrated in Lab 9L2.

9.4.6 *Regulated dc Power Supply*

Full-wave rectifiers, voltage regulators, and the concept of ripple can be used in the analysis of a regulated dc power supply.[18],[19]

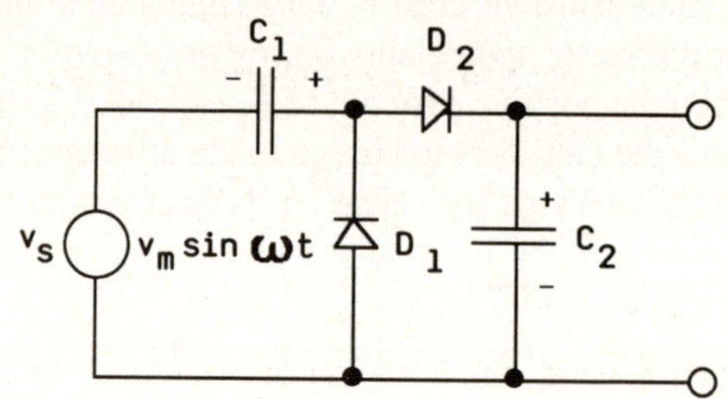

(b) Voltage Doubler

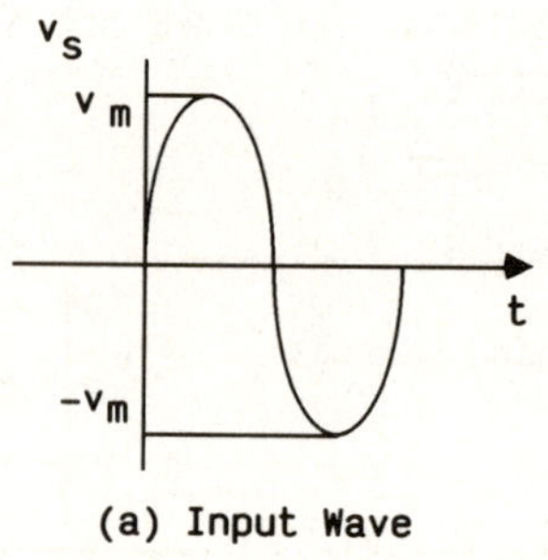

(a) Input Wave

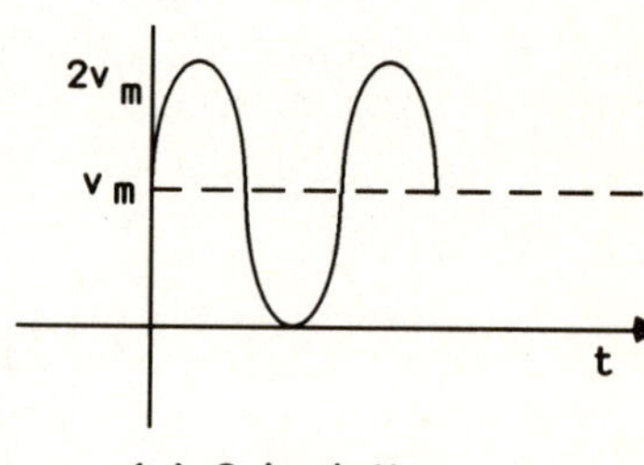

(c) Output Wave

Figure 9.21: Voltage doubler

9.5 Summary

In this chapter, we have studied a number of characteristics and applications of the pn junction diode. First, we examined the current-voltage characteristic of the diode in its three regions of operation: the forward active region, reverse region, and breakdown region. Next, we looked at the diode as a circuit element and considered the various dc and small signal circuit simulation models for the diode. Finally, we considered a number of applications for the diode, viz., current steering; half-wave, full-wave, and peak rectifiers; voltage regulators; and clipping and clamping circuits. We then combined a number of diode subcircuits to design a regulated dc power supply; and we investigated the application of diodes in simple logic circuits.

LAB 9L1

TITLE: The Diode: Basic Operations[20],[21],[22]

OBJECTIVE: To introduce the basic operation and several simple applications of the diode. A design problem is also presented.

EQUIPMENT AND COMPONENTS:
    Digital multimeter
    Oscilloscope
    Function generator
    DC power supply
    Resistors
    Capacitors
    Diodes

PRELABORATORY:

1. Use the equation below, which describes the nonlinear relationship between current and voltage for a diode, to answer the following questions:

$$i_D = I_S \left[e^{q v_D / kT\eta} - 1 \right]$$

a. Define the terms, I_S, q, k, and T.

b. Use the principle of linearity to show that the diode is a nonlinear device. (Let $I_S = 10^{-15}$ amp and T = 300°K.)

c. Quantitatively evaluate the temperature effects of the diode equation by plotting current against voltage on semilog paper for a diode at temperatures 300°K and 350°K. I_S is expressed as a function of temperature. Let v_o vary from 0.1 V to 0.9 V. Use the following values for parameters:

$$
\begin{aligned}
I_S &= 200\ T^{1/4}\ e^{-E_g/kT} \\
E_g &= 1.12\ \text{eV for Silicon} \\
k &= 8.625 \times 10^{-5}\ \text{eV/°K} \\
\eta &= 2
\end{aligned}
$$

2. Analyze the circuits of Figures 9L1.1 and 9L1.2 twice, once for each diode model given in Figures 9(a) and (c). Repeat the procedure for the second value of v_{in}. Your objective is to determine the output voltage for one cycle of the input.

Circuit of Figure 9L1.1:

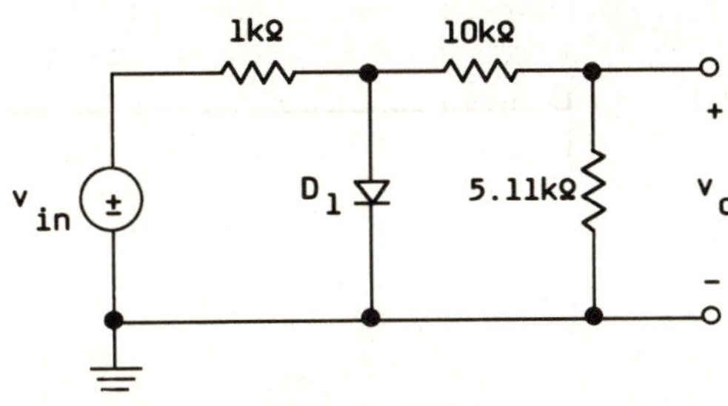

Figure 9L1.1

a. v_{in} is a square wave with a 5-V amplitude and 1-ms period.

b. v_{in} is a square wave with a 0.5-V amplitude and 1-ms period.

Circuit of Figure 9L1.2:

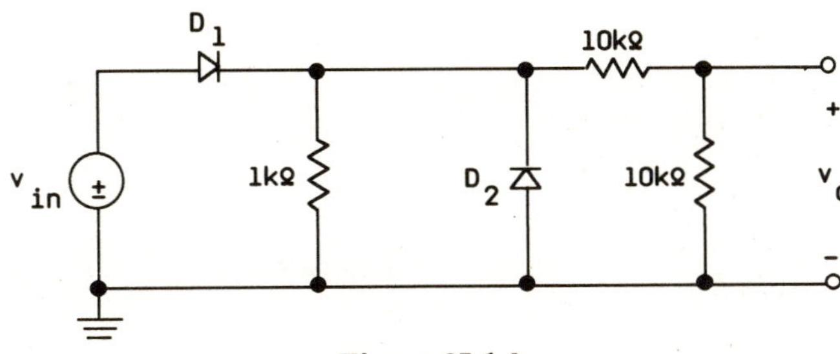

Figure 9L1.2

    a. v_{in} is a square wave with a 10-V amplitude and 1-ms period.

    b. v_{in} is a square wave with a 1-V amplitude and 1-ms period.

3. Complete the graphs for the circuits of Figure 9L1.2, 9L1.3 and 9L1.4. Do the analysis for each diode model given in Figure 9.5(a) and (c).

    a.

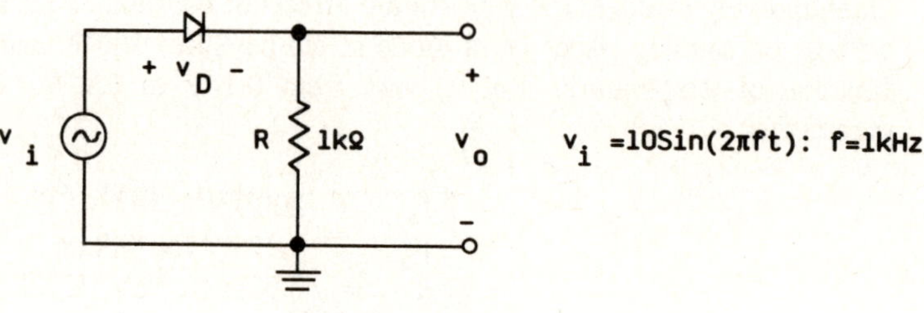

Figure 9L1.3

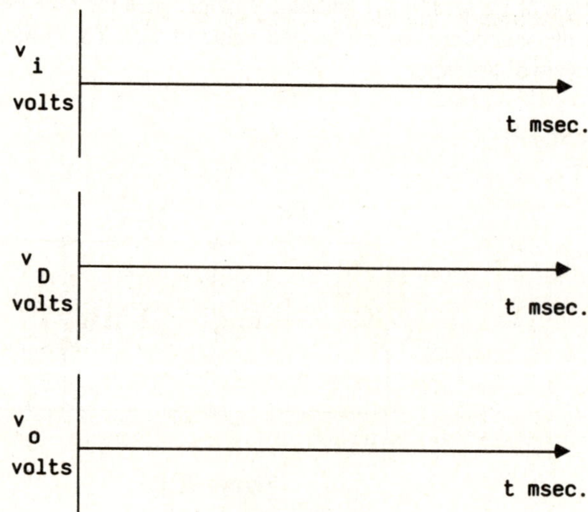

Figure 9L1.4

    b.

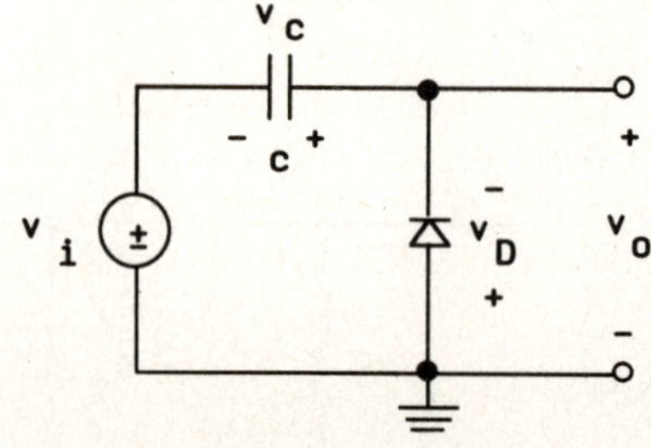

Figure 9L1.5

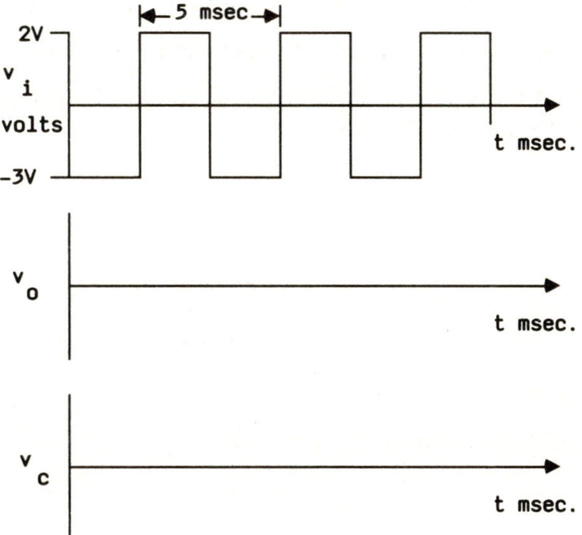

Figure 9L1.6

c.   For T=4 milliseconds, evaluate the following circuit.

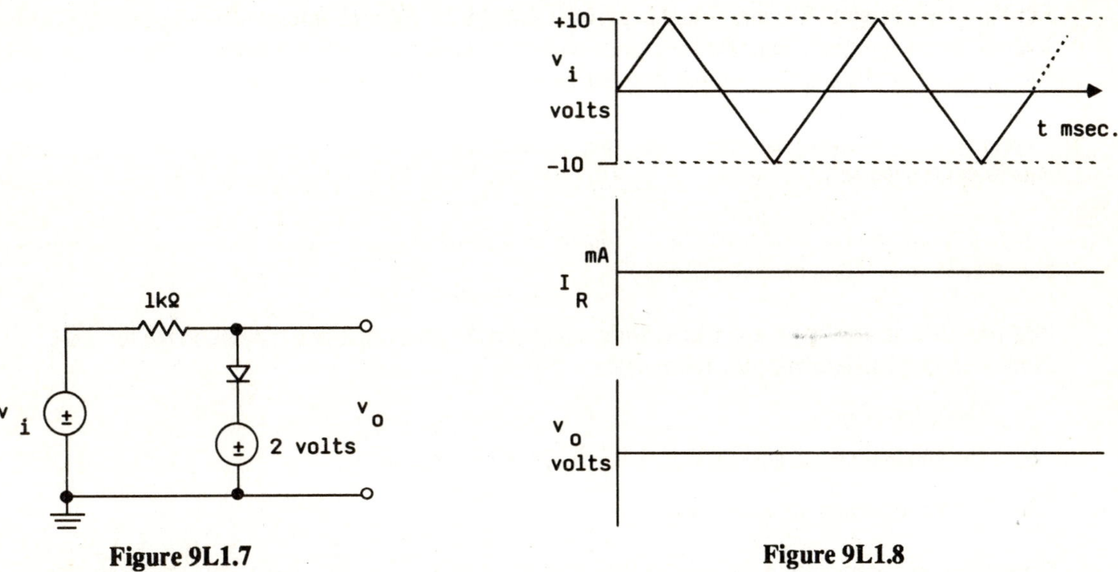

Figure 9L1.7   **Figure 9L1.8**

4.   Design Problem

Given two diodes and a resistor, design a circuit that has the characteristics shown in Figure 9L1.9.
Sketch v_o for $v_i = 10 \sin 2\pi ft$ (f=1 KHz).

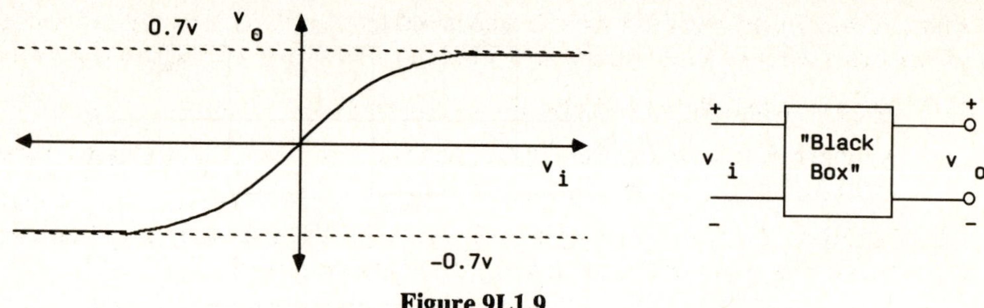

Figure 9L1.9

Laboratory Procedure:

1. Build the circuit shown in Figure 9L1.10.

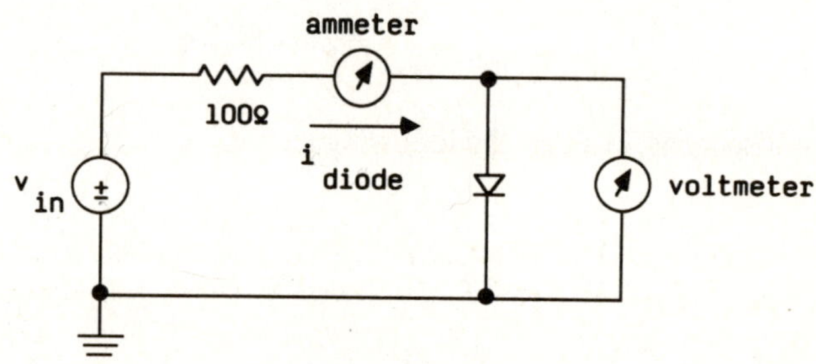

Figure 9L1.10

Let V_{in} vary between -10 and 0 V and between 0 and +10 V. Record current and voltage data points (at least 10 for each region), and plot the results on linear graph paper. (Note: When collecting data points for v_{in} between -10 and 0 V, do not connect the voltmeter during the measurement of current. When collecting data points for V_{in} between 0 and +10 V do not connect the ammeter during the voltage measurement. In each case, the meter that is removed should be replaced by its ideal Thevenin equivalent resistance.)

What is the reason for disconnecting the meters as recommended? Would your results be different if both meters were left in the circuit? Why?

Plot your data on ~~semilog~~ graph paper, with voltage on the linear axis and current on the log axis. From your graphs, determine the following:

a. The value of I_S.

b. The ambient temperature of your laboratory.

c. The region where $\eta=1$ and $\eta=2$.

Repeat the current measurement for two values of v_{in}. Let $v_{in} = -1$ volt, then let $v_{in} = +1$ volt. For each of these readings place a hot soldering iron tip close to but not touching the diode. What happens and does this agree with your prelaboratory results?

2. Build the circuits of Figures 9L1.1 and 9L1.2*. Measure and record the value of v_o for various input signals. Use an oscilloscope and record v_o for each half cycle of each input. For a more accurate measurement, try using a dc power supply to simulate the value for v_o during each half cycle, and measure the resulting v_o with a voltmeter.

Compare your experimental results with that of your prelab. What is the percent error between your prelab results and your measurements for both diode models? Which diode model is more accurate?

3. Build the three circuits shown in Figures 9L1.3, 9L1.5 and 9L1.7.

    a. For Figure 9L1.3, with $v_i = 10\sin(2\pi ft)$ and f=1 KHz, plot v_i, v_D, and v_o on the axes shown in Figure 9L1.11.

    b. For Figure 9L1.5, let v_i be a square wave with a positive peak value of 1 V and a negative value of -3V. Plot v_i, v_c, and v_o on the axes shown in Figure 9L1.12.

    c. For Figure 9L1.7, let v_i is a triangular wave with a 10 V amplitude. Plot v_i, i_R, and v_o on the axes shown in Figure 9L1.13.

*Before you dismantle the circuits, collect all the information needed to complete the plots and answer all questions.

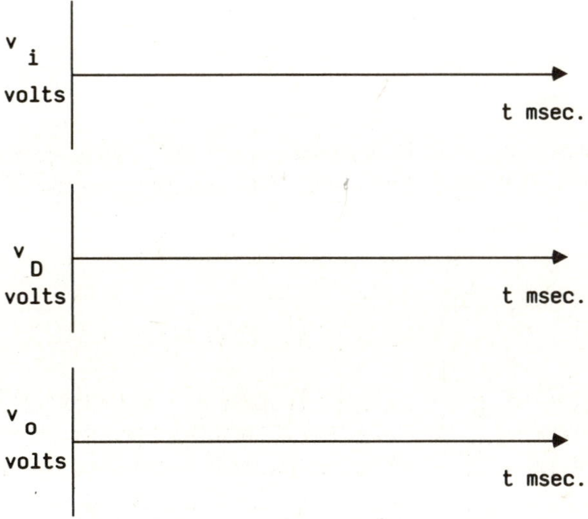

Figure 9L1.11: Plots for Figure 9L1.3

3a. Sketch v_o when a 1000-pf capacitor is connected in parallel with R. Repeat for a 1-μf. capacitor. In terms of diode applications, how would you describe the function of this circuit?

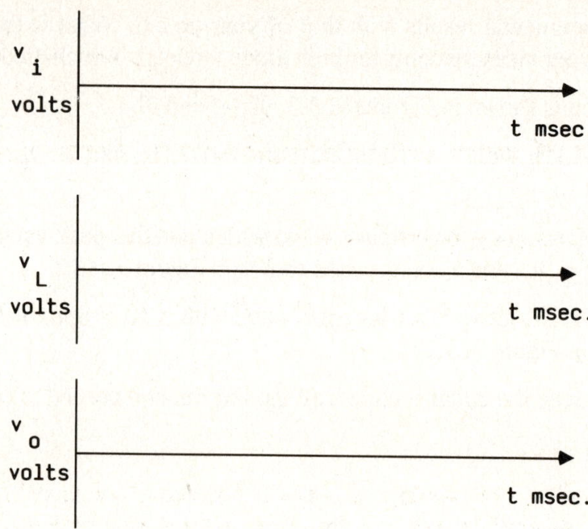

Figure 9L1.12: Plots for Figure 9L1.5

3b. Sketch what happens to v_o and v_c if the diode's anode and cathode are reversed. If v_{in} is changed to a 10-V square wave, what is the voltage across the capacitor?

<u>Circuit 3</u>

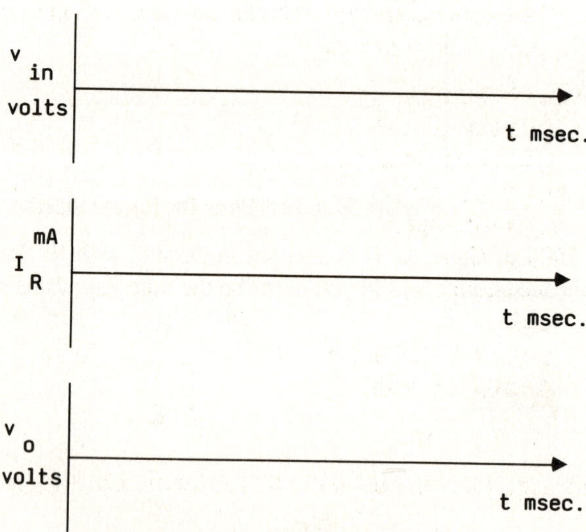

Figure 9L1.13: Plots for Figure 9L1.7

3c. Sketch the resulting v_o if the 2-V supply is changed to -2 V. What happens to v_o when v_R is greater than 10 V? Less than -10 V?

4. Verify your design from part 4 of the prelab. Sketch v_o versus v_i. Vary v_i between -2 and +2 V and collect at least 10 data points.

How would you modify your design to achieve the characteristic shown in Figure 9L1.14?

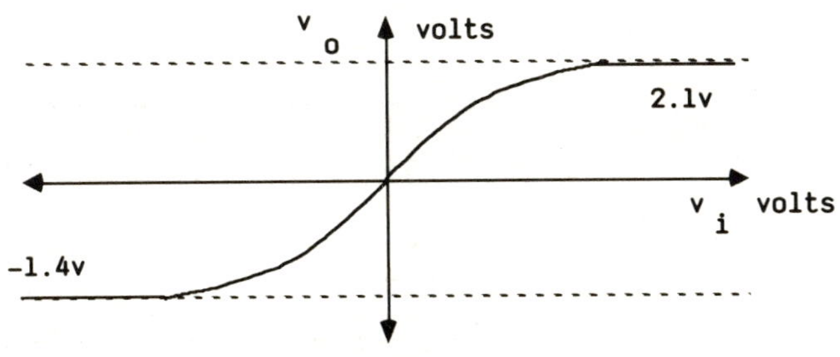

Figure 9L1.14

DISCUSSION: For each section of the lab procedure, calculate the percent error between the theoretical and the experimental results. Include all sources of error in your discussion (e.g., meter accuracy; component tolerance; temperature stability).

CONCLUSIONS:

LAB 9L2

TITLE: The Diode, Temperature Effects and Applications[20],[21],[22]

OBJECTIVE: To introduce a more detailed study of the diode, including the use of a curve tracer. This lab includes diode applications and a design project similar to that of Lab 1, but slightly more complex.

EQUIPMENT AND COMPONENTS:

Digital multimeter
Oscilloscope (dual channel)
Function generator
DC power supply
Resistors
Capacitors
Diodes
  5-V and 10-V Zeners
  4 rectifying diodes

PRELABORATORY:

1. The following equations describe the diode voltage-current relationship:

$$i_D = I_S \left[e^{q v_D / \eta k T} - 1 \right]$$

$$I_S = q A \left[\frac{D_p}{L_p N_d} + \frac{D_n}{L_n N_a} \right] n_i^2$$

$$n_i^2 = 4 M_c \left[\frac{2\pi (m_p^* m_n^*)^{\frac{1}{2}} k T}{h^2} \right]^3 e^{-E_g / kT}$$

Define all the parameters in these equations (see references 2 and 3 for details).

2. Given the three equations of Question 1 and the diode equation

$$i_D = I_S(e^{v_o/v_T\eta} - 1)$$

where $\qquad v_T = kT$

$$k = 8.625 \times 10^{-5}\, eV/{}^\circ K$$

and $\qquad I_S = 200T^{1/3}\, e^{-Eg/kT}$

$$Eg = 1.12\, eV \text{ for silicon}$$

$$\eta = 1$$

a. Plot i_D versus v_D on semilog paper (v_D on the linear scale). Let v_D range from -3V to 0.9V. Also, let $T = 300^\circ K$.

b. Determine the effect of temperature on the diode characteristic. First, find

$$\frac{\partial i_D}{\partial T}\Big|\, v_D = \text{constant}$$

This partial derivative gives the changes in diode current for a small change in diode temperature at a constant diode voltage. Evaluate this expression at room temperature for three values of v_D

$$
\begin{aligned}
1.\ v_D &= -5.0\ V\\
2.\ v_D &= 0.2\ V\\
3.\ v_D &= 0.7\ V
\end{aligned}
$$

Next, the partial derivative

$$\frac{\partial v_D}{\partial T}\Big|\, i_D = \text{constant}$$

gives the change in diode voltage for a small change in diode temperature at a constant diode current. (Assume temperature = $300^\circ K$.)

Evaluate this expression for the following three values of diode current:

$$
\begin{aligned}
&1.\ 1\ \mu A\\
&2.\ 100\ \mu A\\
&3.\ 10\ mA
\end{aligned}
$$

c. Discuss the reverse breakdown by contrasting the avalanche and Zener effect. Compare the breakdown voltages and temperature coefficients for each.

3. Use the assumption of states (is the diode on or off) to analyze the circuit in Figure 9L2.1, using the diode model from Figure 9.5(d).

Let $r_D = 15\ \Omega$ and $V_B = 0.6V$. Do the analysis for three cases:

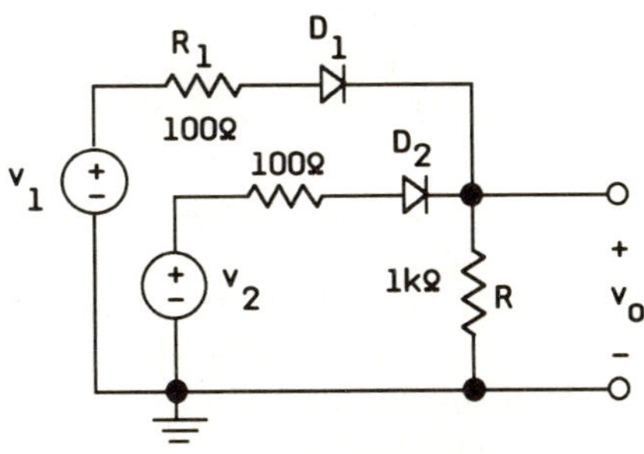

Figure 9L2.1

Case I: $v_1 = 0$ and $v_2 = 0$
Case II: $v_1 = 5$ V and $v_2 = 5$ V
Case III: $v_1 = 5$ V and $v_2 = 10$ V

Using the diode model from Figure 9.5(c), let $V_B = 0.7$ V and analyze the circuit again.

4.   Diode applications. For each of the two circuits in Figure 9L2.2, complete the graphs in Figure 9L2.3 using the diode model from Figure 9.5(c).

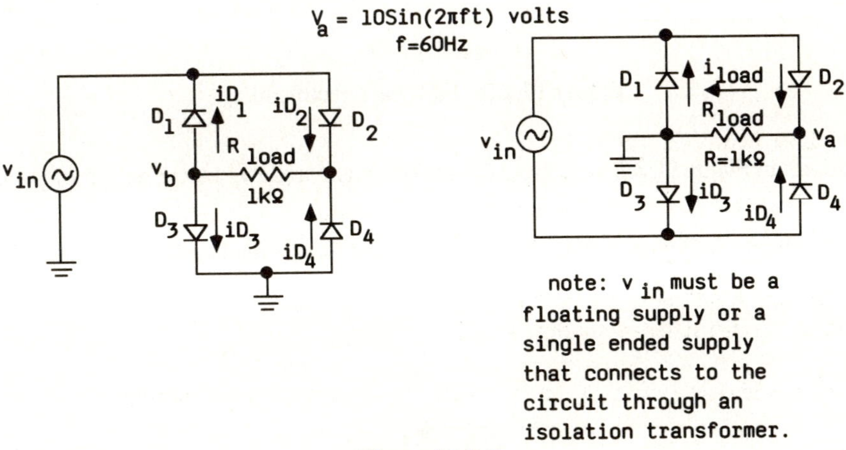

Figure 9L2.2

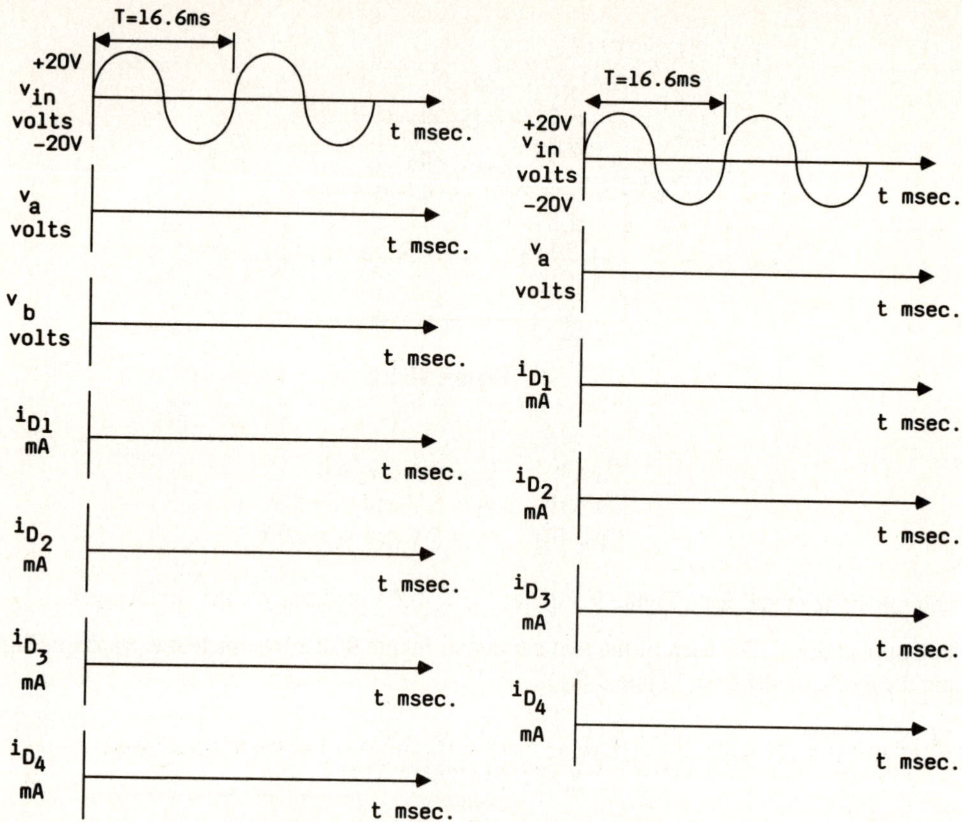

Figure 9L2.3: Full-wave rectification

1. Rectification. In Figure 9L2.2, why can't the ac source and node v_b be connected to ground at the same time?

2. Voltage Doubling. In Figure 9L2.4, assume that C_1 and C_2 are small-valued capacitors.

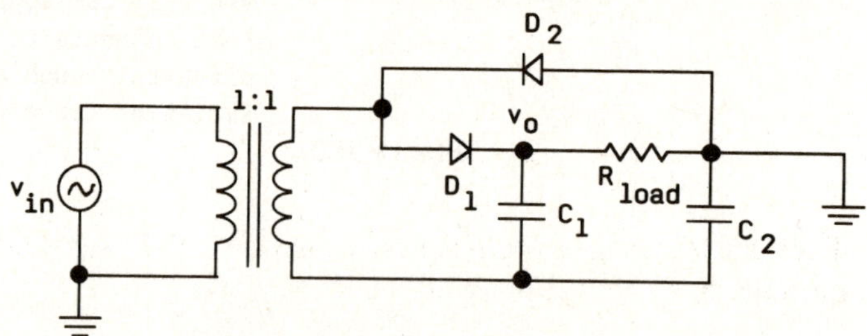

Figure 9L2.4: Voltage doubler

That is, $C_1 = C_2 \approx 1\mu F$. Let $R_{LOAD} \approx 1\,M\,\Omega$. For the sine-wave input v_{in} ($v_{in} = v_1$), plot v_{D1} versus time, v_{D2} versus time, and v_o versus time in Figure 9L2.5.

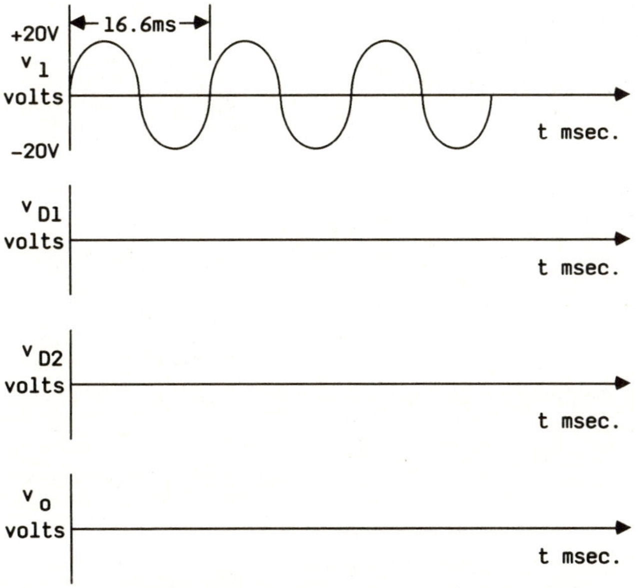

Figure 9L2.5: Plots for Figure 9L2.4

3. **Double-ended clipper.** Figure 9L2.6 shows a circuit called the **double-ended clipper.**

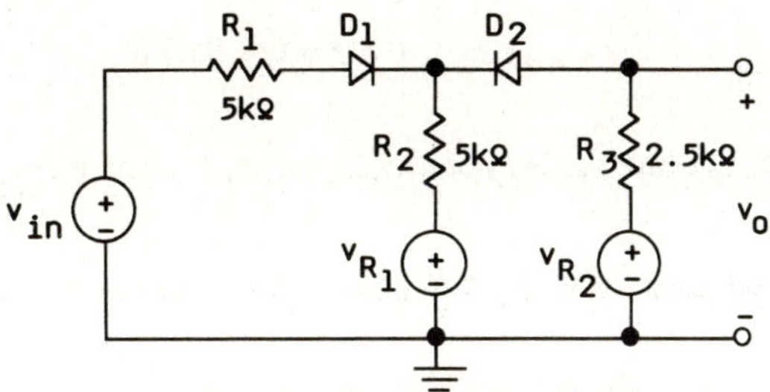

Figure 9L2.6: Double-ended clipper

Let

$$v_{R_1} = 1.5 \text{ V and}$$
$$v_{R_2} = 7.0 \text{ V}$$

a. Use the diode model from Figure 9.5(a) and sketch, in the axes shown in Figure 9L2.7, the relationship between v_o and v_{in}.

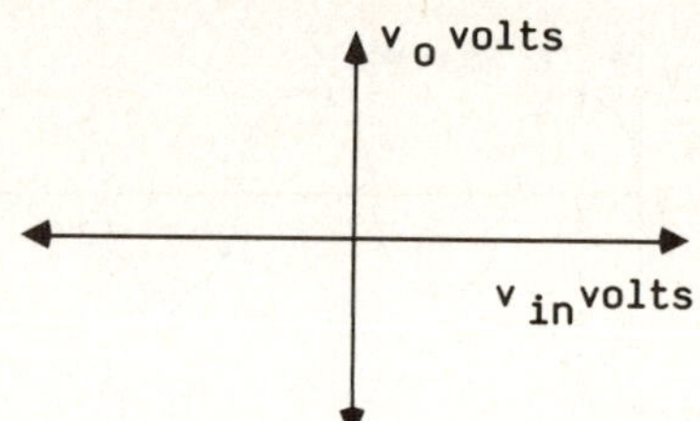

Figure 9L2.7: Plot for Figure 9L2.6

b. Use the diode model from Figure 9.5(c) and sketch the same relationship in the axes given in Figure 9L2.8.

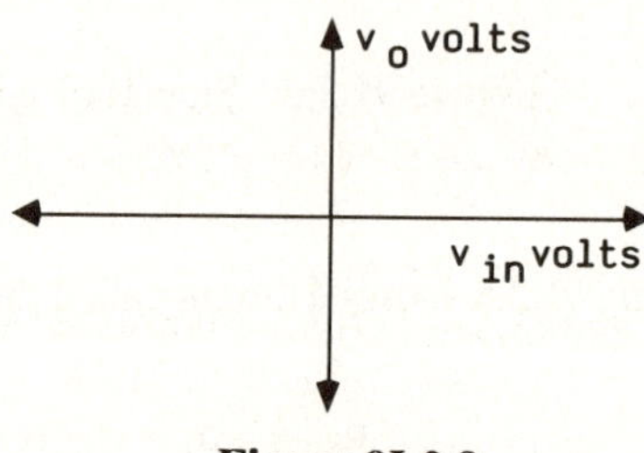

Figure 9L2.8

c. Sketch v_{in} and v_o when v_{in} is equal to $15 \sin(2\pi ft)$, where $f = 1$ kHz. (Use the diode model of Figure 9.5(c).)

4. Another double-ended clipper. For the double-ended clipper circuit shown in Figure 9L2.9:

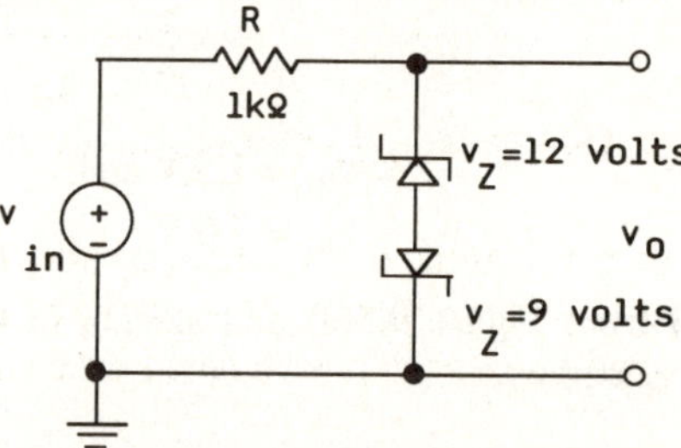

Figure 9L2.9: Double-ended clipper

a. Sketch v_o for $v_{in}=15 \sin(2\pi ft)$, where $f=10$ kHz.

b. Repeat Part a for v_{in} equal to a triangular wave of amplitude 15 V and period T=1 msec.

c. Repeat for v_{in} equal to a square wave of amplitude 10 V and period T=0.1 msec.

5. Design Project. Modify the circuit in Figure 9L2.10(a) (the same as that in Figure 9L2.6, but without the particular values for R_2 and R_3) so that it has the transfer function shown in Figure 9L2.10(b). (Note: The breakpoints occur at $v_{in} = 5$ V and 15 V.)

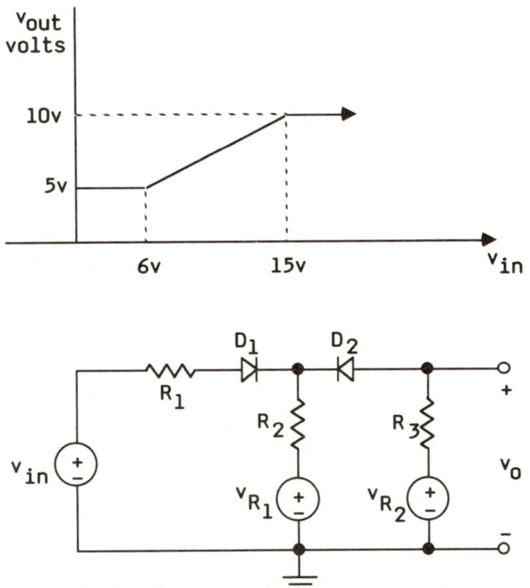

Figure 9L2.10: Circuit and transfer relationship for the design problem

Use the diode model from Figure 9.5(c). You have control over R_1, R_2, R_3, v_{R_1}, and v_{R_2}.

LABORATORY PROCEDURE:

1. Build the following circuit shown in Figure 9L2.11.

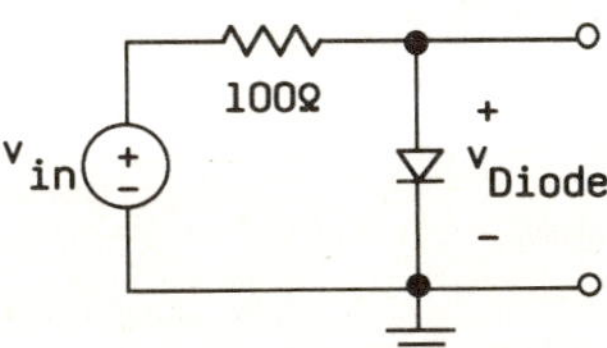

Figure 9L2.11: Characterizing the diode

Vary v_{in} from -10 to +10 V in 1-V increments. Record and plot the diode voltage and current data points on linear graph paper. When collecting the data, do not have the ammeter and voltmeter connected simultaneously. Connecting the two meters at the same time causes an error due to their respective loading effects.

Connect your diode to a curve tracer and display the forward and reverse characteristics of the diode. Photograph or plot the results from the display.

Plot curve tracer data and data collected from the circuit of Figure 9L2.11 on semilog graph paper, with voltage on the linear axis and current on the log axis. From your graph, determine the value of I_S and the temperature of the laboratory. What value of η was assumed?

With material readily available, devise a method for measuring the temperature coefficient

$$\left.\frac{\partial i}{\partial T}\right|_{v_D\,=\,constant}$$ for your diode. (The temperature of boiling and freezing deionized water, or tap water,

can be two stable reference points.)

2.   Build the circuit from part 3 of the prelab.

Measure and record v_o for the three cases listed; that is:

$$
\begin{array}{lll}
\text{Case 1:} & v_1 = v_2 = 0 \\
\text{Case 2:} & v_1 = v_2 = 5\ \text{V} \\
\text{Case 3:} & v_1 = 5\ \text{V}, V_2 = 10\ \text{V}
\end{array}
$$

What happens to v_o if the values for v_1 and v_2 are interchanged in Case 3?

Examine the I-V characteristics of D_1 and D_2 on the curve tracer. Note any discernible difference between the diodes.

3.   Build the four circuits from part 4 of the prelab, shown in Figure 9L2.12. For Part a, build either of the two circuits shown but indicate which circuit is used. Plot v_{in}, v_a, v_b, i_{D1}, i_{D2}, i_{D3} and i_{D4} for the four circuits on the axes given in Figure 9L2.13.

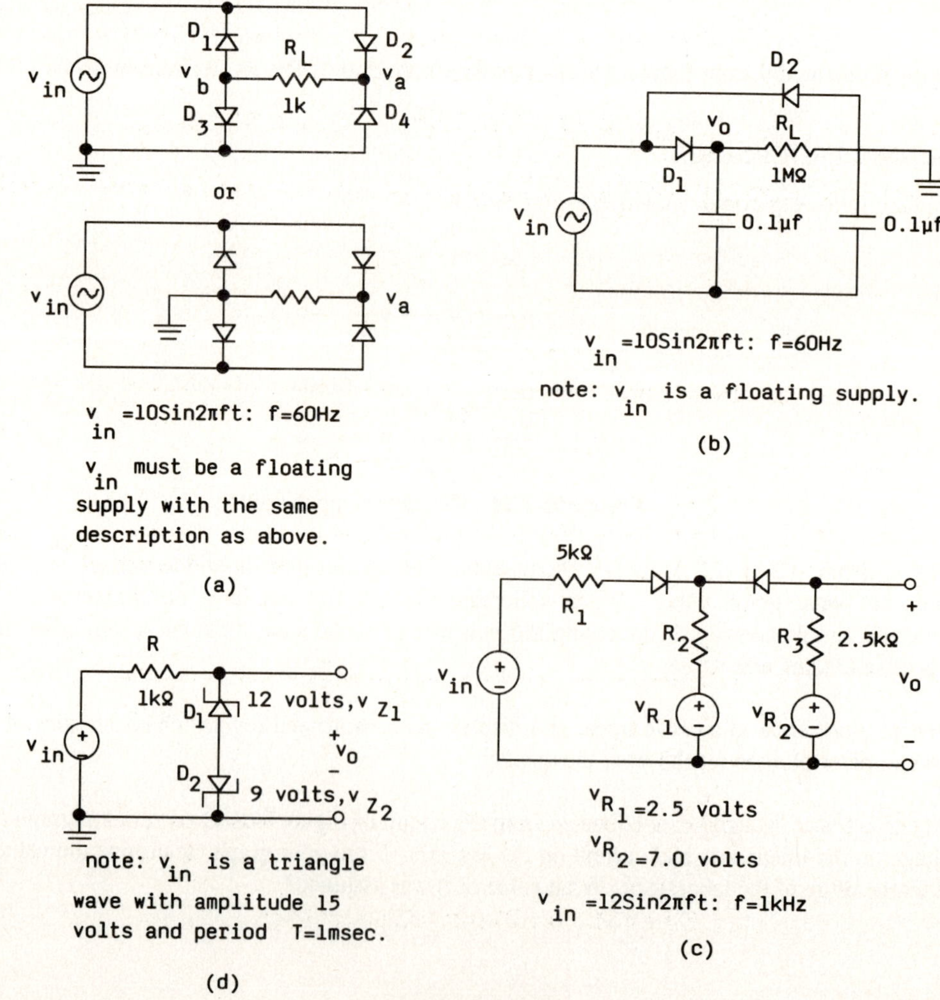

Figure 9L2.12

Part a:

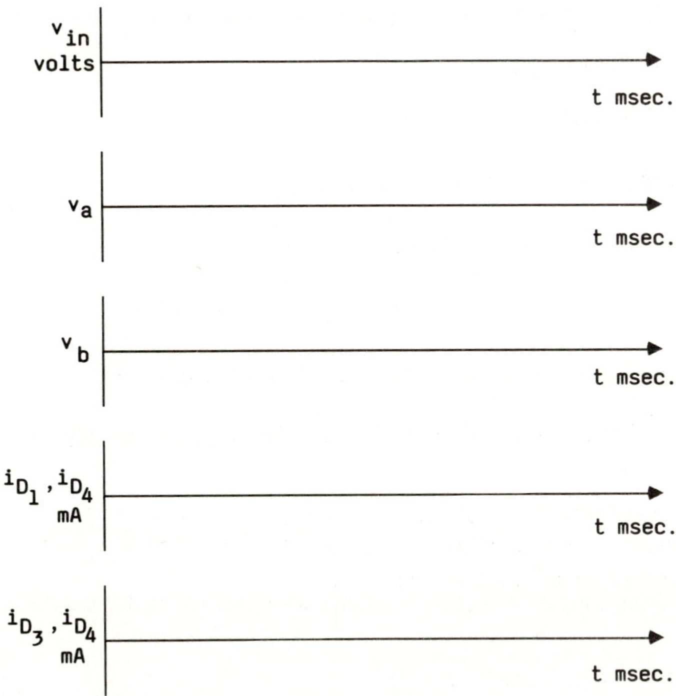

Figure 9L2.13: Plots for circuit in Figure 9L2.12(a)

What is the maximum power dissipated in each diode? What is the average power dissipated in each diode?

Part b:

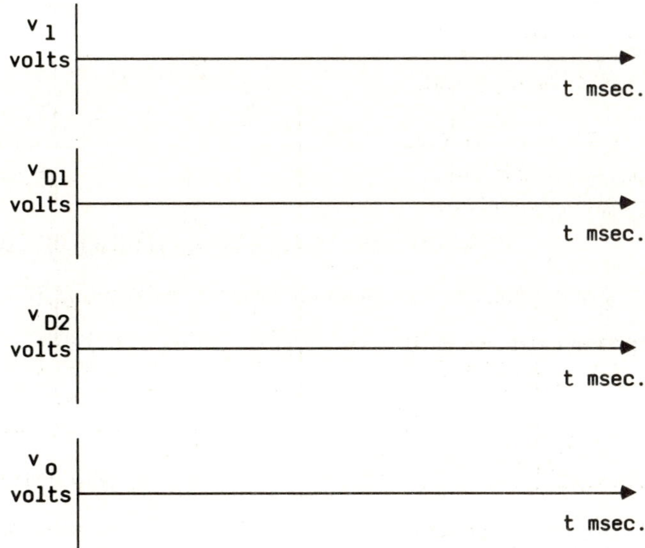

Figure 9L2.14: Plots for circuit in Figure 9L2.12(b)

What happens to v_o as R_L is decreased from 1 MΩ to 100 kΩ? to 10 kΩ? Observe the effect using an oscilloscope. Record the input resistance (impedance) of the scope.

Part c:

Let v_{in} be a dc power supply that you vary in 1-V increments between 0 and 15 V. Record and plot your data.

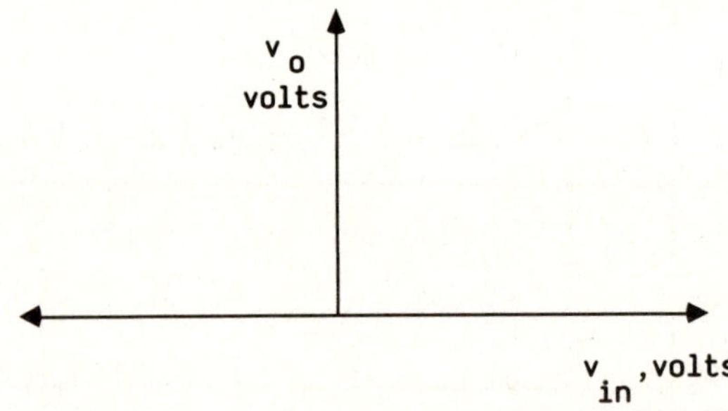

Figure 9L2.15: Plots for circuit in Figure 9L2.12(c)

Let $V_{in} = 15$ V. What happens to v_o as V_{R_1} is increased and V_{R_2} is decreased?

What happens to v_o if $V_{R_1} = V_{R_2} = 0$? Why?

Part d:

   a.   Let $v_{in} = 15\ \text{Sin}\ 2\pi ft$, where $f = 10$ kHz. Sketch v_o.

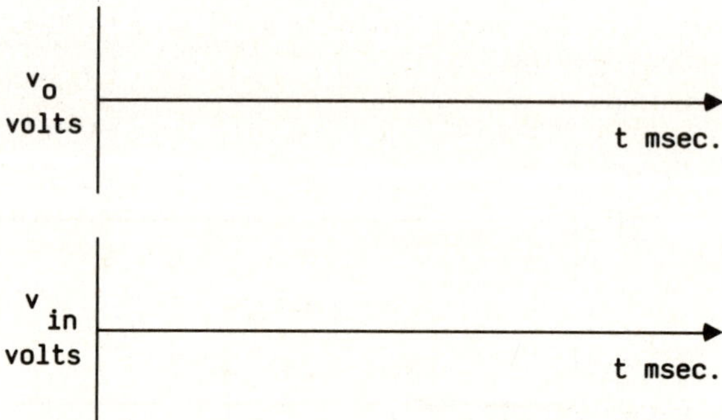

Figure 9L2.16: Plots for circuit in Figure 9L2.12(d)

   b.   Repeat for v_{in} as a triangular wave of amplitude 15 V and a period T=1 msec.

4.   Verify your design from part 5 of the prelab and shown in Figure 9L2.17.

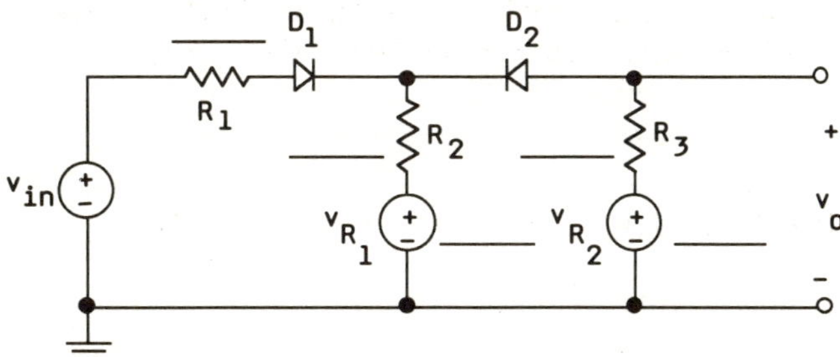

Figure 9L2.17

Record your component values on Figure 9L2.17 and plot the circuit transfer function on the graph shown in Figure 9L2.18.

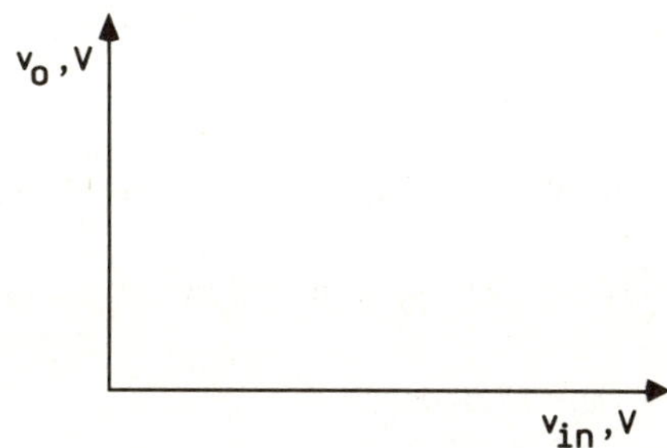

Figure 9L2.18: Plot for the design project

DISCUSSION: For each section of the lab, contrast the experimental results with the theoretical results. Also, consider the accuracy and convenience of the diode models of Figures 9.5(a) and (c).

CONCLUSION: Include explanations of where and why theory and experimental results did not match.

LAB 9L3

TITLE: Diode Current Steering

OBJECTIVE: This laboratory exercise shows the application of diode current steering in the design of a continuously integrating analog to digital (A/D) converter.[23] The A/D converter is discussed in terms of a collection of subcircuits, with the pertinent diode steering section described in greater detail. The student is given specific circuit requirements and asked to design and verify the circuit performance through simulation and laboratory testing.

EQUIPMENT AND COMPONENTS:
Multimeter (capable of measuring up to $10\mu A$)
Power supplies (± 15 V, +20 V)
     low current requirement
Diodes: switching diode (low leakage)
  i.e., Motorola MBD201
      MBD301
Resistors: Quantity of 3, with values determined by each lab group

BACKGROUND   Figure 9L3.1 shows a simplified schematic for the analog section of a continuously integrating analog-to-digital converter. The input to the circuit is a negative voltage, v_{IN}. The circuit produces an output in which the time difference between the positive edge of a periodic external start discharge clock signal running at about 100 Hz and the negative edge of the stop discharge signal is proportional to the analog input voltage.

In this laboratory, you are concerned with the concept of diode current steering. The current steering subcircuit is shown in Figure 9L3.2.

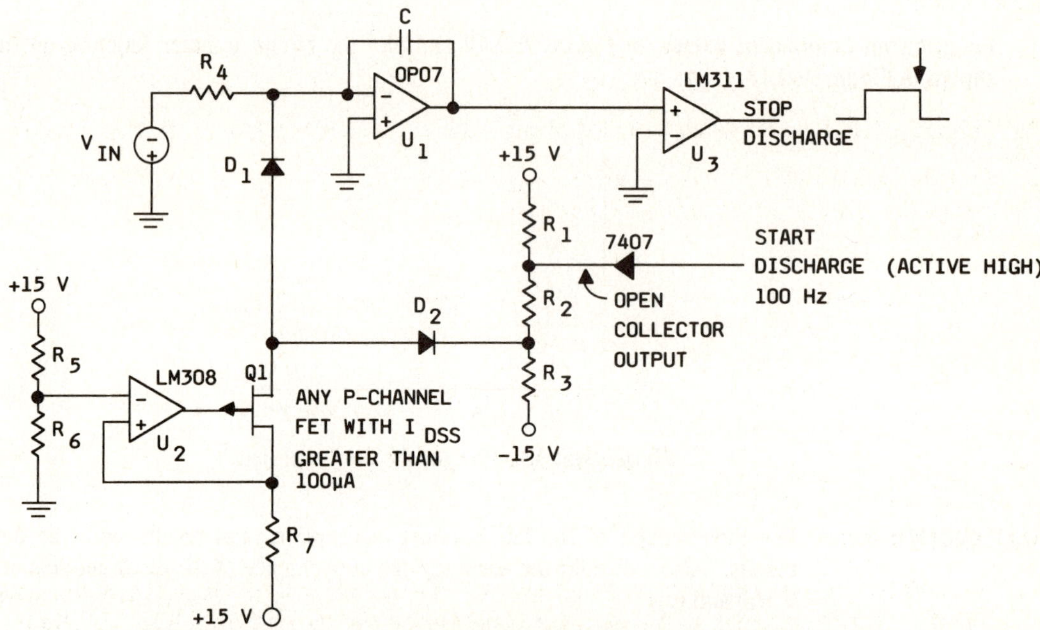

Figure 9L3.1: Simplified continuously integrating A/D converter

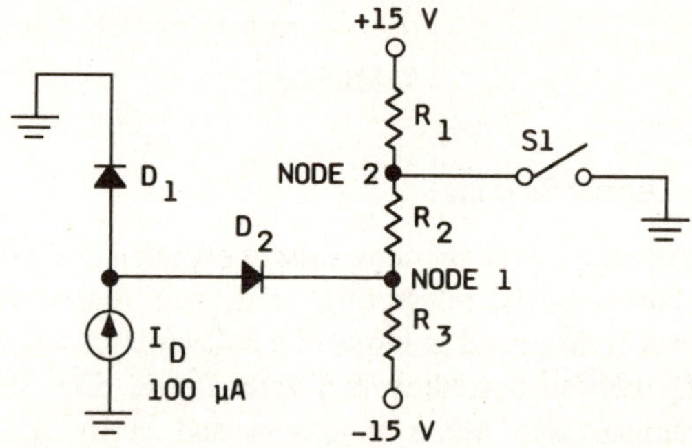

Figure 9L3.2: Current steering section of the integrator

Components R_5, R_6, R_7, U_2, and Q_1 have been replaced by a 100-µA current source, since their combined function is to serve as such a source. Finally, the 7407 open collector buffer is replaced by a switch that either connects node 2 to ground or lets node 2 go to a voltage controlled by the other circuit elements.

As the name implies, the function of this circuit is to steer a current in different directions, in this case two directions. The 100-μA current will flow through D_1 or D_2 depending on the position of S_1. In the complete circuit, when current flows in D_1 it discharges the integrator, and when current flows in D_2 the integrator is allowed to charge.

The objective is to select values for R_1, R_2, and R_3 so that the circuit performs properly.

The design specifications given will ensure the proper operation of the circuit, but remember to examine the circuit's power ratings.

DESIGN SPECIFICATIONS:

1. When S_1 is open, the voltage at node 1 should be 2 V. This will turn off D_2 and turn on D_1, which is the discharge state for the A/D converter.

2. When S_1 is closed, the voltage at node 1 should be -2 volts. This will turn off D_1 and turn on D_2.

3. Use the diode model given in Figure 9L3.3 for your analysis and design. (Hint: When D_2 is conducting, the 100 μA current flows into note 1.)

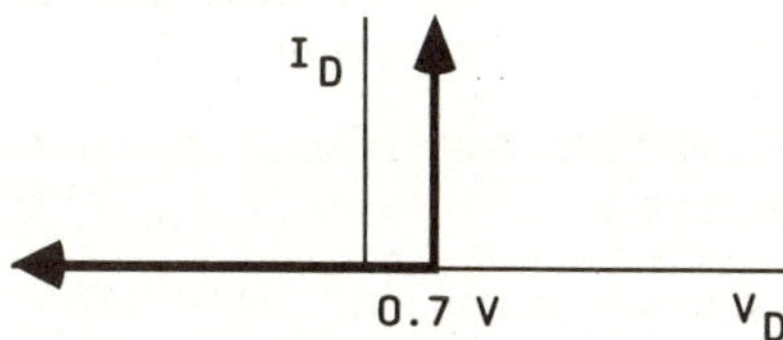

Figure 9L3.3: Diode model

PRELABORATORY:

1. Show three implementations of a 100-μA current source, including the circuit shown in Figure 9L3.4, which acts like a current source for small load resistances. For the circuit of Figure 9L3.4(a),

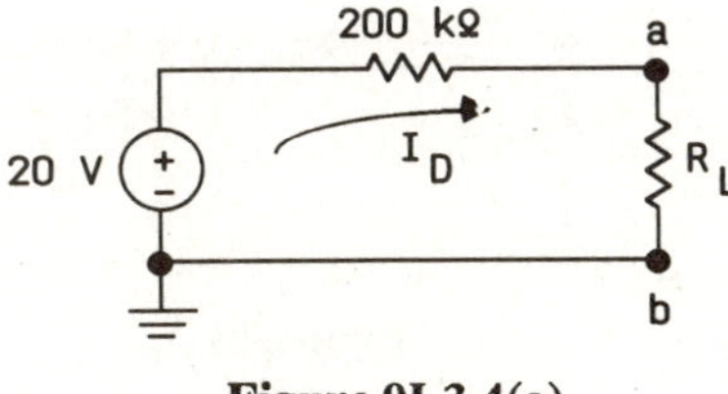

Figure 9L3.4(a)

   a. What range of load resistor R_L values allow this circuit to generate a 100 μA current with less than a 10% error?

   b. Indicate the range of voltage sources that can be placed between nodes a and b that would cause less than a 10% error.

2. Given the circuit of Figure 9L3.2 determine the resistor values R_1, R_2, and R_3 that meet the design specifications given in the "Background" section.

   a. Using your design and resistor values, determine the Thevenin equivalent circuit the current source "sees" under the two specified conditions; viz., that

    1. D_1 conducts and D_2 is nonconducting, and

    2. D_1 is nonconducting and D_2 conducts.

3. Use SPICE or any other circuit simulation program you have available to verify your design. Indicate the voltages at nodes 1, 2, and 3 and the currents in D_1 and D_2 when the switch S_1 is open. Do the same for the case when the switch is closed.

Laboratory Procedure:

1. Build the circuit from Figure 9L3.4(b) of the prelab.

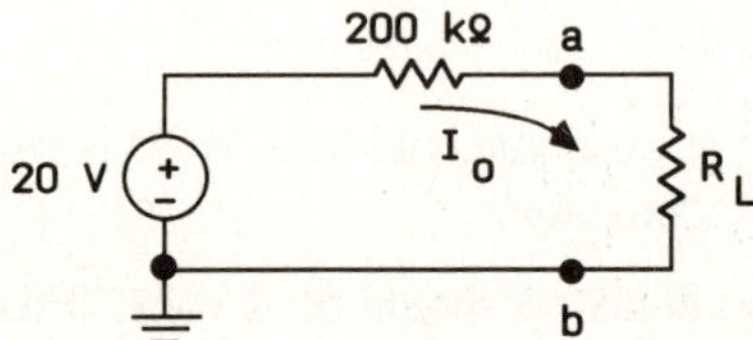

Figure 9L3.4(b) Circuit of Figure 9L3.4 with loadline (I_o, V_o)

    a. Plot the load line (I_o, V_o) of this circuit by varying R_L from a short circuit to an open circuit. (See Figure 9L3.4(b).) Indicate the short circuit current and the open-circuit voltage of your plot.

    b. On the plot from part a, plot the range of load resistances R_L that result in less than a $\pm 10\%$ deviation in the 100-μA current.

    c. On the same plot, show the characteristic of an ideal 100-μA current source.

2. From the design results of part 2 of the prelab, build the circuit and specify the values of R_1, R_2, and R_3 designed in that part of the prelab.

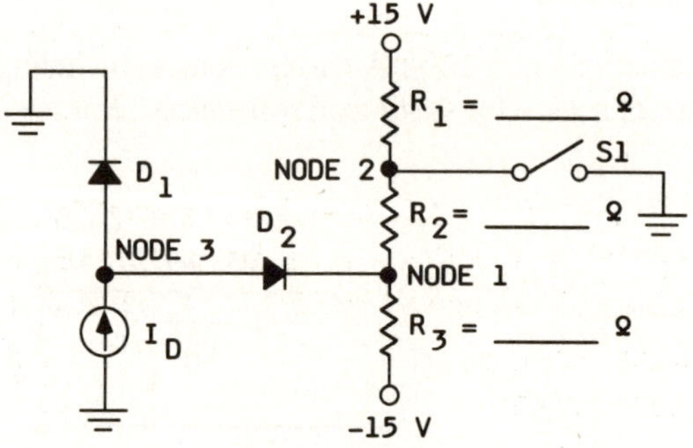

Figure 9L3.5

    a. Using Figure 9L3.5 record the values of R_1, R_2, and R_3.

    b. Verify the performance of the circuit by measuring the current flow in D_1 and D_2 and the voltages at node 1, node 2, and node 3 during the switch-open and switch-closed conditions.

DISCUSSION: Compare the results from your circuit design with those from the computer simulation and the actual laboratory circuit.

CONCLUSIONS:

LAB 9L4

TITLE: The Application of Diodes in Power Supply Design

OBJECTIVE: This laboratory exercise deals with several topics: diode circuit models, load lines, rectifiers, and regulators. We combine these topics in the design of a full-wave Zener-regulated power supply.

EQUIPMENT AND COMPONENTS:

Digital multimeter
Oscilloscope (dual channel)
Function generator
DC power supply: 10 V @ 100-mA capability
Rectifying diodes (bridge or discrete):
  e.g., Motorola IN4001-IN4007
        AR16-AR24
        ER181-ER187
        J1-J2

Zener diode: 13 V, 5% or whatever tolerance is available,
        500 mW
  e.g., Motorola 1N964A (A indicates 10%)
Transformer 120 V rms input
        30-V peak
        250-mA rating
Capacitors
  47 µf, electrolytic with at least a 30-V rating

PRELABORATORY:

1. For each of the the three diode models shown in Figures 9.5(a), (b), and (c) of the text and Equation 9.1, determine the operating point (V_D, I_D) for the diode in the circuit of Figure 9L4.1. Let $I_S = 10^{-9}$ A, $V_T = 26$ mV, $\eta=1$, $r_D = 15\ \Omega$, and $V_B = 0.6$ V.

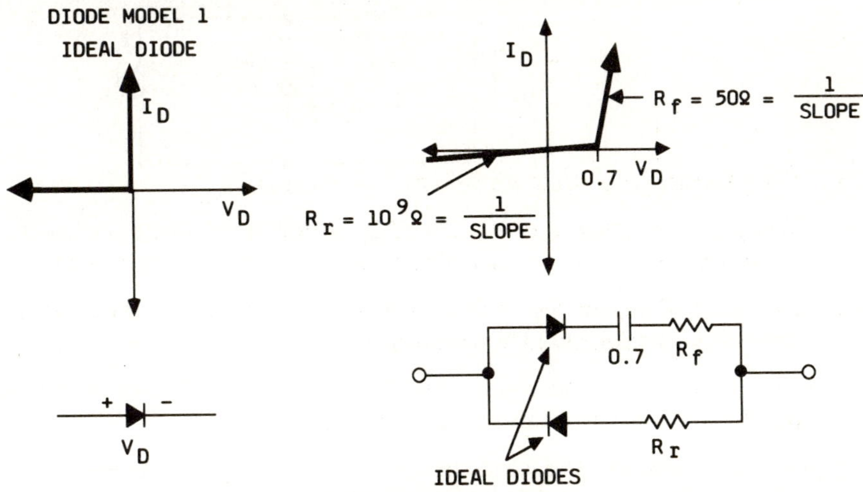

Figure 9L4.1(a)

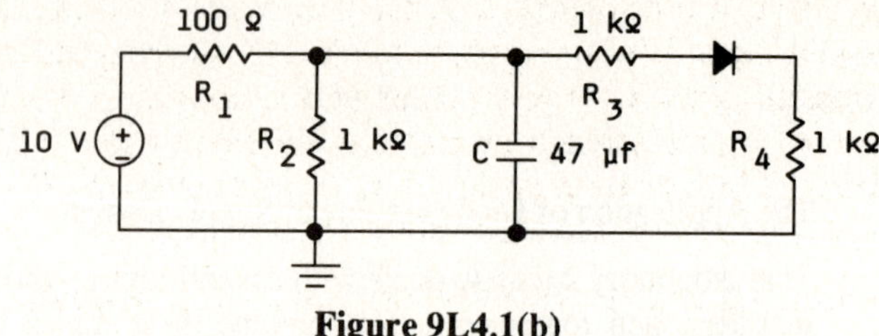

Figure 9L4.1(b)

(Hint: Find the dc Thevenin equivalent of the circuit, excluding the diode.)

2. Given the full-wave rectifier in Figure 9L4.2.

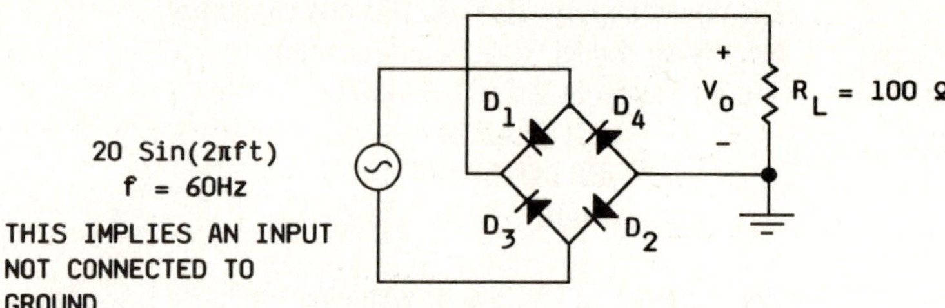

Figure 9L4.2

a. Sketch the input voltage and the voltage across R_L.

b. Connect a 47-µf capacitor across R_L and sketch the output voltage, v_o.

c. Determine the value of C required so that the peak-to-peak ripple voltage is less than 2 V. Repeat this for a ripple voltage less than 100 mV.

d. Repeat parts b and c with R_L = 1 kΩ.

3. In the circuit in Figure 9L4.3; V_{in} has a maximum value of 19 V and a minimum value of 16 V, V_z has a value of 13 V and R_L has a value of 100 Ω; assume that the Zener diode has a power rating of 500 mW and R_L has a 2-w power rating.

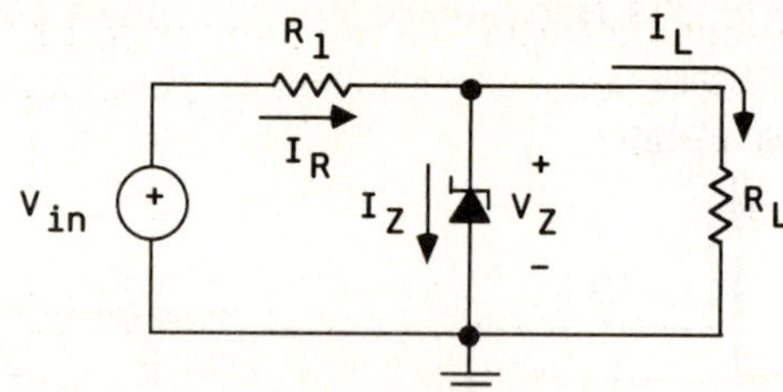

Figure 9L4.3

a. Determine a value for R_1 (include the power rating) that will allow the Zener diode to regulate the voltage across R_L to 13 V without damage to the circuit elements. Assume $I_{zmin} = 0.1\ I_{zmax}$.

b. What happens to the value of R_1 if the Zener has a ±5% tolerance on its voltage? (Obviously, the regulation of the output voltage changes.)

c. Worst case analysis: Use the tolerances of the parts to determine the range of values for V_L given V_{IN} equals 18 V. The value of V_{IN} should be set with a meter having less than a 1% tolerance.

4. The circuit in Figure 9L4.4 is a power supply featuring a full-wave rectifier with Zener voltage regulation. (The circuit is a combination of the previous two circuits shown in Figures 9L4.2 and 9L4.3.) Determine values and power ratings for each unknown component. R_D is a bleeder resistor for the capacitor and can be neglected in your design specifications ($R_D = 1\ M\Omega$). The ac supply must be a floating supply or must pass through an isolation transformer that could also be the output of a stepdown transformer. The voltage from this source should not exceed 30 V peak. The frequency is 60 Hz. Your circuit should meet or exceed the following conditions:

   i. The ripple voltage at node V_c should be less than 1 V peak-to-peak.

   ii. The output voltage V_L should be 13 V, with less than a $\pm10\%$ variation with a load ranging from $100\ \Omega$ to $1\ k\Omega$.

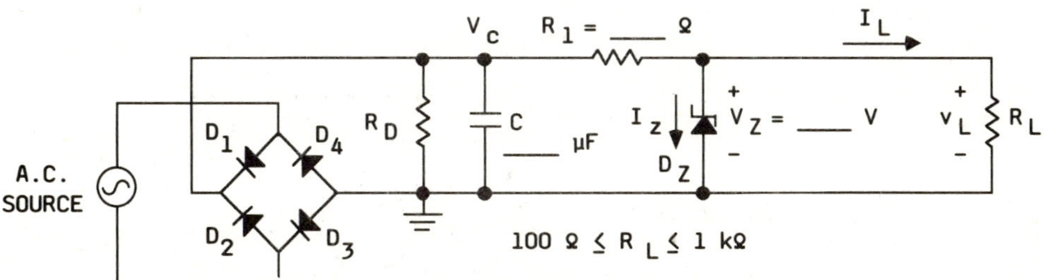

Figure 9L4.4

   a. Determine values for C, R_1, and V_Z that meet the preceding specifications.

   b. Determine minimum power ratings for the components D_1, D_2, D_3, D_4, D_Z, R_1, and R_L.

   c. Determine an acceptable tolerance for D_Z.

5. Verify your design in Question 4 of the prelab using SPICE or another circuit simulation program before going to the laboratory.

   a. Include a sensitivity analysis (.SENS)

   b. Plot the input voltage and V_L.

Laboratory Procedure:

1. Figure 9L4.5 shows the circuit discussed in Part I of the prelab. Build this circuit and record the operating point (I_D, V_D). Compare your results to the results in the prelab. Which model gives the closest answer to the actual circuit?

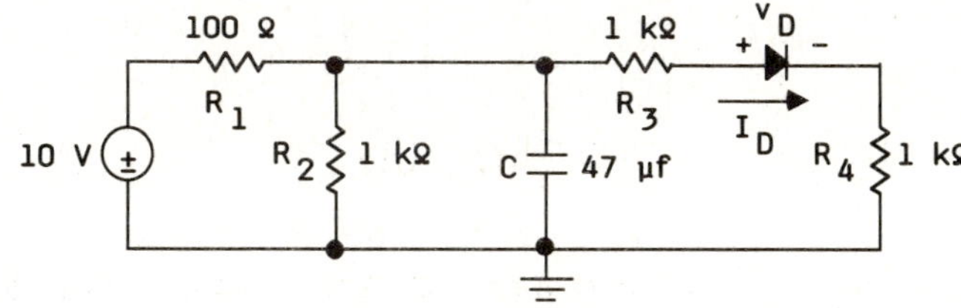

Figure 9L4.5

2. Figure 9L4.6 shows the circuit discussed in Part 2 of the prelab. Build this circuit. Let $v_s = 10\sin 2\pi f t$, where $f = 60 Hz$. The supply v_s must be floating with respect to the ground.

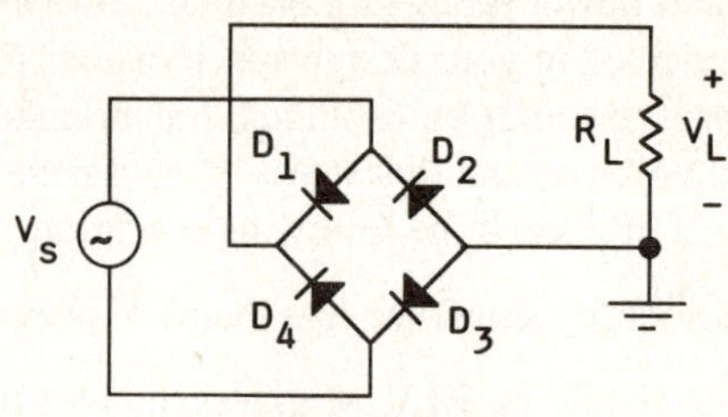

Figure 9L4.6

a. Sketch the following signals with respect to v_S. (If R_L is not referenced to ground then the measurement of v_L and all diode voltages must be measured with a dual channel differential scope.)

   (1) v_L

   (2) v_{D_1}

   (3) v_{D_2}

   (4) v_{D_3}

   (5) v_{D_4}

b. Compare the peak values of v_S and v_L. Are they equal? Why?

c. Connect a 47-µf capacitor across R_L (watch the polarity!), and sketch the voltages from part a again.

d. Use a multimeter and measure the ac voltage across v_S and the ac voltage across R_L (with the capacitor in the circuit). Record your measurements.

e. Use a multimeter to measure and record the dc voltages across R_L and v_S (again, with C in the circuit).

3. Build the circuit of part 3 of the prelab shown in Figure 9L4.3. In Figure 9L4.7, R_1 has been determined by you in the prelab, R_L is 100 Ω and 2 Watt power rating, and the Zener diode is 13 V and has a 500-mW power rating.

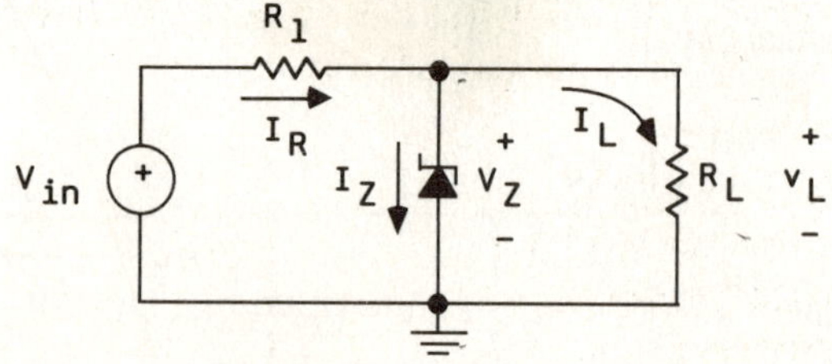

Figure 9L4.7: Circuit of Part 3

a. Sketch the voltage V_L across R_L versus V_{in}. Vary V_{in} from 0-20 V (in 1-V increments). Record and plot your data.

b. Worst case analysis. Does the value you obtain for V_L with $V_{in} = 18$ V fall within the range predicted in your prelab? (Part 3c).

4. Figure 9L4.8 shows the circuit discussed in part 3 of the prelab. Build this power supply circuit for your design and write in the values of the components.

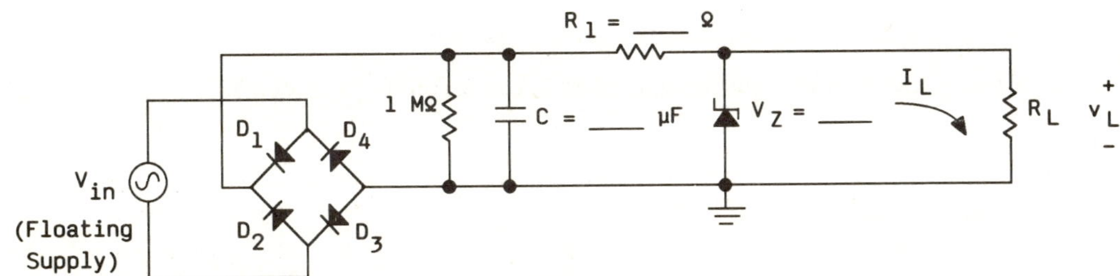

Figure 9L4.8

a.   List the power ratings of all devices.

b.   Let v_{in} be the same as used in the prelab.

c.   Plot the following voltages with respect to time:

   (1) v_{in}

   (2) v_c

   (3) v_L

   (4) v_L (with the Zener removed)

DISCUSSION: Include a comparison of the results from part 4 of the lab with those of the simulation from part 5 of the prelab.

CONCLUSIONS:

LAB 9L5

TITLE: The Diode and the Ideal Operational Amplifiers[24],[25]

OBJECTIVE: This laboratory exercise introduces the student to circuits containing diodes and op amps that can be analyzed using the ideal op amp model and three dc models for the diode. Again, the emphasis is on the various diode models, giving the student an appreciation for the trade-off between simplicity and accuracy. One nonideal op amp characteristic that the student should grasp is that the op amp output voltage cannot exceed the supply voltages.

EQUIPMENT AND COMPONENTS:
                       Digital multimeter
                       Oscilloscope (dual trace)
                       Function generator
                       DC power supply

                       Various resistors
                       Two silicon diodes
                       Operational amplifier (see the data sheet for the pin arrangement of your operational amplifier, or see the instructor)

PRELABORATORY:

1. Analysis of various diode models and their circuit equivalents.

    a. For each diode model, give an equivalent linear circuit for each region.

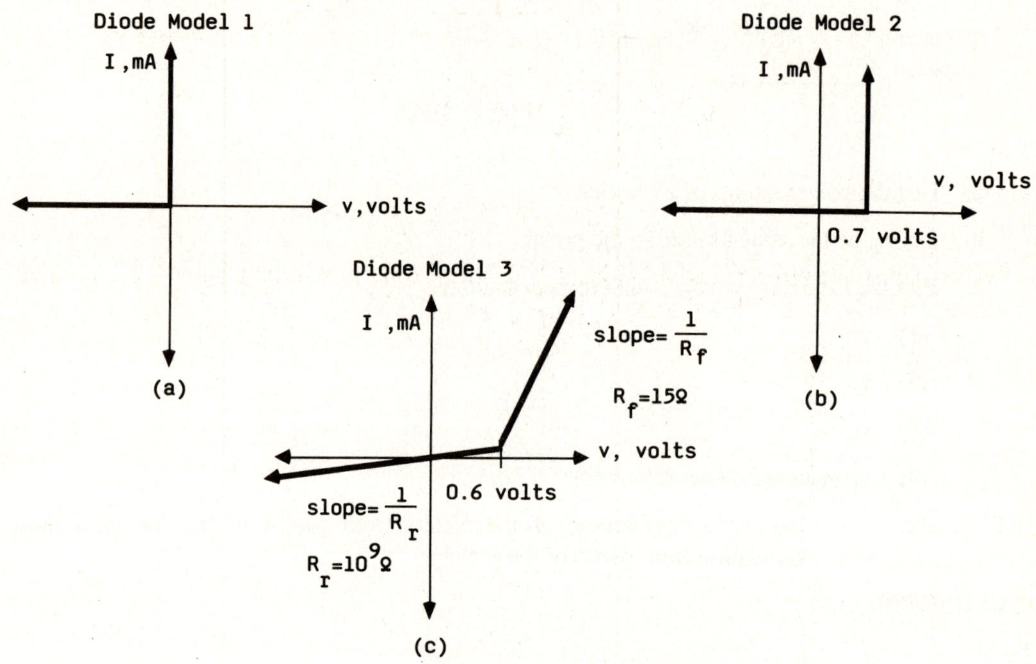

Figure 9L5.1: Diode models

b. Plot diode model 4 described by the following equation. Use semilog paper and plot V_D on the linear axis.

$$I_D = I_S(e^{V_D/\eta V_T} - 1) \tag{9L5.1}$$

$$\text{let } I_S = 10^{-10}\,A$$

$$\eta = 1$$

$$V_T = 26\,mV$$

Assume $\dfrac{v_D}{V_T} \gg 1$ and solve for v_D.

2. Analysis of a diode op amp circuit with three diode models and the diode equation. For the following circuit shown in Figure 9L5.2,

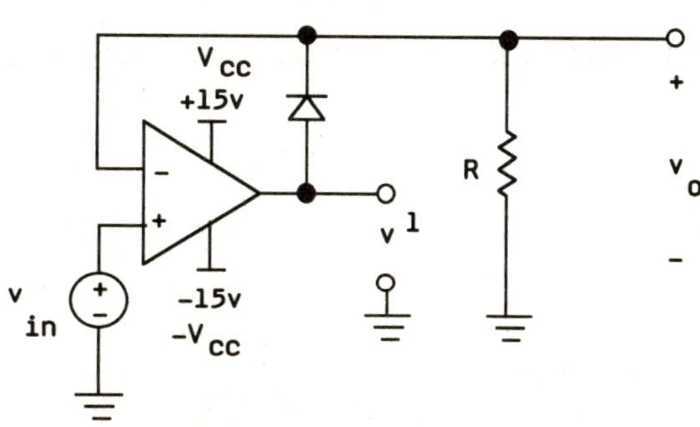

Figure 9L5.2: Precision rectifier

determine the value of v^1 and v_o for all four diode models discussed in Question 1. For diode model 4, assume that $I_S = 10^{-10}$ A and $V_T = 0.026$ V. Let R = 1 kΩ.

   a. $v_{in} = -2.0$ V

   b. $v_{in} = 2.0$ V

3.   Diode Applications. Complete the graphs for each of the following circuits and describe the operation or function of the circuit.

   a.   Circuit 1, Figure 9L5.3. (Use the diode model from Figure 9L5.1(b).)

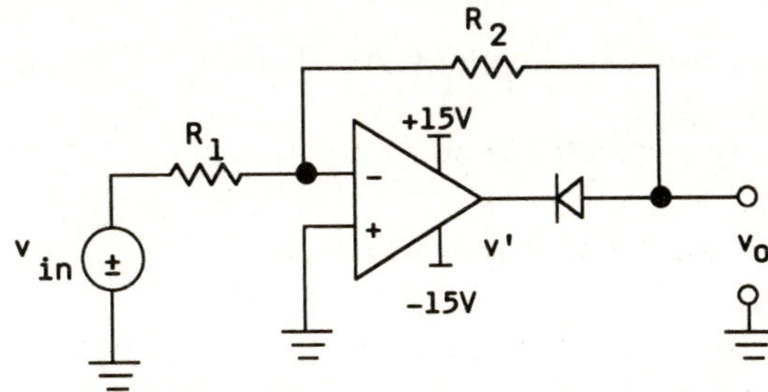

Figure 9L5.3: Asymmetric gain rectifier

On the areas given in Figure 9L5.4, sketch v^1 and v_o, given v_{in}. Assume that $R_2 = 5$ kΩ and $R_1 = 1$ kΩ.

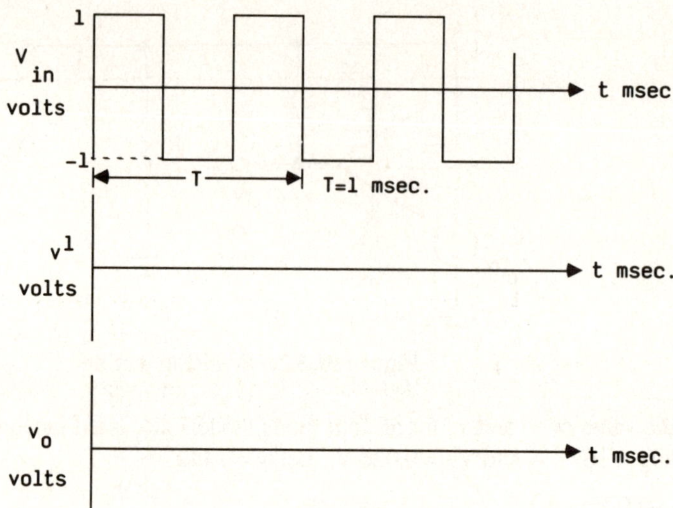

Figure 9L5.4: Plot for circuit in Figure 9L5.3

b.   Circuit 2, Figure 9L5.5. Given the following circuit assume $R = 1\ k\Omega$:

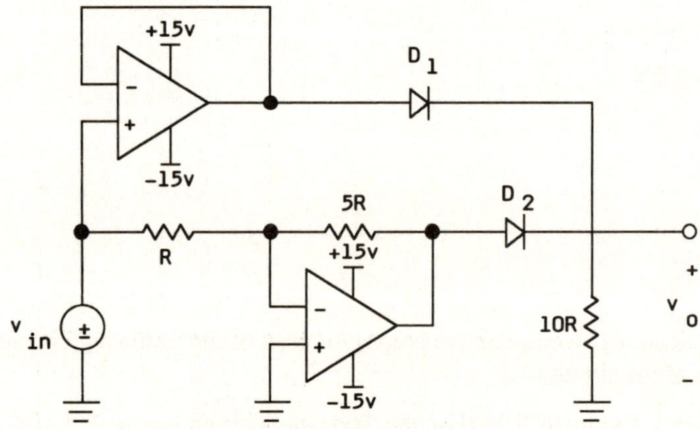

Figure 9L5.5: Full-wave rectifier

Using the diode models of Figures 9L5.1(a) and 9L5.1(b) complete the graphs of v_o on the axes shown in Figure 9L5.6.

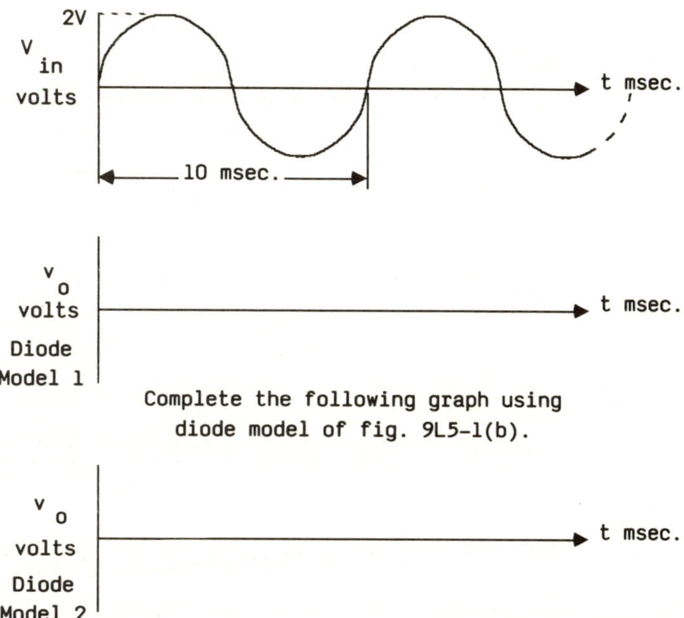

Figure 9L5.6: Plots for circuit in Figure 9L5.5

c.   Circuit 3, Figure 9L5.7.

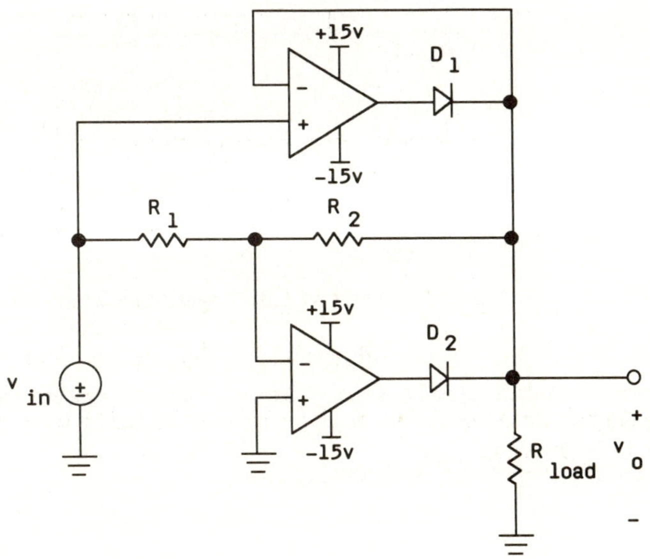

Figure 9L5.7: Improved full-wave rectifier

Assume $R_1 = R_2 = 1\ k\Omega$

Use the diode model of Figure 9L5.1(a) and 9L5.1(b) and complete the graphs of Figure 9L5.8.

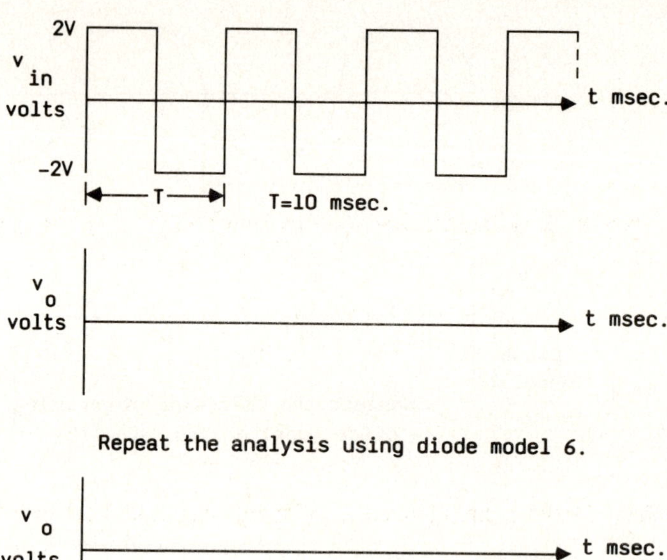

Figure 9L5.8: Plots for circuit in Figure 9L5.7

d. Circuit 4, Figure 9L5.9. Given the following circuit with R = 1 kΩ and R_L = 10 kΩ:

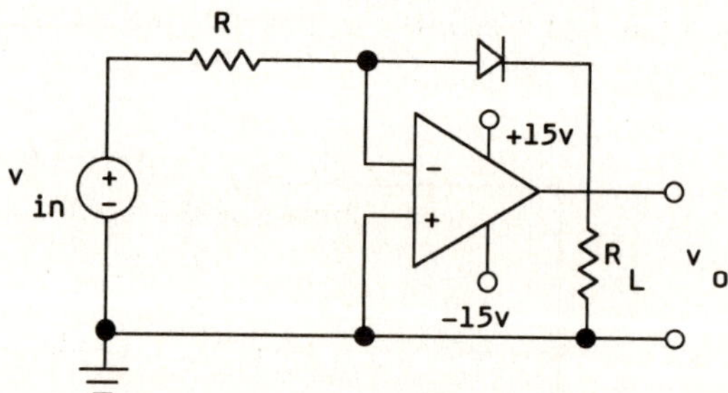

Figure 9L5.9: Logarithmic amplifier

Determine the relationship between v_o and v_{in}. Assume that v_{in} can only have positive values. (R = 1 kΩ)

(1) Use the diode model of Figure 9L5.1(a).

(2) Use the diode model of Figure 9L5.1(b).

(3)   Use the diode model of Figure 9L5.1(c).

(4)   Use the diode model described by Equation (9L5.1). (Assume $I_S = 10^{-10}$ and $V_T = 26$ mV.)

4.   Design Problem. Design a circuit with the following relationship between V_{in} and V_{out}:

$$V_{out} = K_1 \ln \left(\frac{|\, V_{in}\, |}{K_2} \right)$$

$$|\, V_{in}\, | < 10 \text{ volts}$$

K_1 and K_2 are functions of the circuit components.

Laboratory Procedure:

1.   Build the circuit shown in Figure 9L5.10.

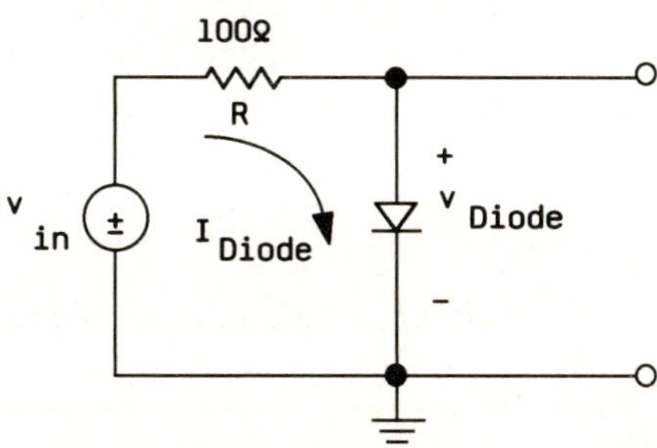

Figure 9L5.10: Characterizing the diode

Vary V_{in} from -10 V to +10 V in 1-V increments. Include V_{in} equal to 10 mV, 20 mV, ...,100 mV,..., Record and plot I_{DIODE} versus V_{DIODE} on linear graph paper. Take into account the nonzero resistance of the ammeter and the noninfinite resistance of a voltmeter.

Plot the data again; this time on semilog graph paper with V_{DIODE} on the linear axis. From your graph, determine the value of I_S.

Save this diode for Circuit 4 in Part 3.

2.   Build the circuit from Figure 9L5.2 of the prelab with R=10 kΩ. Sketch v_o for the following input waveforms:

  a. $v_{in} = 1.5 \sin 2\pi ft$; where $f = 500$ Hz. Sketch v_o on the axes given in Figure 9L5.11.

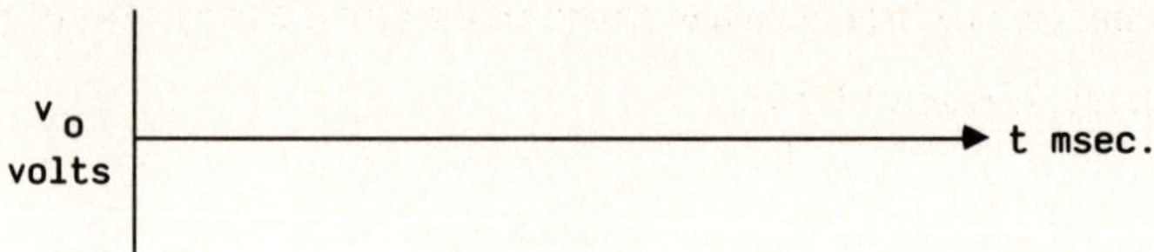

Figure 9L5.11

b. v_{in} is a square wave with amplitude 2 V and period T = 10 ms. Sketch v_o on the axes given in Figure 9L5.12.

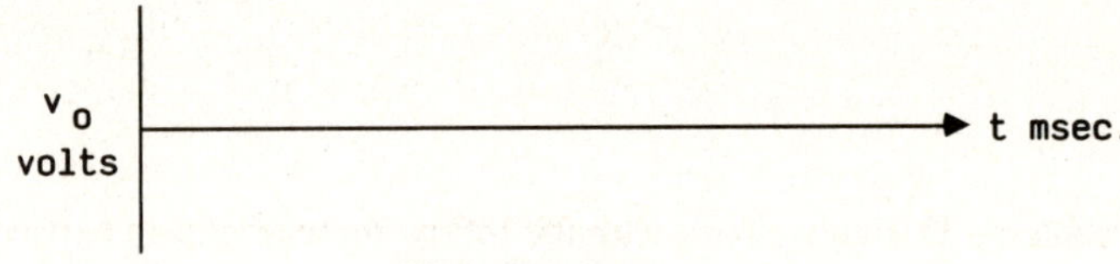

Figure 9L5.12

c. v_{in} is a triangular wave with amplitude 2 V and period T = 10 ms. Sketch v_o on the axes given in Figure 9L5.13.

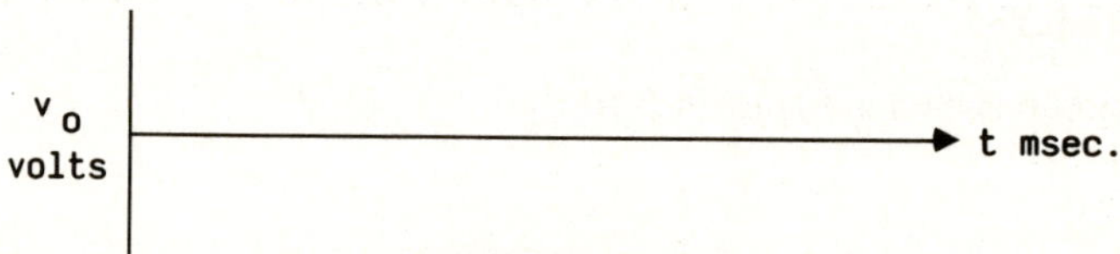

Figure 9L5.13

d. Reverse the polarity of D_1 and repeat part a, sketching v_o on the axes given in Figure 9L5.14.

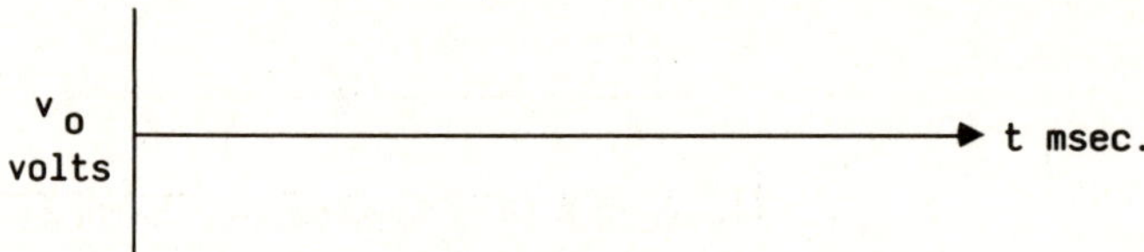

Figure 9L5.14

Connect a 0.1 μf capacitor (0.1×10^{-6} farad) across R, on the axes given in Figure 9L5.15, and sketch v_o. What happens as the frequency is decreased?

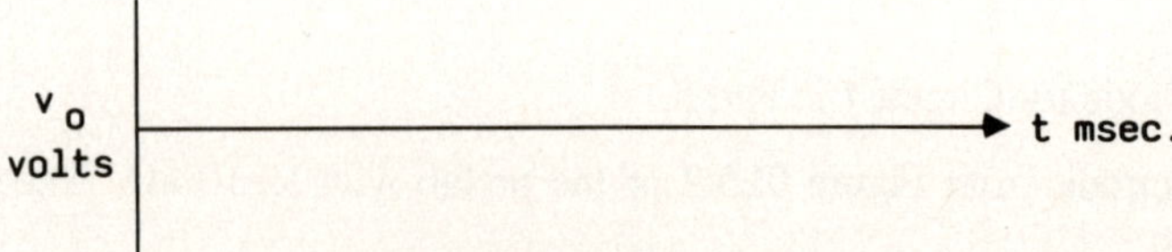

Figure 9L5.15

3. Build the four circuits described below. They are all from section 3 of the prelab. Before dismantling each circuit, collect enough data to complete the plots and answer any questions.

    a. Circuit 1, Figure 9L5.16.

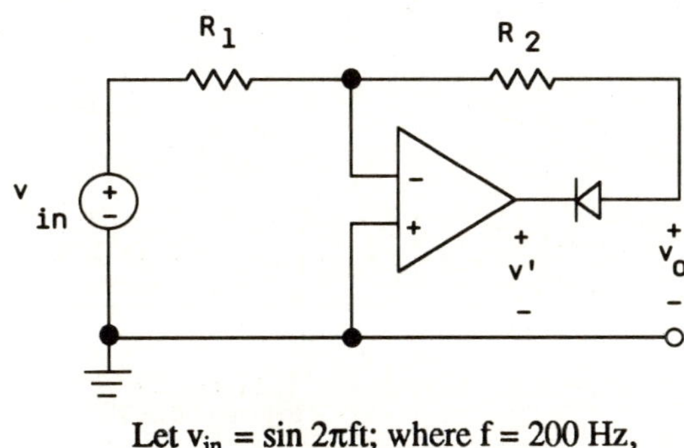

Let $v_{in} = \sin 2\pi ft$; where $f = 200$ Hz,
$R_1 = 1$ kΩ and $R_2 = 5$ kΩ

Figure 9L5.16

Observe and sketch the v' voltage and v_o waveforms on the axes given in Figure 9L5.17.

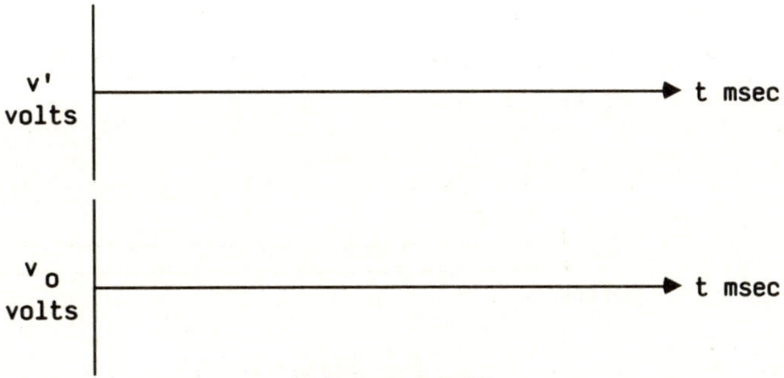

Figure 9L5.17

    b. Circuit 2, Figure 9L5.18.

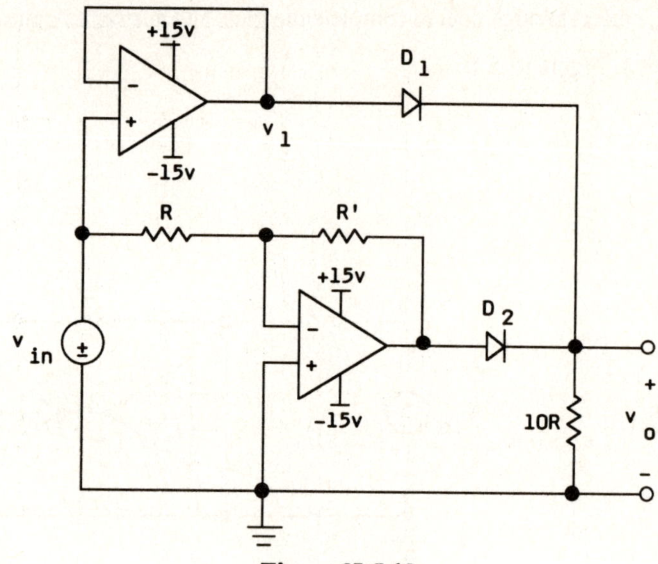

Figure 9L5.18

Let $v_{in} = 4 \sin 2\pi ft$; where f = 200 Hz; R = 1 kΩ and R$'$ = 5R

Sketch the v_1, v_2, and v_o waveforms relative to v_{in} on the axes given in Figure 9L5.19.

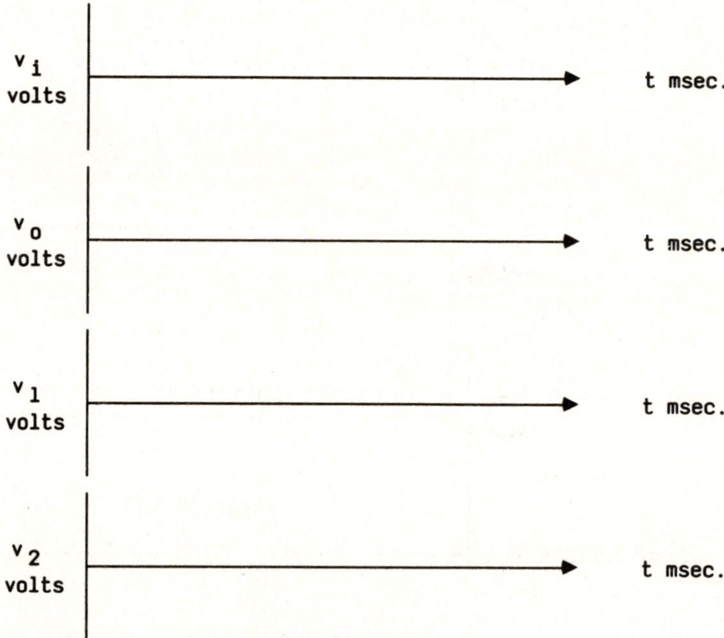

Figure 9L5.19

Now let R$'$=2 kΩ, and repeat the graphs. If the polarity of the diodes is reversed, what happens to v_o? If v_{in} has an amplitude of 1 V what is v_o?

c.   Circuit 3, Figure 9L5.20.

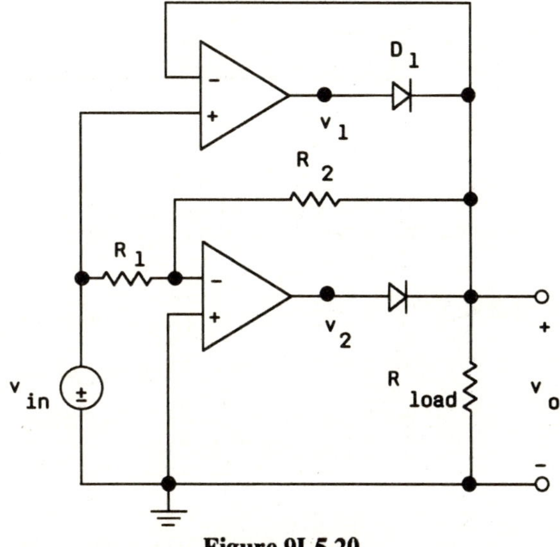

Figure 9L5.20

Let $v_{in} = 5 \sin 2\pi ft$; where $f = 200$ Hz; $R_1 = R_2 = 1$ kΩ, $R_{Load} = 10$ kΩ

Sketch v_1, v_2, and v_o relative to v_{in} on the axes given in Figure 9L5.21.

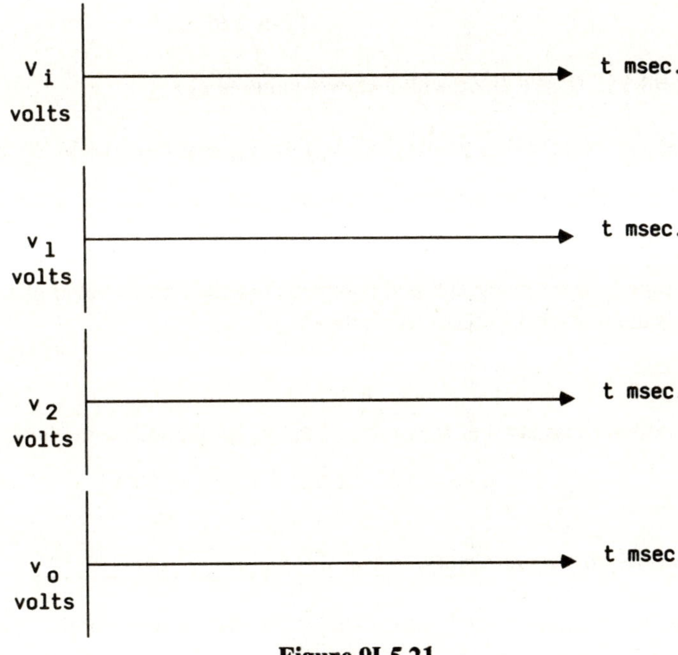

Figure 9L5.21

If the amplitude of v_{in} is changed to 1 V, sketch v_o on the axes given in Figure 9L5.22.

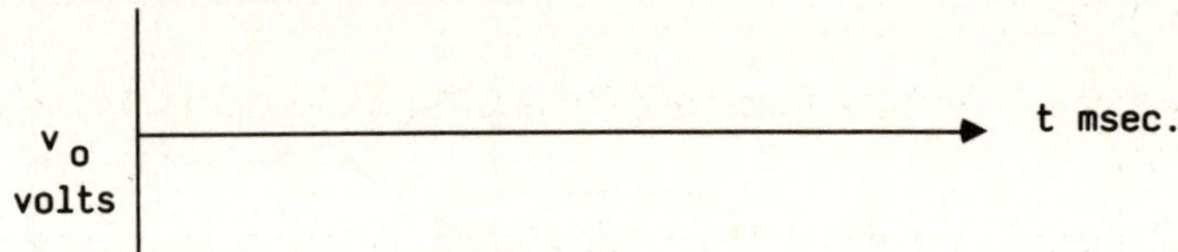

Figure 9L5.22

How does v_o with the amplitude of v_{in} equal to SV differ from v_o with the amplitude of v_{in} equal to 1 V?

    d.   Circuit 4, Figure 9L5.23.

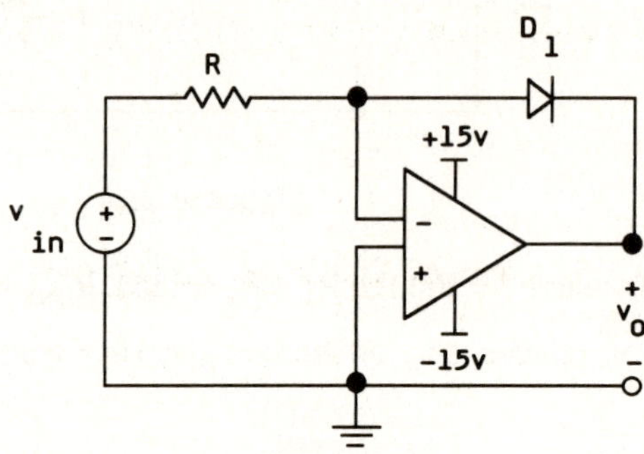

Figure 9L5.23

Let $R = 1\ k\Omega$ (Use a diode with a known value of I_S.)

Record and plot the data points (V_{in}, V_o) for V_{in} equal to 10 mV, 20 mV,..., 100 mV, 200 mV, 1 V, 2 V,...,10 V. Plot the data on semilog paper, with V_o the linear axis. Also, plot $I_{in} = \dfrac{V_{in}}{R}$ versus V_o on this graph.

Determine I_S from this graph and compare it to the known value of I_S. What do you conclude? What is the relationship between V_o and V_{in}?

4.   Design Problem

Verify your design from part 4 of the prelab. Find v_o for the following values of v_{in}:

$$v_{in} = \{\pm 1\ V,\ \pm 2\ V,\ \pm 3\ V,\ \pm 4\ V,\ \pm 5\ V\}$$

Indicate K_1 and K_2 from your theory.

Plot $|v_{in}|$ and v_o on semilog graph paper, with v_{in} on the log axis. From your graph, find K_1 and K_2.

DISCUSSION: Discuss the accuracy of the diode models as compared with the actual experimental results.

CONCLUSIONS:

REFERENCES

[1]   D. H. Navon, *Semiconductor Microdevices and Materials* (New York: Holt, Rinehart and Winston, 1986), pp. 117-123.

[2]   B. G. Streetman, *Solid State Electronic Devices,* 2d ed. (Englewood Cliffs, NJ: Prentice-Hall, 1980), pp. 148-161.

[3]   R. S. Muller and T. I. Kamins, *Device Electronics for Integrated Circuits* (New York: Wiley, 1977), pp. 165-171.

[4]   J. Millman and Arvin Grabel, *Microelectronics,* 2d ed. (New York: McGraw-Hill, 1987), p. 49.

[5]   C. J. Savant, Jr., M. S. Roden, and G. L. Carpenter, *Electronic Circuit Design: An Engineering Approach* (Menlo Park: Benjamin/Cummings, 1987), p. 16.

[6]   D. A. Hodges and H. G. Jackson, *Analysis and Design of Digital Integrated Circuits* (New York: McGraw-Hill, 1983), p. 1160.

[7]   A. S. Sedra and K. C. Smith, *Microelectronic Circuits,* 2d ed. (New York: Holt, Rinehart and Winston, 1987), p. 158.

[8]   Sedra and Smith, p. 160.

[9]   Sedra and Smith, p. 164.

[10] W. H. Hayt, Jr., and G. W. Neudeck, *Electronic Circuit Analysis and Design,* 2d ed. (Boston: Houghton-Mifflin, 1984), p. 9.

[11] Hayt and Neudeck, p. 27.

[12] Sedra and Smith, pp. 215-27.

[13] Savant, Roden and Carpenter, pp. 27-31.

[14] Sedra and Smith, pp. 208-15.

[15] Sedra and Smith, pp. 223-24.

[16] Savant, Roden, and Carpenter pp. 34-6.

[17] Savant, Roden, and Carpenter, pp. 43-6.

[18] S. D. Senturia and B. D. Wedlock, *Electronic Circuits and Applications* (New York: Wiley, 1975), pp. 214-17.

[19] Savant, Roden and Carpenter, pp. 37-42.

[20] Millman and Grabel, pp. 41-70.

[21] Navon, pp. 103-137.

[22] P. R. Gray and R. G. Meyer, *Analysis and Design of Analog Integrated Circuits,* 2d ed. (New York: Wiley, 1984), pp. 1-9.

[23] V. H. Grinch and H. G. Jackson, *Introduction to Integrated Circuits* (New York: McGraw-Hill, 1975), pp. 528-37.

[24] Millman and Grabel, pp. 763-66.

[25] Sedra and Smith, pp. 202-10.

Chapter 10

BIPOLAR AND FIELD EFFECT TRANSISTORS

10.1 Introduction

In this chapter we shall look at the operation of bipolar and field-effect transistors, the form of their current-voltage characteristics, circuit symbols and conventions, and the effects of temperature on important parameters. We shall also discuss dc and small-signal equivalent circuits for bipolar and field-effect transistors.[1],[2],[3]

Equivalent models are used to reduce complicated nonlinear circuits to circuits containing current and voltage sources, resistors, and capacitors. The reduced circuits can then be resolved using standard circuit analysis techniques.

With the exception of some elementary concepts, we shall not become involved in the solid-state theory on which transistor operation is based.

The basic operation of the transistor can be represented by the equivalent circuit shown in Figure 10.1(a).

The diode D_E has a current i_E flowing through it. Here, the collector current is controlled by the voltage v_{BE}. This model is a nonlinear voltage-controlled current source, but it can be converted to a current-controlled current source, as shown in Figure 10.1(b).

Figures 10.1(c) and (d) show two additional equivalent circuit models of the npn BJT. These circuits represent the common-emitter configuration while Figures 10.1(a) and (b) represent the common-base configuration.

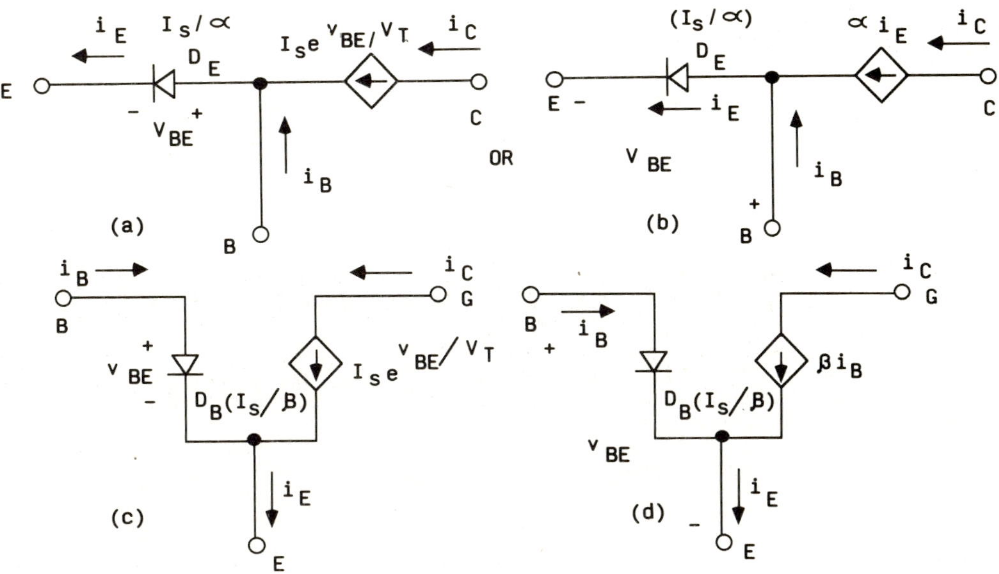

Figure 10.1: npn BJT large-signal circuit models in the active region

Note that we have ignored the small reverse currents I_{EBO} and I_{CBO}. These currents represent the reverse current flowing from emitter to base with the collector open circuited and the reverse current flowing from collector to base with the emitter open, respectively. The value of I_{CBO} is usually in the nanoampere range, while I_{EBO} is even smaller. To include I_{CBO} requires that we write

$$i_C = \alpha i_E + I_{CBO}(e^{v_{BE}/V_T} - 1) \tag{10.1}$$

I_{CBO} depends strongly on temperature and is found experimentally to double for every 10°C rise in temperature.

The pnp transistor operates in a similar way to that of the npn, with some exceptions. Most of the current is conducted by holes injected from the emitter into the base as a result of the forward-bias voltage v_{EB}. The large-signal operation of the pnp transistor can be modeled by any one of four equivalent circuit models that are similar to those representing the npn transistor.

Circuit Symbols and Polarities

The circuit symbols for the npn and pnp BJTs are shown in Figure 10.2. In both symbols, the emitter is distinguished by an arrowhead and the polarity of the transistor is indicated by the direction of the arrowhead on the emitter. The arrow points in the direction of conventional current flow in the emitter.

Figure 10.3 shows the voltage polarities and current flow in transistors biased in the active mode.

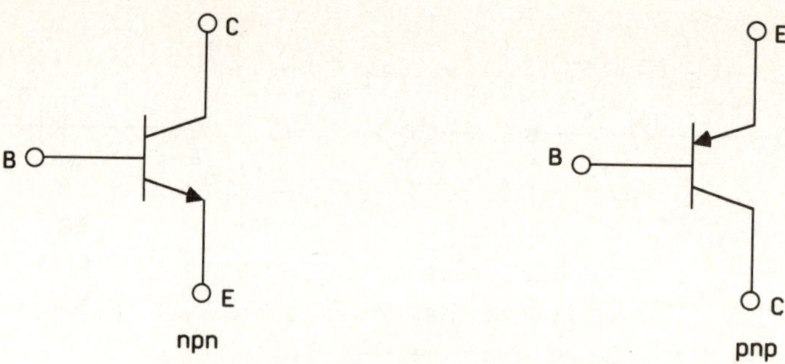

Figure 10.2: Circuit symbols for BJTs

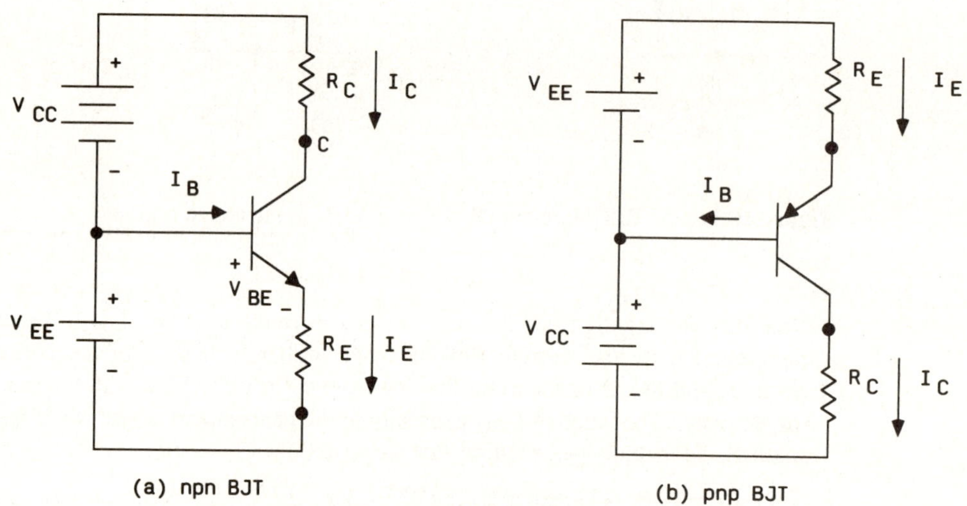

(a) npn BJT

(b) pnp BJT

Figure 10.3: Voltage polarities and current
flow for BJTs operating in the active region

It should be clear from the figure that the npn transistor will operate in the active region
if the emitter base junction is forward biased and the collector is higher in potential
than the base. In a similar manner, the pnp transistor will operate in the active mode if
the potential of the collector is lower than or equal to that of the base. If the collector-
base junction should become forward biased, the transistor enters the mode of
operation called the **saturation region**.

Saturation

The saturated region arises when the forward biased emitter-base junction causes
sufficient collector current to flow in the external circuit such that the collector-base
junction becomes forward biased. The value of this forward-bias collector-base voltage
is about 0.4 to 0.5 V. Any further increase in the base current will result in a very
small increase in the collector current and a corresponding small decrease in the
collector voltage: in saturation, the value of $\dfrac{\Delta i_C}{\Delta i_B}$ is very small, and any increase in
current that is forced into the base terminal flows through the emitter terminal. This
means that the definition of β for a saturated transistor is not the same as that for a
transistor in the active region. In order to be sure that the transistor is in saturation, we
force a base current of at least

$$i_{Bsat} = \frac{i_{C_{SAT}}}{\beta} \qquad (10.2)$$

or

$$\beta_{forced} = \frac{i_{C_{SAT}}}{I_B} \qquad (10.3)$$

where I_B is greater than $I_{B_{SAT}}$ by a factor of 2 to 10 and β_{forced} is a quantity that can be selected by the designer. Figure 10.4 represents the model for the saturated BJT.

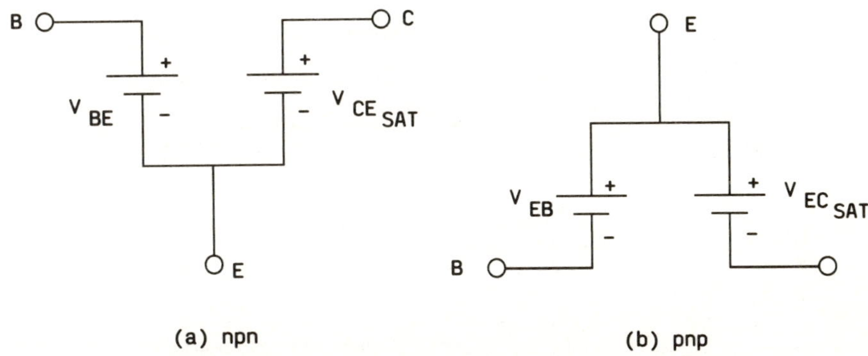

Figure 10.4: Equivalent circuit for the saturated BJT

Cutoff

A third region of operation for the BJT arises when both junctions are reverse biased. In this case, all terminal currents are extremely small and the transistor is said to be off.

Inverted Mode

A fourth mode of operation is the inverted mode, and in this region the collector and emitter are interchanged. The result is that the transistor gain α_R (the inverted mode common base current gain) is much smaller than α_F (the active-mode common-base current gain), since the injection efficiency of the collector is normally much lower than that of the heavily doped emitter.

10.1.1 Junction Field-Effect Transistor (JFET)[1],[2],[3]

A simplified sketch of the n-channel JFET is shown in Figure 10.5.

The channel is formed from lightly doped or low-conductivity n-type silicon with ohmic contacts at either end of the channel. The two gate regions are made of heavily doped or high-conductivity p^+ silicon and are electrically connected.

Regions of Operation

There are two distinct regions of operation for the JFET transistor: the triode region and the pinch-off region. These two regions, labeled in Figure 10.6, are separated by a parabolic boundary marked by a dashed curve.

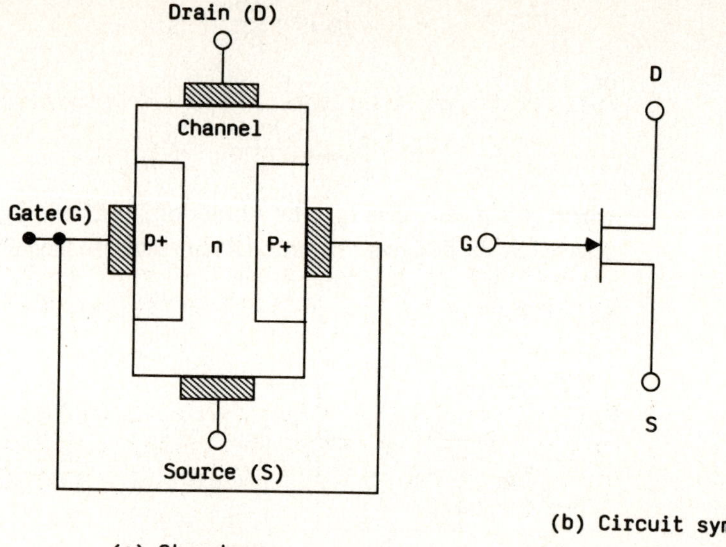

Figure 10.5: Basic structure and circuit symbol of n-channel JFET

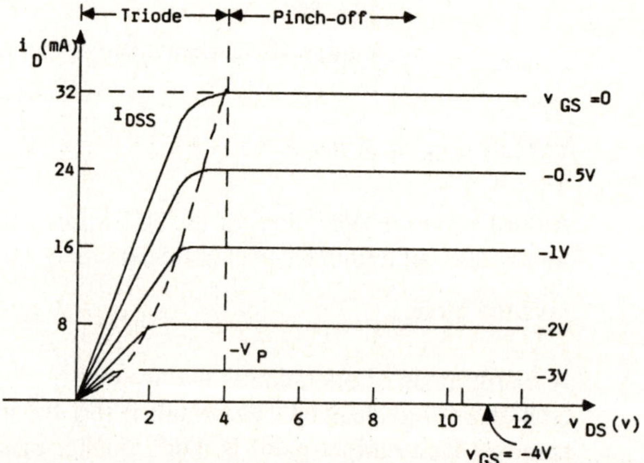

Figure 10.6: i_D–v_{DS} characteristic for an n channel JFET (Reprinted by permission from *Microelectronic Circuits,* Second Edition, by A. S. Sedra and K. C. Smith. Copyright © 1987 by Holt, Rinehart and Winston.)

Triode Region

In the triode region, the JFET acts as a variable resistance (r_{DS}) that is controlled by the gate-to-source voltage v_{GS}. The resistance is linear for small values of v_{DS}.

The i_D-v_{DS} relationship in the triode region is given by

$$i_D = I_{DSS} \left[2(1 - \frac{v_{GS}}{V_P}) \frac{v_{DS}}{-V_P} - (\frac{v_{DS}}{V_P})^2 \right] \tag{10.4}$$

where the values of V_p and I_{DSS} are JFET parameters that are given on the manufacturer's data sheet.

For small values of v_{DS}, Equation (10.4) reduces to

$$i_D = \frac{2I_{DSS}}{-V_p}(1 - \frac{V_{GS}}{V_p})\, v_{DS} \tag{10.5}$$

Equation (10.5) can be rearranged to represent the linear resistance r_{DS}

$$r_{DS} = \frac{v_{DS}}{i_D}\bigg|_{v_{DS_{small}}} = \left[\frac{2I_{DSS}}{-V_p}(1 - \frac{V_{GS}}{V_p})\right]^{-1} \tag{10.6}$$

For an n-channel JFET, V_p is a negative voltage, v_{GS} is always negative, and the gate-to-channel junction is reverse biased at all times. The result is that the gate current-- the leakage current of the reverse-biased pn junction-- is almost zero.

Pinch–off Region

Pinch-off is reached when the reverse-bias voltage at the drain end of the channel is equal to the pinch-off voltage. This condition represents the boundary between the triode region and the pinch-off region. On the boundary between triode region and the pinch-off region,

$$v_{DS} - v_{GS} = -V_P \tag{10.7}$$

Substituting this condition into Equation (10.4) results in

$$i_D = I_{DSS}(\frac{v_{DS}}{V_p})^2 \tag{10.8}$$

Equation (10.8) is the equation of a parabola and is represented by the dashed curve in Figure 10.6.

To operate in the triode region, the JFET drain-to-gate voltage should be less than $-V_p$; that is,

$$v_{DG} < -V_p$$

or

$$v_{DS} < v_{GS} - V_P \quad \text{(triode region)} \tag{10.9}$$

To operate in the pinch-off region, the drain-to-gate voltage must satisfy the condition

$$v_{DG} \geq -V_p$$

or

$$v_{DS} \geq v_{GS} - V_p \quad \text{(pinch–off region)} \tag{10.10}$$

The last condition implies that the drain voltage should be higher than the gate voltage by at least $|V_p|$.

From Figure 10.6, we see that in the pinch-off region the JFET operates as a constant-current source, with the value of current controlled by the voltage v_{GS}. Since the i_D-v_{DS} family of curves are parallel to the horizontal axis or v_{DS}, the constant current source has an infinite resistance.

In addition, the impedance seen looking between the gate-to-source control terminals is infinite, and the equation that relates i_D to the control source v_{GS} can be found from Equations (10.4) and (10.10) to be

$$i_D = I_{DSS}(1 - \frac{v_{GS}}{V_p})^2 \tag{10.11}$$

From these concepts, it is possible to derive an equivalent large signal model for the JFET that applies only to the pinch-off condition. Figure 10.7 shows this model.

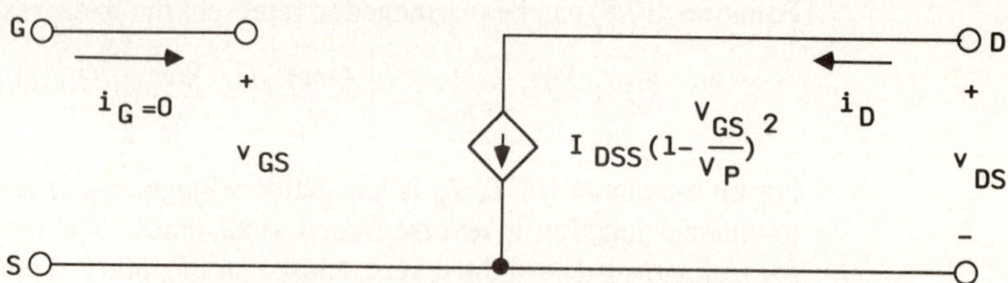

Figure 10.7: Equivalent circuit model for large-signal operation of the n-channel JFET in pinch-off

<u>P–Channel JFET</u>

The p-channel JFET is the exact complement of the n-channel JFET when the n- and p-type regions are interchanged. All voltages and currents are the negative of those for the n-channel case; and the circuit symbol differs from that of the n-channel device only in the direction of the arrowhead on the gate. For the p-channel device, the arrow is pointing outward in the forward direction of the channel (p- type)-to-gate (n-type) junction. The circuit symbol and family of curves are shown in Figure 10.8.

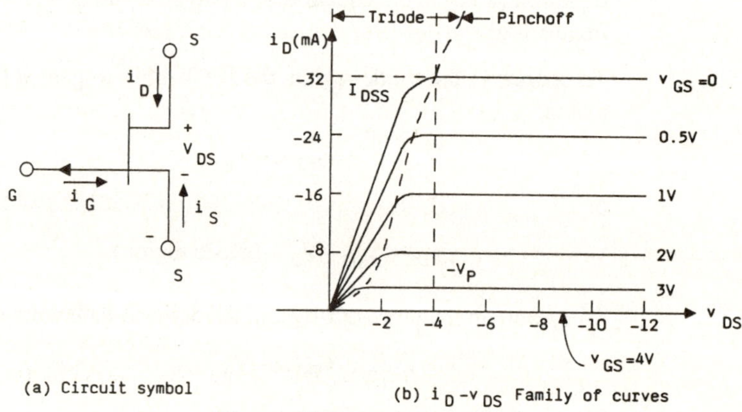

Figure 10.8: P-channel JFET

For the p-channel JFET to operate in pinch-off mode, the drain should be lower than the gate by at least V_p volts; that is,

$$v_{DG} \leq -V_p \tag{10.12}$$

or

$$v_{DS} \leq v_{GS} - V_P \quad \text{(pinch–off region)}$$

To operate in the triode region, the p-channel JFET must satisfy the condition that

$$v_{DG} > -V_p$$

or

$$v_{DS} > v_{GS} - V_p \text{(triode region)} \tag{10.13}$$

The boundary between the two region is given by

$$v_{DG} = -V_p \qquad\qquad (10.14)$$

Equations (10.4) and (10.11), which describe the n-channel JFET, are also valid for the p-channel JFET, except that V_p is positive, v_{GS} is also positive, and v_{DS} is normally negative.

10.1.2 Metal-Oxide-Semiconductor Field-Effect Transistors (MOSFETs)[1],[2],[3]

Unlike the JFET, the MOSFET can be constructed to operate either in the depletion mode, in the enhancement mode, or in both modes (but not simultaneously).

Depletion–Type MOSFET

The physical structure of an n-channel MOSFET of the depletion type is shown in Figure 10.9.

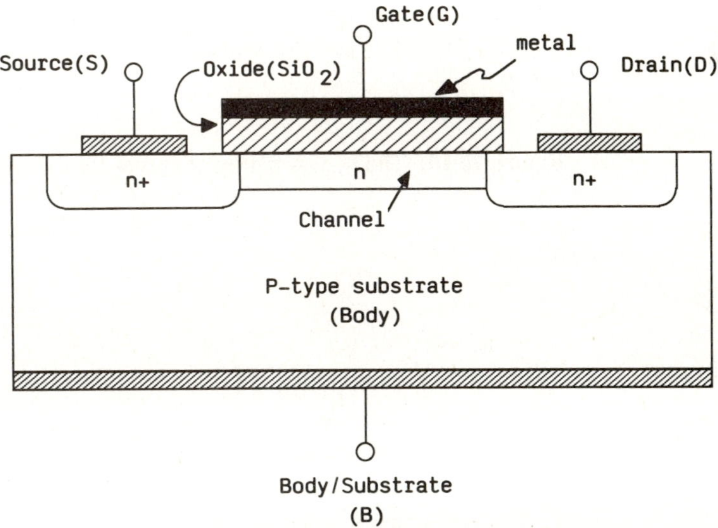

Figure 10.9: n-channel depletion-type MOSFET

The n channel is formed on a p-type silicon substrate, and the two heavily doped n^+ regions are low-resistance connections to the ends of the n channel and the metal contacts of the source (S) and drain (D).

The physical operation of the depletion mode MOSFET is much like that of the JFET, except that the JFET channel is depleted by reverse-biasing the gate-to-channel junction while the MOSFET channel is depleted by an electric field applied between the gate and the channel. The family of curves is shown in Figure 10.10.

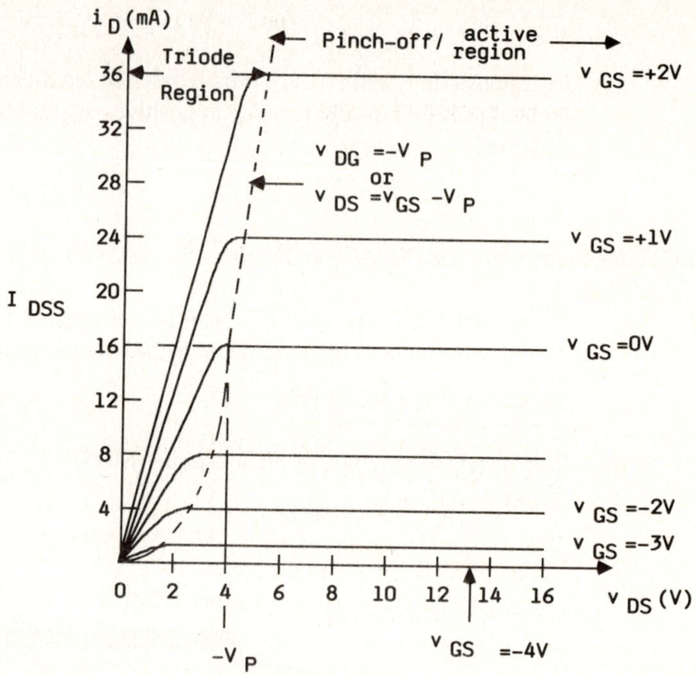

Figure 10.10: The i_D-v_{DS} family of curves for a depletion-type n-channel MOSFET

The circuit symbol for the depletion-type, n-channel MOSFET is shown in Figure 10.11(a).

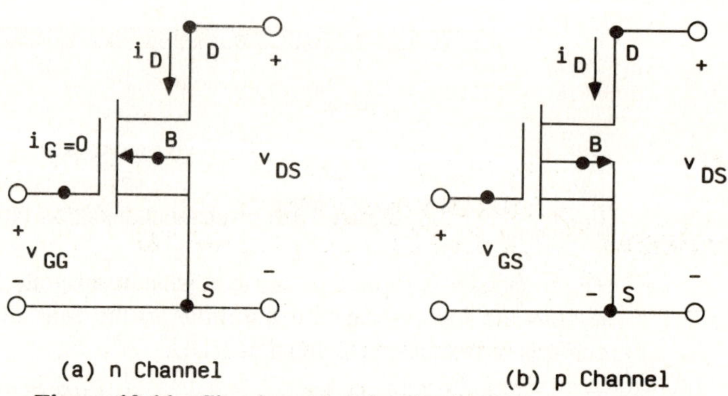

Figure 10.11: Circuit symbol for the depletion-type MOSFET

Comparisons can be made between the n-channel, depletion-type MOSFET and the JFET. In fact, the family of curves for the two devices are identical, except that positive values of v_{GS} are allowed for the MOSFET. A depletion-type MOSFET can therefore be biased to operate in the enhancement mode. The equations describing the characteristic curves of the depletion-type MOSFET are identical to those for the JFET. That is, in the triode region,

$$v_{DG} < -V_p$$

or

$$v_{DS} < v_{GS} - V_p \tag{10.15}$$

and

$$i_D = I_{DSS} \left[2\left(1-\frac{v_{GS}}{V_p}\right)\left(\frac{v_{DS}}{-V_p}\right) - \left(\frac{v_{DS}}{V_p}\right)^2 \right] \qquad (10.16)$$

In the pinch-off region,

$$v_{DG} \geq -V_p$$

or

$$v_{DS} \geq v_{GS} - V_p$$

and

$$i_D = I_{DSS} \left(1 - \frac{v_{GS}}{V_p}\right)^2 \qquad (10.17)$$

The depletion-type p-channel MOSFET operates the same as the n-channel MOSFET, except that all the polarities of the voltages and currents are reversed. Equations (10.16) and (10.17) apply, with the pinch-off voltage V_p taken as positive. Circuit techniques that apply to the JFET also apply to the depletion-type MOSFET. The circuit symbol for the p-channel depletion-type MOSFET is shown in Figure 10.11(b).

The Enhancement–type MOSFET

Figure 10.12(a) shows the physical structure of an enhancement-mode MOSFET, and Figure 10.12(b) shows the circuit symbols. As can be seen,

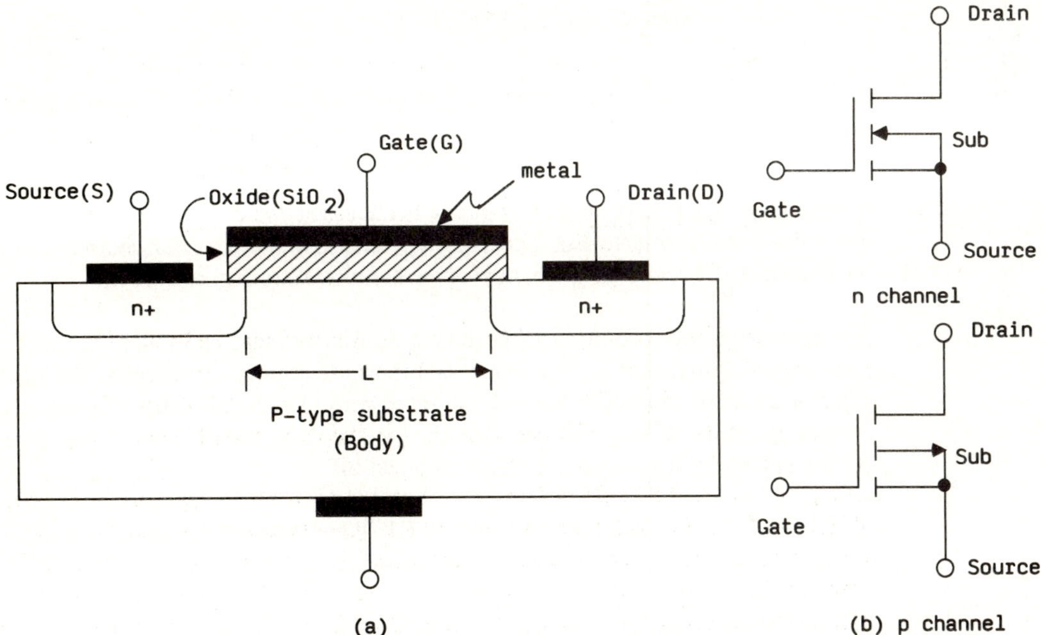

Figure 10.12: Enhancement-type MOSFET and circuit symbols

this structure has no built-in channel. So if the gate is floating, or if $v_{GS} = 0$ as we move from the source to the drain, we encounter two back-to-back series diodes. Since no current can flow under such circumstances, an n channel has to be created. If a positive voltage is applied to the gate relative to the source, the positive voltage will attract electrons from the substrate to the surface regions under the oxide. The voltage at the gate must be large enough so that an n channel can be formed. Therefore, for measurable drain current to flow, the value of v_{GS} must be greater than some threshold voltage V_t. The value of V_t is a function of the particular type of MOSFET used.

Figure 10.13 shows the family of curves for an n-channel, enhancement-type MOSFET.

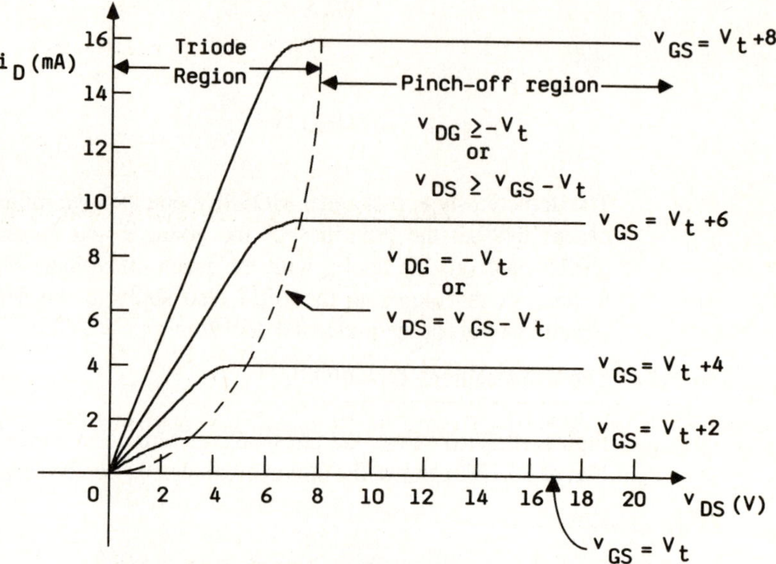

Figure 10.13: Ideal family of curves for an n-channel, enhancement-type MOSFET

When the values of v_{DS} are small (< 500 mV) and the values of v_{GS} are increased above V_t, the induced channel becomes deeper and the channel resistance decreases. These voltages create the triode region of the enhancement type MOSFET.

If the voltage v_{GS} is kept constant, but v_{DS} is increased, the channel depth at the source end will remain constant. But increasing v_{DS} means that v_{GD} has decreased, and thus the channel will become more shallow at the drain end (see Figure 10.14).

As the voltage v_{DS} continues to be increased, the channel resistance will increase, resulting in a nonlinear i_D-v_{DS} curve in the triode region. Eventually, the channel depth at the drain end of the channel will reach zero. Pinch-off occurs when $v_{GD} \leq V_t$. Further increases in v_{DS} will not change the shape of the channel, so i_D remains constant at the value reached at the start of pinch-off.

As in the JFET and depletion-mode MOSFET, there are two regions of operation for the enhancement-mode MOSFET, viz., the triode region and the pinch-off region.

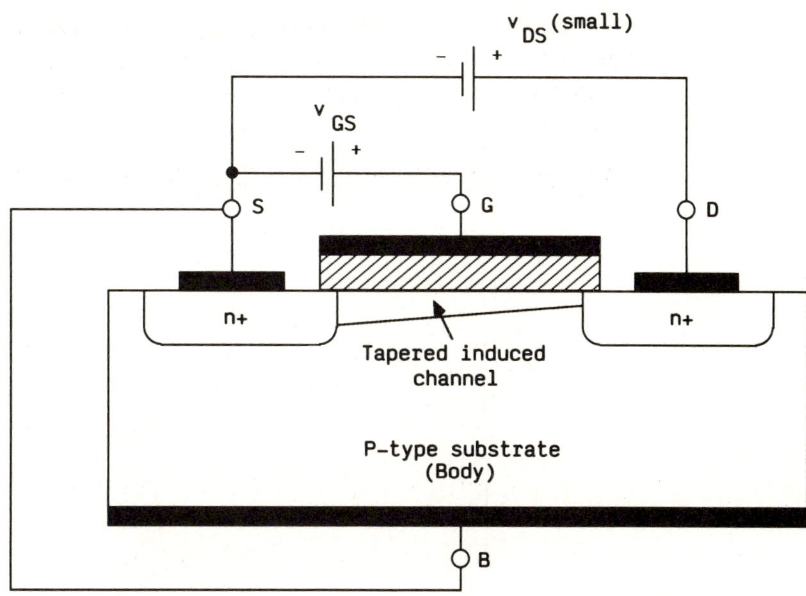

Figure 10.14: Enhancement-mode n-channel MOSFET
with $v_{GS}\,|_{\text{const}}$ but v_{DS} increased

Operation in the triode region can be expressed mathematically as

$$v_{GS} \geq V_t \quad \text{(onset of channel)} \tag{10.18}$$

and

$$v_{GD} \geq V_t \quad \text{(continuous channel)}$$

or

$$v_{DS} < v_{GS} - V_t$$

The i_D–v_{DS} characteristic curves in the triode region are described by

$$i_D = K\left[2(v_{GS} - V_t)v_{DS} - v_{DS}^2\right] \tag{10.19}$$

where

$$K = \frac{1}{2}\mu_n C_{ox}\left(\frac{W}{L}\right),\ A/V^2 \tag{10.20}$$

$$
\begin{aligned}
\mu_n &= \text{electron mobility} \\
C_{ox} &= \text{oxide capacitance between the gate and} \\
&\quad\ \text{the channel, in capacitance per unit area} \\
L &= \text{channel length} \\
W &= \text{channel width}
\end{aligned}
$$

Operation in the pinch-off region/saturation region is described by the conditions

$$
\begin{aligned}
& v_{GS} \geq V_t \\
\text{and} \quad & v_{GD} \leq V_t \text{ or } v_{DS} \geq v_{GS} - V_t
\end{aligned} \tag{10.21}
$$

The boundary between the triode region and the pinch-off region is given by

$$v_{DS} = v_{GS} - V_t \tag{10.22}$$

Substituting Equation (10.22) into Equation (10.19) gives the i_D–v_{DC} relationship at pinch-off, viz.,

$$i_D = K(v_{GS} - V_t)^2 \qquad (10.23)$$

Equation (10.23) indicates that in pinch-off the enhancement MOSFET models a voltage-controlled current source. A plot of this equation is shown in Figure 10.15.

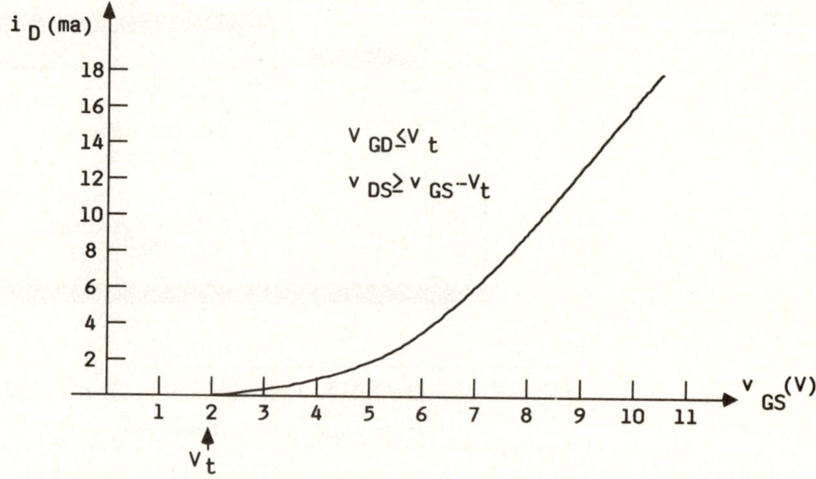

Figure 10.15: i_D-v_{GS} characteristic for an enhancement-mode MOSFET

From the figure, it is clear that for the enhancement-mode transistor to conduct, v_{GS} must exceed the threshold voltage V_t. However, the depletion-mode transistor conducts the current I_{DSS} at $v_{GS}=0$.

10.2 Temperature Effects[1],[2],[3]

10.2.1 Bipolar Transistors

For bipolar silicon transistors, the voltage across the emitter-base junction decreases by about 2 mV for each rise of 1°C in temperature, provided that the current through the junction is constant. The relationship between the collector current and v_{BE} is given by Equation (10.1).

The value of I_S for silicon is in the range of 10^{-12} to 10^{-15} A, is a function of temperature, and doubles for every 5°C rise in temperature.

The reverse current across the collector-base junction, I_{CBO}, is the reverse current flowing from the collector to the base with the emitter open circuited. I_{CBO} depends strongly on temperature, approximately doubling for every 10°C rise.

The common-emitter current gain $\beta_{dc}(h_{FE})$ or β_{ac} (h_{fe}) is a function of the collector current i_C and temperature. These dependencies are shown in Figure 10.16.

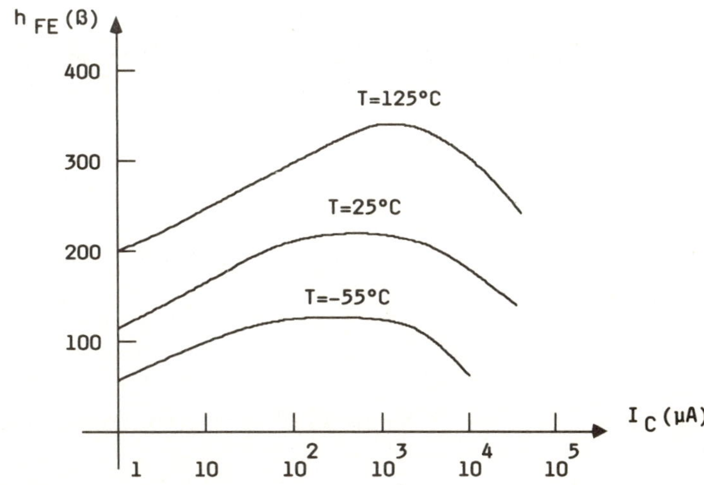

Figure 10.16: Typical BJT dependence of β on i_C and temperature (Reprinted by permission from *Microelectronic Circuits*, Second Edition, by A. S. Sedra and K. C. Smith. Copyright © 1987 by Holt, Rinehart and Winston.)

10.2.2 Junction Field-Effect Transistors

As mentioned, the gate leakage current increases with temperature, roughly doubling for every 10°C rise in temperature. The conductivity of the channel and the turn-on threshold voltage of the gate-channel junction are also functions of temperature. However, with proper circuit design, these factors can be minimized.

10.2.3 Depletion and Enhancement MOS Transistors

For the enhancement mode MOSFET, the threshold voltage V_t and the constant K from Equation (10.23) are functions of temperature. The magnitude of V_t decreases by about 2 mV for every 1°C rise in temperature. The decrease in the magnitude of the threshold voltage gives rise to a corresponding increase in drain current as the temperature is increased, but the dominant effect is a decrease in K with temperature which results in a decrease in drain current. Temperature effects for the depletion mode MOSFET are similar to those of the JFET.

10.3 Output Resistance/Early Voltage[1],[2],[3]

10.3.1 Bipolar Transistors

Practical BJTs show some dependence of the collector current on the collector-to-base voltage, with the result that the i_C-v_{CB} family of curves are not horizontal straight lines (see Figure 10.17). When extrapolated, the characteristic lines meet at a point $v_{CE} = -V_A$ on the negative v_{CE} axis. The voltage V_A is positive and has typical values of 50-125V. It is called the **Early** voltage, after the scientist who first studied it.

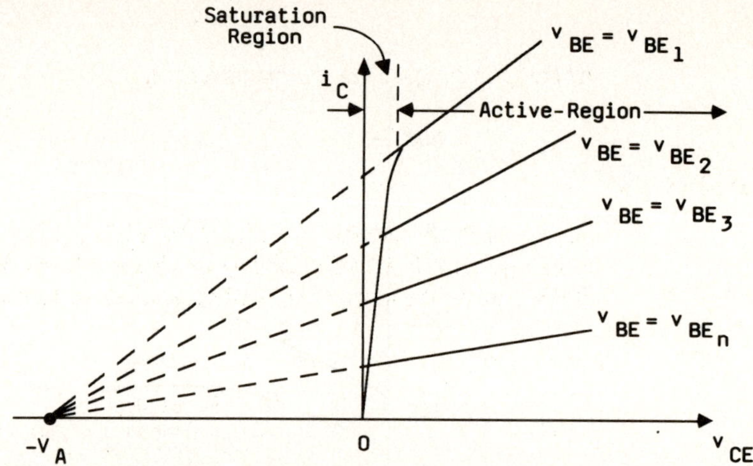

Figure 10.17: The i_C-v_{CE} characteristics of a real BJT

The non-infinite value of the output resistance is

$$r_o \approx \frac{V_A}{I_C} \tag{10.24}$$

The collector current I_C is the current corresponding to the value of v_{BE}. The significant factor in circuit designs is the resistance r_o.

10.3.2 Junction Field-Effect Transistors

The i_D-v_{DS} family of curves of real JFETS also show the finite, nonzero slope in pinch-off (see Figure 10.18).

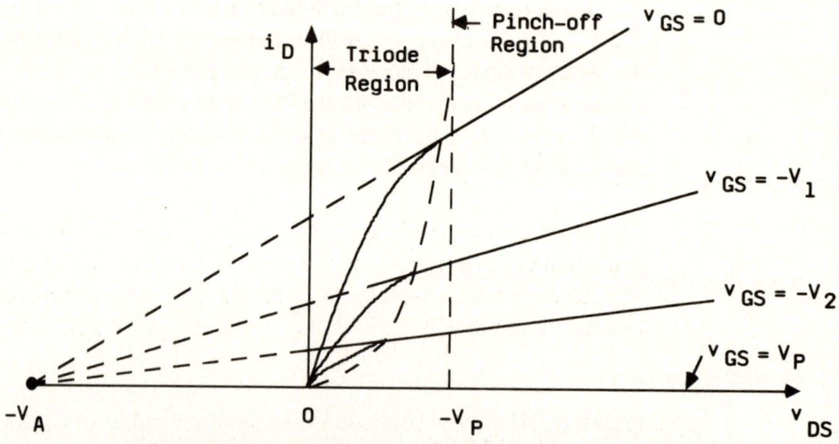

Figure 10.18: The i_D=v_{DS} characteristic of a real JFET

The characteristic curves, in pinch-off, can be extrapolated to a point along the v_{DS} axis. This point, as in the case for the BJT, is labeled $-V_A$, where V_A is a positive voltage. The resistance is actually the inverse of the slope of the i_D-v_{DS} characteristic lines and can be written as

$$r_o \approx \frac{V_A}{I_D} \bigg|_{V_A \gg V_P} \tag{10.25}$$

where I_D is the drain current for a fixed v_{GS}.

10.3.3 MOSFETS

As with JFETS and BJTs, practical MOSFETS have finite output resistance. This finite resistance is accounted for by considering the channel length modulation. Figure 10.19 displays the i_D-v_{DS} family of curves for a real enhancement-mode MOSFET.

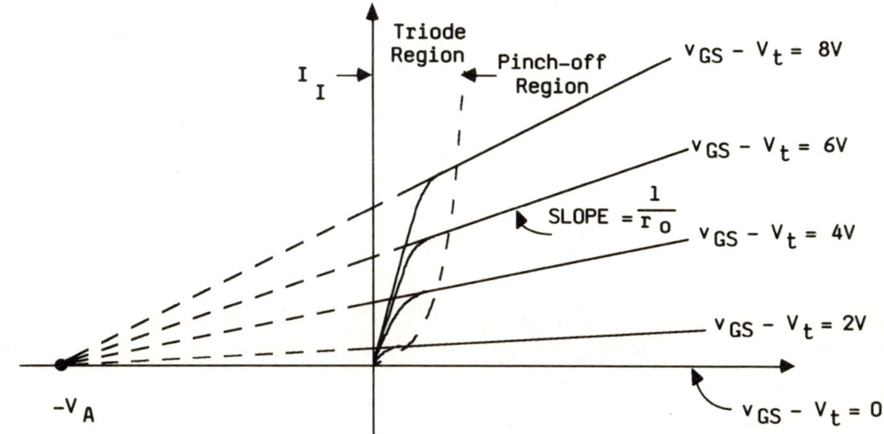

Figure 10.19: The i_D-v_{DS} family of curves for a real enhancement-mode MOSFET

Note that in pinch-off the i_D-v_{DS} straight lines project to a point $v_{DS} = -V_A$, where V_A is a positive voltage. The output resistance r_o can be expressed as

$$r_o = [K(V_{GS} - V_t)^2/V_A]^{-1} \tag{10.26}$$

which can in turn be approximated by

$$r_o = \frac{V_A}{I_D} \bigg|_{v_{GS} = \text{constant}} \tag{10.27}$$

10.4 Small Signal Models[1],[2],[3]

The small signal models to be presented in this section all have in common the fact that each element of the equivalent circuit is based on a physically measurable parameter such as collector current, transconductance, or gate-to-source voltage. Therefore, each element of the circuit model can be measured in the laboratory.

10.4.1 Junction Field Effect Transistor

The small-signal JFET model shown in Figure 10.20 has three features that are directly based on the theory of its operation:

1. Since the drain current is relatively independent of the drain-source voltage in the cut-off region, its effect can be modeled by a current generator. It can also be observed (see Figure 10.6) that i_D is a strong function of v_{GS}. Therefore, the

model contains a voltage-controlled current generator between the drain and source.

2. The nonzero slope of the curves in cut-off can be modeled by a resistor r_o, in parallel with the current generator.

3. The reverse-bias junction between the gate and the source yields a very high input resistance, so the input is modeled as an open circuit.

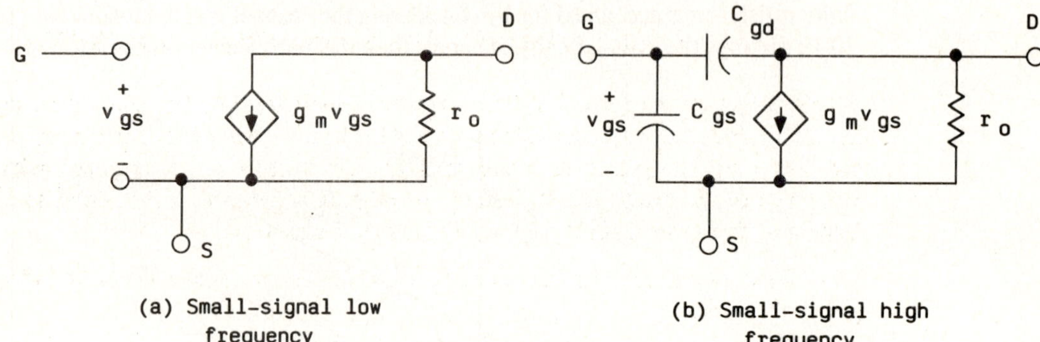

(a) Small-signal low
frequency

(b) Small-signal high
frequency

Figure 10.20: Small-signal models for the JFET

The relationship between the output current and the input voltage is given as a transconductance g_m. From Equation (10.11),

$$g_m \equiv \left. \frac{\partial i_D}{\partial v_{GS}} \right|_{Q\ point} = \left. \frac{\partial}{\partial v_{GS}} I_{DSS}(1 - \frac{v_{GS}}{V_P})^2 \right|_{Q\ point}$$

$$= g_{mo}(1 - \frac{v_{GS}}{V_P}) \tag{10.28}$$

where
$$g_{mo} = \frac{2I_{DSS}}{|V_P|}$$

The quantity g_{mo} is the maximum possible value of the transconductance, and with $|V_P|$ present in the denominator, Equation (10.28) applies to both n- and p-channel devices. The small-signal low-frequency models, shown in Figure 10.20(a), fails to model circuit conditions adequately at high frequencies. Since the gate-to-channel junction is reverse biased, a depletion capacitance exists between gate and source and gate and drain. Including these two capacitances results in the model of Figure 10.20(b). Both n-channel and p-channel JFETS can be described by the equivalent circuits of this model.

10.4.2 *MOS Depletion and Enhancement Mode*

Figure 10.21 shows the small signal equivalent circuit model for both the depletion- and enhancement-mode MOSFET.

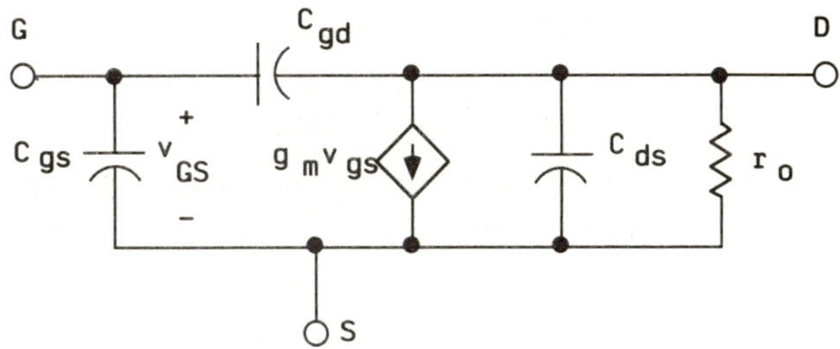

Figure 10.21: Small-signal equivalent circuit model of the MOSFET

In calculating the gain of MOSFET circuits at low frequencies, the capacitances can be ignored -- i.e., their impedances are such that they can be considered open circuits. For the high-frequency case, the total effective input capacitance, together with the resistance of the signal source feeding the amplifier, produces a first order low-pass filter that reduces the gain of the amplifier at high frequencies.

From Equation (10.23), the transconductance -- i.e., the quantity relating i_D and v_{gs} in pinch-off is given by

$$g_m = \left. \frac{\partial i_D}{\partial v_{gs}} \right|_{Q \, point} = \left. \frac{\partial}{\partial v_{gs}} [K(v_{GS} - V_t)^2] \right|_{Q \, point}$$

$$= 2K(V_{GS} - V_t) \tag{10.29}$$

From Equations (10.20), (10.23) and and (10.29), g_m can be written as

$$g_m = \sqrt{2\mu_n C_{ox}} \sqrt{W/L} \sqrt{I_D} \tag{10.30}$$

This equation shows that the transconductance is proportional to the square root of the dc bias current, and at a given bias current g_m is proportional to the square root of the aspect ratio W/L of the device.

10.4.3 Bipolar Junction Transistors

For small signals, the hybrid-π equivalent circuit model shown in Figure 10.22 can represent the BJT. The equivalent circuit represents both the npn and pnp transistors.

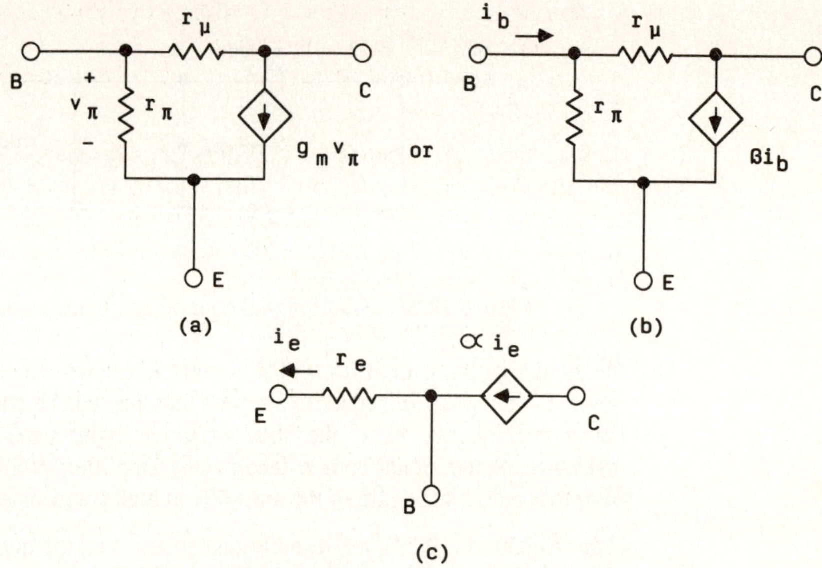

Figure 10.22: Hybrid-π small signal equivalent circuit

The transconductance term g_m the input resistance r_π at the base, the collector-base resistance r_μ, and the emitter resistance r_e, are given respectively by

$$g_m = I_c / \frac{kT}{q} = I_C / V_T \tag{10.31}$$

where $V_T = \dfrac{kT}{q}$ = thermal voltage

$$r_\pi = \frac{V_T}{I_B} = \frac{\beta}{g_m} \tag{10.32}$$

$$r_e = \frac{V_T}{I_E} = \frac{\alpha}{g_m} \approx \frac{1}{g_m} \tag{10.33}$$

and

$$r_\mu = \beta_o r_o$$

All the equivalent circuits of Figure 10.22 apply at a particular bias point, since all of the parameters depend on the value of I_C or I_E. Figure 10.23 represents the small-signal high-frequency hybrid-II equivalent circuit of the BJT.

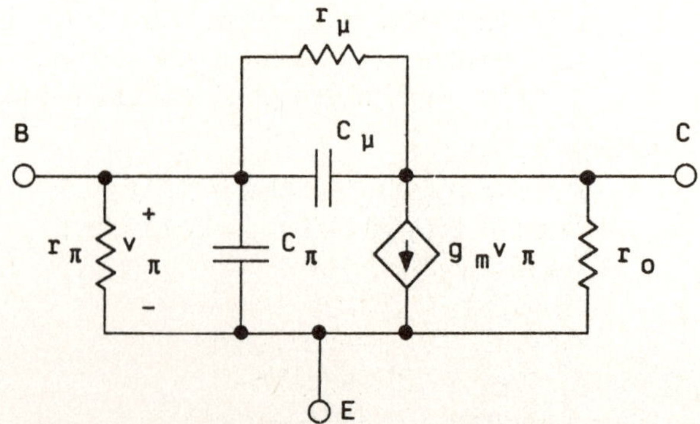

Figure 10.23: High frequency hybrid-Π equivalent circuit

C_π, the emitter-base capacitance, consists of two parallel capacitances, viz., a depletion capacitance and a diffusion capacitance. C_μ, the collector base junction capacitance, is a depletion capacitance because the collector-base junction is reverse biased in the active mode.

Both C_π and C_μ determine the high frequency response of transistor amplifiers, but at low to middle frequencies these capacitors have capacitive reactances that are much smaller than the impedance of the resistors with which they are associated and can be ignored. At low frequencies, the blocking and bypass capacitors determine the response of transistor amplifiers.

The mid-frequencies are defined as those frequencies for which the blocking and bypass capacitors have capacitive reactances that are much smaller than the impedances of their associated resistances, while capacitors C_π and C_μ have reactances much larger than impedances of resistors with which they are associated. Therefore, the blocking and bypass capacitors are replaced by short circuits, and C_π and C_μ are replaced by open circuits.

The equivalent circuit of Figure 10.23 also includes the output resistance, r_o, previously defined in Equation (10.24).

10.5 Summary

In this chapter, we have discussed three types of transistors: the bipolar junction transistor, the junction field effect transistor, and the metal oxide semiconductor field-effect transistor. Very brief discussions of the fundamental principles of operations of each of these devices were presented. Also examined were the regions of operation, circuit symbols, temperature effects, output resistance in pinch-off, and large- and small-signal equivalent circuits. The purpose of including these concepts is to give the student a convenient reference for some of the fundamental concepts of transistors needed to do the prelabs and laboratory experiments that follow.

LAB 10L1

TITLE: A Single-Stage Amplifier Using the Bipolar Junction Transistor (BJT): Low Frequency

OBJECTIVE: The objective of this laboratory exercise is to fortify, by demonstration, the concepts of large- and small-signal models, to determine dc biasing, and to determine small-signal parameters at low frequencies. The common-collector, common-emitter, and common-base transistor circuit configurations will be examined and the following characteristics of each circuit determined: voltage gain v_o/v_i; current gain i_o/i_i; input resistance r_{in}; and output resistance r_{out}. The various regions of biasing will be observed, with major emphasis on the active region.

EQUIPMENT AND MATERIALS:
Power supply (25 V at 100 mA), adjustable
Function generator (with offset capability)
Oscilloscope (dual trace)
Transistor curve tracer
General-Purpose npn transistor
    e.g., 2N2222 or 2N3904
Assortment of resistors (see lab)
Transistor curve tracer

COMMENTS: Several times during the lab, you are asked to set an ac voltage to zero. Some function generators, however, cannot be adjusted to zero. So, for example, if when measuring the output resistance of a circuit, you need to set the ac input signal to zero, we suggest that you replace the function generator (set to the proper dc offset) with a dc power supply set to the dc offset voltage. Also, some function generators cannot have their ac component set to a smaller amplitude than several hundred millivolts. If you encounter this problem, set the ac component as small as possible and observe the output signal with an oscilloscope. If any part of the output is clipped due to saturation or cutoff of your transistor, you may need to increase V_{CC} and/or move your operating point.

PRELABORATORY:

1. Use the 2N3904 data sheet to study the family of curves shown on the sheet (I_C versus V_{CE} with I_B varied). From this family of curves, do the following:

    a. Identify five operating points -- one in the cut-off region, three in the active region, and one in the saturation region. Record I_C and V_{CE} for each point.

    b. Determine β_F (forward β), and g_m for the three active region points.

    c. Determine α_F (forward α) for the three active region points.

    d. Determine the early voltage V_A from the curves.

    e. Determine the input resistance looking into the base of the BJT for the three active-region points.

    f. Determine the input resistance looking into the collector for the three active-region points.

    g. Determine the input resistance looking into the emitter for three active-region points.

    h. Show a large signal model for the transistor in the active region, saturated region, and cut-off region.

    i. For the three active-region points, show the simplified hybrid-Π model, including r_π, gm, and r_o.

2. Analyze the circuit of Figure 10L1.1.

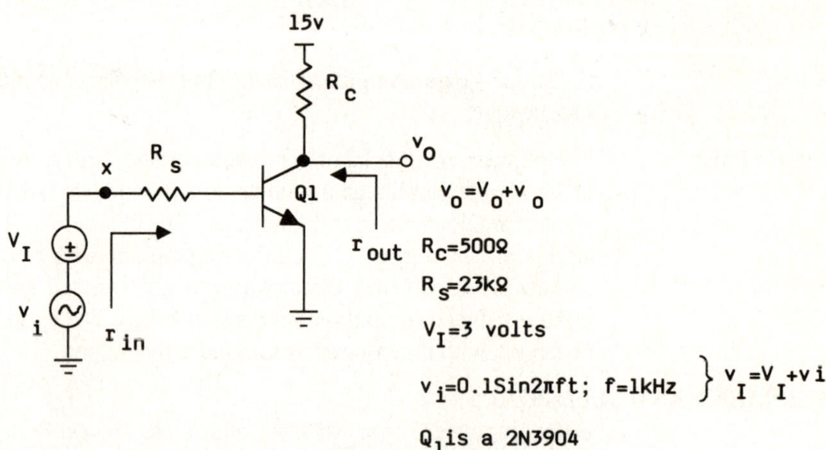

Figure 10L1.1: Precision rectifier

a. Is Q_1 in the active mode? Why?

b. What is the operating point for Q_1? Find V_C, V_{BE}, V_{BC}, I_B, I_C, and I_E.

c. Find the hybrid-Π model for Q_1 and redraw the circuit containing the hybrid-Π model. Make sure r_π, g_m, and r_o reflect your operating point.

    d. Looking into the circuit at point x, determine the input resistance r_{in}. (Remember to use the hybrid-Π model.)

    e. Looking into the circuit at v_o, determine the resistance r_{out}.

    f. Determine the voltage gain v_{o/v_i} (small signal).

    g. Determine the current gain i_{o/i_i} (small signal).

3. Analyze the circuit of Figure 10L1.2.

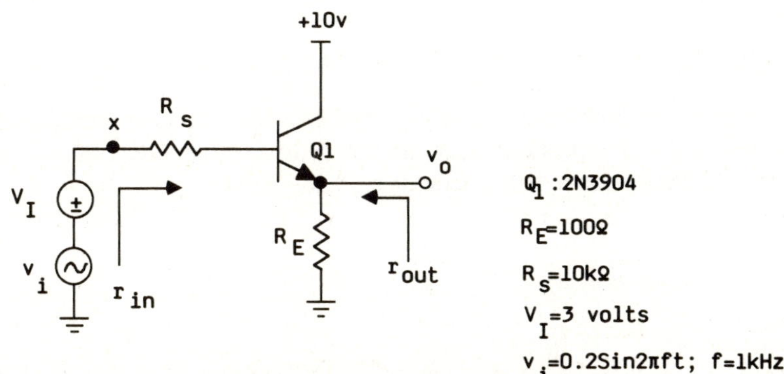

Figure 10L1.2: Circuit for Section 3

    a. Is Q_1 in the active region? Why?

    b. What is the operating point? Find V_{CE}, V_{BE}, V_{BC}, I_C, I_B, and I_E.

    c. Redraw the circuit using the simplified hybrid-Π model. Indicate the values of r_o, gm, and r_π.

    d. Determine the resistance r_{in} seen looking into the circuit at point x.

    e. Determine the resistance r_{out} looking into the circuit at v_o.

    f. Determine the voltage gain v_o/v_i using small signals.

    g. Determine the current gain i_o/i_i, where i_o is the small-signal current in R_E.

4. For the circuit in Figure 10L1.3:

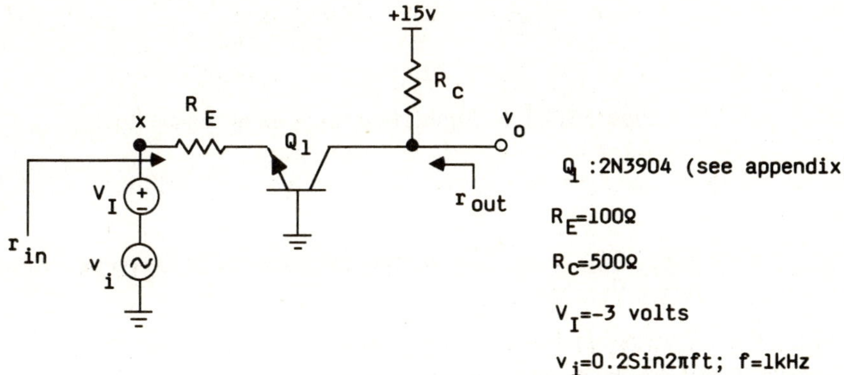

Figure 10L1.3: Circuit for Part 4

   a. Is Q_1, in the active mode? Why?

   b. Determine the operating point for Q_1. Find V_{CE}, V_{BE}, V_{BC}, I_C, I_B, and I_E.

   c. Redraw the circuit using the modified version of the hybrid-Π model called the T-model, (including r_e and g_m).

   d. Determine the small-signal input resistance r_{in}.

   e. Determine the small-signal output resistance r_{out}.

   f. Determine the small-signal voltage gain v_o/v_i.

   g. Determine the small-signal current gain i_o/i_i. (i_o is the current in R_C and i_i is the current into the emitter.)

Laboratory Procedure:

1. Use a curve tracer to display the family of curves I_C versus V_{CE} for various values of I_B. Plot these curves and identify an operating point in the active region. Record this point graphically and numerically. At this operating point, determine the following:

   a. B_F

   b. g_m

   c. r_π

   d. r_e

   e. Early Voltage, V_A

   f. r_o

Show the simplified hybrid-Π model for this transistor at the operating point identified.

2. Figure 10L1.4 shows the circuit of Figure 10L1.1 with particular values given for the resistors. Build this circuit.

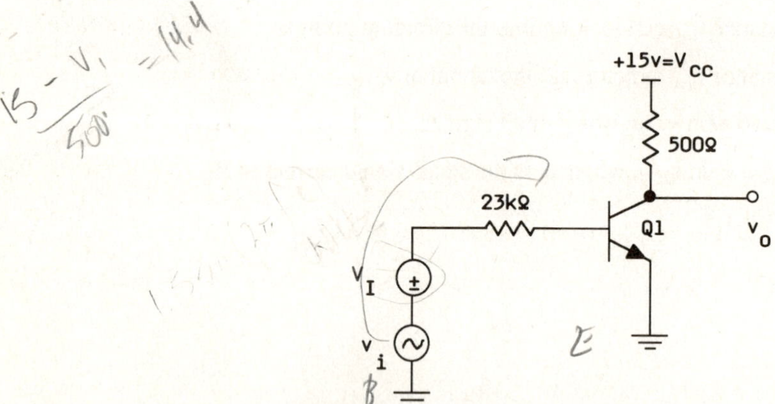

Figure 10L1.4: Application of circuit of Figure 10L1.1

The biasing voltage V_I and the small-signal voltage v_i can be obtained from a function generator with offset.

   a. Set up the biasing of Q_1 in the same manner as described in the prelab. Record the operating point of Q_1. Adjust V_I so that $V_o = 5$ V.

   b. Measure the ac voltage and current gain of the circuit. That is, measure v_o / v_i and i_o / i_i, where i_o is the current flowing in the 500-Ω resistor.

   c. Change V_I to -3.0 V. What happens to V_o? Why?

d. Pick one of the active-region operating points from your prelab, and adjust V_I and V_{CC} (15 V initially) to bias your circuit at this point. Then measure the voltage gain, current gain, output resistance, and input resistance of the circuits, and compare them to the corresponding values obtained in your prelab. Refer to Chapter 4 for help in measuring r_{out} and r_{in}.

e. Let v_i be a triangular wave with a peak to peak value of 0.2 V. Observe the output on an oscilloscope as the amplitude of v_i is increased, and determine the maximum value of v_i that results in no observable distortion. Indicate whether your distortion is caused by the saturation region or the cut-off region.

3. Build the circuit part three of the prelab and repeated in Figure 10L1.5:

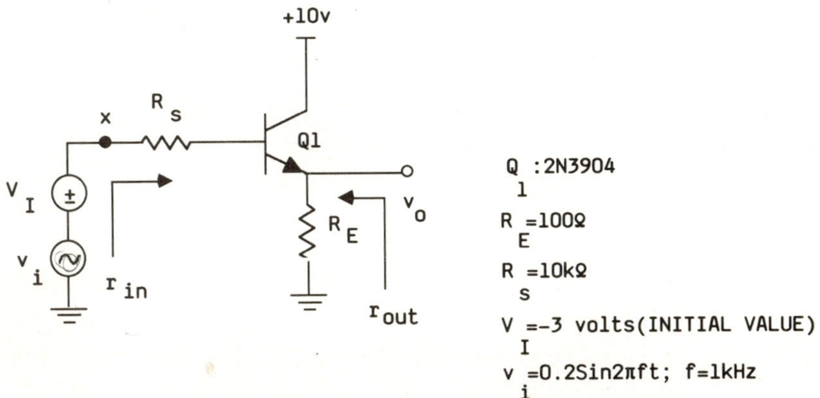

Figure 10L1.5: Circuit for Section 3

The input voltage $V_I + v_i$ is developed the same way as in part 1 of the lab.

a. Start with $V_I = 3$ V and then adjust it so $V_o = 2$ V, and record the operating or Q-point of the transistor.

b. Measure the small-signal current i_o/i_i and the voltage gain v_o/v_i (i_o is measured in R_E).

c. Measure r_{in} for the circuit by measuring the ac voltage and current at node x.

d. Measure the circuit's output resistance looking into node v_o (refer to Chapter 4).

e. Observe the output voltage as the input is increased from an amplitude of 0.2 V to 2 V. Sketch the output for both extremes of the input. What happens at the output for the 2-V amplitude sinusoidal input? Remember to keep V_I at the value set in part a.

4. Build the circuit from part four of the prelab and shown in Figure 10L1.6:

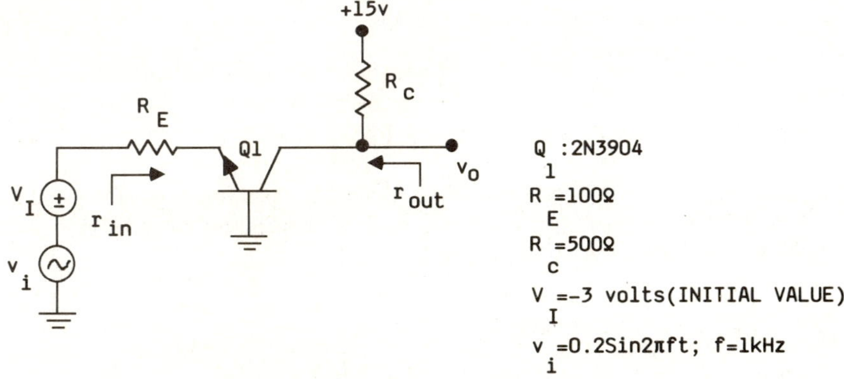

Figure 10L1.6: Circuit for Section 4

The input voltage is developed in the same manner as in previous sections, except that the offset has the opposite polarity.

   a. Set all voltages as described in the figure, and determine the transistor's operating point. Measure V_{CE}, V_{BE}, V_{BC}, I_B, I_E, and I_C.

   b. Measure the voltage and current gain for the circuit. That is, measure v_o/v_i and i_o/i_i, where i_o is the current flowing into the collector of the BJT.

   c. Measure the input resistance r_{in} and output resistance r_{out} in a manner similar to that made in previous section.

5. Take one of the circuits of Figures 10L1.1 through 10L1.3 and analyze it using SPICE. Find the operating point, voltage gain, current gain, input impedance, and output impedance.

DISCUSSION: Compare the various configurations in terms of voltage gain, current gain, input resistance, and output resistance. Put the results in a table. If you used the same transistor in the lab as in the prelab, compare your results.

CONCLUSIONS:

LAB 10L2

TITLE: A Single-Stage Amplifier Using the Junction Field-Effect Transistor (JFET): Low Frequency

OBJECTIVE: The objective of this laboratory exercise is to expose the student to the large signal characteristics of the JFET, including the triode, active and cutoff regions. After identifying several operating points in the active region, a small signal model of the JFET is used to analyze the small-signal gain of a JFET amplifier in the common-source and common-drain configurations. Self-biasing is used, and the effect of a source resistance bypass capacitor is demonstrated. Included is a circuit simulation using SPICE.

EQUIPMENT AND COMPONENTS:

        Power supply
        Function generator with dc offset control
        Curve tracer to give JFET characteristics
        Oscilloscope (dual trace)
        Multimeter
        Resistors
        JFET: n-Channel depletion mode
        e.g., 2N5459, 2N5485, 2N5640
        Capacitor: 10 μF

COMMENTS:

PRELABORATORY:

1. For the circuit in Figure 10L2.1, set $R_S=0\ \Omega$ and find the operating point for the JFET if $I_{DSS} = 16$ mA and $V_P = -8$ V. Set $v_i = 0$ initially and let $V_G = -2$ V. Repeat this calculation for $V_G = -4$ V and $V_G = -6$ V. Redo the analysis for $I_{DSS} = 4$ mA and $V_p = -2$ V.

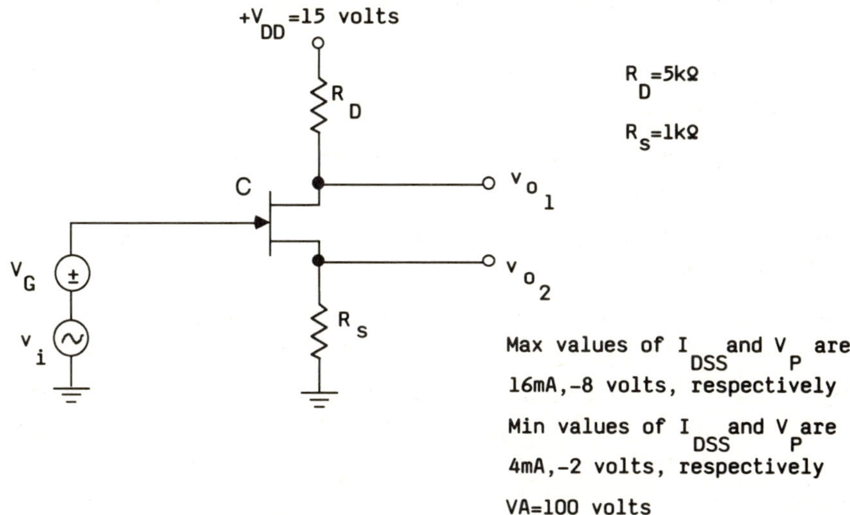

Figure 10L2.1: Circuit for Part 1

    a.   Find the small-signal parameters g_m and r_o that are in the active or pinch-off region.

    b.   Find the voltage gain v_{o1}/v_i for operating points in the active or pinch-off region. Find the input resistance R_{in} and the resistance R_o seen looking into the output for all pinch-off region operating points.

2.   Set $R_s = 1\ k\Omega$ and let $R_D = 0\ \Omega$. Repeat the previous calculation of Part 1, but determine v_{o2}/v_i as the voltage gain.

3.   Set $R_S = 1\ k\Omega$ and $R_D = 5\ k\Omega$. Repeat the previous calculation of Part 1, but determine both v_{o1}/v_i and v_{o2}/v_i.

Laboratory Procedure:

1.   Characterize your JFET.

    a.   Use a curve tracer to display the I_D versus V_{DS} curves for a series of gate voltages. Plot this family of curves.

    b.   From the curves, determine I_{DSS}, V_P, and the Early voltage, V_A.

    c.   Pick an operating point in the active region and determine g_m and r_o.

    d.   Pick an operating point in the triode region and determine r_{DS}.

2.   Connect your JFET as shown in Figure 10L2.2, to bias it.

Set $R_S = 0$ (short R_S) and let $v_i = 0$. Then adjust V_G so that Q_1 is in the active region. By measuring V_{DG}, how can you determine whether Q_1 is in the active region? Make sure that V_{DS} is at least 3 V away from the triode region.

    a.   For this operating point, determine g_m and measure the voltage gain v_{o1}/v_i for $v_i = 0.1 \sin 2\pi ft$. Let f vary from 1 kHz to 10 kHz.

    b.   Is there any variation in the gain as the frequency goes from 1 kHz to 10 kHz?

    c.   Let $R_S = 1\ k\Omega$, let $R_D = 0$ (short R_D), and let $v_i = 0$. Is the JFET still in the active mode? Why? For this operating point, determine g_m and measure the voltage gain v_{o1}/v_i. Vary f from 100 Hz to 10 kHz and record the gain for several frequencies.

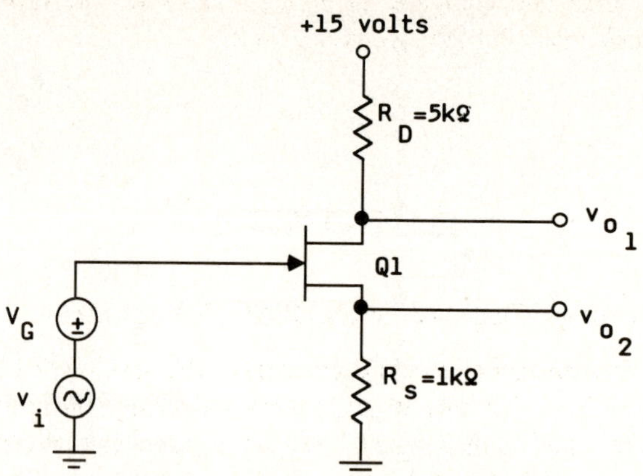

Figure 10L2.2: Circuit for Part 2

3. Measuring voltage gain.

    a. Let $R_S = 1$ kΩ and $R_D = 5$ kΩ, as shown in Figure 10L2.2.

       (1) Adjust V_G so that the JFET operating point is at least 3 to 5 V away from the triode region.

       (2) Measure the voltage gains v_{o1}/v_i and v_2/v_i for an input signal of

$$v_i = 0.1\sin 2\pi ft$$

Let f vary from 100 Hz to 1 MHz. Plot the data on semilog graph paper, with frequency on the log axis. Collect at least 2 data points per decade of frequency.

    b. Add a 1-μF capacitor across R_S, as shown in Figure 10L2.3.

       (1) Does the capacitor have any effect on the operating point of the JFET? Why?

       (2) Repeat the voltage gain measurements as a function of frequency from Part 3.a.2 for the circuit in Figure 10L2.3. Plot the data on the same graph as in the previous section.

       (3) With R_S and R_D both in the circuit, pick an operating point in the active region. What is the voltage gain, $|v_{o1}/v_i|$? If a 1-μf capacitor is placed in parallel with R_S, what is the magnitude of the impedance seen by the input source at very low frequencies (less than 100 Hz). At 1 kHz? At 10 kHz? At 100 kHz? Calculate the gain v_{o1}/v_i for each of these frequencies.

       (4) For the situation just described in Part (3), run a SPICE simulation for the circuit and determine $|v_o/v_i|$ as a function of frequency.

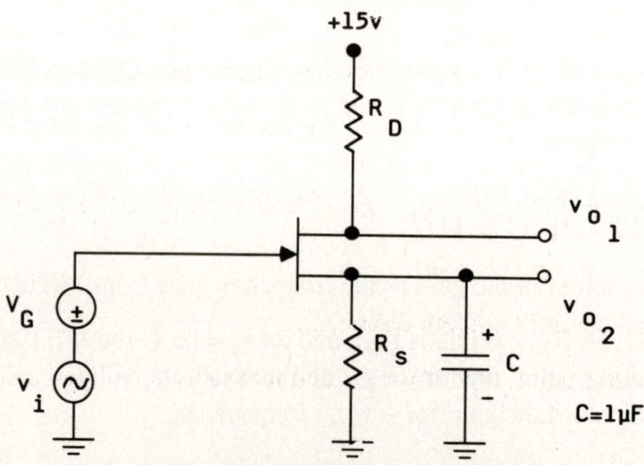

Figure 10L2.3: Figure for Part 3.B

DISCUSSION: Contrast and compare the voltage gains for all variations of the JFET amplifier considered. Put the results in a table.

CONCLUSIONS:

LAB 10L3

TITLE: The Single-Stage n-Channel MOSFET Amplifier

OBJECTIVE: In this laboratory exercise we explore the basic operation of a single-stage MOSFET amplifier. We use depletion- or enhancement-type n-channel MOSFETS. The large- and small-signal, low-frequency characteristics are examined for the common-source and source-follower topologies. The effect of a source resistor bypass capacitor is also observed. The lab concludes with a comparison of hand calculations, experimental results, and a SPICE simulation.

EQUIPMENT and COMPONENTS:
Function generator with offset adjustment
   dc to 1-MHz frequency range.
Oscilloscope: dual trace
Resistors: 1 kΩ and 4 kΩ
Single power supply: 10 V @ 10 mA
Capacitor: Electrolytic 1 μF, 20 V rating
Multimeter
3N128
2N4351

COMMENTS:

PRELABORATORY:

1. Given the circuit of Figure 10L3.1, find an appropriate value of R such that it is possible to observe the break frequency (of 3db) with a maximum generator frequency of 1 MHz. Use one-half the maximum value of C_{iss} for each input.

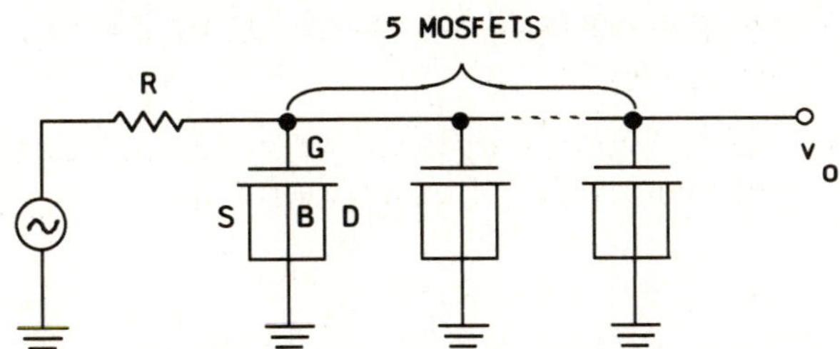

(Note: It is suggested that the 3N128 MOSFET be used.)

Figure 10L3.1: Circuit for Part 1

2. Using information only from the tables and plots of the data sheets, answer the following questions:

   a. Is the 3N128 an n- or p-channel device?

   b. Is it an enhancement or depletion device?

   c. What are the minimum, typical, and maximum values for V_{GS} cutoff voltage?

    d.   What is the maximum value of the gate input capacitance C_{iss}?

    e.   What are the minimum, maximum, and typical values for I_{DSS}?

3. Using the data sheets for the 2N4351 transistor, answer the following questions.

    a.   Is the 2N4351 an n- or p-channel device?

    b.   Is it an enhancement or depletion device?

    c.   What is the minimum and maximum threshold voltage?

    d.   What is the maximum gate leakage current (at 25°C)?

    e.   Consider the equations

$$i_D = K(v_{GS} - V_T)^2 \qquad (10L3.1)$$

$$K = \frac{1}{2} C_{ox} \mu_n W/L$$

where

$$
\begin{aligned}
i_D &\equiv \text{drain current} \\
v_{GS} &\equiv \text{voltage between the gate and the source} \\
V_T &\equiv \text{threshold voltage (use } V_T = 3 \text{ V)} \\
\mu_n &\equiv \text{electron mobility} = 600 \text{ cm}^2/\text{V} \cdot \text{s} \\
 &\quad \text{(in silicon @ 300°K)} \\
C_{ox} &\equiv \frac{\text{gate capacitance}}{\text{unit area}} \text{ use } 3.5 \times 10^{-8} \text{ F/cm}^2 \\
W &\equiv \text{channel width} \\
L &\equiv \text{channel length}
\end{aligned}
$$

Use these equations to plot I_d against V_{GS}. Estimate W and L from the drawing on the data sheet.

    f.   Determine g_m using the following equation and the value of K in part e:

$$g_m = \frac{\partial i_D}{\partial v_{GS}} \Big|\, V_{DS} = \text{constant} = 2K(v_{GS} - V_T) \qquad (10L3.2)$$

    g.   Compare g_m with $|y_{fs}|$ on the data sheet.

4. Analyze the circuit of Figure 10L3.2. Assume Q_1 is a 2N4351.

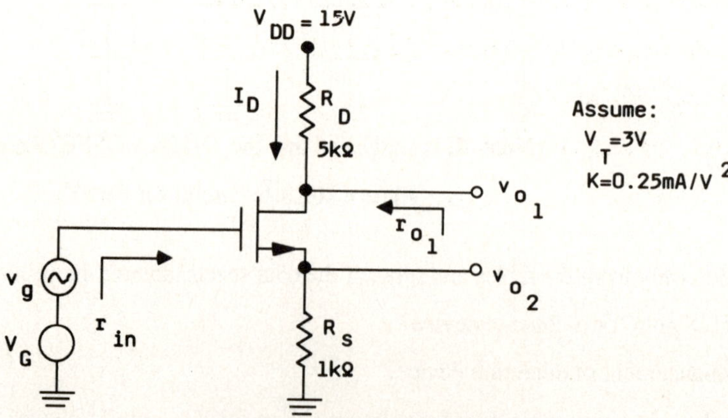

Figure 10L3.2: Circuit for Part 4

a. If $R_S = 0\ \Omega$, then what polarity and magnitude must V_G be to start the drain current i_D?

b. What value of V_G yields a 1-mA drain current?
(Again, assume that $R_S=0$.)

c. For $R_S = 0$ and $I_D = 1$ mA, determine the following:

    (i) g_m

    (ii) r_{in}, r_{o_1}

    (iii) v_{o1}/v_g

d. Let $R_S = 1$ kΩ and $I_D = 1$ mA. Repeat Part c.
Also, find v_{o2}/v_g.

e. Connect a 1 μF capacitor across R_S and repeat Part d. Also, find v_{o1}/v_g and v_{o2}/v_g at dc, 100 Hz, 10 kHz, and 100 kHz.

f. Run a SPICE simulation on the circuit in Part e. Perform a frequency analysis on this circuit for v_{o1}/v_g and v_{o2}/v_g. Let $v_g = 0.1$ sin wt.

Laboratory Procedure:

1. Measure input capacitance

    a. Set up the circuit of Figure 10L3.3. Let R equal the value you determined in the prelab.

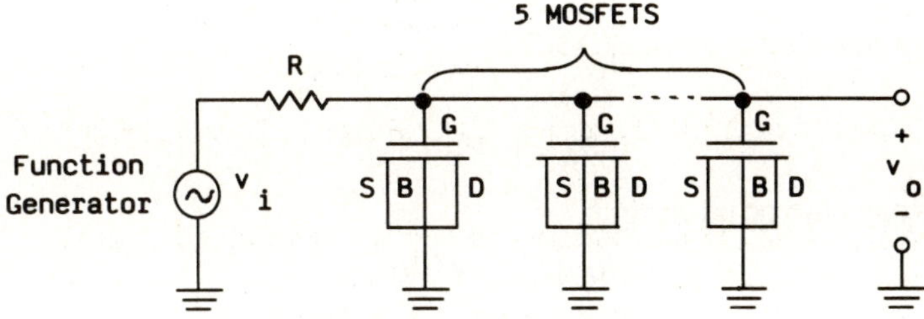

Figure 10L3.3: Circuit for Part 1

    b. Increase the frequency of v_i until the amplitude v_o decreases to 0.71 of its low frequency amplitude. Record this frequency, and use it to calculate the average gate capacitance C_{iss}. Remember, $f_{3db} = \dfrac{1}{2\pi RC_T}$; where $C_T \equiv$ total capacitance. Remove all MOSFETs, and repeat the measurement to determine the amount of stray capacitance.

    c. Compare your value of C_{iss} to that on the data sheet.

2. Common-source MOSFET amplifier.

    a. Set up the circuit of Figure 10L3.4.

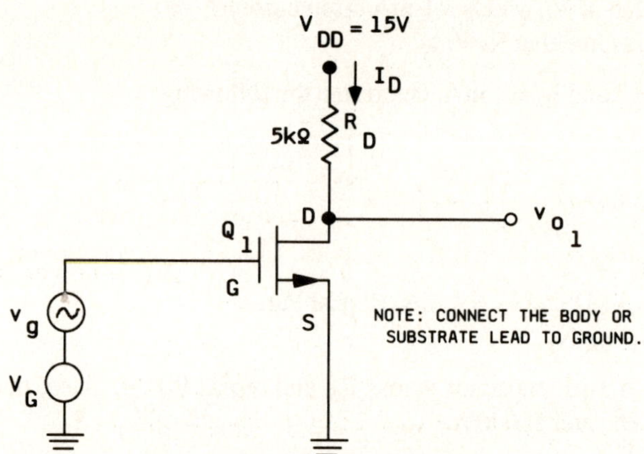

(**Note:** Q_1 can be a depletion- or enhancement-type n-channel MOSFET.)

Figure 10L3.4: Circuit for Part 2.a

b. Let $v_g = 0$ and let $V_G = 0$. Measure I_D, and use this information to determine whether Q_1 is an enhancement- or a depletion-type MOSFET. Depending on the type of device, you need to determine K and V_T or I_{DSS} and V_p. If Q_1 is a depletion-type MOSFET,

$$I_D = I_{DSS}(1 - V_{GS}/V_P^2)$$
(10L3.3)

If Q_1 is an enhancement-type MOSFET,

$$I_D = K(V_{GS} - V_T)^2$$
(10L3.4)

For the depletion-type transistor,

$$I_D = I_{DSS} \text{ if } V_{GS} = 0$$

and

$$I_D = 0 \text{ if } V_{GS} = V_P$$

For the enhancement-type transistor,

$$I_D = 0 \text{ if } V_{GS} = 0$$

$$I_D \neq 0 \text{ if } V_{GS} > V_T \text{ (let } I_D = 10\,\mu A).$$

c. Adjust V_G so $I_D = 1$ mA. Record V_{DS}, I_D, V_{GS}, and V_G, and then calculate g_m. Remember, for the enhancement-type MOSFET,

$$g_m = 2K(V_{GS} - V_T)$$

and for the depletion-type MOSFET,

$$g_m = \left[\frac{I_{DSS}}{|-V_P|}\right](1 - V_{GS}/V_P) = \frac{I_{DSS}}{-V_P}\sqrt{\frac{I_D}{I_{DSS}}}$$
(10L3.5)

d. Let $v_g = 0.1 \sin \omega t$; with $\omega = 6{,}280$ rad/sec.

(1) Sketch v_{o1} and v_g.

(2) Measure the voltage gain v_{o1}/v_g.

3. Biasing with negative feedback. Build the circuit of Figure 10L3.5.

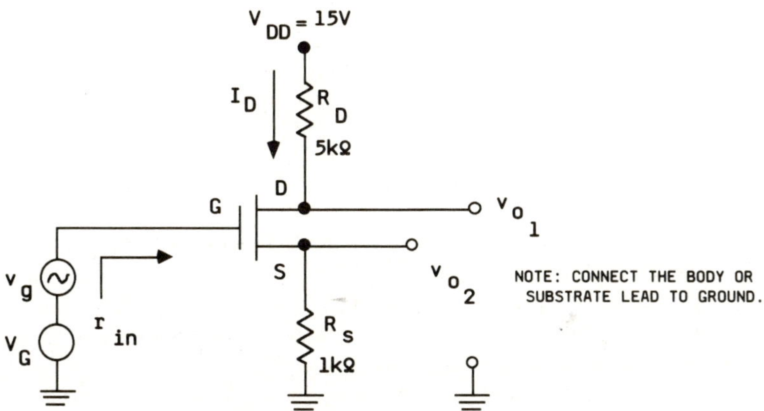

Figure 10L3.5: Circuit for Part 3

a. Adjust V_G (let $v_g = 0$) so that $I_D = 1$ mA. Record V_G, V_D, and V_S.

b. Measure the ac input resistance r_{in}.

c. Measure the ac output resistance r_o looking into the output v_{o1}. Record this value.

d. Measure the ac output resistance r_s looking into the output v_{o2}. Record this value.

e. Let $v_g = 0.2 \sin \omega t$; with $\omega = 6{,}380$ Hz. Sketch v_g, v_{o1}, and v_{o2}. Use this sketch to determine the voltage gains v_{o1}/v_g and v_{o2}/v_g.

4. Effect of a source resistor bypass capacitor.

    a. Add a bypass capacitor to the circuit of Figure 10L3.5, as shown in Figure 10L3.6.

    b. Record the operating point (I_D, V_{DS}). Compare it to the operating point in part 3.

    c. Let $v_g = 0.2 \sin \omega t$: where $f = (\omega/2\pi)$, varies from dc to 1 MHz in decades (10 Hz, 100 Hz, 1 kHz, ...,1 MHz). Record the amplitude of v_{o1} and v_{o2} for each frequency.

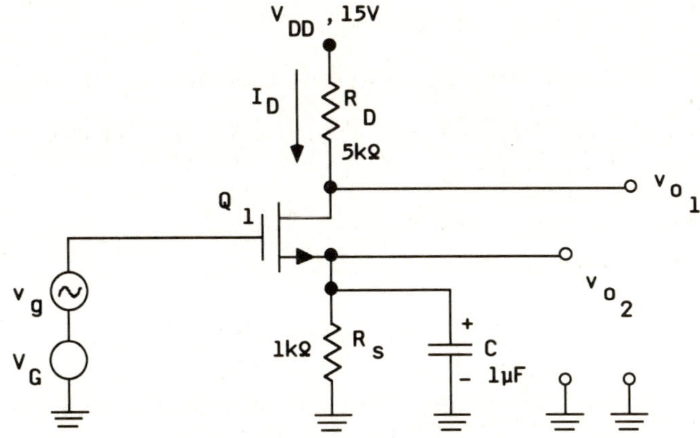

Figure 10L3.6: Circuit for Part 4a

DISCUSSION: Include a comparison of the frequency response of the circuit in Figure 10L3.6 for v_{o1}/v_g and v_{o2}/v_g using the three methods of the lab (hand calculation, experimentation, and simulation) for generating data.

CONCLUSIONS:

LAB 10L4

TITLE: The Current Source[2],[4]

OBJECTIVE: The current source, as well as the current sink, are used extensively in the design of integrated circuits. Their functions are twofold: to bias amplifiers-- normally, differential amplifiers-- and to act as loads for amplifiers, thus taking advantage of the high impedance of a current source. This laboratory exercise examines current sources implemented with BJTs. A comparison is made between these current sources and the ideal case with infinite output impedance. This comparison includes hand calculations, experimentation, and computer simulation.

EQUIPMENT AND MATERIALS:
Power supply, ±15 V and 10 V (adjustable)
Function generator
Curve tracer
Discrete npn transistor
Capacitor (1 μF)
Resistors (1/8 W; values determined in lab)
Matched pair of npn transistors

PRELABORATORY:

1. Analysis of a simple current mirror that functions as a current sink.

    a. Consider the circuit of Figure 10L4.1.

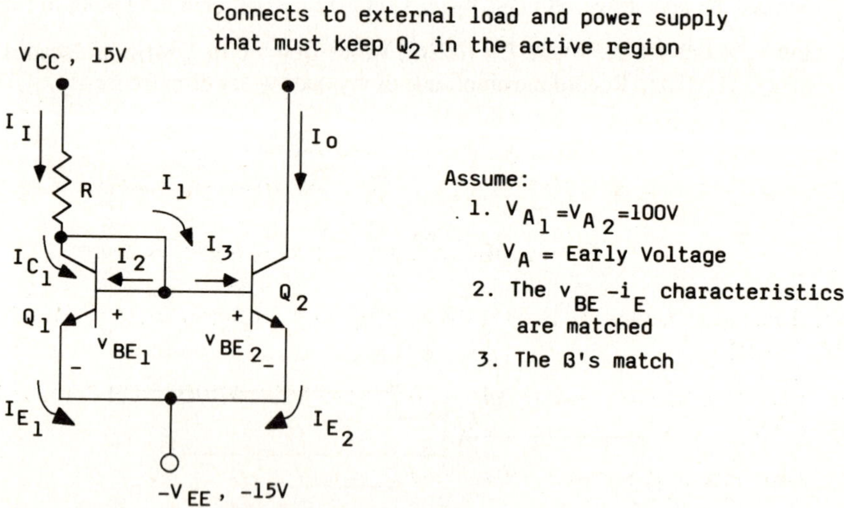

Figure 10L4.1: Simple current mirror

Since the V_{BE}-versus-I_E characteristics match and since the β's match, the Ic's must also match. Using KVL around the loop, we see that $V_{BE1} = V_{BE2}$, and since the V_{BE}-versus-I_E characteristics match, $I_{E1} = I_{E2}$. Answer the following questions:

(1) Write I_2 and I_3 in terms of I_E (remember, $I_{E1} = I_{E2}$).

(2) Write I_1 in terms of I_2 and I_3.

(3)   Determine I_{C1} in terms of I_2.

(4)   Determine I_I in terms of I_{C_1} and I_2. Also, what is the relationship between I_o and I_{C_1}?

(5)   Show that $I_o / I_I = \dfrac{\beta}{(\beta + 2)}$.

(6)   What is the minimum voltage that could appear at the collector of Q_2 and allow the circuit to function as a current source?

(7)   What value of R gives an I_o of 1 mA? (Assume that $\beta=200$ and $V_{BE} = 0.7$ V at $I_E = 1$ mA.)

(8)   Draw the small signal model (low-frequency hybrid-π model) for Q_2. Assume that its base is connected to ground. What is the small-signal output resistance of the circuit? The reason we assume that the base of Q_2 is grounded is because looking into the base of Q_1, we see a small resistance. What is this resistance?

(9)   Draw the Norton and Thevenin equivalent circuits for the current source.

(10)   Determine the mismatch between I_I and I_o for various values of β. Take $\beta = 10, 50$, and 200.

2.   The constant current source. Analyze the circuit of Figure 10L4.2.

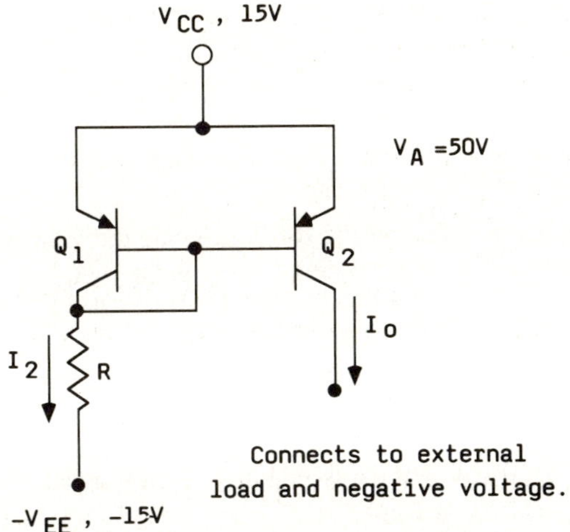

Figure 10L4.2: Circuit for constant-current source

Make the same assumptions for this circuit as we made for the circuit of Figure 10L4.1 (except for the early voltage).

a.   Find the ratio I_o/I_2. Evaluate I_o/I_2 for β equal to 10, 50, 100, 200. Assume R=5 KΩ and $V_{BE} = 0.7$ V.

b.   Determine the value of R that results in $I_o = 1$ mA.

c.   What is the resistance seen looking into the collector of Q_2?

d.   Draw the Norton and Thevenin equivalent circuits for the current source.

3.   Evaluation of the Widlar current source (sink).

Given the circuit of Figure 10L4.3, find I_o/I_2 and the output impedance of the current sink.

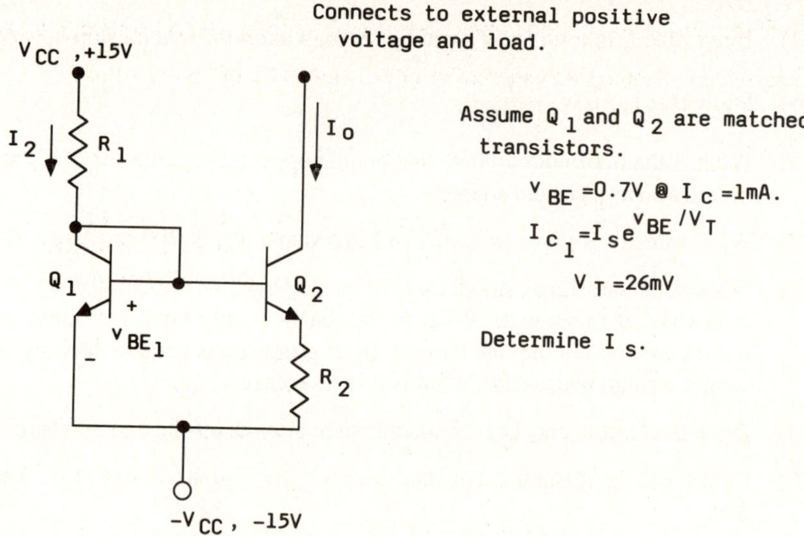

Figure 10L4.3: Widlar current source sink

It is known that

$$I_{C_1} = I_{S_1} e^{V_{BE1}/V_T} \text{ and } I_{C_2} = I_{S_2} e^{V_{BE2}/V_T}.$$

Also, since the Q's are matched, $I_{S_1} = I_{S_2} = I_S$.

a. Write a loop equation including the two V_{BE}'s and the voltage drop of R_2. Assume that $\beta \gg 1$, which allows you to assume also that $I_{C_1} = I_{E_1}$ and $I_{C_2} = I_{E_2}$.

b. Arrange the equation from Part a in the form

$$I_o R_2 = V_T \ln\left(\frac{I_2}{I_o}\right) \tag{10L4.1}$$

c. If I_o is equal to 10 µA and I_2 is equal to 2 mA, what values of R_1 and R_2 are needed? Remember, $V_{BE} = 0.7$ V at 1 mA.

d. Use the small-signal equivalent circuit of Figure 10L4.4 for Q_2 of Figure 10L4.3. The base load of Q_2 is approximately $\dfrac{1}{g_m}$ from Q_1. Determine the output impedance r_{out} looking into the collector of Q_2.

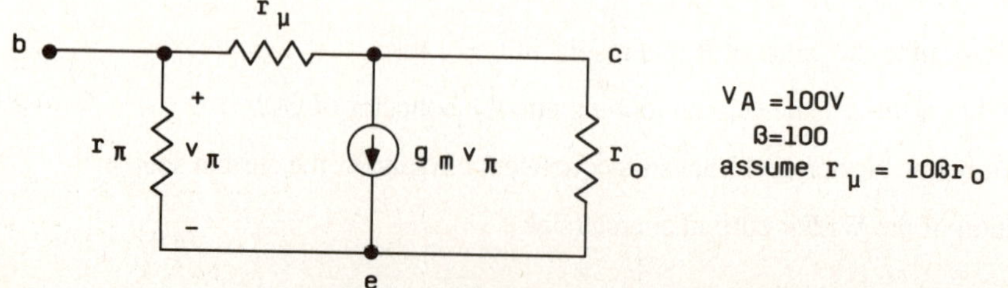

Figure 10L4.4: Small-signal equivalent circuit

   e.   Draw the Norton and Thevenin equivalent circuits for Figure 10L4.3.

   f.   Draw the schematic for pnp implementation of a Widlar current source (instead of sink).

4.   Evaluation of the Wilson current source. Analyze the circuit of Figure 10L4.5.

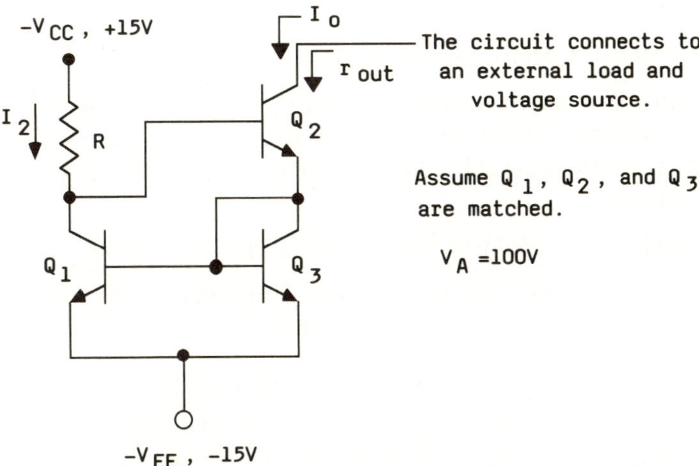

Figure 10L4.5: Wilson current source

   a.   Use the analysis of Reference [4] to determine the Norton and Thevenin equivalent circuits for the Wilson current source. Show that

$$I_o = I_2 \left[1 - \frac{2}{\beta^2 + 2\beta + 2} \right] \text{ and } r_{out} \approx \frac{\beta r_o}{2} \tag{10L4.2}$$

where r_o is the collector-to-emitter small-signal resistance.

Laboratory Procedure:

1.   Build a current mirror with two discrete npn transistors (not made on the same piece of silicon).

   a.   Use two discrete npn transistors, and measure the β's for each transistor (use a curve tracer). Also, plot the v_{BE} - i_E characteristic for both transistors using the curve tracer.

   b.   Build the circuit of Figure 10L4.6. Pick R_I so that I_{R_I} equals 0.1 mA.

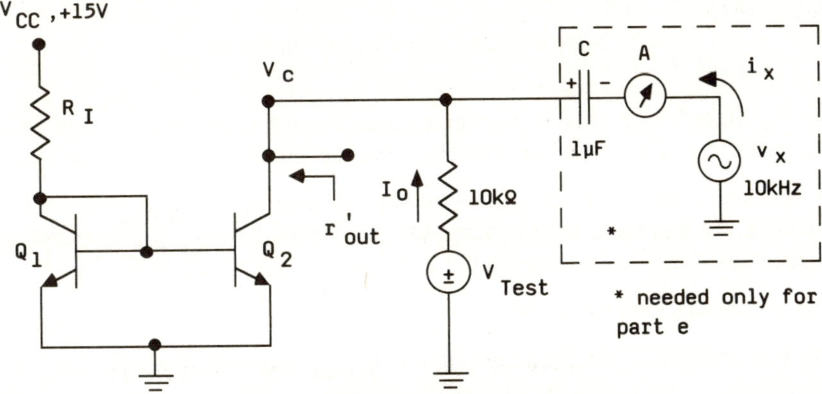

Figure 10L4.6: Circuit for Part 1

    c. Vary V_{test} from 0 V to 10 V in 0.5-V increments. Record I_o and V_C for each value of V_{test}. Plot I_o/I_I versus V_{test}.

    d. Determine r'_{out} by measuring the collector-to-ground small-signal output resistance. This is done by ac-coupling a test voltage v_x to the collector of Q_2 and measuring the small-signal current i_x. Notice that the resistance looking into the collector of Q_2 is in parallel with the 10 kΩ resistor. Do this procedure for V_{test} = 1 V, 5 V, and 10 V. Where could you move the ammeter so that this measurement is made directly?

2. Build a current mirror with a dual npn transistor.

    a. Use one of the dual npn transistors suggested in the equipment list, and determine β for each transistor. Use either a curve tracer or a common-emitter amplifier configuration. Be sure to measure the β's at the same quiescent collector current I_C. Also, plot the v_{BE} - i_E characteristic for both transistors.

    b. Build the current mirror of Figure 10L4.6 using the dual npn transistor of Part a.

    c. Vary V_{test} from 0 V to 10 V in 0.5-V increments. Record I_o and V_C for each value of V_{test}. Plot I_o/I_I versus V_{test}. (Use the same graph as in Part 2.)

    d. Repeat the measurement of r'_{out} for the dual-transistor implementation of the current source.

    e. Compare your results with discrete devices to those with matched devices.

3. Build the Widlar current mirror with a dual npn transistor.

    a. Using the same dual transistor from part 2, build the current source shown in Figure 10L4.7. Let R_1 equal 3 kΩ. Adjust V_{CC} so that I_I equals 10 mA. Determine R_2 so that I_o equals 40 μA.

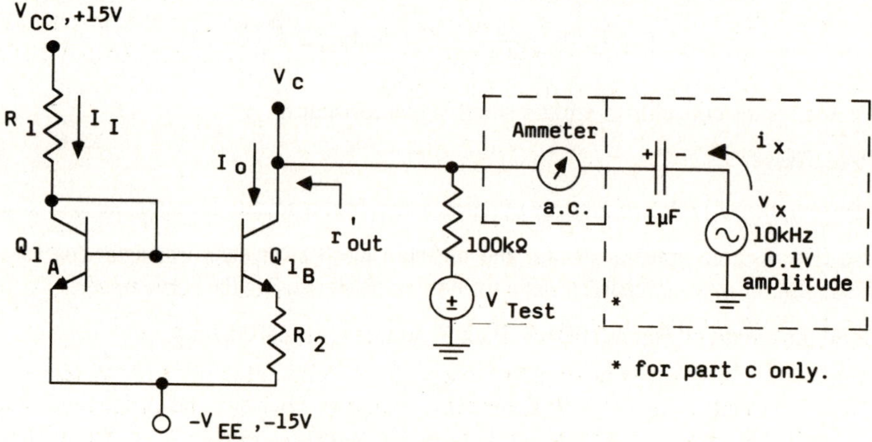

Figure 10L4.7: Widlar current mirror

    b. Vary V_{test} from 0 V to 10 V in 1-V increments. Record I_o and V_C for each value of V_{test}. Plot I_o/I_2 versus V_{test}. Over what range of V_{test} does the circuit function as a current mirror?

    c. Use the same method as in Part 1 and measure r'_{out}, the small-signal resistance of the collector. Remember, the resistance seen looking into the collector of Q_{1B} is in parallel with the 100 kΩ resistor. Repeat this measurement for V_{test} equal to 1 V, 5 V, and 10 V.

4. Build a Wilson current source (optional).

    a. Build two versions of the circuit of Figure 10L4.8, one with three discrete transistors and the other with one of the multitransistor building blocks.

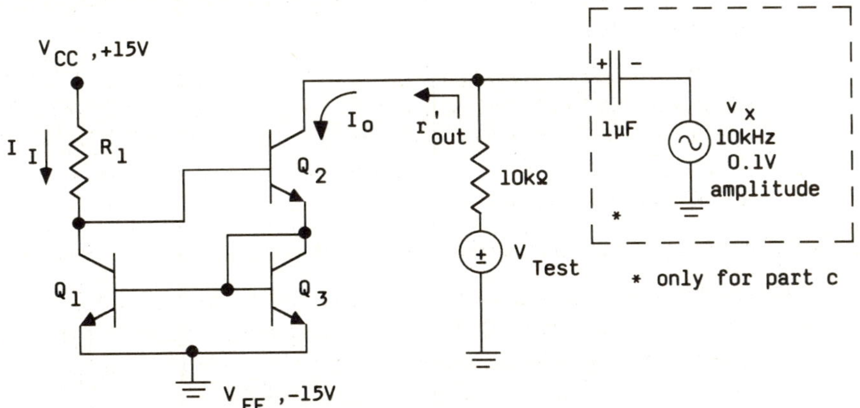

Figure 10L4.8: Wilson current source

b.   Use Equation (10L4.2) in the prelab, and determine R_1 so that I_o equals 0.1 mA. Record and plot I_o/I_2 versus V_{test} as V_{test} ranges from 0 V to 10 V in 1-V increments.

c.   Use the technique of Sections 1 and 2, and determine the output resistance seen looking into the collector of Q_2. Measure r'_{out} for V_{test} equal to 1 V, 5 V, and 10 V.

5.   Run a SPICE simulation on the circuits of Part 1, 3, and 4. Repeat the conditions of the laboratory in the simulation. Use the .TF option to obtain the output impedance.

DISCUSSION: Include a comparison of each circuit's performance as a current source. Compare your prelab results with the experiment results. Discuss the range of voltages at the collector of the output transistor that allowed good current-source performance and why such performance was obtained.

CONCLUSIONS:

LAB 10L5

TITLE: The Differential Amplifier[2],[4]

OBJECTIVE: The differential amplifier is a fundamental building block of virtually all integrated analog circuits. This laboratory exercise examines the differential amplifier as an evolving device. We start with the simplest form of the differential amplifier and systematically include circuit modifications that improve its performance. Each variation is characterized by calculating and measuring the following parameters: differential-mode gain v_{OD}/v_{ID}; common-mode gain, v_{OCM}/v_{ICM}; common-mode rejection ratio expressed as $20\log(v_{OCM}/v_{OD})$; input resistance, both differential and common-mode, R_{in_D} and R_{in_c}, respectively; and common-mode input voltage range.

Since the differential amplifier can be directly coupled to another differential amplifier without the use of capacitors, it is an excellent amplifier for integrated circuits. Components such as capacitors and resistors require large areas of silicon, so the designer's motivation is to devise ways of minimizing the use of these passive components in favor of transistors. The evolution of the lab's differential amplifier is toward this end.

The input stage of the operational amplifier is the differential amplifier. The operational amplifier's specifications give several characteristics that are determined predominately by the input differential amplifier. Among these parameters are the offset voltage V_{os}; bias current I_{bias}; offset current I_{os}; and range of input common-mode voltage. These parameters are calculated and measured for each phase of the lab.

Again, simulation of the differential amplifier is explored using SPICE, with emphasis on the effects of component mismatch on circuit performance.

EQUIPMENT AND MATERIALS:

Dual-trace oscilloscope (capable of a differential measurement)
Digital multimeter
Function generator (with ± 12-V offset)
Triple power supply (± 15 V and $+6$ V)
Discrete and dual matched transistors
Resistors
Capacitor (1 μF), 20-V rating

PRELABORATORY:

1. Given the circuit of Figure 10L5.1, assume that Q_1 and Q_2 are matched transistors. (The base-emitter characteristics and the β's match.) Therefore, $I_{E_1} = I_{E_2}$, $I_{C_1} = I_{C_2}$, $v_{cm} = v_d = 0$. This means that $I_{E_1} = I_{E_2} \simeq (1/2)I$ (assuming negligible current flows in R_E). Also, assume that β is not infinite (therefore, $\alpha \neq 1$).

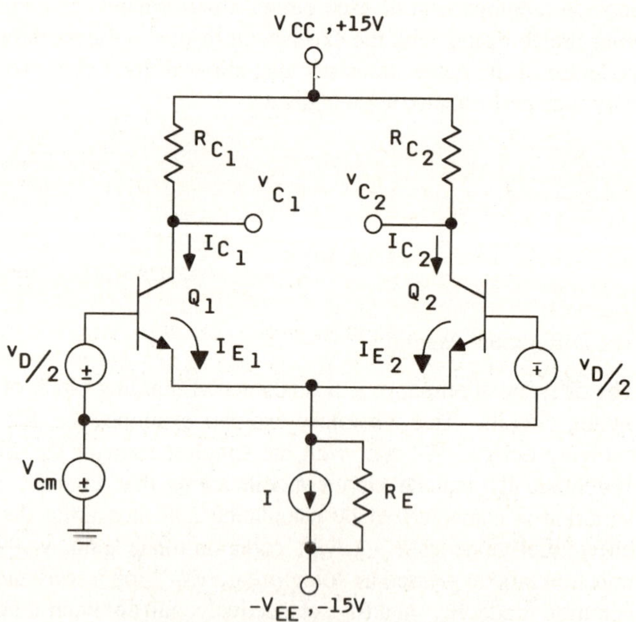

Figure 10L5.1: Differential amplifier

A single-ended output is v_{C_1} or v_{C_2} and the differential output is $v_{Od} = v_{C_2} - v_{C_1}$ where v_D is the differential input and v_{cm} is the common-mode input voltage.

a. Explain the half circuit technique, and show that the two common-mode half-circuits ($v_{cm} \neq 0$ and $v_D = 0$) and the differential-mode half circuits ($v_{cm} = 0$ and $V_D \neq 0$) are given by Figure 10L5.2(a) and (b), respectively.

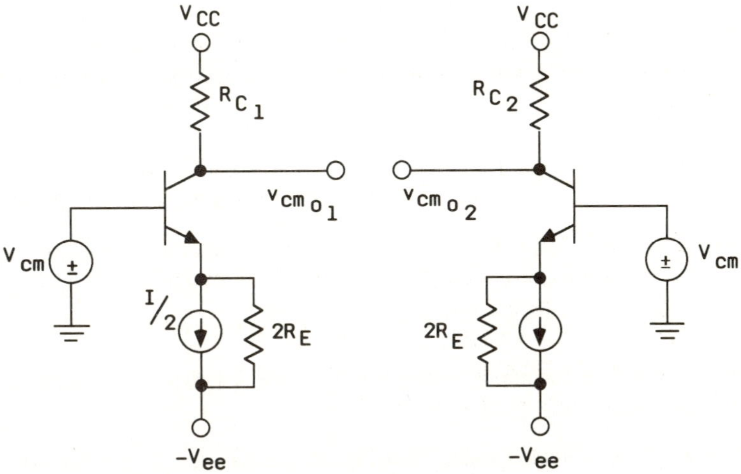

(a) Common–mode Half circuits

Figure 10L5.2 (a): Common-mode half circuits

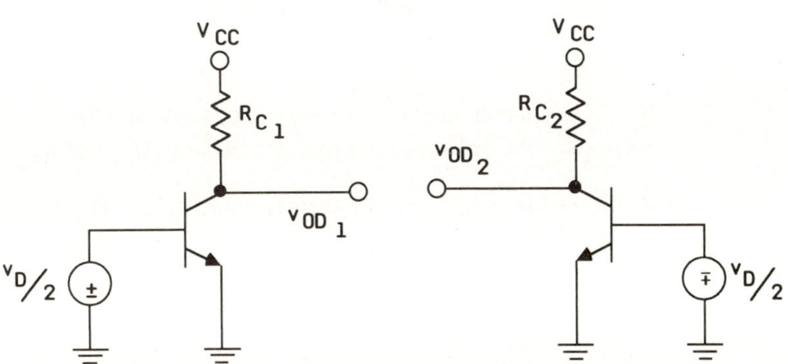

(b) Differential–Mode Half circuits

Figure 10L5.2(b): Differential mode half circuits

Use the small-signal hybrid-π model for low frequencies, assuming $r_o \rightarrow \infty$, $r_\mu \rightarrow \infty$, and $2R_E >> r_e$, where $r_e = \dfrac{r_\pi}{(\beta + 1)}$. Also, assume $R_{C_1} = R_{C_2}$.

b. Derive the single-ended common-mode voltage gain of the circuit of Figure 10L5.2(a). That is, find v_{cmo1}/v_{cm} or v_{cmo2}/v_{cm}. Show that the gain is $A_{cm} \approx \dfrac{\alpha R_C}{2R_E}$.

c. Show that the single-ended differential-mode gain A_d, $A_d \equiv v_{OD_2} / v_D$ is equal to $-1/2\, g_m R_C$ and the differential output gain, $\dfrac{v_{OD_2} - v_{OD_1}}{v_D} = -g_m R_C$.

d. Show that the common-mode rejection ratio CMRR, defined as $-20\log(\,|\, \dfrac{A_{cm}}{A_D} \,|\,)$ is equal to $-20\log(g_m R_C)$.

e. Again, use the half-circuits of Figure 10L5.2 to show that the common-mode input resistance

$$R_{icm} \approx \frac{r_\mu}{2} \; \| \; (\beta + 1)R_E \; \| \; (\beta + 1)\frac{r_o}{2} \tag{10L5.1}$$

Obviously, the small-signal model must include r_μ and r_o to yield this equation. Remember, the common-mode signal is applied to both inputs, and both of the half-circuits of Figure 10L5.2(a) are in parallel; thus the 1/2 for r_μ and r_o. In addition, it is assumed that the common-mode gain is small, so the ac voltage at the collector is very small, and thus it can be assumed that the collector is at ac ground.

f. Again, using the half circuits of Figure 10L5.2(b), show that the differential input resistance

$$R_{in_D} = \frac{v_D}{i_D} = (\beta + 1)\, 2r_e = R_{in_D} \tag{10L5.2}$$

Remember, the differential input voltage v_D is applied across both transistor inputs, which are differentially in series.

g. If the transistors and collector resistors of Figure 10L5.1 are closely, but not perfectly, matched, then an offset voltage or nonzero differential voltage will appear ($V_{c_2} - V_{c_1} \neq 0$) when $v_{cm} = v_D = 0$. This offset voltage is usually referred to as the input differential voltage required to make the differential output voltage $v_{c_2} - v_{c_1} = 0$. It can be shown that the offset voltage V_{OS} is given by Equation 10L5.3.

$$V_{os} = V_T \left[-\frac{\Delta R_C}{R_C} - \frac{\Delta I_S}{I_S} \right] \tag{10L5.3}$$

where

$$
\begin{array}{lll}
V_T & \equiv & \text{the thermal voltage, equal to 26 mV at } 25°C \\
\Delta R_C & \equiv & \text{the difference in resistance between } R_{C_1} \text{ and } R_{C_2} \\
R_C & \equiv & (\dfrac{R_{C_1} + R_{C_2}}{2}), \text{ the average value of } R_{C_1} \text{ and } R_{C_2}, \\
\Delta I_S & \equiv & \text{the difference in saturation currents of } Q_1 \text{ and } Q_2 \\
I_S & \equiv & (\dfrac{I_{S_1} + I_{S_2}}{2}), \text{ the average saturation current}
\end{array}
$$

If R_{C_1} is 1,010 Ω, R_{C_2} is 995 Ω, I_{S_1} is 1.1×10^{-13} A, and I_{S_2} is 9.95×10^{-14} A, determine the offset voltage V_{os}. Also, if $v_D = 0$ and $v_{cm} = 0$, what does $V_{C_2} - V_{C_1}$ equal?

h. For a nonideal operational amplifier, there is an input current referred to as the **input bias current**, I_{BIAS}. Calculate this input bias current for the circuit of Figure 10L5.1 if $v_{cm} = v_D = 0$, $\beta_1 = \beta_2 = \beta = 200$, $I = 1$ mA, $R_E = 100$ kΩ, and $V_{BE_1} = V_{BE_2} = 0.7$ V.

2. Design of a current source/sink.

a. Given the circuit of Figure 10L5.3, show that the relationship between I_I and I_o is given by

$$I_o R_2 = V_T \ln \left[\frac{I_I}{I_o} \right] \tag{10L5.4}$$

Hint: Write a loop equation including the two base-emitter voltages. Use the fact that $V_{BE} = V_T \ln (\frac{I_C}{I_S})$, where I_C is the collector current. $V_{CC} = +15$ V and $V_{EE} = -15$ V.

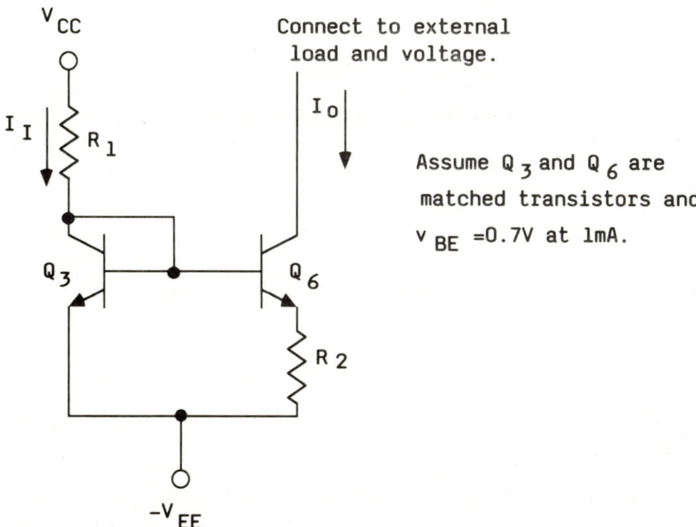

Figure 10L5.3: Widlar current source

b.   Determine R_1 and R_2 so that the current I_I equals 1 mA and I_o = 50 μA. Find the Norton and Thevenin equivalent models for the circuit of Figure 10L5.3 as one looks into the collector of Q_6. To find the Norton and Thevenin equivalent resistance of the current source, assume that the base of Q_6 is connected to ground. Show that the R_T (the Thevenin resistance) is given by

$$R_T \approx R_2 \parallel r_{\pi_3} + (1 + gm_6(r_{\pi_3} \parallel R_2))r_{o_6} \tag{10L5.5}$$

where

$$R_T \approx [1 + g_{m_6}(r_{\pi_3} \parallel R_2)]r_{o_6} \tag{10L5.6}$$

Since the first term of Equation 10L5.5 is much smaller than the second term, Equation 10L5.6 results. This analysis neglects r_μ.

3.   The current source. In a current source, the constant current is pushed into a load connected to a more negative voltage. Analyze the circuit of Figure 10L5.4.

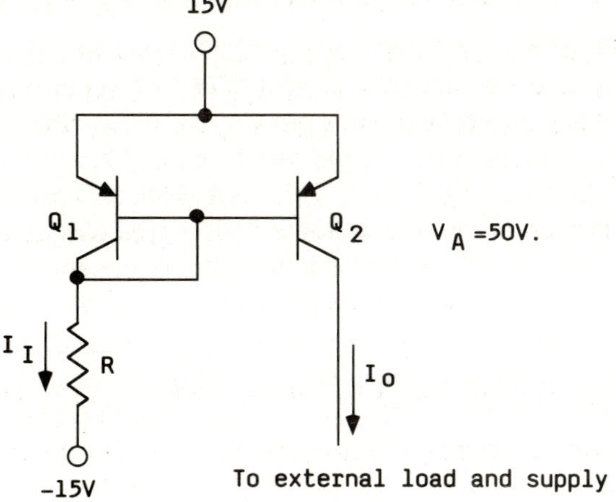

Figure 10L5.4: pnp current source

a. Find the ratio of I_o to I_I. Evaluate I_o / I_I for β equal to 10, 50, 100, 200. (Assume R=5 KΩ and $V_{BE} = 0.7$ V.) What is the value of I_o for each β?

b. For $\beta = 100$, determine the value of R that results in $I_o = 1$ mA.

c. Draw the Norton equivalent circuit for the current source.

4. The current source as an active load. The analysis of the differential amplifier in Part 1 of the prelab shows that the differential gain, single ended or differentially, is proportional to the collector resistor R_C. Because resistors fabricated on silicon for integrated circuits consume large areas of silicon, a substitute for a passive resistor must be found. One approach is to take advantage of the large value of the collector-to-emitter resistance r_o for a bipolar junction transistor.

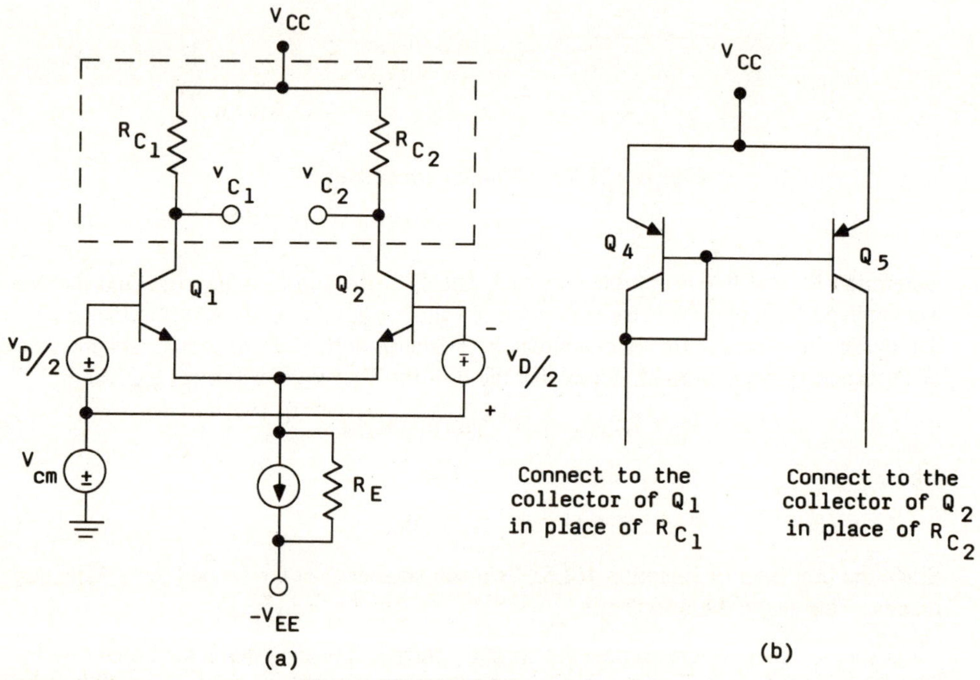

Figure 10L5.5: Differential amplifier

In the load circuit of Figure 10L5.4(a), let the current mirror of Figure 10L5.5(b) replace R_{C_1} and R_{C_2}. The current source is connected in two steps. First, replace R_{C_1} with Q_4. Once the collector of Q_4 is connected to the collector of Q_1, the collector current of Q_1 will set the current mirror's input current (see Figure 10L5.4). This small-signal current is proportional to the differential input voltage v_d. If v_d increases, then the collector current of Q_1 will increase by $g_m v_D / 2$, and by symmetry, if v_D decreases, the collector current will decrease by $g_m v_D / 2$. This decrease in collector current can be viewed in the small-signal sense as an increase in the current flowing out of the collector of Q_2. Since Q_4 and Q_5 are matched transistors, the collector of Q_5 will match the collector current of Q_4 and the collector current will increase by $g_m v_D / 2$. Now, if the collector of Q_5 connects to the collector of Q_2, replacing R_{C_2}, and a load resistor R_L is connected to the collector of Q_2 and ground, then a small-signal current of $g_m v_D$ will flow in R_L. The small-signal output voltage will then increase by $g_m R_L v_D$. We shall see in Part 5 that if R_L is replaced with an open circuit, the voltage at the collector of Q_2 will increase and therefore the voltage gain of the circuit will also increase. These increases are due to the relatively large value of the collector-to-emitter resistance of Q_5 compared to the value of R_C that could be fabricated on a small area of silicon.

a. Show that the output resistance R_o looking into the collector of Q_5 is equal to the r_o of Q_5, where r_o is the small-signal resistance between the collector and emitter of a BJT. (Assume $\beta \gg 1$.)

b. Show that when the collectors of Q_2 and Q_5 are connected together, the circuit's output resistance, when considering a small-signal differential input, v_D, is given by Equation 10L5.7

$$R_o = r_o(Q_2) \parallel r_o(Q_5) \qquad\qquad (10L5.7)$$

where $r_o(Q_2)$ and $r_o(Q_5)$ are the output resistances looking into the collectors of Q_2 and Q_5, respectively.

5. Integrated differential amplifier. Consider the circuit of Figure 10L5.6.

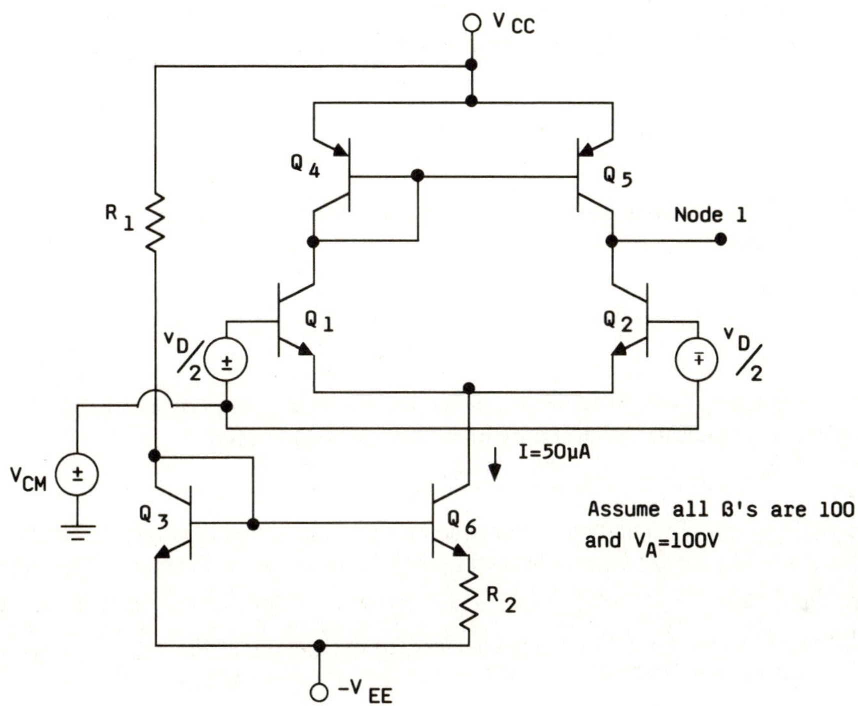

Figure 10L5.6: Integrated differential amplifier

a. Identify the three circuits of the previous sections included in the circuit of Figure 10L5.6, and identify which transistors should be matched.

b. If $v_d = v_{cm} = 0$, what are the emitter currents of Q_1 and Q_2? What are the collector currents of Q_1 and Q_2, and what is the voltage at the emitters?

c. Let $v_d = 0$ and $v_{cm} \neq 0$. What is the range of v_{cm} that allows all transistors to operate in the active mode?

d. Let $v_{cm} = 0$ and $v_d = 1$ mV.

   (1) By what amount will the collector current of Q_1, I_{C_1} increase?

   (2) By what amount will the collector current of Q_2 I_{C_2} decrease?

   (3) What is the difference between the collector currents of Q_1 and Q_2?

   (4) Draw the small-signal circuit at node 1 by replacing Q_2 and Q_5 by the hybrid-π model of Figure 10L5.7.

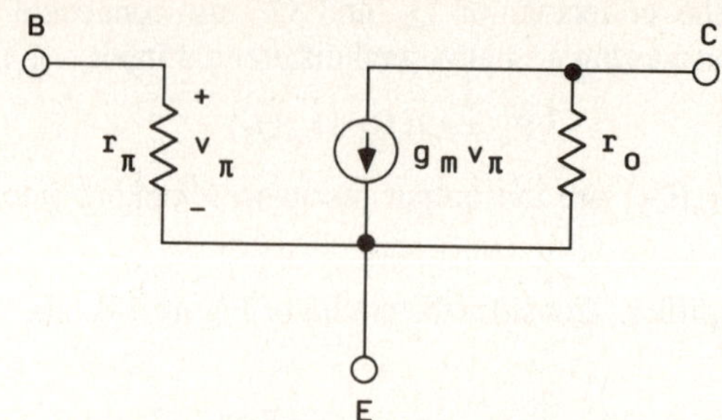

Figure 10L5.7: Hybrid-π model

(5)   What is the small-signal output resistance looking into node 1 of the circuit of Figure 10L5.6?

(6)   What is the change in output voltage at node 1 due to the connection of v_d at the input? What is the single-ended differential voltage gain? What is the differential input resistance?

e.   Let $v_d = 0$ and $v_{cm} \neq 0$. What is the common-mode input resistance seen by v_{cm}? Assuming that all devices are matched, what is the common-mode voltage gain?

f.   Determine the common-mode rejection ratio.

g.   Let $v_{cm} = v_d = 0$. Determine the input bias current (the current flow into the base of Q_1 and Q_2). If the β's of Q_1 and Q_2 are mismatched by 10%, what is their difference in base currents? (Note: $I_{B_1} - I_{B_2} = I_{O_s}$.)

6.   Use SPICE and compare the following parameters for the circuits of Figure 10L5.1 and 10L5.6. Use the values of v_A and β given in Figure 10L5.6. Let $R_{C_1} = R_{C_2} = 1 \text{ k}\Omega$, $R_E = 100 \text{ k}\Omega$, $I = 1 \text{ mA}$ $V_{CC} = 15 \text{ V}$, and $V_{EE} = -15 \text{ V}$.

a.   Common-mode voltage gain
b.   Single-ended differential gain
c.   Differential gain
d.   Common-mode rejection ratio
e.   Common-mode input resistance
f.   Input bias current $\dfrac{I_{B_2} + I_{B_1}}{2}$
g.   Input offset current $| I_{B_1} - I_{B_2} |$
h.   Input offset voltage V_{os}

Laboratory Procedure:

1.   The discrete differential amplifier.

a.   Build the circuit of Figure 10L5.8.

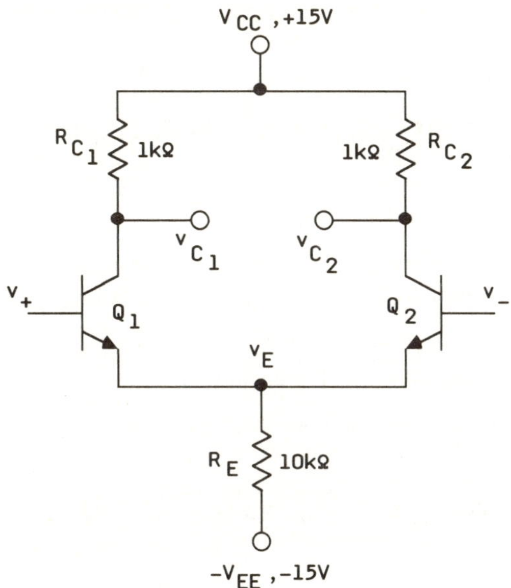

Figure 10L5.8: Discrete differential amplifier

b. Connect v_+ and v_- to ground and measure v_{C_1} v_{C_2}, and v_E. Calculate the differential voltage gain given by Equation 10L5.8. Note: $v_D/2 = v_+$ and $- v_D/2 = v_-$.

$$\frac{v_{C_2} - v_{C_2}}{v_D} = - g_m R_C \tag{10L5.8}$$

Determine the input offset voltage v_{os}.

c. Let $v_+ = 25$ mV and $v_- = -25$ mV. Measure and record v_{C_1}, v_{C_2}, and v_E. Determine the differential voltage gain $(v_{C_2} - v_{C_1})/v_D$ and the single-ended voltage gain v_{C_2}/v_D. Note: $v_D = 50$ mV and v_{C_1}, v_D, and V_{C_2} are the small-signal changes from their quiescent zero input values, found in Part B.) Complete Table 10L5.1.

Table 10L5.1: Differential and single-ended voltage gain

v_+	v_-	V_{C_1}	V_{C_2}
0	0		
25 mV	-25 mV		
50 mV	-50 mV		
75 mV	-75 mV		
100 mV	-100 mV		
150 mV	-150 mV		
200 mV	-200 mV		

d. From the table, plot the differential voltage gain $(v_{c_2} - v_{C_1})/v_D$ and the single-ended differential voltage gain v_{c_2}/v_D, versus v_D. Note $v_D = (v_+ - v_-)$. Over what range of v_D are the gains constant?

e. Let $v_+ = v_- = v_{cm}$. Vary v_{cm} from +15 V to -15 V. Over what range of v_{cm} do Q_1 and Q_2 remain in the active mode? Let $v_{cm} = \sin(2\pi ft)$, with f=1 kHz; and measure the common-mode voltage gain $(v_{c_2})/v_{cm}$. Record the common-mode voltage gain A_{cm}.

f. Calculate the common-mode rejection ratio CMRR. (Use the definition given in the prelab.)

g. Let $v_+ = v_- = 0$ V, and measure the bias current I_B flowing into the base of each transistor Q_1 and Q_2. Determine the offset current $| I_{B_1} - I_{B_2} |$.

2. The differential amplifier using a matched pair of npn transistors. Replace the discrete transistors Q_1 and Q_2 of Figure 10L5.8 with a matched pair. Then repeat all measurements of Part 1, b, c, d, e and f and record the data.

3. The differential amplifier using an active current source.

    a. Replace the 10 kΩ resistor used in Part 2 with an active current sink shown in Figure 10L5.9.

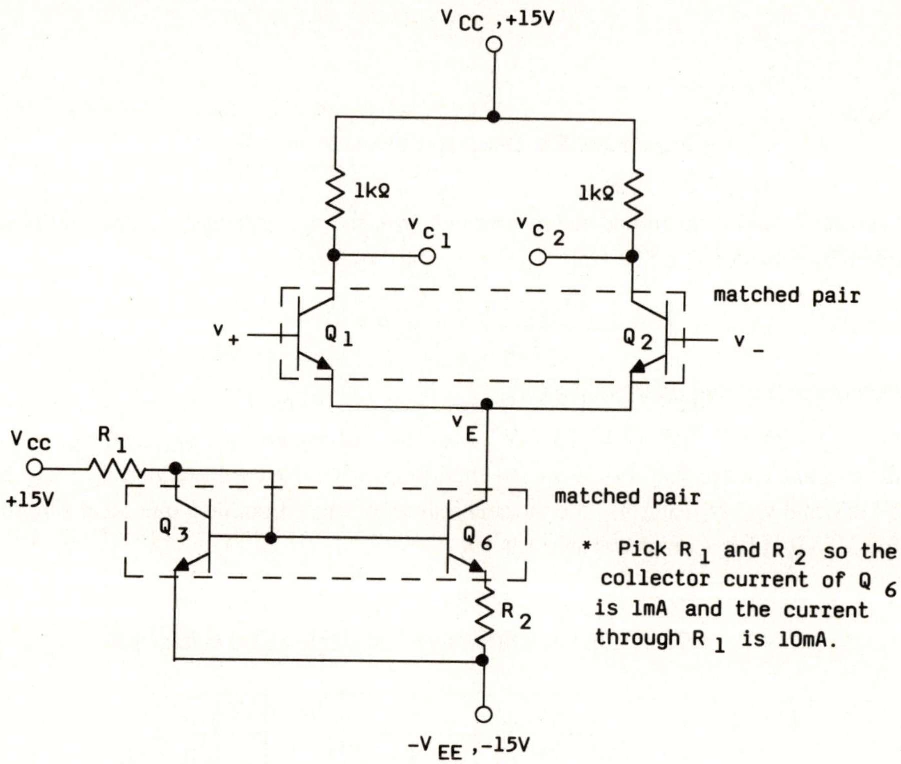

Figure 10L5.9: Differential amplifier with active current source

b. Repeat the measurement of Part 2.a using the circuit of Figure 10L5.9.

4. The integrated differential amplifier.

    a. Replace the 1 KΩ resistors used in Part 3 with the active loads shown in Figure 10L5.10.

    b. Repeat the measurements of Part 1, b, c, d, e, and f, and record all data.

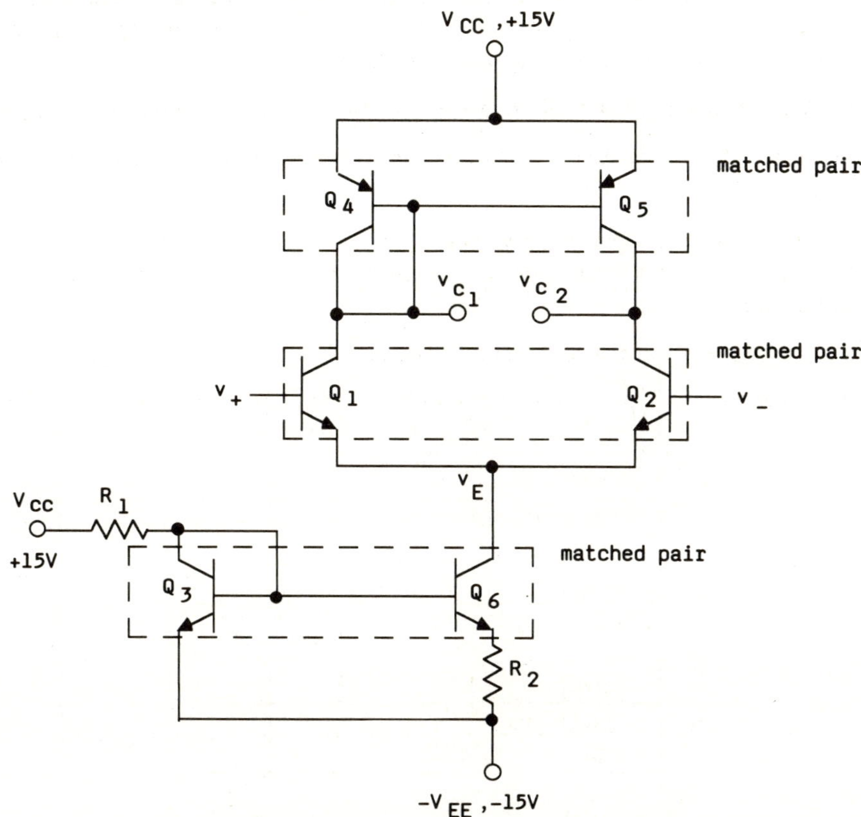

Figure 10L5.10: Integrated differential amplifier

DISCUSSION: Use the data collected in parts 1 through 4, and complete Table 10L5.2.

Table 10L5.2: Summary of experimental data

Laboratory Section					
Parameter	Part 1	Part 2	Part 3	Part 4	SPICE
Input bias current $\dfrac{I_{B_1} + I_{B_2}}{2}$					
Input common-mode voltage range					
Input offset voltage V_{os}					
Input offset current $\mid I_{B_1} - I_{B_2} \mid$					
Common-mode input resistance, R_{icm}					
Single-ended differential voltage gain v_{oc_1}/v_{i_D}					
Common-mode voltage gain v_{ocm}/v_{icm}					
Common-mode rejection ratio CMRR					

CONCLUSIONS:

LAB 10L6

TITLE: The MOSFET Integrated Differential Amplifier[2],[4]

OBJECTIVE: This laboratory exercise examines the use of MOSFETs to implement analog differential amplifiers. From their conception in the 1940s until the 1970s, MOSFETs were almost exclusively found in the design of digital systems, including memories and microprocessors. Most analog designs used BJTs. Bipolar transistors have a larger transconductance and closer parameter matching than MOSFETs. One advantage of MOSFETs over BJTs is their very large input resistance.

To make systems more reliable and less costly, one can put both analog and digital functions on a single chip, a monolithic design. BJTs can be used to implement digital functions (TTL), but the high density of MOSFETs cannot be matched by BJTs because of the complexity of BJTs and their larger power dissipation.

Therefore, an alternative is to implement analog functions using MOSFET technology. The differential amplifier is well suited for monolithic design because of its ability to couple stages without capacitors. Also, all the supporting devices for a differential amplifier, such as matched load resistors and current sources, can be implemented with the MOSFET. These "all-MOSFET differential amplifiers" can be made very small because of the ability to scale down the size of the device. This laboratory exercise evaluates the MOSFET differential amplifier as it evolves from a discrete design to a monolithic one. Each of the three stages of evolution of the MOSFET differential amplifier exercise is evaluated for the following parameters:

1: Differential gain, $\dfrac{v_o}{v_D}$

2: Common-mode gain, $\dfrac{v_o}{v_{cm}}$

3: Common-mode rejection ratio, CMRR

4: Common-mode voltage swing

5: Input resistance

6: Effect of component mismatch on offset voltage.

The laboratory includes the use of SPICE to aid in complex calculations.

EQUIPMENT AND MATERIALS:

Dual-channel oscilloscope capable of making a differential measurement

Triple power supply ($\pm$15 V and +12 V)

Function generator (with offset)

Dual MOSFETS:

n-channel enhancement MOSFET (2N4351)

p-channel enhancement MOSFET (3N161)

BACKGROUND: Consider the circuit of Figure 10L6.1.

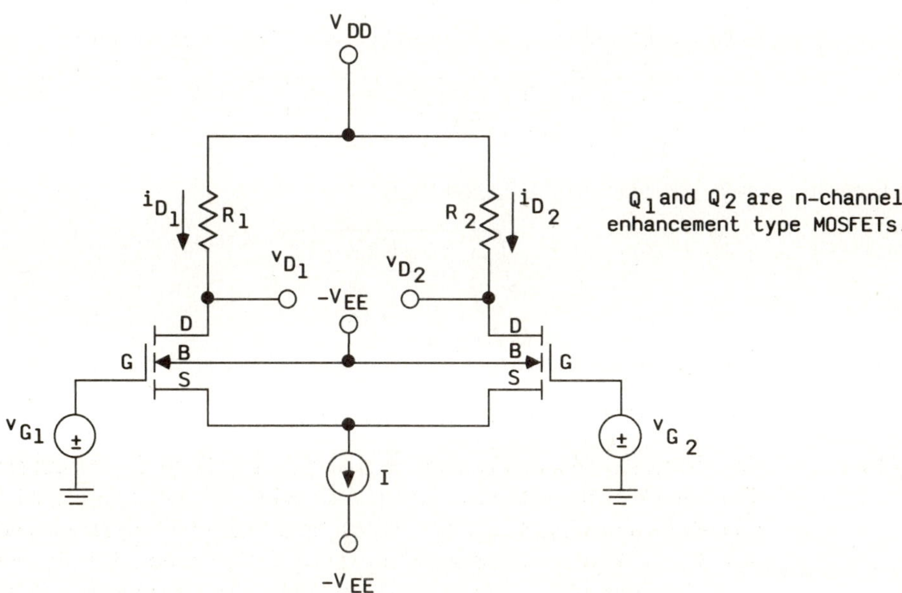

Figure 10L6.1: Differential amplifier

For this circuit, we assume that Q_1 and Q_2 are matched-- that is, that their threshold voltage V_t and $K=\frac{1}{2}\mu_n C_{ox}\dfrac{W}{L}$ are identical. In this equation,

μ_n =   electron mobility,

C_{ox} =   gate oxide capacitance per unit area,

W/L =   channel aspect ratio (width to length).

As long as Q_1 and Q_2 are in the active region, their drain currents are respectively given by Equations 10L6.1(a) and (b)

$$i_{D_1} = K(v_{GS_1} - V_t)^2 \qquad\qquad (10L6.1(a))$$

and

$$i_{D_2} = K(v_{GS_2} - V_t)^2 \qquad\qquad (10L6.1(b))$$

If $v_{GS_1} = v_{GS_2}$ and the devices remain in the active region, then $i_{D_1} = i_{D_2}$.

Now, applying KCL at the node labeled "S" in Figure 10L6.1, we get

$$i_{D_1} + i_{D_2} = I \qquad\qquad (10L6.2)$$

Our objective is to solve for i_{D_1} and i_{D_2} by using Equations (10L6.1(a)), and (b) and (10L6.2). We start by taking the square root of each of Equations (10L6.1 (a) and (b)). Then we subtract the latter from the former to yield

$$\sqrt{i_{D_1}} - \sqrt{i_{D_2}} = \sqrt{K}\left[v_{GS_2} - v_{GS_1}\right] \qquad\qquad (10L6.3)$$

Now, define $v_{GS_1} - v_{GS_2}$ as v_{ID}. Substituting v_{ID} and Equation (10L6.2) into Equation (10L6.3), we obtain

$$i_{D_1}^2 - i_{D_1}\,I + \frac{1}{4}(Kv_{ID}^2 - I)^2 = 0 \qquad\qquad (10L6.4)$$

This quadratic equation can be solved to yield Equation 10L6.5.

$$i_{D_1} = \frac{I}{2} \pm \sqrt{(\frac{I}{2})^2 - (\frac{Kv_{ID}^2 - I}{2})^2} \qquad\qquad (10L6.5)$$

Recognizing that the expression under the square root is the difference of two squares, we obtain

$$i_{D_1} = \frac{I}{2} \pm (\frac{v_{ID}}{2})\sqrt{2KI}\sqrt{1 - \frac{v_{ID}^2 K}{2I}} \qquad\qquad (10L6.6)$$

Since $i_{D_1} + i_{D_2} = I$ and v_{ID} is positive, it follows that $v_{GS_1} > v_{GS_2}$. Therefore, i_{D_1} will increase, and the only solution that makes sense is

$$i_{D_1} = \frac{I}{2} + (\frac{v_{ID}}{2})\sqrt{2KI}\sqrt{1 - \frac{v_{ID}^2 K}{2I}} \qquad\qquad (10L6.7)$$

By a similar derivation, one obtains

$$i_{D_2} = \frac{I}{2} - (\frac{v_{ID}}{2})\sqrt{2KI}\sqrt{1 - \frac{v_{ID}^2 K}{2I}} \qquad\qquad (10L6.8)$$

At the quiescent point, $v_{ID}=0$ and $i_{D_1}=i_{D_2}=\dfrac{I}{2}$. This corresponds to $v_{GS_1} = v_{GS_2} = v_{GS}$, where $\dfrac{I}{2}=K(v_{GS} - V_t)^2$. From this, it is possible to find an equivalent expression for $\sqrt{2KI}$ as follows:

$$2K = \frac{I}{(v_{GS} - V_t)^2}$$

$$2KI = \frac{I^2}{(v_{GS} - V_t)^2}$$

$$\sqrt{2KI} = \frac{I}{(v_{GS} - V_t)} \tag{10L6.9}$$

Substituting this equation into Equations (10L6.7) and (10L6.8) results in

$$i_{D_1} = \frac{I}{2} + \frac{I}{(v_{GS} - V_t)}\left(\frac{v_{ID}}{2}\right)\left[1 - \left(\frac{\dfrac{v_{ID}}{2}}{v_{GS}-V_t}\right)^2\right]^{\frac{1}{2}} \tag{10L6.10}$$

and

$$i_{D_2} = \frac{I}{2} - \frac{I}{(v_{GS} - V_t)}\left(\frac{v_{ID}}{2}\right)\left[1 - \left(\frac{\dfrac{v_{ID}}{2}}{v_{GS}-V_t}\right)^2\right]^{\frac{1}{2}} \tag{10L6.11}$$

If $V_{ID/2}$ is much smaller than $(v_{GS} - V_t)$, then the expression $\dfrac{\dfrac{v_{ID}}{2}}{v_{GS} - V_t} \approx 0$ and Equations (10L6.10) and (10L6.11) reduce to

$$i_{D_1} \approx \frac{I}{2} + \left[\frac{I}{(v_{GS} - V_t)}\right]\left[\frac{v_{ID}}{2}\right] \tag{10L6.12}$$

and

$$i_{D_2} \approx \frac{I}{2} - \left[\frac{I}{(v_{Gs} - V_t)}\right]\left[\frac{v_{ID}}{2}\right] \tag{10L6.13}$$

It can easily be shown that $\dfrac{I}{(v_{GS} - V_t)}$ is equivalent to the transconductance of a MOSFET. Therefore, Equations (10L6.12) and (10L6.13) can be written as

$$i_{D_1} \approx \frac{I}{2} + g_m \frac{v_{ID}}{2} \tag{10L6.14}$$

and

$$i_{D_2} \approx \frac{I}{2} + g_m\left[\frac{v_{ID}}{2}\right] \tag{10L6.15}$$

(See Figure 10L6.1.) The differential output voltage is the difference between

$$v_{D_2} = V_{DD} - i_{D_2}R_D \tag{10L6.16}$$

and

$$v_{D_1} = V_{DD} - i_{D_1}R_D \tag{10L6.17}$$

Substituting Equations (10L6.14) and (10L6.15) into Equations (10L6.16) and (10L6.17) yields

$$v_{D_2} - v_{D_1} = g_m R_D v_{ID} \tag{10L6.18}$$

where

$$g_m = \frac{I}{v_{GS} - V_t}$$

Also, the single-ended small-signal voltage v_{D_2} is given by

$$v_{D_2} = g_m R_D \frac{v_{ID}}{2} \tag{10L6.19}$$

Now, consider the common-mode gain ($v_{ID} = 0$ and $v_{G_1} = v_{G_2}$). Qualitatively, this can be analyzed by looking at Figure 10L6.1. If $v_{ID} = 0$, then as long as the current source is functioning, I will be the total current, and it must split equally between Q_1 and Q_2. Therefore v_{D_1} and v_{D_2} should not change when a common-mode voltage is applied. This means that as v_G increases, V_S must increase to keep v_{GS} constant and, therefore, i_D constant. You will determine a quantitative expression for the common-mode voltage gain in the prelab.

PRELABORATORY:

1. If the ideal current source of Figure 10L6.1 is replaced by a current source with finite impedance, the circuit of Figure 10L6.2 is obtained.

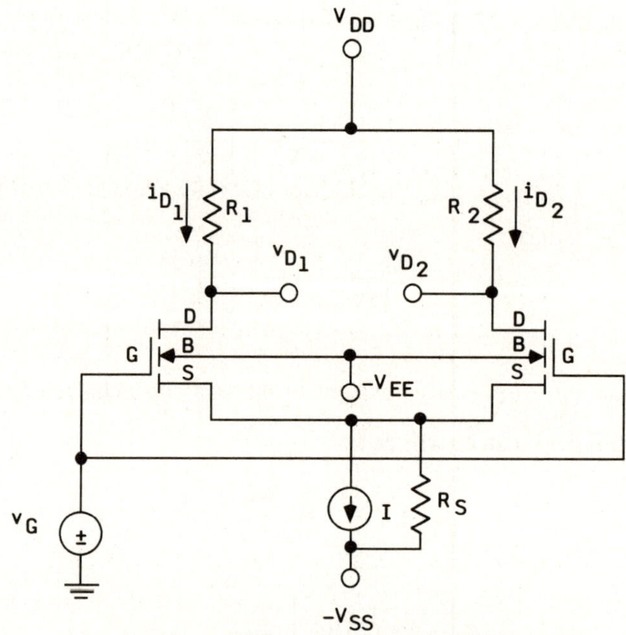

Figure 10L6.2: Differential amplifier

The common-mode gain can be determined by the application of the half-circuit technique, shown in Figure 10L6.3.

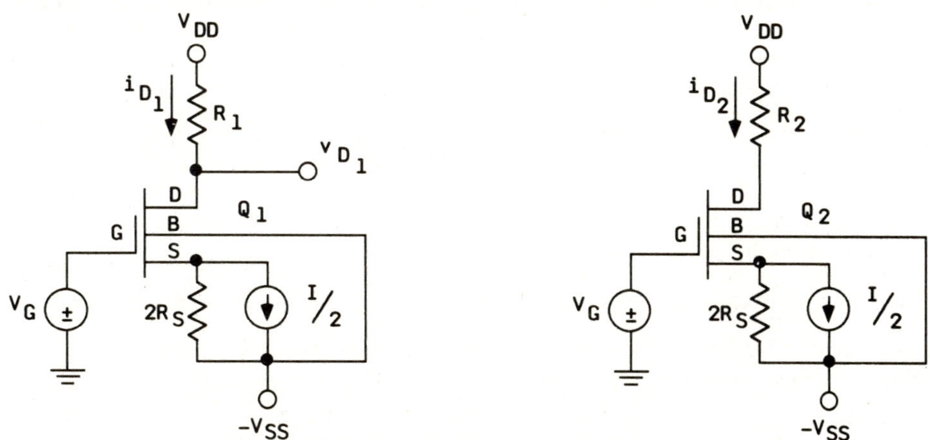

Figure 10L6.3: Half circuit technique

a. Draw the small-signal equivalent circuit for either half-circuit; also, show that the common-mode voltage gain

$$A_{cm} \equiv \frac{v_{D_1}}{v_g} = \frac{v_{D_2}}{v_g} = \frac{-g_m R_d}{1 + 2 g_m R_s} \tag{10L6.20}$$

where v_g is the small-signal component of v_G.

If $g_m 2 R_S > 1$, then

$$A_{cm} \approx \frac{R_D}{2 R_s} \tag{10L6.21}$$

For this analysis, use the MOSFET model shown in Figure 10L6.4. Neglect the body effects of the device.

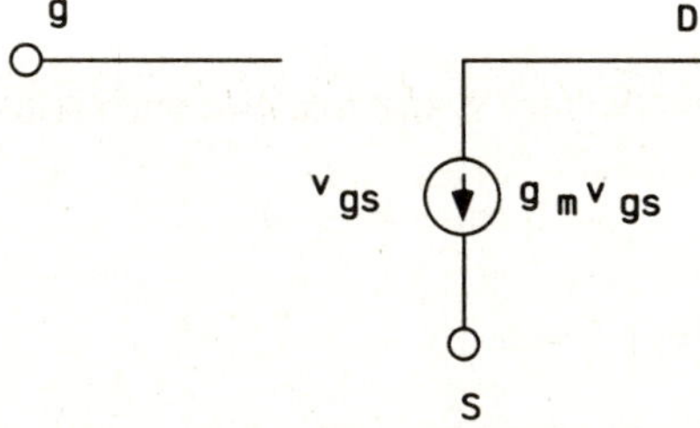

Figure 10L6.4: MOSFET equivalent circuit model

b. Determine the common-mode rejection ratio, defined as

$$20 \log \left| \frac{A_{cm}}{A_d} \right| \ \text{dB}$$

c. Let the following be the values of the parameters for the components in the circuit of Figure 10L6.2:

$$V_t = 2 \text{ V} \qquad R_S = 1 \text{ M}\Omega \qquad -V_{SS} = -15 \text{ V}$$
$$I = 1 \text{ mA} \qquad K = 10 \text{ }\mu\text{A/V}^2$$
$$R_1 = 1 \text{ k}\Omega \qquad V_{DD} = 15 \text{ V}$$

Determine the voltage range of the common-mode and differential signal that allows Q_1 and Q_2 to operate in the active region.

d. Find the input resistance looking into either gate. Use the model shown in Figure 10L6.4.

e. Source of offset voltage, v_{OS}. Consider the circuit of Figure 10L6.5.

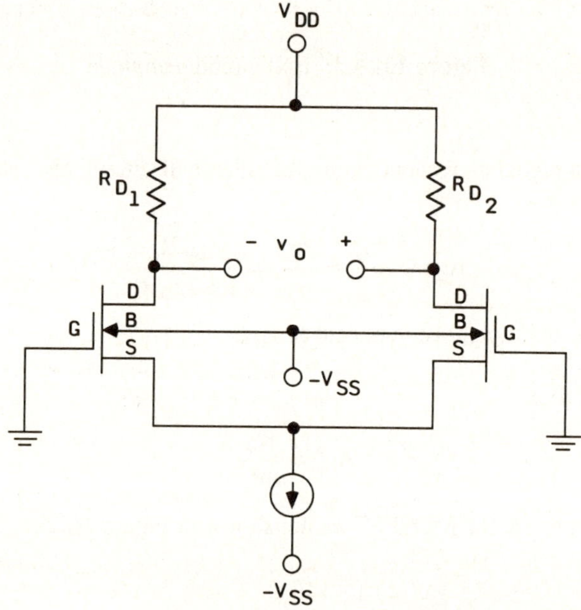

Figure 10L6.5: Differential amplifier

(i) Show that the output voltage v_o due to a mismatch in the drain resistors R_{D_1} and R_{D_2} is

$$| \, v_o \, | = \frac{I}{2} \, \Delta R_D \tag{10L6.22}$$

where $\Delta R_D = | \, R_{D_1} - R_{D_2} \, |$

Also, show that the effective offset voltage at the input is given by

$$\text{Input offset voltage, } V_{OS} = \left[\frac{V_{GS} - V_t}{2} \right] \frac{\Delta R_D}{R_D} \tag{10L6.23}$$

(ii) The differential output voltage v_o can also be caused by a mismatch in the aspect ratio $\frac{W}{L}$, where W is the width and L the length of the MOSFET channel. This results in a mismatch in K's, ΔK, and is given by the offset voltage

$$V_{OS} = \frac{(v_{GS} - V_t)}{2} \left[\frac{\Delta K}{K} \right] \tag{10L6.24}$$

(iii)   The offset voltage due to the effect of a mismatch between the two threshold voltages

$$V_{OS} = \Delta V_t \qquad (10L6.25)$$

If we assume that all three sources of offset voltage are statistically independent, then the standard deviation of the total offset voltage is found by taking the square root of the sum of the squares of the standard deviations of each source of offset voltage. That is,

$$\sigma_{\text{total offset voltage}} = \sqrt{\sigma_1^2 + \sigma_2^2 + \sigma_3^2} \qquad (10L6.26)$$

where σ_1, σ_2, and σ_3 represent the standard deviations in offset voltage caused by each source of offset.

Is the assumption of independence for the three sources of offset valid? Why?

2.   The MOSFET differential amplifier with a MOSFET current source.

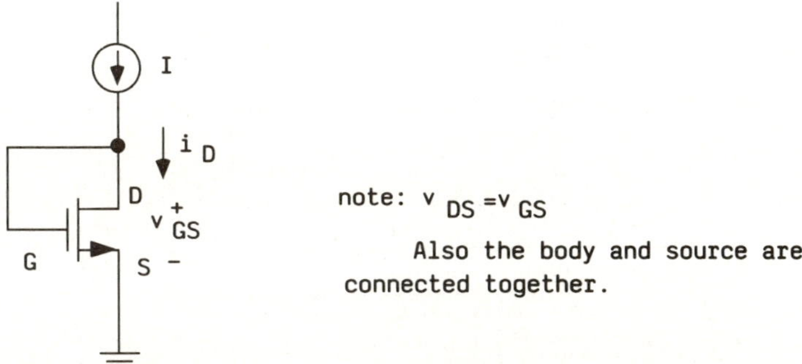

Figure 10L6.6: n-channel enhancement MOSFET

The circuit shown in Figure 10L6.6 is an n-channel enhancement-type MOSFET. The current-voltage characteristic is:

$$i_D = K(v_{GS} - V_t)^2 \qquad (10L6.27)$$

Observe that $v_{DS} = v_{GS}$ and the drain current equals I. Equation (10L6.27) is plotted in Figure 10L6.7.

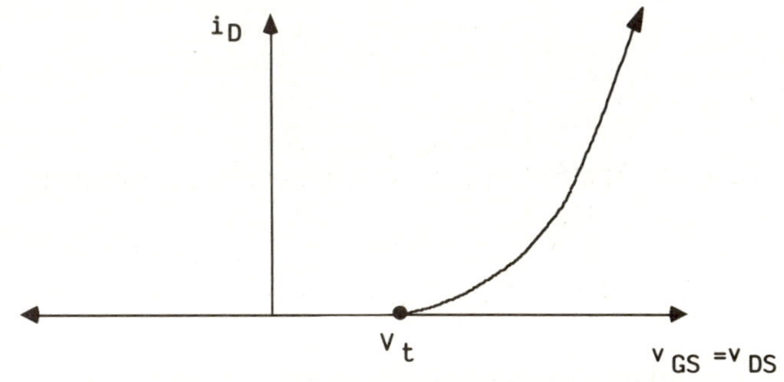

Figure 10L6.7: Graph of Equation 10L6.27

Now consider two matched n-channel enhancement MOSFETs connected as shown in Figure 10L6.8.

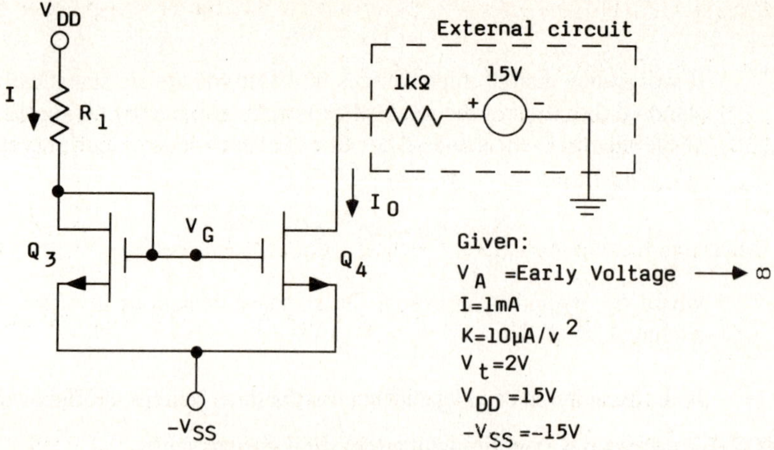

Figure 10L6.8: Current mirror

a. Determine the voltage v_G, resistor, and the output resistance seen looking into the drain of Q_4. Also, find the drain current of Q_4, labeled I_o.

b. Determine the output resistance seen looking into the drain of Q_4 if $V_A = 60$ V.

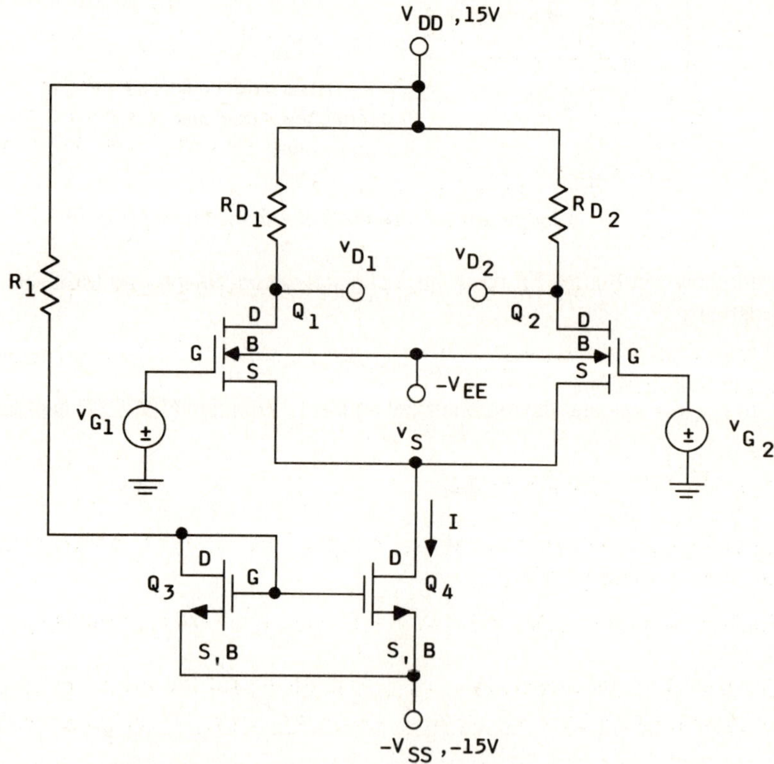

Figure 10L6.9: Differential amplifier with current source

For $v_{G_1} = v_{G_2} = 0$V, determine the value of I, v_S, and g_m for Q_1 and Q_2 of Figure 10L6.9. Let K $(Q_3 \& Q_4) = 10 \, \mu A/V^2$, K $= (Q_1 \& Q_2) \, 60 \, \mu A/V^2$, $V_t = 2$ V, $V_A = 100$ V, $R_{D_1} = R_{D_2} = 150 \, k\Omega$, $R_1 = 248 \, k\Omega$ and $-V_{EE} = -15$ V.

c.   If $v_{G_1} \neq v_{G_2}$ and $v_{G_1} - v_{G_2} \equiv v_{ID}$, determine the differential voltage gain $\left| \dfrac{v_{D_2} - v_{D_1}}{v_{iD}} \right|$. Also find the single-ended differential gain, v_{D_2}/v_{id}.

d.   Replace the current source composed of Q_3 and Q_4 by its Norton equivalent model. Then use the half-circuit technique presented in Part 1 of the lab to determine the common-mode voltage gain, $A_{cm} = V_{D_2}/v_{cm}$, where $v_{cm} = v_{G_1} = v_{G_2}$.

e.   Determine the common-mode rejection ratio, CMRR $= -20 \log \left| \dfrac{A_{cm}}{A_d} \right|$; where $A_d = v_{D_2}/v_{id}$.

f.   Determine the common-mode voltage range over which all transistors remain in the active region.

g.   Determine the differential mode voltage range over which Q_1 and Q_2 remain in the active mode.

3.   The MOSFET differential amplifier with active loads. In the circuit of Figure 10L6.10, Q_5 and Q_6 are p-channel enhancement-mode devices.

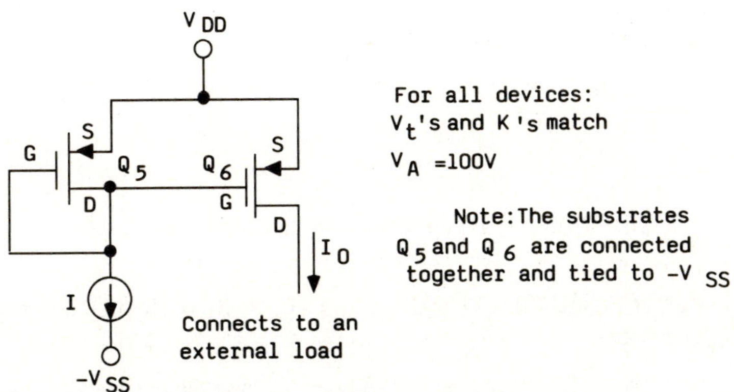

Figure 10L6.10: Active load for differential amplifier

If the voltage drop across Q_5 is at least V_t, then Q_5 will conduct the current I, and the voltage between the gate and source is

$$I = K(V_{GS} - | V_t |)^2 \tag{10L6.28}$$

where V_{GS} and V_t are negative voltages.

Since Q_5 and Q_6 are matched and the v_{GS} voltages are the same for both devices, the drain voltage of Q_6 equals the drain current of Q_5.

a.   Find the output resistance and Norton equivalent circuit of the drain of Q_6.

Consider the replacement of R_{D_1} and R_{D_2} in Figure 10L6.9 by the circuit of Figure 10L6.10. The resulting circuit is shown in Figure 10L6.11. Q_4 acts as the current source I in Figure 10L6.10.

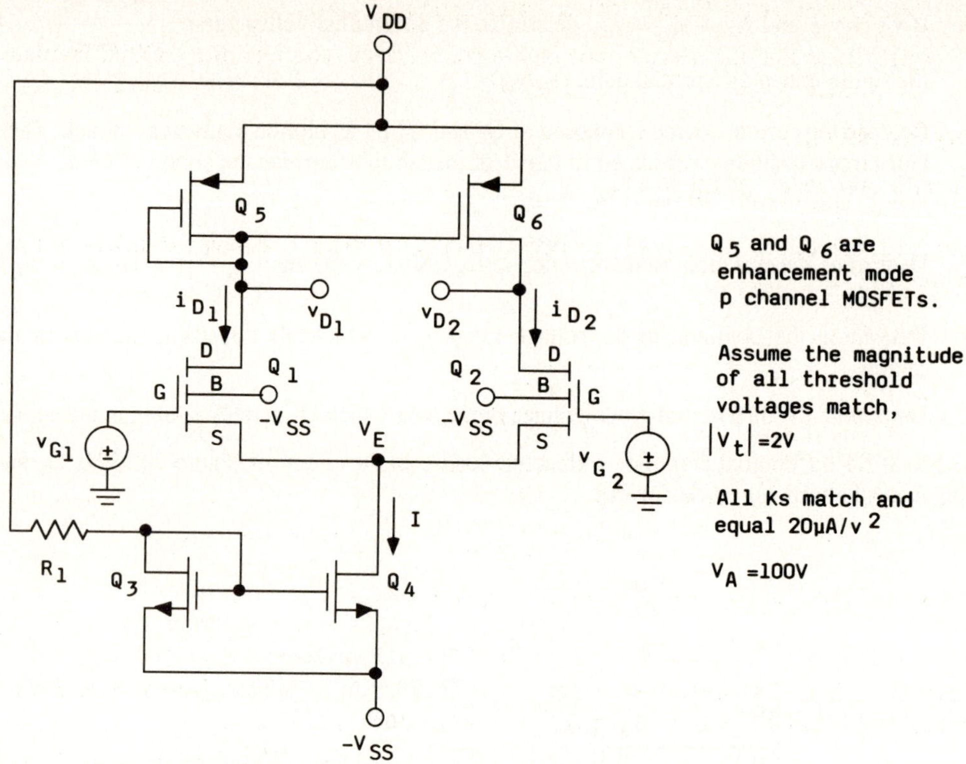

Figure 10L6.11: Differential amplifier with active loads

b.   For the circuit of Figure 10L6.11 with $-V_{SS}$ = -15 V, R_1 = 256 kΩ, and I = 0.1 mA, answer the following questions.

   (i)   Let v_{G_1} = v_{G_2} = 0. Determine the voltage, V_E, and the current flowing through Q_1 and Q_2.

   (ii)   Let v_{G_1} - v_{G_2} ≡ v_D ≠ 0. Find the resulting increase in the drain current of Q_1, and the resulting decrease in drain current of Q_2. Since Q_5 and Q_6 act as a current mirror, find the current flowing out of the drain of Q_6.

   (iii)   Draw the small-signal circuit around the node labeled v_{D_2}. Use the small-signal model of Figure 10L6.12 to model Q_6 and Q_2. Determine the small-signal single-ended differential voltage gain, $A_d \equiv \dfrac{v_{D_2}}{v_d}$.

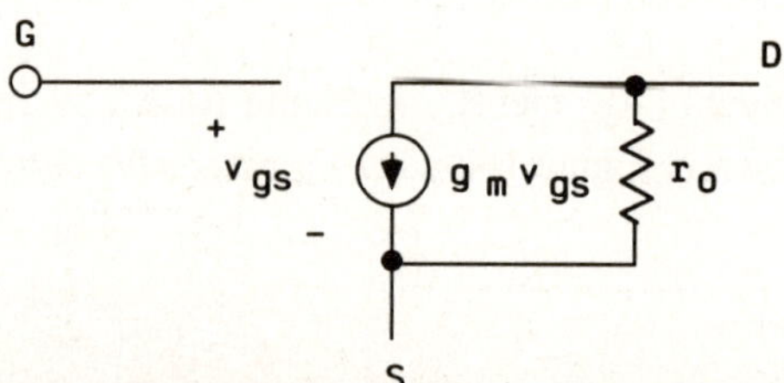

Figure 10L6.12: Hybrid-π equivalent circuit

   (iv)   Use the half-circuit technique of part 1 and the small-signal model of Figure 10L6.12 to determine the common-mode voltage gain of the circuit of Figure 10L6.11.

(v) Determine the CMRR for the circuit of Figure 10L6.11.

(vi) Determine the range of common mode voltages for which the MOSFETs remain in the active range.

Laboratory Procedure:

1. Component parameter measurement.

Build the circuit of Figure 10L6.13. Use this circuit to determine K and V_t for each n- and p- channel enhancement MOSFET.

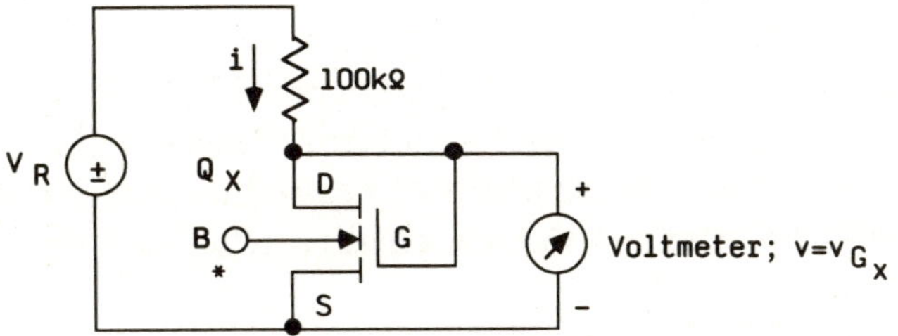

Figure 10L6.13: Circuit of Part 1

(Note: For n-channel devices, connect the body to the source and connect V_R as shown. For p- channel devices, connect the body to the source and reverse the polarity of V_R.)

a. From the data sheet, determine the current level at which V_t is specified. Increase V_R until Q_x conducts this amount of current. Record V_{G_x} and i for each MOSFET as you vary V_R from 0 V to 10 V. Record at least six data points for each FET.

b. From the data collected in Part 1a, determine K for each FET. Record and label each FET for future identification.

2. Discrete differential amplifier

a. Build the circuit of Figure 10L6.14.

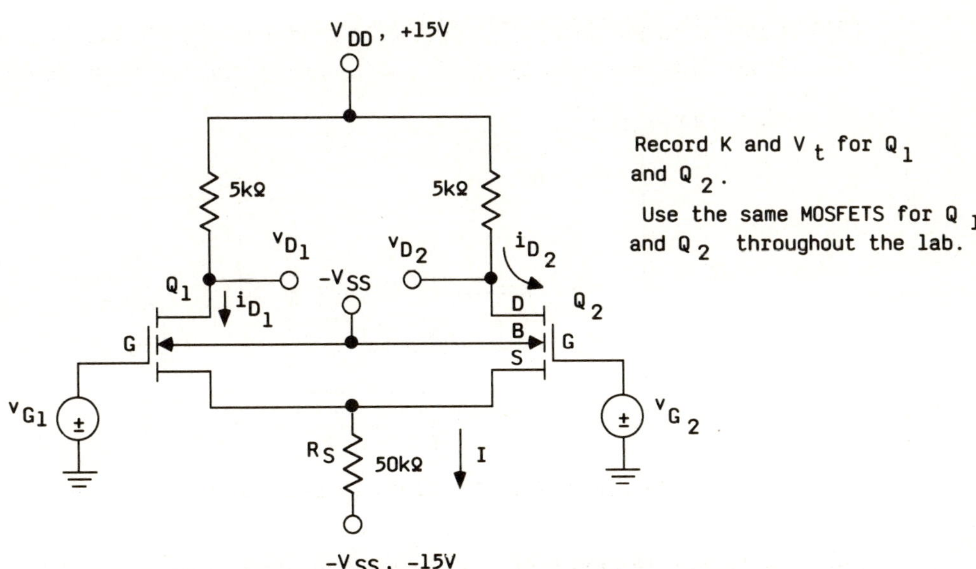

Figure 10L6.14: Discrete differential amplifier

(1) Let $v_{G_1} = v_{G_2} = 0$ V. Measure the voltages v_{D_1}, v_{D_2}, and v_S (source voltage), and determine the value of I along with that of drain currents i_{D_1} and i_{D_2}. Verify that the FETs are operating in the active region. Also, record the offset voltage $v_o = v_{D_1} - v_{D_2}$ and calculate g_m $\left[\dfrac{g_m - I}{V_{Gs} - V_t} \right]$

(2) Let $v_{G_1} = v_{G_2} = 0.5\sin(2\pi ft)$; f=1kHz. Measure the voltages v_{D_1} and v_{D_2}. Sketch the input voltage v_G, the two output voltages, V_{D_1} and V_{D_2}, and the voltage $v_{D_2} - v_{D_1}$. (Use the differential mode on your oscilloscope to measure $v_{D_2} - v_{D_1}$.) Be sure to preserve the phase relationship of the indicated voltages. Determine the common-mode voltage gain $A_{cm} = v_{D_2}/v_{cm}$. What range of common-mode voltage allows Q_1 and Q_2 to remain in the active region?

(3) Let $v_{G_2} = 0$ and $v_{G_1} = 0.1\sin(2\pi ft)$ with f=1 kHz. Sketch v_{G_1}, v_{D_1}, v_{D_2} and v_S. Also include $v_{D_2} - v_{D_1}$, measured as in Part 2. Determine the differential voltage gain $A_d = v_{D_2}/v_D$. Increase the amplitude of v_{G_1} until the output voltage v_{D_2} is distorted. What causes the distortion? Can you adjust the offset (dc component) voltage of the input to eliminate the distortion? Why?

(4) Use the results from Parts (2) and (3) to determine

$$\text{CMRR} = 20 \log \left(\frac{A_{cm}}{A_d} \right)$$

(5) Measure the input resistance for either input.

b. Measurement of input offset voltage.

(1) Measure the offset voltage at the input by applying a dc input voltage that forces $| v_{D_1} - v_{D_2} | = 0$. Record this voltage as V_{os}.

(2) Based on the measured mismatch of R_{D_1} and R_{D_2}, V_{t_1}, and V_{t_2}, and K_1 and K_2 (for Q_1 and Q_2, respectively), do the equations from the prelab account for the measured value of V_{os}?

3. Evaluation of an n- Channel MOSFET current mirror.

a. Build the circuit of Figure 10L6.15 using dual (monolithic matched n- channel enhancement-mode) MOSFETs. Use the same MOSFETs for Q_3 and Q_4 throughout the lab.

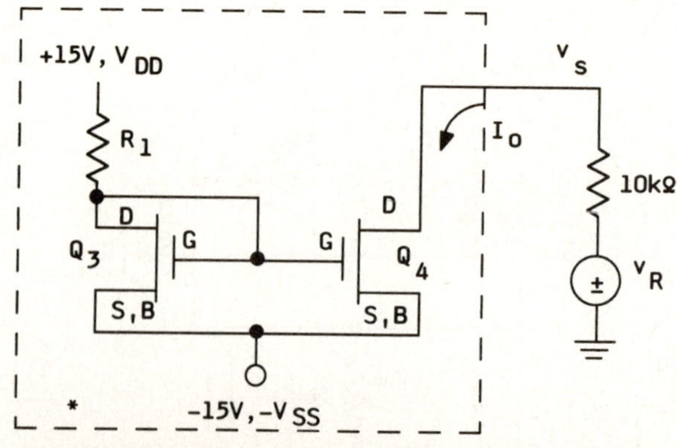

Figure 10L6.15: N-Channel MOSFET current mirror

(1)   Record V_t and K of Q_3 and Q_4.

(2)   Determine R_1 so that the channel current of Q_3 is $\approx 500\ \mu A$.

(3)   Vary V_R from -15 V to +15 V in 2 V increments. Record v_R, v_S, and I_o for each value of v_R. Over what range of V_R is the CUT operating as a current source?

b.   Measuring output resistance. Next, ac-couple a function generator to the node labeled V_s. Use a 1 μF capacitor, and set the function generator to a 1 kHz sine wave with an amplitude of 0.5 V (1 V peak to peak). This is shown in Figure 10L6.16. Determine the output resistance of the CUT by dividing the voltage v_x by the measured current i_x. Let V_R equal 15 V for this measurement.

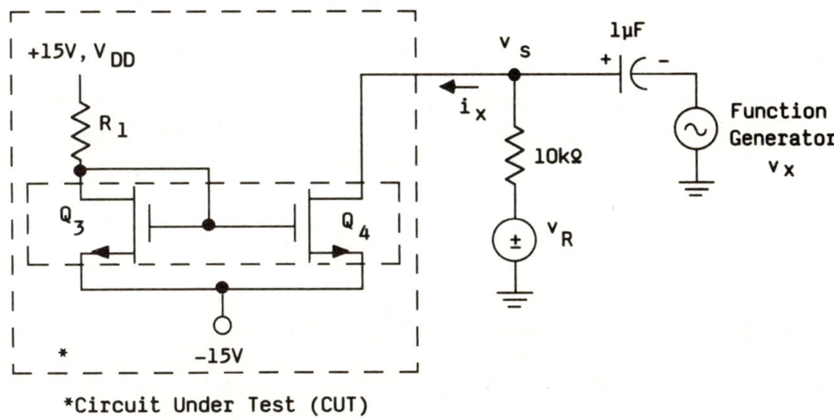

Figure 10L6.16: Circuit for Part 3.b

4.   Application of the MOSFET current mirror to the MOSFET differential amplifier.

a.   Build the circuit of Figure 10L6.17. Be sure to use the same Q_1-Q_2 pair and Q_3-Q_4 pair from parts 1 and 2 of the lab. R_1 should be the same value you calculated in Part 2.

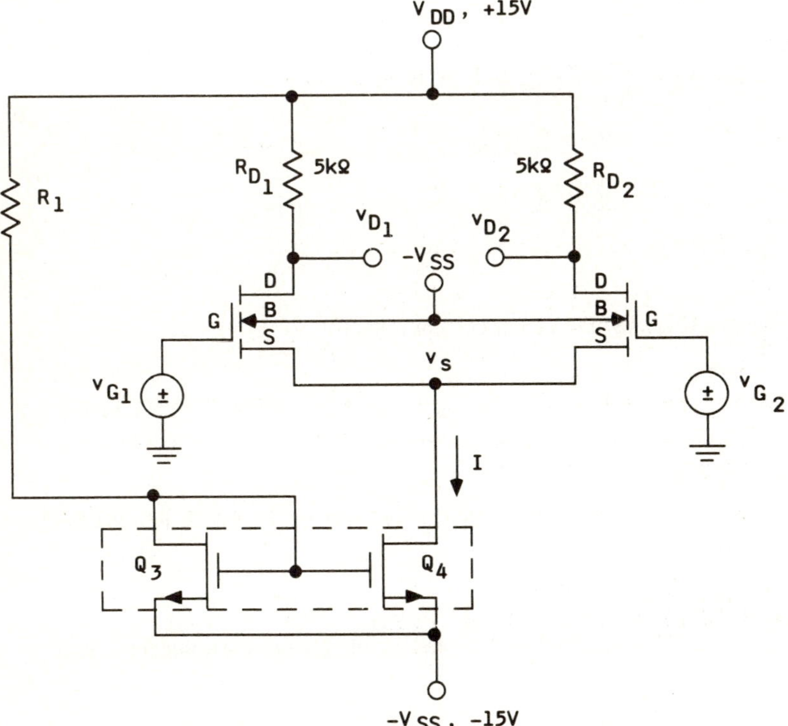

Figure 10L6.17: Circuit for Part 4.a

b. Make the following measurements:

(1) Let $v_{G_1} = v_{G_2} = 0$ V. Measure and record v_{D_1}, v_{D_2}, and v_S. Verify that the channel currents for Q_1 and Q_2 are equal, at $I/2 \approx 250$ μA. Calculate $g_m = \dfrac{I}{(v_{GS}-V_t)}$. Let $v_{G_2} = v_{G_1} = 1 \sin(2\pi f t)$; with $f = 1$ kHz. Measure the voltages v_{D_1} and v_{D_2}. Sketch v_G, v_D, v_S, and v_{D_2}. Use the scope, in differential mode, to display and sketch $v_{D_2} - v_{D_1}$. Be sure to use internal triggering so that all phase relationships are observed. Determine the common-mode voltage gain, $A_{cm} = \dfrac{v_{D_2}}{v_{cm}}$. What magnitude of common-mode voltage allows all MOSFETs to remain in the active mode?

(2) Let $v_{G_2} = 0$ and $v_{G_1} = 0.1 \sin(2\pi f t)$; where $f = 1$kHz. Sketch v_{G_2}, v_{D_1}, v_{D_2}, and v_s. Also sketch $v_{D_2} - v_{D_1}$. Determine the differential voltage gain, $A_d = \dfrac{v_{d_2}}{v_d}$. Increase the input voltage amplitude until there is distortion observed in the output waveform v_{D_2}. What is the cause of the distortion? Can you adjust the offset of the input signal to eliminate this distortion? Why?

(3) Use the results from Part (2) and Part (3) to determine the common-mode rejection ratio. $\text{CMRR} = 20 \log \left| \dfrac{A_{cm}}{A_d} \right|$ dB.

5. Evaluation of a P-channel current mirror.

A. Build the circuit of Figure 10L6.18.

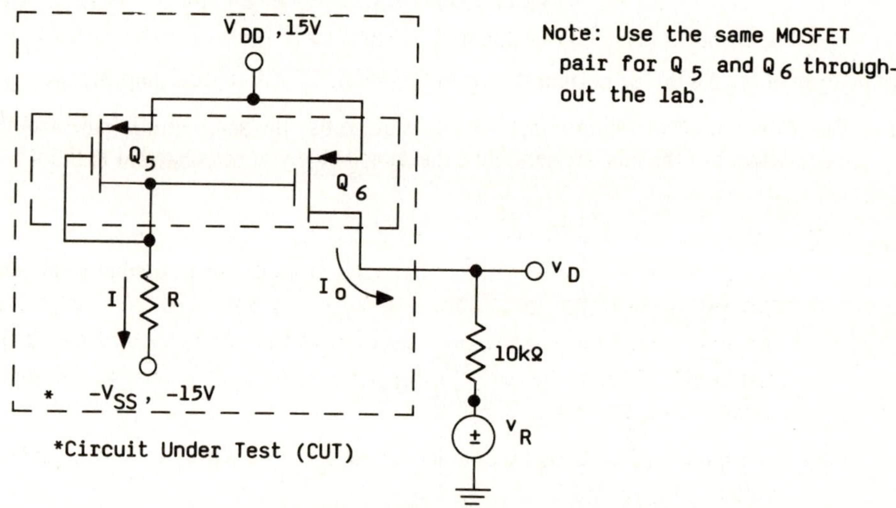

Figure 10L6.18: P-channel current mirror

a. Pick R so that $I = 250$ μA. Measure I_o and v_D for V_R varied from -15 V to +15 V in 2 V increments. Determine the range of voltages at node v_D that allow the CUT to operate as a current mirror. Plot I_o versus v_D.

b. Use the technique of Part 2b to determine the output resistance of the CUT. Record the output resistance.

6. Monolithic-type MOSFET differential amplifier

a. Build the circuit of Figure 10L6.19. The components coupled by dotted lines are matched dual MOSFETs.

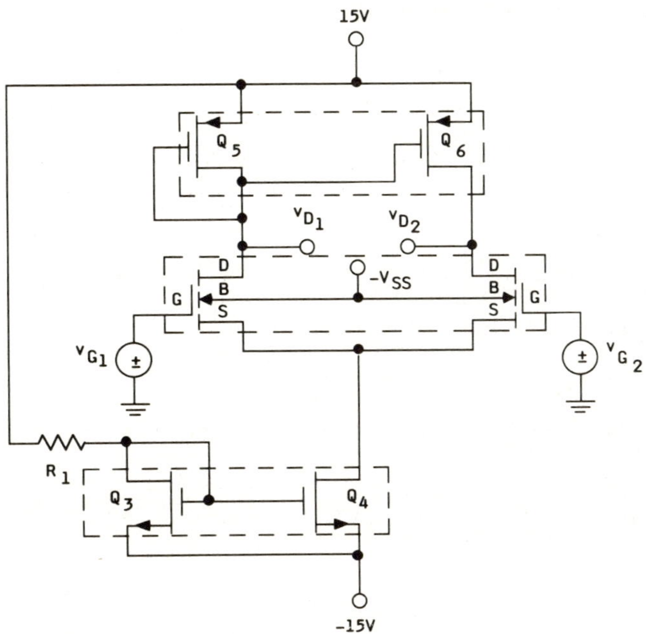

Figure 10L6.19: Monolithic differential amplifier

b. Answer the following questions pertaining to the circuit of Figure 10L6.19:

   (1) Let $v_{G_1} = v_{G_2} = 0$ V. Measure v_{D_1}, v_{D_2}, v_S, I, and the drain currents i_{D_1} and i_{D_2} for Q_1 and Q_2, respectively. Verify that all MOSFETs are operating in the active region.

   (2) Let $v_G = v_{G_1} = v_{G_2} = 0.5\sin(2\pi ft)$, with f = 1 kHz. Determine the common-mode voltage gain v_{d_2}/v_{D_2}. Sketch v_{D_2} and v_G as functions of time. Note the phase relationship of the signals. What is the maximum range of the common-mode voltage that allows Q_1 and Q_2 to operate in the active region?

   (3) Let $v_{G_1} = 0.1\sin(2\pi ft)$, with f = 1 kHz, and $v_{G_2} = 0$ V. Sketch v_{G_1}, v_{D_1}, v_{D_2}, and v_S. Also, sketch $v_{D_2} - v_{D_1}$. (Use the differential measurement capability of the oscilloscope.) Determine the differential gain $A_d = v_{D2}/v_d$, where $v_d = v_{G_1} - v_{G_2}$. Increase the input amplitude until there is visible distortion in the output v_{D_2}. What is the cause of this distortion? (Sketch the waveform.) Can you adjust the offset of the input signal to eliminate the distortion? Why?

   (4) Use the results of Parts (3) and (4) to determine the common-mode rejection ratio CMRR = $20 \log \left| A_{cm}/A_d \right|$.

   (5) Measure the input resistance of either input.

c. Measure the input offset voltage.

   (1) Measure the input offset voltage by applying a dc voltage between the two inputs that causes $| v_{D_2} - v_{D_1} | = 0$. Record this voltage as V_{os}.

   (2) Why is this offset voltage less than the value found in Part 2b, since the same MOSFETs are used for Q_1 and Q_2?

7. Perform a SPICE simulation of the circuit of Figure 10L6.19. Use the values of K and V_t found in Part 1. Verify the results of Part 5 with this simulation.

DISCUSSION: Compare the results you obtained in the prelab, lab experiment, and SPICE simulation.

CONCLUSIONS:

REFERENCES

[1] C. J. Savant, Jr., M. S. Roden, and G. O. Carpenter, *Electronic Circuit Design* (Menlo Park, CA: Benjamin/Cummings, 1987).

[2] A. S. Sedra and K. C. Smith, *Microelectronic Circuits,* 2d ed. (New York: Holt, Rinehart and Winston, 1987).

[3] D. H. Navon, *Semiconductor Microdevices and Materials* (New York: Holt, Rinehart and Winston, 1987), Chapters 8, 9 and 10.

[4] P. R. Gray and R. G. Meyer, *Analysis and Design of Analog Integrated Circuits, 2d ed.* (New York: John Wiley, 1984), Chapters 4 and 12.

Chapter 11

OPERATIONAL AMPLIFIERS

11.1 Introduction[1],[2],[3]

In Chapter 5, we introduced the ideal operational amplifier (op amp). For many op amp applications, assuming the op amp is ideal is a good assumption. However, it is important that the student know the characteristics of a real op amp and appreciate the limitations a real op amp has on the performance of op amp circuits.

The purpose of the next section is to discuss the most significant characteristics of the nonideal op amp. From these characteristics, it is possible to develop a model or equivalent circuit of the practical op amp.

The most significant difference between ideal and real op amps is the open loop voltage gain. The ideal op amp has an infinite voltage gain, but the actual op amp has a finite gain that decreases with frequency. This decrease occurs because of parasitic junction capacitance and minority-carrier charge in devices making up the op amp circuit. In addition, the ideal op amp has infinite input impedance and zero output impedance, but the practical op amp has a large but finite input impedance and a small output impedance.

The equivalent circuit for the ideal op amp is shown in Figure 11.1.

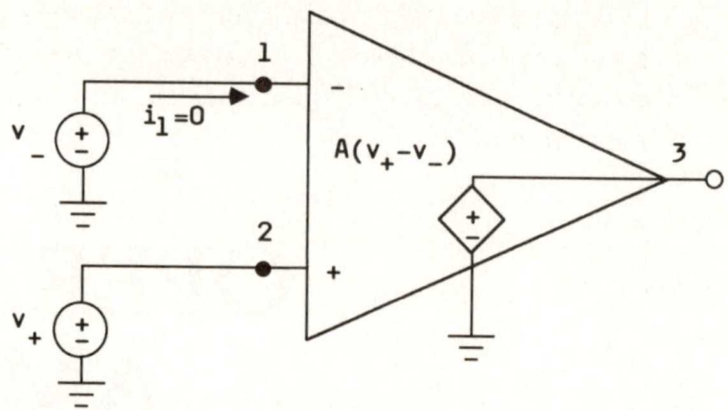

Figure 11.1: Equivalent circuit of the ideal op amp

A low-frequency equivalent circuit for the practical op amp that includes the effects of input bias current and input offset voltage is shown in Figure 11.2.

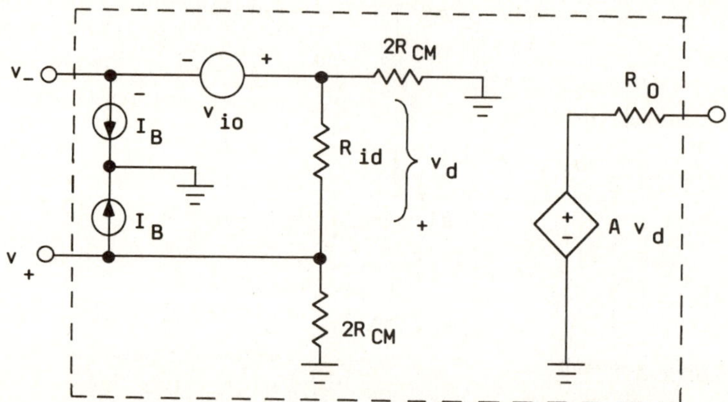

Figure 11.2: An equivalent circuit for a nonideal op amp

V_{io} is the input offset voltage, R_o is the output resistance, and R_{id} is the input resistance measured between the inverting and noninverting terminals. The model also contains a resistor between each of the two inputs and ground. These resistors are the common-mode resistances and are equal to $2R_{cm}$ each. If the two inputs are connected, these resistors are in parallel and the combined resistance to ground is R_{cm}. If the op amp is ideal, R_{cm} is equal to infinity. An equivalent circuit of the type shown in Figure 11.2 can be used to do SPICE analysis of practical op amp circuits.

The current source I_B represents the input bias current. Although ideal op amp inputs draw no current, bias current does enter each input terminal in the practical op amp. I_B is the base current of the input transistor.

In the next section, we discuss a number of the characteristics that describe the practical op amp. These include input offset voltage, input bias current, output and input resistance, power supply rejection ratio, common mode rejection, slew rate, full-power band width, frequency compensation, and, of course, open loop voltage gain and bandwidth.

A number of practical operational amplifier circuits are discussed in the remainder of the chapter, including generalized impedance converters and negative impedance circuits, active voltage regulators, active filters, and nonlinear applications such as the logarithmic amplifier. The laboratory exercises were developed to illustrate the characteristics of real op amps, as well as a number of practical op amp circuit applications.

11.2 Nonideal Parameters[1]

11.2.1 Open-Loop Voltage Gain and Bandwidth

The open-loop voltage gain of an op amp is the ratio of the change in output voltage to a change in the input voltage without feedback. Voltage gain is a unitless quantity, and we shall use the symbol A to indicate the open-loop voltage gain.

Op-amps have high voltage gain for inputs from dc to low frequencies. However, the gain decreases with increasing frequency. Figure 11.3 shows a plot for the open-loop gain as a function of frequency for a typical op amp.

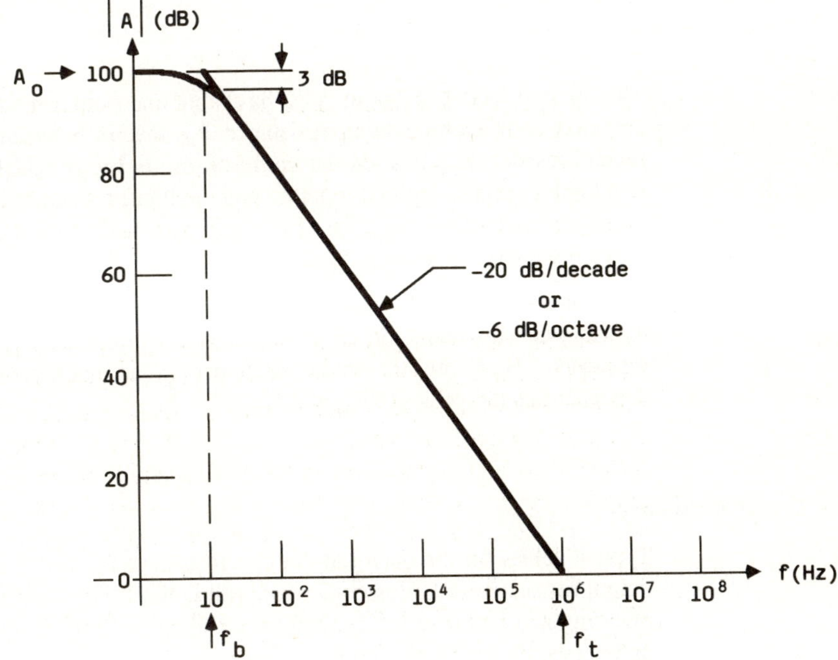

Figure 11.3: Open loop gain of a typical op amp

11.2.2 Finite Bandwidth

Let the open-loop gain A(jω) of the op amp be expressed as

$$A(j\omega) = \frac{A_o}{1 + j\dfrac{\omega}{\omega_b}} \tag{11.1}$$

where A_o denotes the dc gain and ω_b is the 3-dB or break-frequency. The break-frequency or corner-frequency is the frequency at which the magnitude of the open-loop gain is 3 dB down from its dc magnitude. For frequencies $\omega \gg \omega_b$, Equation (11.1) becomes

$$A(j\omega) \approx \frac{(A_o \omega_b)}{j\omega} \tag{11.2}$$

Let $\omega_t = A_o \omega_b$. Then Equation (11.2) can be written as

$$A(jw) \approx \frac{\omega_t}{jw} \tag{11.3}$$

where ω_t is called the **unity gain bandwidth** and is specified on the data sheets of op amps. It should be clear from Equation (11.3) that the magnitude of the open-loop gain, $|A|$ becomes unity or 0 dB when $\omega = \omega_t$.

11.2.3 Input Resistance

The op amp model of Figure 11.2 has input and output resistances. The practical op amp has a differential input resistance R_{id} that is between the input terminals. A second resistor, R_{icm}, called the **common-mode input resistance**, is measured if the two input terminals are tied together and the input resistance to ground is measured. It is convenient to divide R_{icm} into two equal parts and connect each resistor between one of the input terminals and ground.

Sedra and Smith [1] show that for an inverting op amp, taking R_{id} and R_{cm} into account by using the equivalent circuit of Figure 11.2 will have a negligible effect on the input resistance. However, the input resistance of an noninverting op amp is strongly dependent on the values of R_{id} and R_{cm}.

11.2.4 Output Resistance

Typical values for the open-loop output resistance R_o are 75 Ω to 100 Ω. To find the output resistance of a closed-loop amplifier, short the signal source, which makes the inverting and noninverting configurations identical, and apply a test voltage V_x to the output (see Figure 11.4).

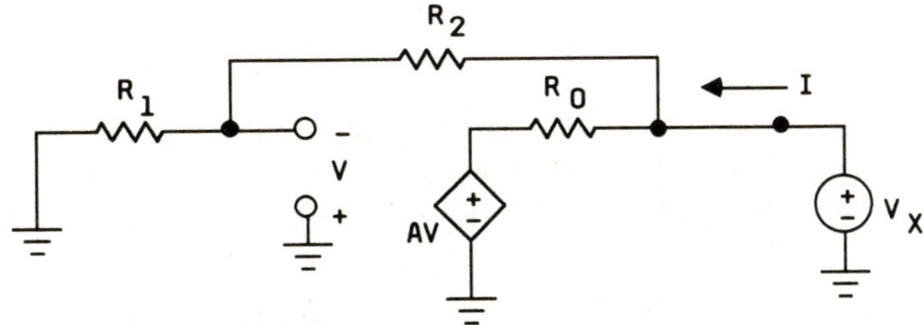

Figure 11.4: Derivation of the output resistance R_o

The output resistance $R_{out} \equiv \dfrac{V_X}{I}$ can be found by a straightforward analysis of the circuit. The result is

$$R_{out} = [R_1 + R_2] \parallel \left[\frac{R_o}{(1 + A\beta)}\right]$$
(11.4)

where

$$\beta \equiv \frac{R_1}{R_1 + R_2}$$

If $R_o << R_1 + R_2$, then

$$R_{out} \approx \frac{R_o}{1 + A\beta}$$
(11.5)

Normally, $A\beta >> 1$; so

$$R_{out} \approx \frac{R_o}{A\beta}$$
(11.6)

At low frequencies, A is real and large, and thus R_{out} is very small.

If A is frequency dependent, take $A = \omega_{t/s}$ (ω represents a finite op amp bandwidth). Then

$$Z_{out} = \frac{R_o}{1 + \beta\dfrac{\omega_t}{s}}$$
(11.7)

and

$$Y_{out} = \frac{1}{R_o} + \frac{\beta\omega_t}{sR_o}$$
(11.8)

If we let $L = \dfrac{R_o}{\beta\,\omega_t}$, then an equivalent circuit would be R_o in parallel with an inductance (see Figure 11.5).

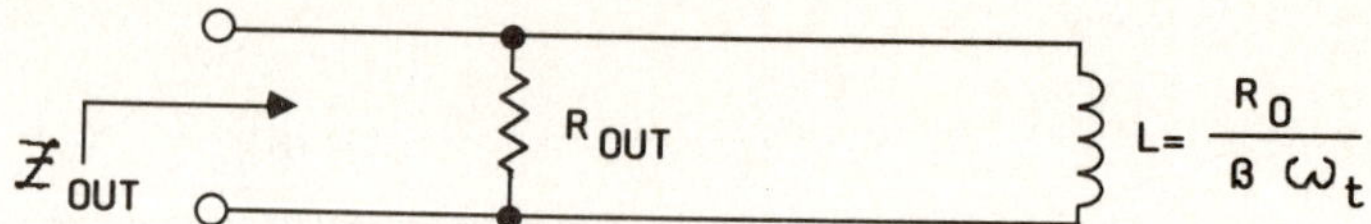

Figure 11.5: Equivalent circuit of the closed-loop output impedance of an op amp

11.2.5 Input offset voltage (V_{io}, V_{os}, or V_{off})

For an ideal op amp, when the input voltage is zero the output voltage is also zero. This is not the case for a real op amp. The input offset voltage V_{io} is defined as the input voltage required to make the output voltage equal to zero. The value of V_{io} may be of either polarity; thus, the input offset voltage may be written as $\pm V_{io}$. The V_{io} value is usually caused by an unbalance in input-stage currents for a bipolar-transistor circuit, or by mismatches between pinchoff or threshold voltages for JFET or MOS transistors. The offset voltage will drift with temperature as $\dfrac{\Delta V_{io}}{\Delta T}$ $\left(\dfrac{\mu V}{C^\circ}\right)$. For bipolar transistor op amps, V_{io} is typically in the range $\pm 10\,\mu V$ to 10 mV, while for some FETs it can be greater than ± 50 MV.

The input offset voltage may be measured by setting the input to zero and measuring the output (see Figure 11.6). This output voltage is known as the **output dc offset voltage.** If the output dc voltage is divided by the open-loop gain of the op amp, V_{io} will be obtained.

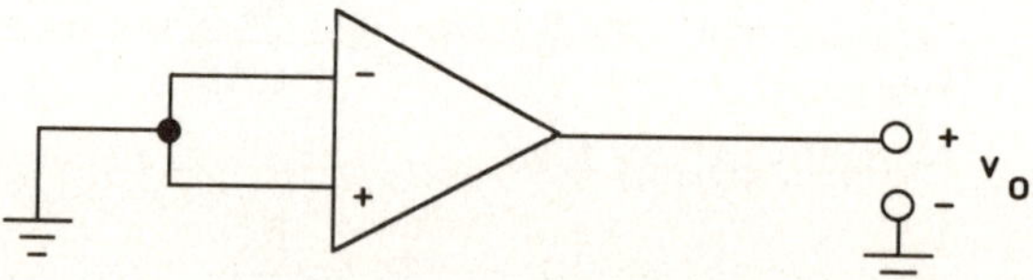

Figure 11.6: Measuring V_{io}

The effects of input offset voltage can be incorporated into the op amp model (see Figure 11.7). Many op amps are provided with two extra terminals and a prescribed technique for reducing the offset voltage to zero.

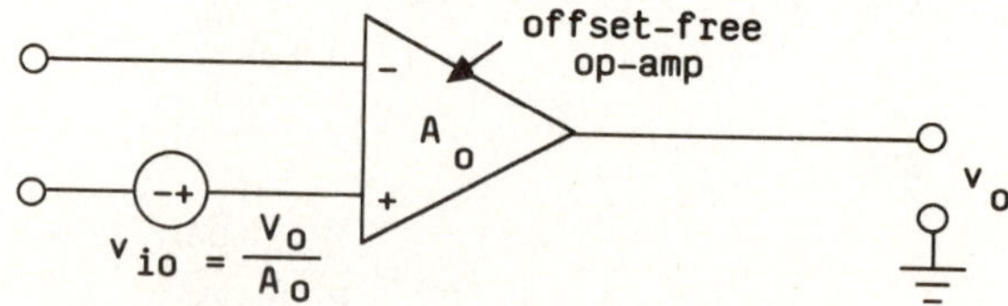

Figure 11.7: Input offset voltage V_{io}

11.2.6 Input Bias Current (I_B)

In order for the practical op amp to operate, its two input terminals have to be supplied by finite dc currents. These currents are called the **input bias currents.** (see Figure 11.8).

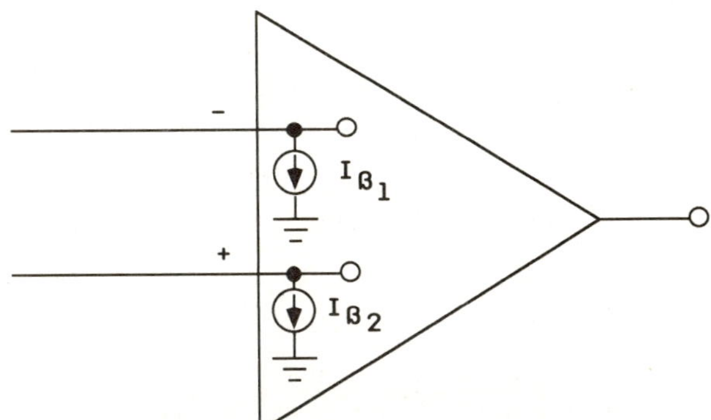

Figure 11.8: Input bias current I_B

The values of the current sinks, I_{B_1} and I_{B_2}, are independent of the source impedance. The bias current is defined as the average value of the two current sinks. That is,

$$I_B = \frac{I_{B_1} + I_{B_2}}{2} \tag{11.9}$$

The difference between the two sink currents is called the **input offset current** I_{io} or I_{os} and is expressed as

$$I_{io} = \mid I_{B_1} - I_{B_2} \mid \tag{11.10}$$

For bipolar-transistor inputs, I_B is in the range of 10 nA to 1 μA, while FETs have I_B values less than 100 pA. The value of I_B will decrease with increasing temperature for bipolars, while for JFETS the input current will double for every 8°-12°C increase of temperature.

For a well-matched different-input stage, the input offset current is found to have a value about one-fifth that of I_B. The value I_{io} is temperature-dependent, with $\dfrac{\Delta I_{io}}{\Delta T}$ expressed in $\dfrac{pA}{C°}$.

To minimize the effect of the input bias currents, a resistor that is equal to the dc resistance seen by the inverting terminal should be placed in the noninverting terminal.[1] If the amplifier is ac-coupled, the resistors should be selected such that $R_3 = R_2$ (see Figure 11.9).

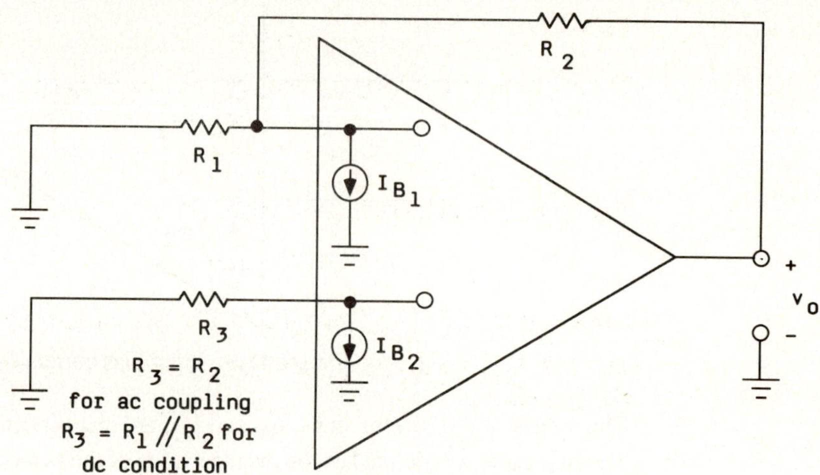

Figure 11.9: Reducing the effect of input bias current

11.2.7 Power Supply Rejection Ratio (PSRR)

The power supply rejection ratio (PSRR) is a measure of the ability of an op amp to ignore changes in the power supply voltage.

The PSRR is defined as the ratio of the change in v_o to the total change in power supply voltage. PSRRs is usually specified in $\dfrac{\mu V}{V}$ or dB. Typical op amps have PSRRs of $\sim 30 \, \dfrac{\mu V}{V}$.

To minimize changes in supply voltage, the power supply for each group of op amps should be decoupled from those of other groups. The power supply lines should be capacitively bypassed to ground. This arrangement will ensure that varying currents drawn from the power supply in response to signal variations will not spread throughout the system.

11.2.8 Common-Mode Rejection (CMRR)

If the two input terminals of an op amp are tied together, and a signal v_{ICM} is applied, the output will not be zero (see Figure 11.10).

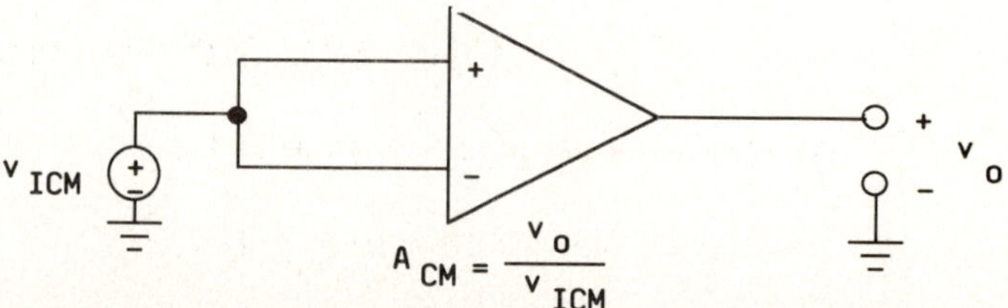

Figure 11.10: Definition of A_{cm} for a nonideal op amp

The ratio of the output voltage v_o to the input voltage v_{ICM} is defined as the common-mode gain A_{CM}.

If an op amp has v_1 and v_2 applied to its inverting and noninverting input terminals, respectively, the difference between the two input signals is the differential-mode signal.

$$v_{ID} = v_2 - v_1 \tag{11.11}$$

The average of the two input signals is the common-mode signal

$$v_{ICM} = \frac{v_1 + v_2}{2} \tag{11.12}$$

The output voltage can be expressed as

$$v_o = Av_{ID} + A_{CM}v_{ICM} \tag{11.13}$$

where A is the differential gain and A_{CM} is the common-mode gain.

The CMRR is a measure of the ability of an op amp to reject common-mode signals and is defined as

$$CMRR = \frac{|A|}{|A_{CM}|} \tag{11.14}$$

or

$$CMRR = 20\log\frac{|A|}{|A_{cm}|} \tag{11.15}$$

For the case of the inverting op amp, the positive input terminal is grounded, and therefore, the common-mode input signal is practically zero. For the case of the noninverting op amp, the common-mode input signal is approximately equal to the applied input signal, and therefore, the finite CMRR of the op amp has to be considered.

Sedra and Smith[1] offer a procedure for taking into account the effect of the finite CMRR in calculating closed-loop gain. The real op amp is replaced with an ideal op amp, but with a voltage generator that takes into account the effect of the finite CMRR (see Figure 11.11).

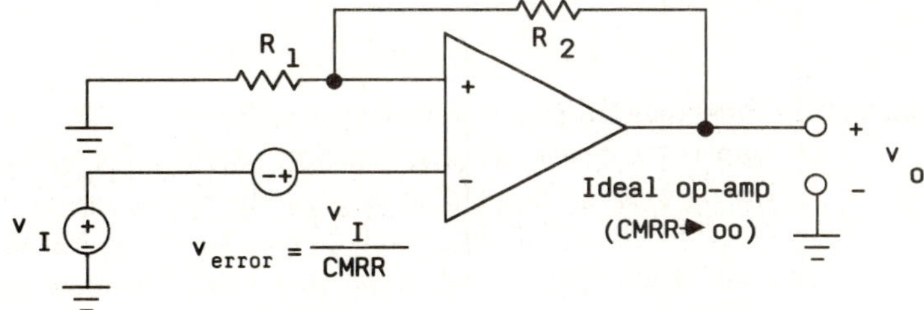

Figure 11.11: CMRR for nonideal op amp

The output voltage can then be expressed as

$$v_o = v_I(1 + \frac{1}{CMRR})(1 + \frac{R_2}{R_1}) \tag{11.16}$$

Actually, the polarity of the error-voltage generator can be positive or negative.

The CMRR is a function of frequency, decreasing as the frequency is increased. Typical values of CMRR range from 80 to 100 dB. It is desirable to have the CMRR as high as possible.

11.2.9 Slew Rate

The real op amp has a frequency-dependent response; that is, it does not respond uniformly to all frequencies. Therefore, if a step function (0-5 V) is applied to the input of a real op amp, the response is not an ideal step (see Figure 11.12).

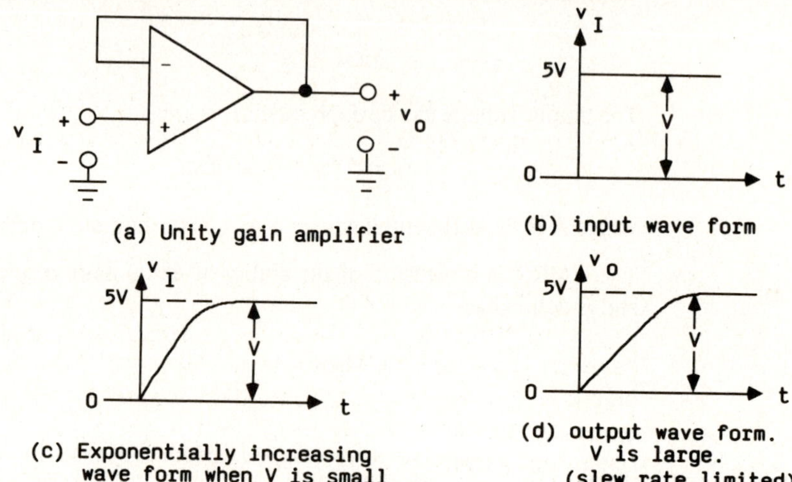

Figure 11.12: Illustration of slew-rate limiting using a unity gain amplifier

For example, if the circuit has a single-pole transfer function given by

$$\frac{v_o}{v_i} = \frac{A}{1 + s\tau} \tag{11.17}$$

with

$$\tau = \frac{1}{2\pi f_o}$$

where f_o is the 3dB frequency, and the input is a step function written as $v_i = \frac{5}{s}$, then the response should be

$$v_o = 5(1 - e^{-t/\tau}) \tag{11.18}$$

This predicted response is shown in Figure 11.12(c). The actual response of the op amp is the linear ramping function shown in Figure 11.12(d). This linear ramping response shows that the op amp output cannot rise at the rate predicted by Equation (11.18). The actual response is termed **slew-rate limiting**, and the slope of the linear ramp at the output is called the **slew rate**. The slew rate is defined as the maximum possible rate of change of the op amp output voltage:

$$SR = \frac{dv_o}{dt} \Big|_{max} \tag{11.19}$$

Slew rate is specified on the op amp data sheet in units of $\frac{V}{\mu s}$.

11.2.10 Full-Power Bandwidth

Op-amp slew-rate limiting can cause nonlinear distortion in sinusoidal waveforms. This means that above a certain frequency, the amplifier can no longer keep up with the maximum slope of a sinusoid. This limiting frequency can be expressed as

$$f_m = \frac{SR}{2\pi V_{o_{max}}} \tag{11.20}$$

where SR is the slew rate and $V_{o_{max}}$ is the maximum output voltage. The output sinusoids of amplitudes smaller than $V_{o_{max}}$ will show slew-rate distortion at frequencies higher than f_m.

11.2.11 Frequency Compensation

The shape of the curve in Figure 11.3 is typical of internally compensated op amps. **Internal compensation** refers to the RC network that causes the open-loop gain amplitude response to fall off at higher frequencies and to ensure that the op amp circuit will be stable with respect to signal frequencies. That is, the op amp will not burst into spontaneous oscillation when a signal is applied. The addition of the compensation network, however, results in a decreased small-signal bandwidth, slow slew rate, and reduced-power bandwidth. If the internal frequency-compensating circuit were removed, the result would be a higher slew rate and greater power bandwidth, but the op amp would probably oscillate.

To prevent oscillation, manufacturers of op amps bring out a number of frequency-compensating terminals. These terminals give the user the option of choosing the best allowable combination of stability and bandwidth.

11.3 Operational Amplifier Applications

11.3.1 Negative Impedance Circuit (NIC)[1],[2]

The circuit shown in Figure 11.13 results in a negative input impedance.

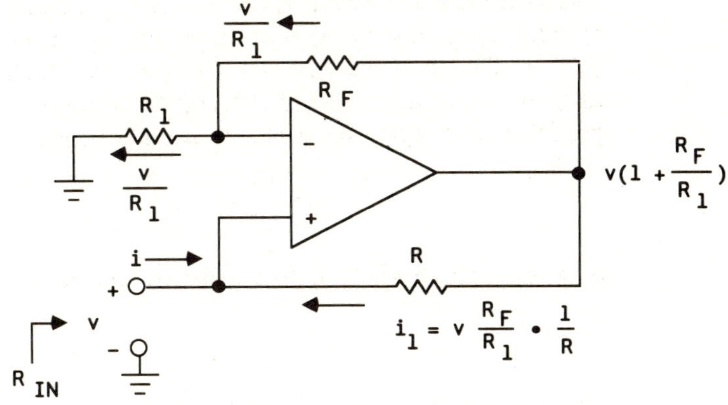

Figure 11.13: Negative impedance circuit

The input resistance is found from the definition

$$R_{IN} = \frac{v}{i} \tag{11.21}$$

Because of the virtual short circuit between the op amp input terminals, the voltage at the inverting or (-) terminal will be v. The current through R_1 will then be $\frac{v}{R_1}$. If we assume that the input impedance of the op amp is quite large compared to R_F, then the current through R_F will also be $\frac{v}{R_1}$. Then, the voltage at the op amp output will be

$$v + \frac{v}{R_1} R_F = (1 + \frac{R_F}{R_1})v \tag{11.22}$$

The current through resistor R

$$i_1 = \frac{v(1 + \frac{R_F}{R_1}) - v}{R} = v \frac{R_2}{R_1} \frac{1}{R} \tag{11.23}$$

Again, we assume that the input impedance of the op amp is very large and no current flows into the positive input terminal of the op amp. Then

$$i = -i_1 = -\frac{v}{R} \frac{R_F}{R_1} \tag{11.24}$$

Equation 11.24 can be rewritten as

$$R_{IN} = -R \frac{R_1}{R_F} \tag{11.25}$$

The input resistance is negative, with a value equal to the resistance in the positive-feedback path, $\frac{R_1}{R_F}$. The circuit is called a **negative impedance converter** (NIC), where R may be replaced by an arbitrary impedance.

11.3.2 Generalized Impedance Circuit (GIC)

The impedance converter is an extension of the NIC. The object is to calculate the input impedance Z_{IN} of the circuit of Figure 11.14. The two op amps are considered ideal, so that in each amplifier, $v_+ = v_i$.

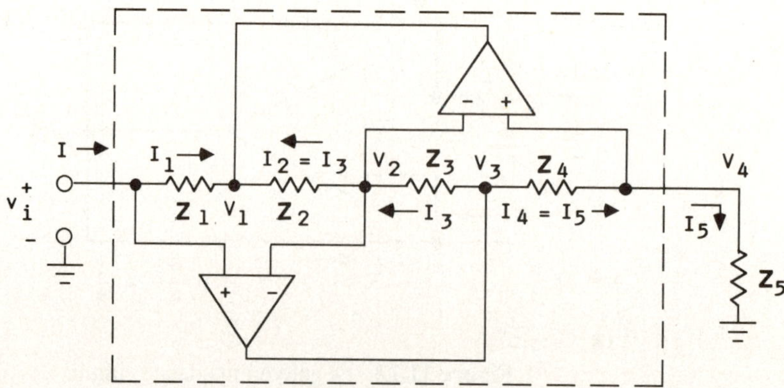

Figure 11.14: Impedance converter

Therefore,

$$V_i = V_2 = V_4 \tag{11.26}$$

The current through Z_4 and Z_5 is given by

$$I_5 = \frac{V_4}{Z_5} = \frac{V_i}{Z_5} \tag{11.27}$$

The voltage V_3 is

$$V_3 = V_4 + I_5 Z_4 + V_i + \frac{Z_4}{Z_5} V_i \tag{11.28}$$

$$= V_i(1 + \frac{Z_4}{Z_5}) \tag{11.29}$$

The current in Z_3 is then

$$I_3 = \frac{V_3 - V_2}{Z_3} = \frac{V_i(1 + \frac{Z_4}{Z_5}) - V_i}{Z_3} = \frac{V_i Z_4}{Z_3 Z_5} \tag{11.30}$$

The voltage V_1 is given by

$$V_1 = V_2 - I_3 Z_2 = V_1 - \frac{V_i Z_4 Z_2}{Z_3 Z_5} \tag{11.31}$$

The current is

$$I_1 = \frac{V_i - V_1}{Z_1} = \frac{V_i Z_2 Z_4}{Z_1 Z_3 Z_5} \tag{11.32}$$

Finally, the input impedance is

$$Z_{IN} = \frac{V_i}{I} = \frac{Z_1 Z_3 Z_5}{Z_2 Z_4} \tag{11.33}$$

The GIC is the circuit within the box of Figure 11.14. Impedance Z_5 is the terminating impedance. The GIC contributes a general conversion factor $\dfrac{Z_1 Z_3}{Z_2 Z_4}$. The result is that Z_5 is multiplied by this factor.

With the proper selection of the impedances Z_1 through Z_5, it is possible to obtain a wide variety of Z_{IN} functions. For example, the GIC finds applications in the area of active inductorless filter design. If we let $Z_1 = R_1$, $Z_2 = \dfrac{1}{sC_2}$, $Z_3 = R_3$, $Z_4 = R_4$ and $Z_5 = R_5$, then Equation (11.33) becomes

$$Z_{IN} = sC_2 \frac{R_1 R_3 R_5}{R_4} \tag{11.34}$$

In Equation (11.34), let

$$L = C_2 \frac{R_1 R_3 R_5}{R_4} \tag{11.35}$$

then

$$Z_{IN} = sL \tag{11.36}$$

Thus, the input impedance looks inductive. For integrated circuits, inductive characteristics can be simulated without the use of inductive elements.

11.3.3 Active Voltage Regulators[3],[4]

Voltage regulators accept an unregulated or varying dc input voltage and produce a regulated or constant output voltage that can be used as a supply voltage for other circuits. Thus, fluctuations in the supply voltage are eliminated and the performance in circuits using this supply is improved.

Figure 11.15 illustrates an unregulated dc power supply.

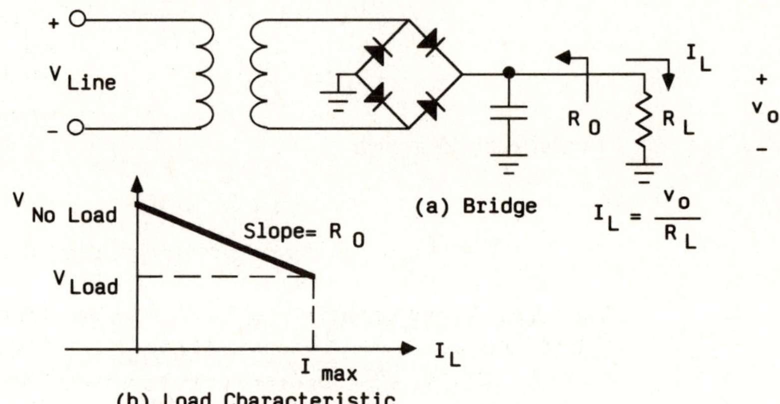

Figure 11.15: Unregulated dc power supply

Power supply regulation is defined by

$$V_R = \frac{V_{NO\,Load} - V_{Load}}{V_{NO\,Load}} \times 100 \qquad (11.37)$$

From Figure 11.15 (b), it is clear that one characteristic of a good power supply is a low output resistance R_o.

A voltage regulator is inserted between the output filter C and the load in Figure 11.16. The load characteristic then improves to that shown in Figure 11.17.

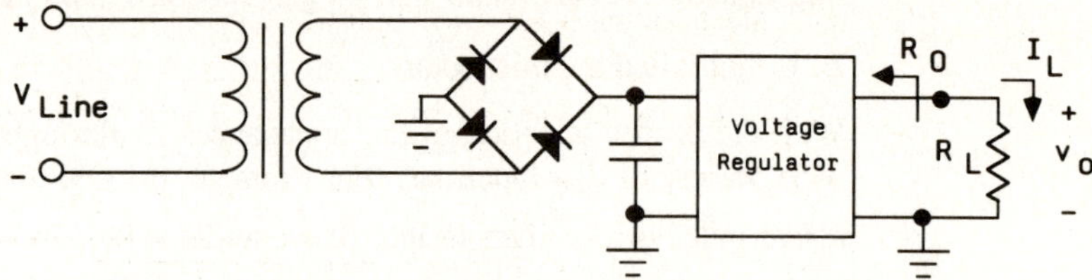

Figure 11.16: Full-wave bridge with voltage regulator circuit

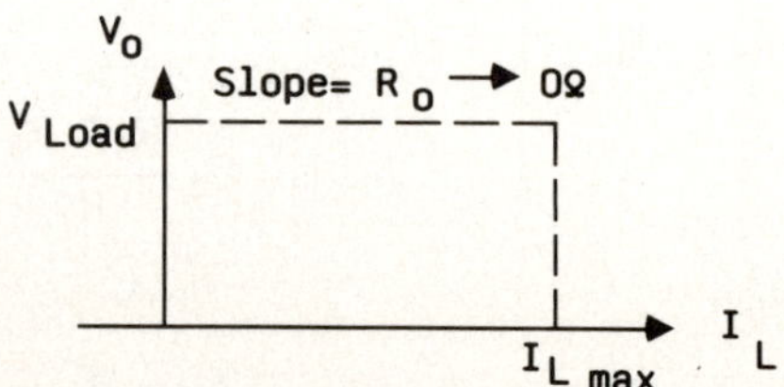

Figure 11.17: Load characteristic with voltage regulator

The most common type of voltage regulator is the "series" regulator, shown in Figure 11.18.

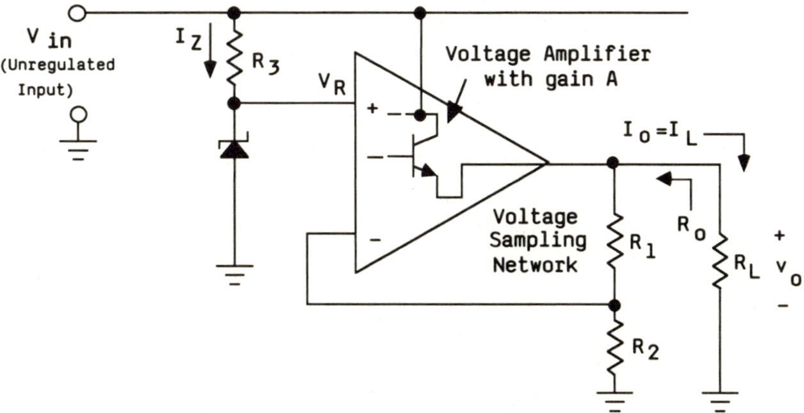

Figure 11.18: Schematic of a series voltage regulator

In the figure, a stable reference voltage V_R is obtained using a Zener diode. V_R could also be obtained by using a band gap reference.[5] The reference voltage is then fed to v_+, or the noninverting terminal of the high-gain amplifier, where it is compared with a sample of the output voltage taken by resistors R_1 and R_2.

From general feedback concepts,[6,7] it should be recognized that the circuit of Figure 11.18 is a series-shunt feedback circuit, and for large loop gain,

$$V_o = V_R \left[\frac{R_1 + R_2}{R_2} \right]$$

(11.38)

Therefore, over the operating range of the circuit, the output voltage is determined by a stable avalanche-diode reference voltage and the ratio of two resistors.

A series-shunt feedback amplifier configuration is shown in Figure 11.19.[6] The equivalent circuit is given in Figure 11.20.

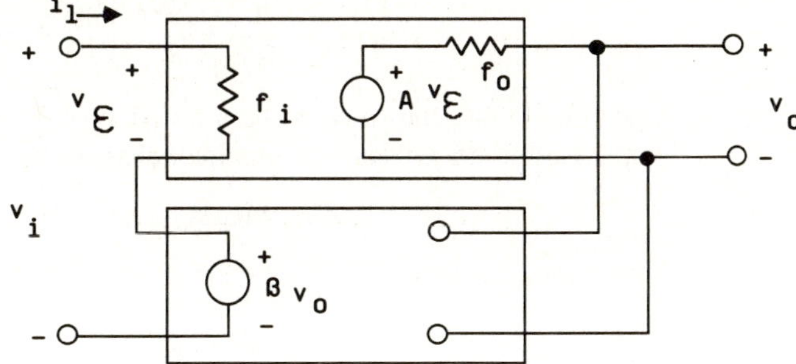

Figure 11.19: Series-shunt configuration

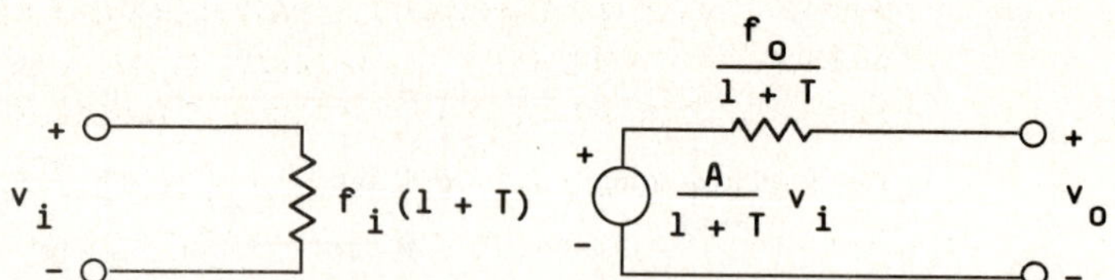

Figure 11.20: Equivalent circuit of a series-shunt feedback amplifier

The input impedance Z_i with feedback applied is

$$Z_i = \frac{v_i}{i_1} = (1 - T)z_i \tag{11.39}$$

The output impedance Z_o with feedback applied is

$$Z_o = \frac{z_o}{1 + T} \tag{11.40}$$

T is called the **loop gain** and is written as $T = A\beta$. The quantity f in Figures 11.19 and 11.20 is called the **feedback transfer function**. In terms of the circuit of Figure 11.18, Equation (11.40) can be written as

$$R_o = \frac{r_{oa}}{1 + T} \tag{11.41}$$

where $T = A\,\beta$ and $r_{oa} = z_o$, the output resistance of the amplifier without feedback. But the feedback transfer function β for a series-shunt configuration can be expressed as

$$f = \frac{R_2}{R_1 + R_2} \tag{11.42}$$

Then the loop gain T is written as

$$T = A\frac{R_2}{R_1 + R_2} \tag{11.43}$$

If it is assumed that V_R is constant and $T \gg 1$, then Equation (11.41) can be written as

$$R_o = \frac{r_{oa}}{AV_R}\,V_o \tag{11.44}$$

Equation (11.44) shows R_o to be a function of V_o if A, V_R, and r_{oa} are fixed. If the output current drawn from the regulator changes by ΔI_o, then V_o changes by

$$\Delta V_o = R_o \Delta I_o \tag{11.45}$$

Substituting Equation (11.44) into Equation (11.45) results in

$$\frac{\Delta V_o}{V_o} = \frac{r_{oa}}{AV_R}\Delta I_o \tag{11.46}$$

From this equation, the load regulation of the regulator can be calculated. Equation (11.46) gives the percentage change in V_o for a specified change in I_o. The objective is for $\dfrac{\Delta V_o}{V_o}$ to be as small as possible.

11.3.4 Active Filters[8],[9]

A filter is a two-port network that passes signals within a specified frequency range but attenuates signals outside this frequency range. Active filters contain amplifiers, which facilitate the design of a variety of transfer functions.

Four of the most commonly used types of filters are low-pass, high-pass, band-pass, and band-stop filters. In addition to examining these, we shall consider the Butterworth filter as well.

The ideal low-pass filter allows frequencies up to a given limit to pass and attenuates frequencies above that limit. The ideal high-pass filter passes frequencies above a given frequency limit and attenuates those below that limit. The ideal band-pass filter allows only a particular band of frequencies to pass and attenuates the remaining frequencies. Ideal band-stop filters pass frequencies outside of a particular frequency band, but reject those frequencies within the band. Butterworth filters produce no ripple in the pass band and attenuate unwanted frequencies outside of this band.

Active filters produce gain and are composed of op amps, resistors, and capacitors. The combination of op amps, capacitors, and resistors can simulate the performance of LC filters.

The general configuration of the operational amplifier circuit used for active-network design is shown in Figure 11.21.

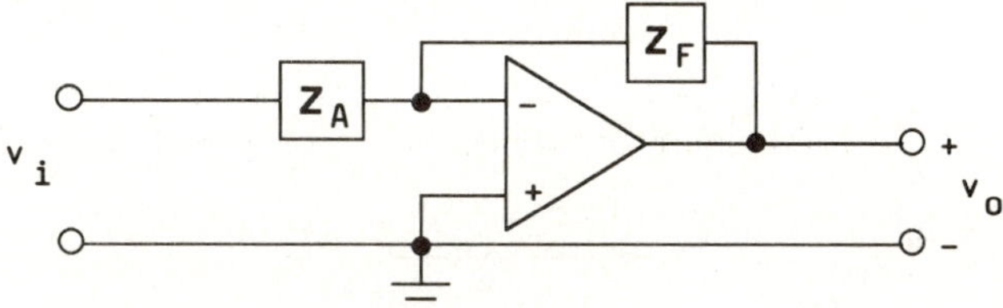

Figure 11.21: General form of the active network

The closed-loop gain of the system is given by

$$\frac{v_o}{v_i} = -\frac{Z_F}{Z_A}$$

(11.47)

The impedances Z_A and Z_F can be the result of combinations of capacitors and resistors. Table 11.1 illustrates some common two-terminal configurations and the resulting impedance functions.

Table 11.1: Common two-terminal circuit configurations and their respective impedance functions

(Reprinted by permission from *Electronic Circuit Design* by C. J. Savant, Jr.,
M. S. Roden, and G. L. Carpenter. Copyright © 1987 by Benjamin/Cummings.)

Circuit	Impedance Function
(1)	$\dfrac{s + \dfrac{1}{RC}}{C_1 s\left(s + \dfrac{1}{R}\dfrac{C_1 + C}{C_1 C}\right)}$
(2)	$\dfrac{RR_1\left(s + \dfrac{1}{RC}\right)}{(R + R_1)\left[s + \dfrac{1}{C(R + R_1)}\right]}$
(3)	$\dfrac{Rcs + 1}{Cs}$
(4)	$\dfrac{(C + C_1)\left[s + \dfrac{1}{R(C + C_1)}\right]}{CC_1 s\left[s + \dfrac{1}{RC}\right]}$
(5)	$\dfrac{R_1\left(s + \dfrac{1}{C}\dfrac{R + R_1}{R_1 R}\right)}{s + \dfrac{1}{RC}}$
(6)	$\dfrac{1}{C\left(s + \dfrac{1}{RC}\right)}$

A low-pass filter has a frequency response that is nonzero for $\omega = 0$ and zero as $\omega \to \infty$ (see Figure 11.22). By contrast, a high-pass filter has a frequency response that is near zero for $\omega = 0$ and nonzero as $\omega \to \infty$ (see Figure 11.23).

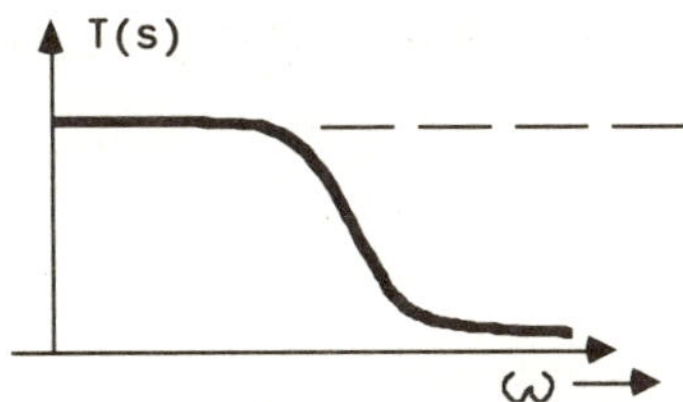

Figure 11.22: Low-pass filter

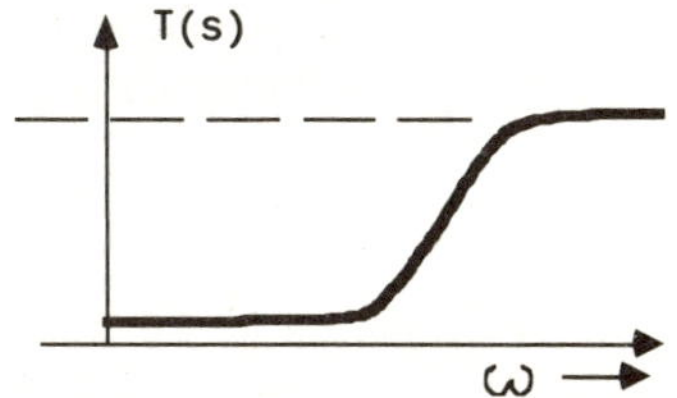

Figure 11.23: High-pass filter

An example of a transfer function for a low-pass filter is

$$T(s) = \frac{10}{s+5} \tag{11.48}$$

Here the gain at $\omega=0$ is 4, and it decreases at -20 dB/decade for frequencies larger than the corner frequency of $\omega = 5\,rad/s$. Another example is shown by

$$T(s) = \frac{5s - 10}{s^2 + s + 100} \tag{11.49}$$

For $\omega = 0$, the dc term is $-\frac{1}{10}$. The magnitude decreases with increasing frequency at a rate of -20 dB/decade and approaches zero for large values of ω. Equation (11.49), then, represents a low-pass filter.

An example of a low-pass filter is shown in Figure 11.24.

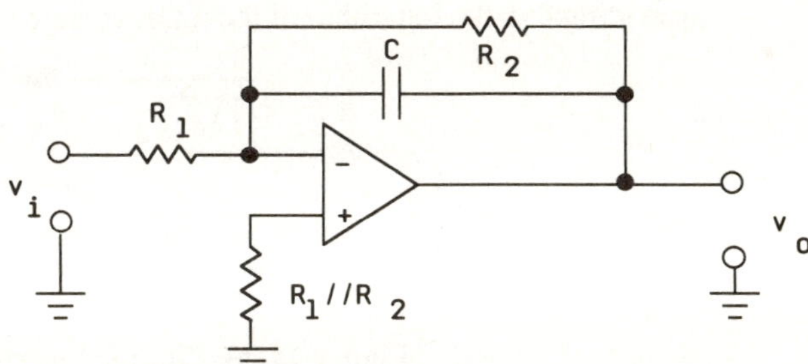

Figure 11.24: Example of a low-pass filter

The transfer function for this circuit, from Table 11.1 (circuit 6), is

$$T(s) = -\frac{R_2}{R_1} \frac{1}{1 + s\,R_2 C} \tag{11.50}$$

The dc gain is $\dfrac{R_2}{R_1}$ and the corner frequency is $\dfrac{1}{R_2 C}$.

For Butterworth filters the transfer function is chosen so that the magnitude response curve is as flat as possible within the passband of the filter. The transfer function of the Butterworth filter is expressed as the reciprocal of a polynomial in s. If $T(s)$ is the transfer function and $B_n(s)$ is the polynomial for the nth-order filter, then

$$T(s) = \frac{1}{B_n(s)} \tag{11.51}$$

The magnitude of the transfer function is unity within the passband, and the break frequency is unity. This is called the **normalized filter**. The transfer function for the normalized filter has a corner frequency at $\omega = 1$. The magnitude at this corner frequency is 0.707.

The normalized polynomials for Butterworth filters are as follows:

$$
\begin{aligned}
B_1(s) &= s + 1 \\
B_2(s) &= s^2 + 1.414s + 1 \\
B_3(s) &= s^3 + 2s^2 + 2s + 1 \\
B_4(s) &= s^4 + 2.61s^3 + 3.41s^2 + 2.61s + 1 \\
B_5(s) &= s^5 + 3.24s^4 + 5.24s^3 + 5.24s^2 + 3.24s + 1 \\
B_6(s) &= s^6 + 3.86s^5 + 7.46s^4 + 9.14s^3 + 7.46s^2 + 3.86s + 1
\end{aligned}
$$

For example, a third-order Butterworth low-pass filter with a dc gain of 15 and a cutoff frequency of 1,500 Hz is defined by the transfer function

$$
T(s) = \frac{15}{(\frac{s}{2\pi \times 1,500})^3 + 2(\frac{s}{2\pi \times 1,500})^2 + 2(\frac{s}{2\pi \times 1,500}) + 1} \tag{11.52}
$$

11.4 Nonlinear Applications

11.4.1 Logarithmic Amplifiers[10]

When a diode is used in place of the feedback resistor R_F, the output voltage is proportional to the logarithm of the input voltage (see Figure 11.25).

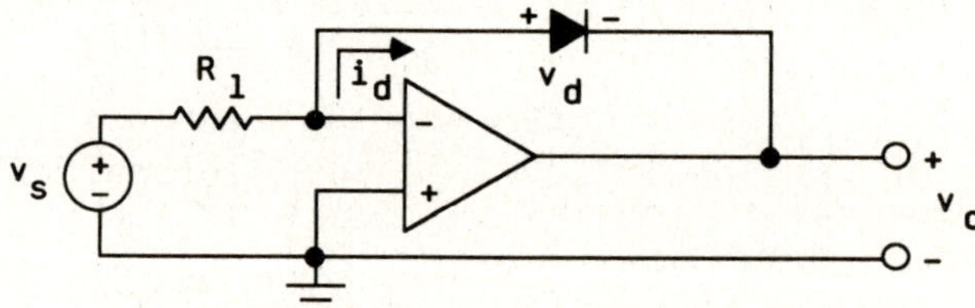

Figure 11.25: Simple logarithmic amplifier

If

$$
i_d = I_o \left[\exp\left(\frac{qv_d}{nkt}\right) - 1 \right] \tag{11.53}
$$

I_o = saturation current

and

$$
n \approx 1
$$

$$
\frac{kT}{q} \approx 25 \text{ mV @ room temperature}
$$

then

$$v_o = -v_d = -\frac{nkt}{q}\ \ln\ \frac{v_s}{R_1 I_o}$$

(11.54)

$$\frac{v_d}{\frac{nkt}{q}} \gg 1$$

The usefulness of the diode is limited because v_o is temperature dependent as a result of the factor $\frac{nkT}{q}$ and the saturation current I_o, which are both temperature dependent. The factor n depends on the diode current. It can be eliminated by replacing the diode with a grounded-base transistor (see Figure 11.26).

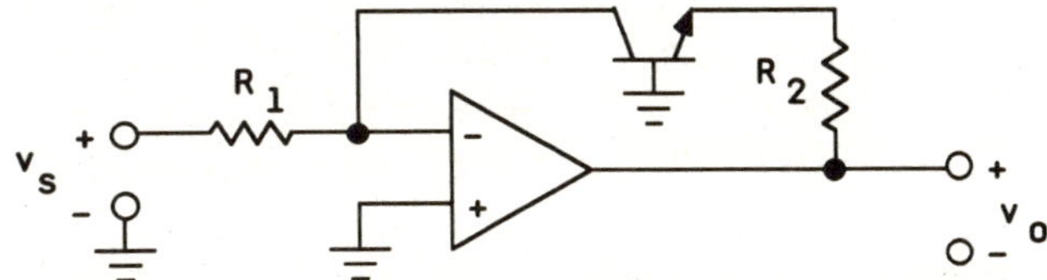

Figure 11.26: Logarithmic amplifier using a transistor

For the current of this circuit, the coefficient remains at unity over a collector current range from $<10^{-12}$A to $\sim 10^{-4}$A. Thus, for $v_o = V_{BE} > 760$ mV, the output voltage v_o becomes

$$v_o = -V_{BE} \approx -\frac{kT}{q}\ \ln\left(\frac{v_s}{R_1 I_o}\right)$$

(11.55)

11.5 Summary

In this chapter, we began by illustrating the differences between the ideal op amp and the practical op amp. We then defined the parameters used to describe practical op amps. Among these parameters are: finite gain and bandwidth, input offset voltage, offset and bias current, common mode rejection, power supply rejection, slew-rate and full-power bandwidth, and finite input and output resistance. On the basis of these parameters, we can alter the idealized equivalent circuit of the op amp to a more practical equivalent circuit. Laboratory 1 is designed to give the student experience measuring these nonideal op amp parameters.

The remainder of the chapter was devoted to a discussion of practical circuit applications of the op amp. The remaining laboratory experiments of this chapter are designed to give the student experience with some of these circuit applications.

LAB 11L1

TITLE: Measuring Nonideal Operational Amplifier Characteristics

OBJECTIVE: This laboratory exercise is designed to give the student experience in measuring the electrical parameters that characterize nonideal or real operational amplifiers. Among these parameters are finite dc gain (A_o), common-mode rejection ratio (CMRR), power supply rejection ratio (PSRR), full-power bandwidth, slew rate, input offset voltage (V_{io}), input bias current (I_B), and input offset current I_{OS}.

To make the measurements of these parameters more interesting, you are given several op amps. Each has particular parameters for which it is optimized. Your instructor will give you the data sheets that characterize your unknown op amps. Your objective is to match each op amp with the correct data sheet.

EQUIPMENT AND COMPONENTS:

Oscilloscope (dual trace and delay sweep)
dc power supplies: ±15 V and 10 V
Digital multimeter (DMM)
Function generator (square and sine wave)
Three unknown op amps labeled u_1, u_2, and u_3

Resistors	2	100-kΩ	1/8-W	resistors
	2	10-kΩ	1/8-W	resistors
	1	10-Ω	1/8-W	resistors

Breadboard
1,000-pf capacitor

PRELABORATORY:

1. Define the following terms:

    a. Input bias current (I_B).

    b. Input offset current (I_{OS}).

    c. Input offset voltage (V_{OS}).

    d. Power supply rejection ratio (PSRR).

    e. Common mode rejection ratio (CMRR).

    f. Open-loop gain (A_d).

    g. Slew rate.

    h. Unity-gain closed-loop bandwidth.

2. Voltage offset compensation. Most of the test circuits used to measure op amp characteristics require using an additional op amp that has a very low offset voltage or using external compensation. Determine the op amp you will be using, and refer to an appropriate data book for the voltage offset compensation.

3. Measuring input offset voltage (V_{io}). Consider the circuit of Figure 11L1.1. The label "OUT" stands for "op amp under test" and represents one of the unknown op amps. The buffer is the offset-compensated op amp and will be used throughout the lab.

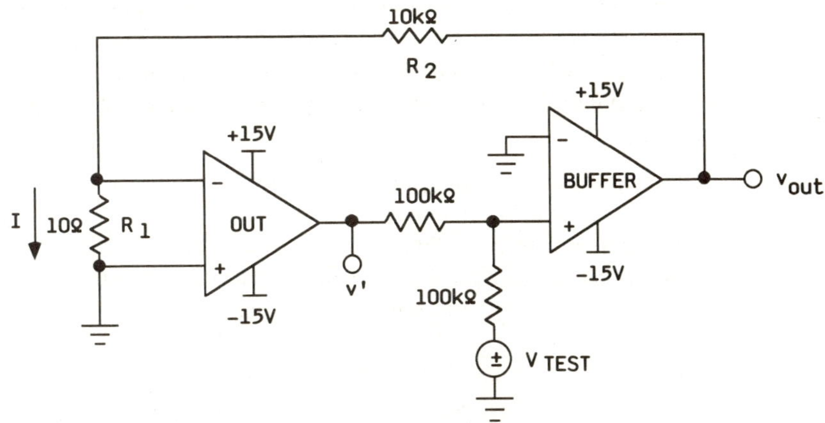

Figure 11L1.1: Measuring input offset voltage

By setting V_{TEST} equal to zero, the feedback forces the output of the OUT to zero. The best way to see the feedback action qualitatively is to assume that the output of the OUT is not zero; let us say that it is positive. Then follow the effect of this positive voltage around the loop. If V' is 1 V, then V_+ for the BUFFER equals 0.5 V and the output swings positive. This forces the V_- voltage at the input of the OUT to be more positive than V_+. This in turn causes the output of the OUT to swing to a negative voltage. Following the same line of reasoning forces V' back positive. So, assuming the loop is stable, the only stable operating point is $V' \approx 0$. This means that V_{out} is forced to the value of voltage necessary to place a voltage drop across R_1, so the voltage IR_1 compensates for the offset voltage V_{io} of the OUT. So V' is approximately zero.

Therefore, one concludes that the voltage drop across R_1

$$IR_1 = V_{io} \qquad\qquad\qquad (11L1.1)$$

and

$$V_{out} = IR_1 + IR_2 \qquad\qquad\qquad (11L1.2)$$

Substituting Equation (11L1.1) into (11L1.2) results in

$$V_{out} = V_{io} + R_2 \frac{V_{io}}{R_1} = V_{io}\left[1 + \frac{R_2}{R_1}\right] \approx 1,000 V_{io} \qquad\qquad (11L1.3)$$

Figure 11L1.2 shows the nonideal characteristics that appear at the input of the op amp. The analysis assumes that the bias currents, I_{B_1} and I_{B_2}, do not contribute a significant voltage to V_{out}. If $I_{B_1} = I_{B_2} = 100$ nA, what voltage is generated at V_{out}? Show that the bias current needs to be 5 µA in order to generate a 10% error in the measurement of V_{io} if V_{io} is actually 500 µV?

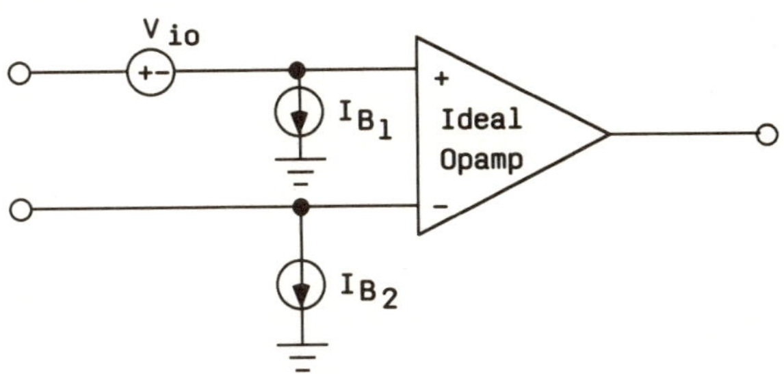

Figure 11L1.2: Input nonideal characteristics

4. Measuring the power supply rejection ratio (PSRR). Use the same circuit as given in part 2 of the prelab. How can you determine the PSRR? (Hint: Change V_{CC} (the +15-V supply) by 1 V and measure the change in V_{io}.) The OUT and the BUFFER should have separate power supplies. Also, this procedure should be repeated for the negative supply. The supply that produces the poorer result defines the PSRR.

5. Measuring the open-loop differential gain (A_d). Consider the circuit of Figure 11L1.3. Notice that V_{TEST} is equal to -10 V. By an analysis similar to that of part 2 of the prelab, V' will equal 10 V. This means that a signal is generated at the differential input to the OUT to make $V' = 10$ volts. The input signal, across R_1, produces a current that also flows in R_2. The current through R_1 is given by

$$I = \frac{V_i + V_{io}}{R_1} \tag{11L1.4}$$

and

$$V_{OUT} = \left[\frac{V_{io} + V_i}{R_1}\right]\left[R_1 + R_2\right] \tag{11L1.5}$$

The open-loop gain A_d is equal to the ratio of the output voltage to the input voltage. The output voltage of the OUT is equal to 10 V. The input voltage can then be solved for, using Equation 11L1.5.

Write an expression for the OUT's open-loop gain A_d in terms of V_{io}, R_1, and R_2.

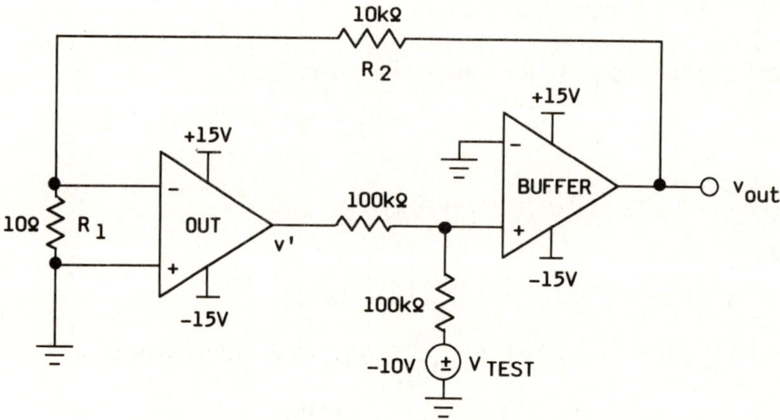

Figure 11L1.3: Circuit for determining A_d

6. Measuring input bias and offset currents I_{B_1}, I_{B_2}, and I_{OS}.

Consider the circuit of Figure 11L1.4.

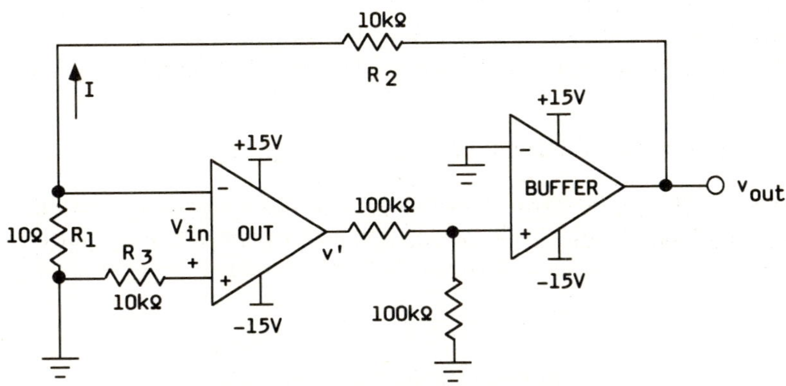

Figure 11L1.4: Circuit for measuring I_{B_1}

Because V_{TEST} is equal to zero, the feedback causes V' also to equal zero. This means that the voltage drop in R_1 must compensate for voltages generated by the offset voltage and bias currents of the OUT. Using the model of Figure 11L1.2 to represent the OUT, we can write the loop equation

$$IR_1 + V_{io} - V_{in} - R_3 I_{B_1} = 0 \qquad (11L1.6)$$

Since V' equals zero, V_{in} equals zero, and Equation 11L1.6 reduces to

$$IR_1 = R_3 I_{B_1} + V_{io} \qquad (11L1.7)$$

The current I also flows in R_2 and therefore V_{out} is given by

$$-V_{out} = IR_1 + IR_2 = \left[\frac{R_3 I_{B_1} + V_{io}}{R_1}\right](R_1 + R_2) \qquad (11L1.8)$$

Equation 11L1.8 can be solved for I_{B_1} to yield

$$I_{B_1} = -\frac{1}{R_3}\left[\frac{R_1 V_{out}}{R_1 + R_2} + V_{io}\right] \qquad (11L1.9)$$

To measure I_{B_2}, move R_3 to the inverting terminal, as shown in Figure 11L1.5.

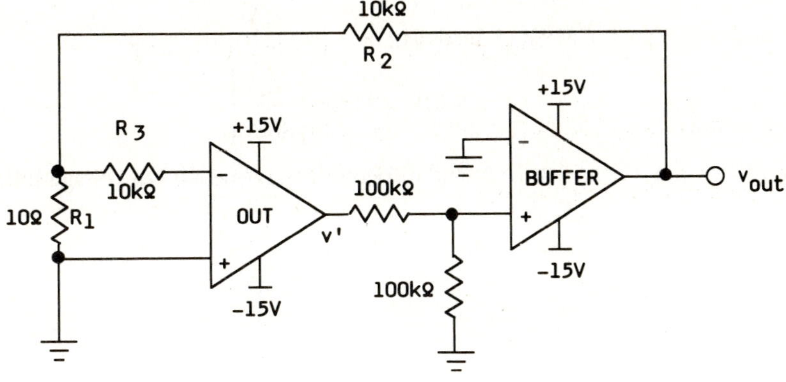

Figure 11L1.5: Circuit for measuring I_{B_2}

Show, by a similar analysis, that I_{B_2} is given by

$$I_{B_2} = +\frac{1}{R_3}\left[\frac{R_1 V_{OUT}}{R_1 + R_2} + V_{io}\right] \tag{11L1.10}$$

The offset current I_{os} is the difference between I_{B_1} and I_{B_2}.

7. Measuring common mode rejection ratio (CMRR). Since we determined the differential gain A_d in Part 4, we need the common-mode gain A_{cm} to find the CMRR. Accordingly, consider the circuit of Figure 11L1.6. Since V_{cm} is common to both inputs, any signal generated at V_{out} due to V_{cm} is caused by a common-mode gain.

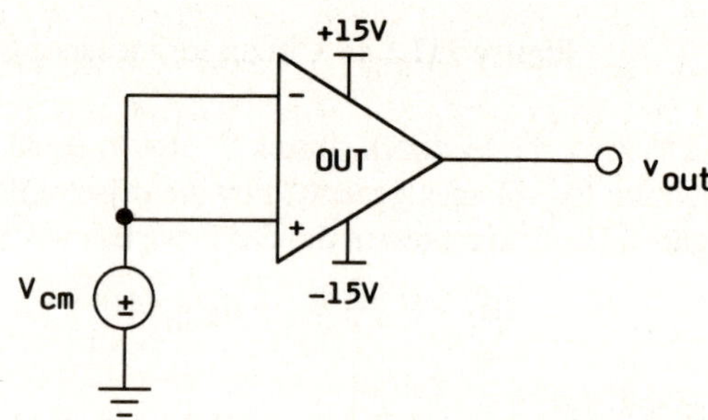

Figure 11L1.6: Circuit for measuring CMRR

First, we need to subtract the part of V_{out} caused by the offset voltage V_{io}. The easiest way to do this is to set V_{cm} equal to zero and measure V_{out} and then set V_{cm} equal to 10 V and measure V_{out}. The common-mode gain then is

$$A_{cm} = \frac{V_{out}\Big|V_{cm}=10 - V_{out}\Big|V_{cm}=0}{10} \; V \tag{11L1.11}$$

Finally,

$$CMRR = \left|\frac{A_d}{A_{cm}}\right| \tag{11L1.12}$$

Most data sheets express CMRR in decibels:

$$CMRR = 20\log\left|\frac{A_d}{A_{cm}}\right| \tag{11L1.13}$$

8. Measuring slew rate. Consider the circuit of Figure 11L1.7 which is a voltage follower.

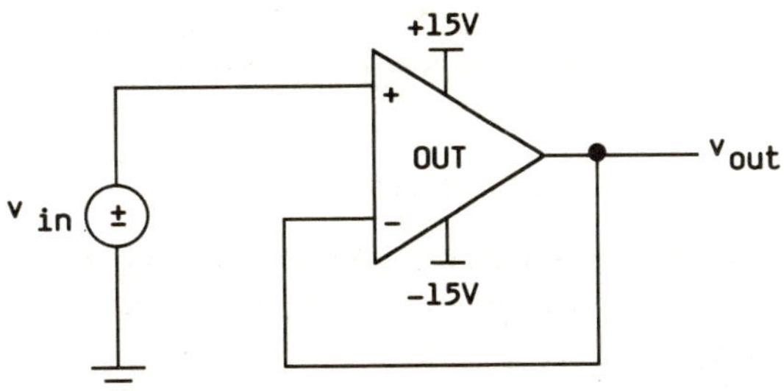

Figure 11L1.7: Circuit for measuring slew rate

Let v_{in} be a pulse train with a 50% duty cycle, 1-millisecond period, and 5-V amplitude. From the definition, describe how one can determine the slew rate of an op amp through the use of the oscilloscope. The use of the delay sweep feature of the oscilloscope will prove useful.

9. Measuring unity-gain closed loop bandwidth. In the circuit of Figure 11L1.7 v_{in} is a sine wave with an amplitude of 0.5 V. What will the amplitude of V_{OUT} be when the frequency is equal to the unity gain closed loop bandwidth?

Laboratory Procedure:

1. Compensating the offset voltage of the BUFFER

    a. Construct the BUFFER by connecting the appropriate voltage offset compensation circuit for the op amp you are given. The paper implementation should have been done in the prelaboratory.

    b. Adjust the variable resistor in your compensation circuit so that the output is zero or passes through zero when the input is set to zero.

2. Measuring the input offset voltage V_{os}

    a. Build the circuit of Figure 11L1.1 and repeated in Figure 11L1.8, which is essentially that of Figure 11L1.1 with a 1,000-pF capacitor attached across R_2.

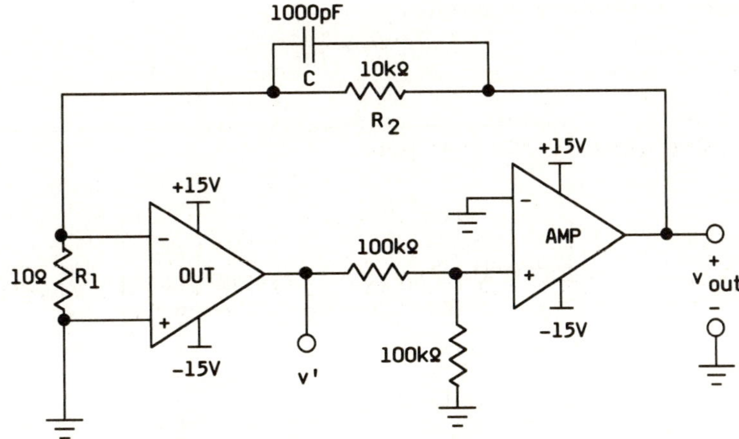

Figure 11L1.8: Circuit for measuring V_{OS}

    b. Measure and record the voltages V' and V_{out}.

    (1) Is $V' \approx 0$?

    (2) Calculate the input offset voltage for the OUT and record it in Table 11L1.1.

    (3) Repeat this procedure for three different op amps under test.

3. Measuring the power supply rejection ratio.

    a. Build the circuit shown in Figure 11L1.9. Note that this is the same circuit as that given in Figure 11L1.8 except that the positive 15-V supply for the OUT must be separate from that of the BUFFER.

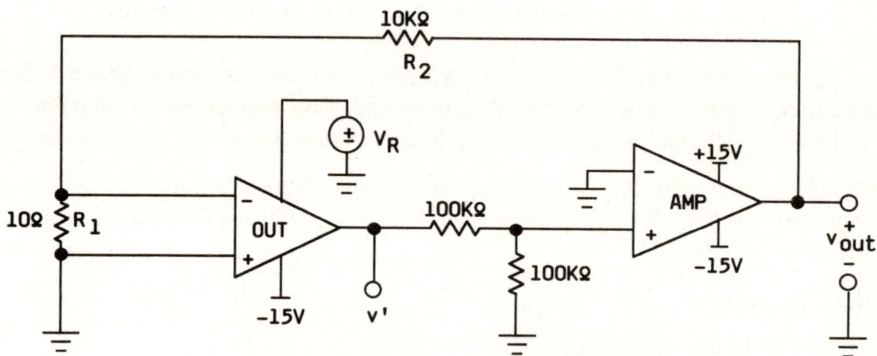

Figure 11L1.9: Circuit for measuring PSRR

    b. Let V_R equal 15 V, and measure and record V_{out}. (This is the same as in Part 2 of the lab.)

    c. Let V_R equal 14 V, and measure and record v_{out}. Then use your definition of PSRR to determine its value from the collected data. Record your PSRR in Table 11L1.1.

    d. Repeat Part c for each op amp being tested.

4. Measuring the open-loop differential gain A_d.

    a. Build the circuit shown in Figure 11L1.10, which is essentially the same as that given in Figure 11L1.3. Verify that v^1 equals 10 V.

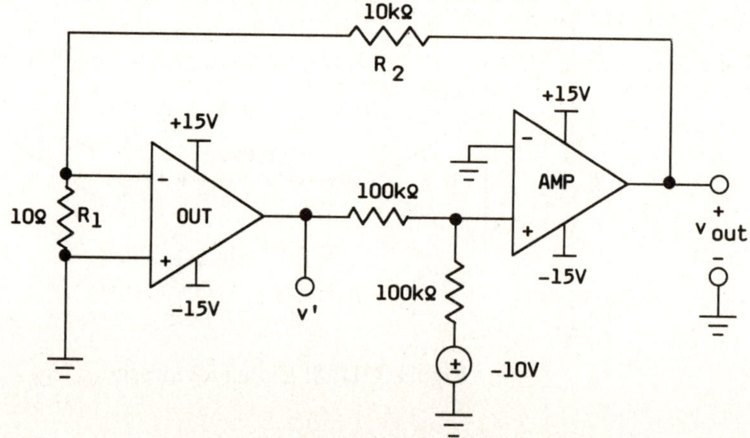

Figure 11L1.10: Circuit for measuring A_d

b. Measure and record V_{out}. Use your result from Part 5 of the prelab, and calculate the open-loop differential gain A_d. Record A_d in Table 11L1.1.

c. Repeat Part b for each op amp being tested.

5. Measuring input bias and offset currents: I_{B_1}, I_{B_2}, and I_{os}.

a. Build the circuit of Figure 11L1.11. This is essentially the same circuit as that shown in Figure 11L1.4.

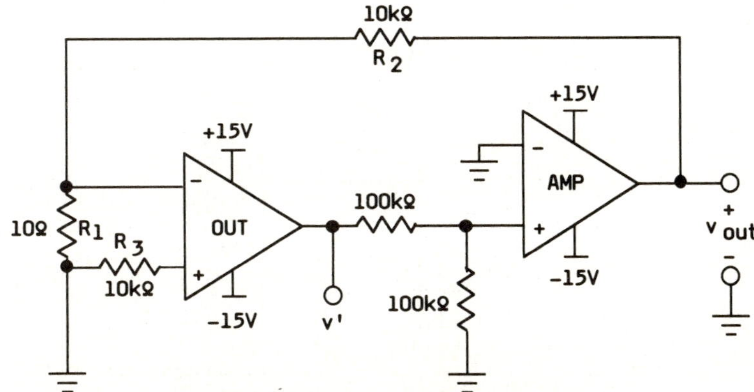

Figure 11L1.11: Circuit for measuring I_{B_1}

b. Verify that V' equals zero volts. Measure and record V_{out}. Use Equation (11L1.9) to determine I_{B_1} (the bias current for the noninverting terminal of the OUT). Record I_{B_1} in Table 11L1.1.

c. Repeat Part b for each op amp being tested.

d. Move R_3 to the inverting terminal and the opposite side of R_1. Then ground the noninverting terminal as shown in Figure 11L1.5.

e. Measure and record V_{out}. Use Equation (11L1.10) to determine I_{B_2}. Record I_{B_2} in Table 11L1.1. Find and record the offset current I_{os}.

f. Repeat Part e for each op amp being tested.

6. Measuring the common-mode rejection ratio CMRR.

a. Build the circuit of 11L1.12. This is essentially the same circuit as that shown in Figure 11L1.6.

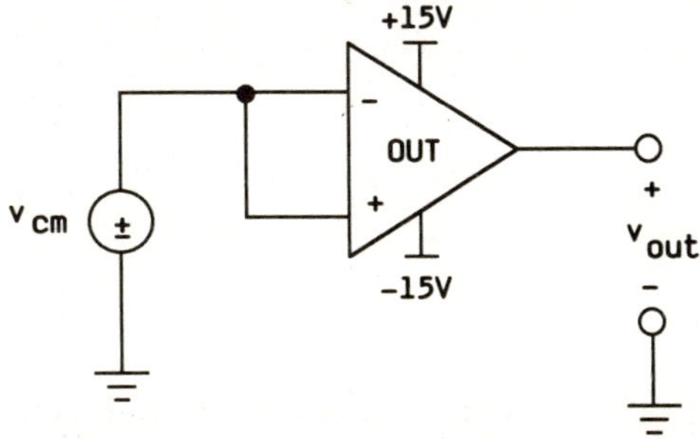

Figure 11L1.12: Circuit for measuring CMRR

    b. Let V_{cm} equal zero. Then measure and record V_{out}.

    c. Let V_{cm} equal 10 V. Then measure and record V_{out}.

    d. Use Equation 11L1.11 to determine the common-mode gain A_{cm}.

    e. Calculate the CMRR expressed in decibels. Record A_{cm} in Table 11L1.11.

    f. Repeat the above procedure for each op amp being tested.

7. Measuring slew rate SR.

    a. Build the circuit of Figure 11L1.13. This is essentially the same circuit as that shown in Figure 11L1.7.

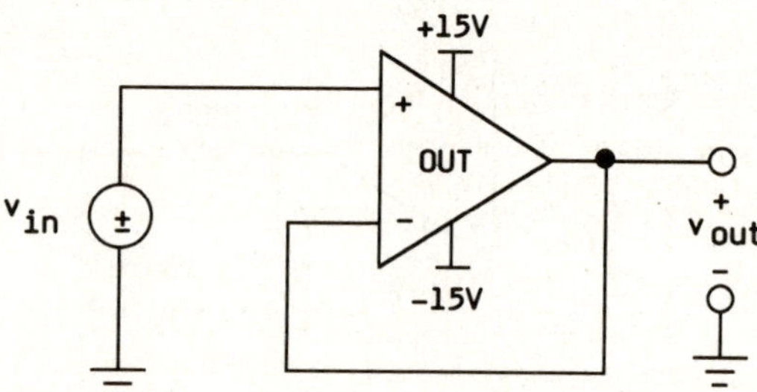

Figure 11L1.13: Circuit for measuring slew rate

Let v_{in} be a pulse train as described in the prelab Part 8.

    b. Observe v_{out} using the delayed sweep capability of your oscilloscope. Sketch the oscilloscope's display (show the positive and negative edge), and calculate the slew rate (express slew rate in V/μsec). Use the smaller number as the slew rate and record it in Table 11L1.1.

    c. Repeat Part b for each op amp being tested.

8. Measure the unity gain closed-loop bandwidth A_{CL}.

    a. Use the circuit of Figure 11L1.13, but let v_{in} be a 1-kHz sine wave with an amplitude of 0.5 V. Increase the frequency until the amplitude of v_{out} decreases to 0.71 of its 1-kHz amplitude.

    b. Record the results of Part a on Table 11L1.1 and repeat this procedure for each op amp being tested.

Table 11L1.1: Summary of experimental data

Op amp Parameters	OP AMPS UNDER TEST		
	OUT #1	OUT #2	OUT #3
V_{OS} (μV) Input Offset Voltage			
PSRR (μV/V) Power Supply Rejection Ratio			
A_d (V/mV) Open Loop Diff. Gain.			
I_{B_1} (nA) Input bias current			
I_{B_2} (nA) Input bias current			
I_{OS} (nA) Input offset current			
CMRR (dB) Common-mode rejection ratio			
SR (V/μs) Slew Rate			
A_{CL} (MHz) Unity-Gain Closed-loop BW			

DISCUSSION: Determine which op amp being tested corresponds to which data sheet. Justify your selection.

CONCLUSIONS:

LAB 11L.2

TITLE: Introduction to Active Filters

OBJECTIVE: This laboratory exercise gives an introduction to the active filter. A single-pole low pass and high-pass filter is presented for analysis. Band-pass and band-stop filters are developed by connecting the low-pass and high-pass filter in cascade and cascode configurations. Also, the filters' transfer functions are plotted on semilog paper.

EQUIPMENT AND COMPONENTS:
Power supply, ± 15 V
Function generator
Oscilloscope (dual trace)
Resistors
Capacitors
General-purpose operation amplifier
(unity-gain compensated)

PRELABORATORY:

1. Given the circuit of Figure 11L2.1, find the transfer function $\dfrac{v_{out}(j\omega)}{v_{in}(j\omega)}$. Assume that the operational amplifier is ideal.

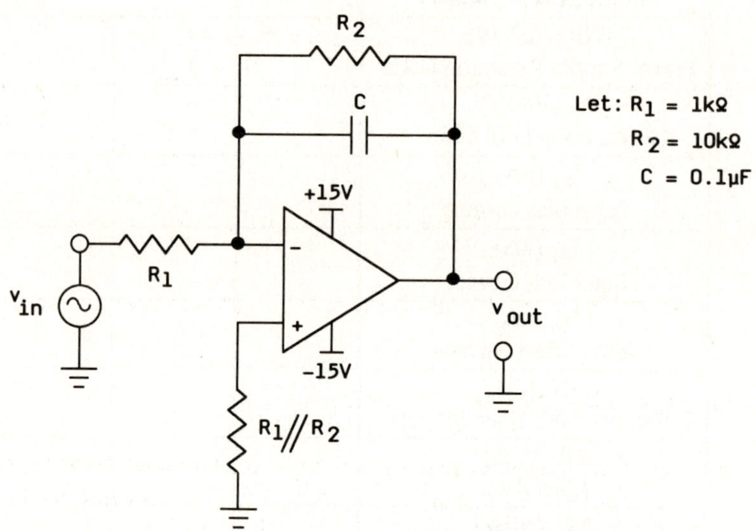

Figure 11L2.1: Active filter using an inverting amplifier

   a. Find the magnitude of the transfer function, $\left|\dfrac{v_{out}(j\omega)}{v_{in}(j\omega)}\right|$

   b. Plot the magnitude versus frequency ω on at least 4 decade semilog paper. Plot the magnitude on the linear axes and the frequency on the log axes. Start with ω equal to 10 rad/sec.

   c. Express the magnitude in dB's, where $\left|\dfrac{v_{out}(j\omega)}{v_{in}(j\omega)}\right| dB = 20 \log \left|\dfrac{v_{out}(j\omega)}{v_{in}(j\omega)}\right|$ Plot the magnitude in dB's on the linear axis versus frequency, ω, on the log axis.

   d. Is the circuit a low-pass or high-pass filter?

2. Given the circuit of Figure 11L2.2 find the transfer function, $\dfrac{v_{out}(j\omega)}{v_{in}(j\omega)}$. Assume that the operational amplifier is ideal.

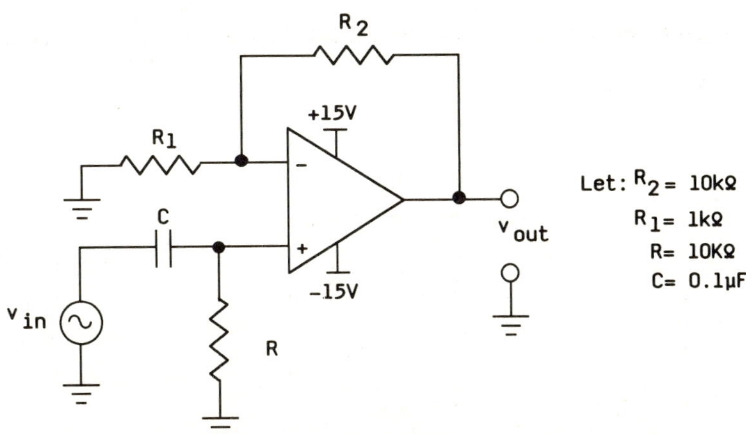

Figure 11L2.2: Active filter using a noninverting amplifier

a. Find the magnitude of the transfer function $\dfrac{v_{out}(j\omega)}{v_{in}(j\omega)}$.

b. Plot the magnitude versus frequency ω on four decade semilog graph paper. Plot the magnitude on the linear axes and the frequency on the log axes. Start with ω equal to 1 rad/sec and cover 1, 10, 100, 1,000, and 10,000 rad/sec.

c. Express the magnitude in dB's, where $\left|\dfrac{v_{out}(j\omega)}{v_{in}(j\omega)}\right|$ dB = 20 log $\left|\dfrac{v_{out}(j\omega)}{v_{in}(j\omega)}\right|$. Relabel the linear axes in dB's, and plot the magnitude in dB's versus frequency ω on semilog paper.

d. Is the circuit a low-pass or high-pass filter?

3. Given the circuit of Figure 11L2.3, determine the transfer function between the output and the two inputs.

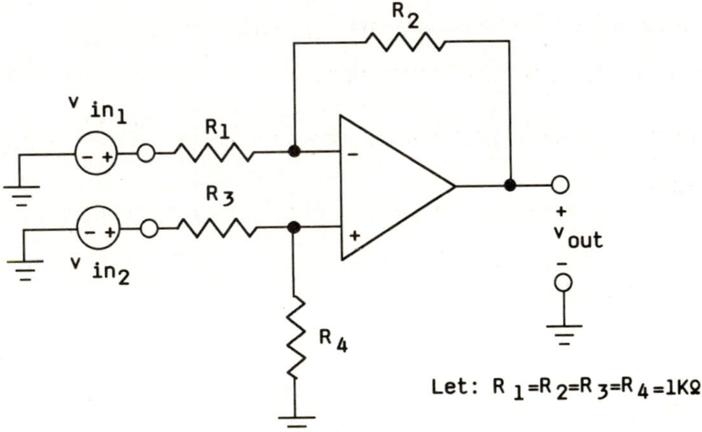

Figure 11L2.3: Circuit of section 3

4. What is the resulting transfer function when the circuits of Figure 11L2.1 and 11L.2 are cascaded together?

5. What is the resulting transfer function when the circuits of Figure 11L2.1 and 11L2.2 are cascoded together? Note that the outputs of the two circuits cannot be connected together. Instead, the outputs of the circuits of Figure 11L2.1 and Figure 11L2.2 are connected to the inputs of the circuit of Figure 11L2.3, as shown in Figure 11L2.4.

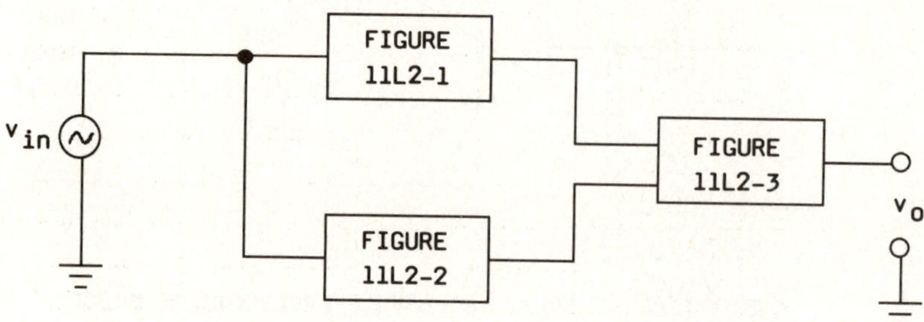

Figure 11L2.4: Circuits of Figure 11L2.1 and Figure 11L2.2 connected in cascode

Laboratory Procedure:

1. Build the circuit of Figure 11L2.5.

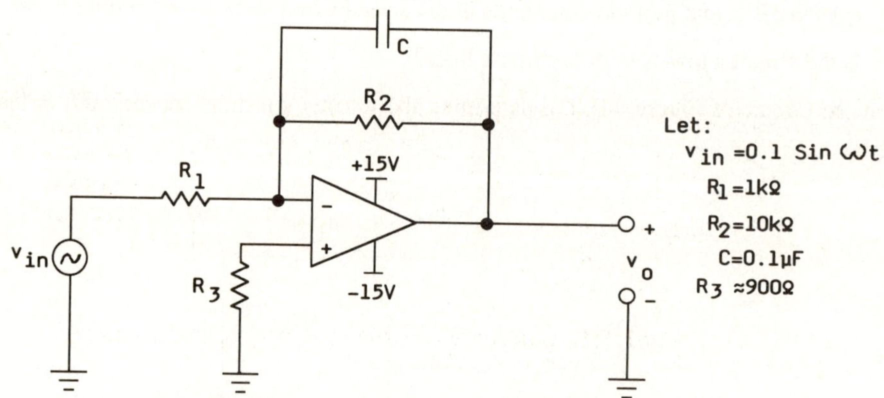

Figure 11L2.5: Active filter in inverting configuration

a. Display v_o and v_{in} in Table 11L2.1.

Table 11L2.1: Data for the circuit of Figure 11L2.5

ω rad/sec	$\mid v_{in} \mid$ volts	$\mid v_o \mid$ volts	$\mid \dfrac{v_o}{v_{in}} \mid$ dB	$\mid \dfrac{v_o}{v_{in}} \mid$ degrees
dc				
10				
20				
100				
200				
1,000				
2,000				
10,000				
20,000				
100,000				

    b.   Plot the magnitude of the gain (in dB) and plot the magnitude and phase angle (in degrees) versus frequency on semilog paper.

2.   Build the circuit of Figure 11L2.6. The operational amplifier should be the same as that used in Part 1.

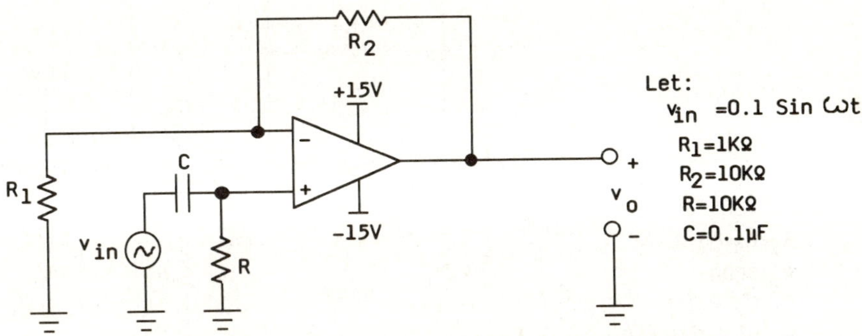

Figure 11L2.6: Active filter in noninverting configuration

    a.   Display v_o and v_{in} on a dual-trace oscilloscope. Be sure that the mode of displaying the waveforms preserves their phase relationship. Record the magnitude and phase angle between v_o and v_{in} in Table 11L2.2.

Table 11L2.2: Data for circuit of Figure 11L2.6

ω rad/sec	$\mid v_{in} \mid$ volts	$\mid v_o \mid$ volts	$\mid \dfrac{v_o}{v_{in}} \mid$ dB	$\mid \dfrac{v_o}{v_{in}} \mid$ degrees
dc				
10				
20				
100				
200				
1,000				
2,000				
10,000				
20,000				
100,000				

b.   Plot the magnitude in dB and phase angle (in degrees) versus frequency on semilog paper.

3.   Build the circuit of Figure 11L2.7. This circuit is the circuit of Figure 11L2.5 and the circuit of Figure 11L2.6 cascaded together.

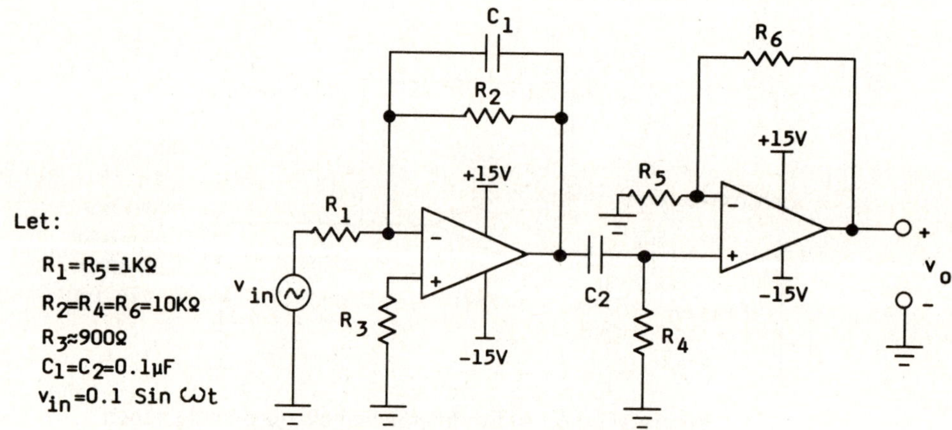

Figure 11L2.7: Cascade configuration

a.   Display v_o and v_{in} on a dual-trace oscilloscope. Be sure that the waveforms are displayed in a way that preserves their phase relationship. Record the magnitude and phase angle between v_o and v_{in} in Table 11L2.3.

Table 11L2.3: Data for cascade configuration

ω rad/sec	$\|v_{in}\|$ volts	$\|v_o\|$ volts	$\|\dfrac{v_o}{v_{in}}\|$ dB	$\|\dfrac{v_o}{v_{in}}\|$ degrees
dc				
10				
20				
100				
200				
1,000				
2,000				
10,000				
20,000				
100,000				

b. Plot the magnitude (in dB) and phase angle (in degrees) versus frequency on semilog paper.

4. Build the circuit of Figure 11L2.8. This circuit is the circuits of Figures 11L2.5 and 11L2.6 cascoded together, using the circuit of Figure 11L2.4.

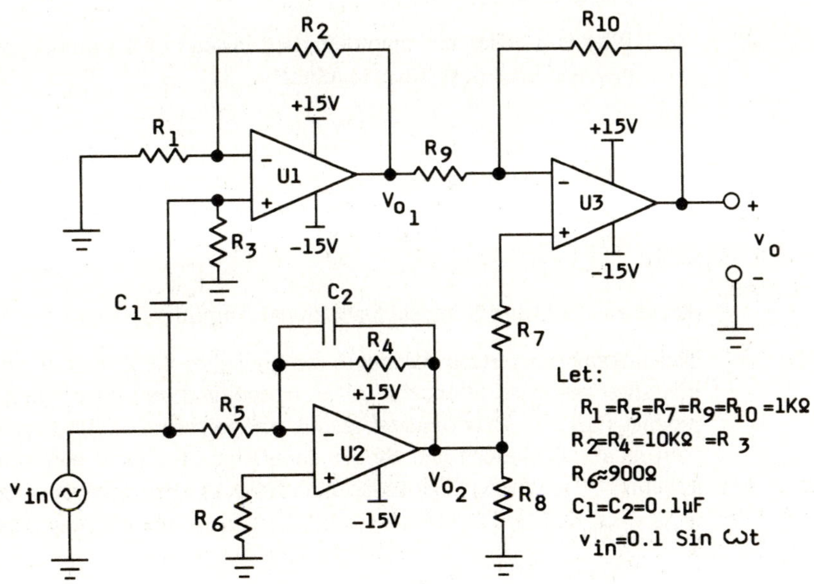

Figure 11L2.8: Cascode configuration

a. Display v_o and v_{in} on a dual-trace oscilloscope. Be sure that the mode of displaying the waveforms preserves their phase relationship. Record the magnitude and phase angle between v_o and v_{in} in Table 11L2.4.

Table 11L2.4: Data for cascade configuration

ω rad/sec	$\mid v_{in} \mid$ volts	$\mid v_O \mid$ volts	$\mid \dfrac{v_O}{v_{in}} \mid$ dB	$\mid \dfrac{v_O}{v_{in}} \mid$ degrees
dc				
10				
20				
100				
200				
1000				
2000				
10,000				
20,000				
100,000				

b. Plot the magnitude (in dB) and phase angle (in degrees) versus frequency on semilog paper.

c. Display and sketch v_{01} and v_{02}. Be sure to preserve their phase relationship, and be sure to use units of rad/sec as the input frequency for your sketch.

DISCUSSION: In your discussion, include the answers to the following questions:

1. If the low pass and high pass circuits were interchanged, would the transfer function change? Why?

2. Why is a difference amplifier used instead of a summer to cascade the low-pass and high-pass filters together?

CONCLUSIONS:

LAB 11L3

TITLE: Active Filters Using Nonideal Operational Amplifiers

OBJECTIVE: This laboratory exercise introduces the student to the design of active filters and the limitation on these filters due to an operational amplifier's finite gain-bandwidth product (GBW). Experimental results are presented in Bode plots and compared with hand calculations and SPICE simulations for ideal and nonideal operational amplifiers. Low-pass, High-pass, and band-pass filters are examined.

EQUIPMENT AND COMPONENTS:

Power supply, ±15 V
Function Generator (sine waves needed)
Oscilloscope (dual trace)
Resistors (1/8 W)
Capacitors (30-V rating)
General-purpose operational amplifiers
    (preferably unity gain compensated)

PRELABORATORY:

1. Given the circuit of Figure 11L3.1 find the transfer function $\dfrac{v_o(j\omega)}{v_{in}(j\omega)}$. Assume the operational amplifier is ideal.

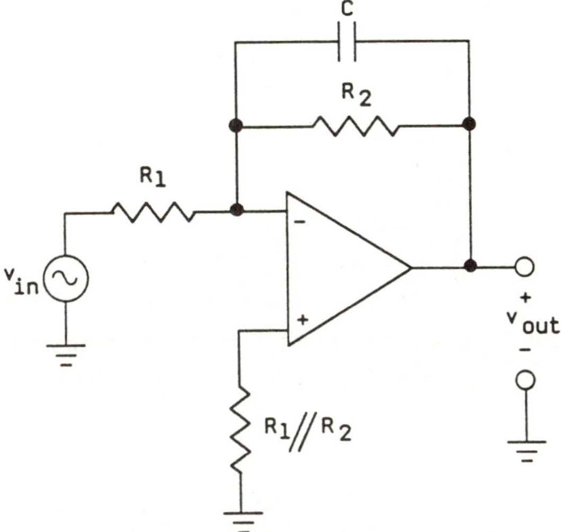

Figure 11L3.1: Low-pass filter

Next, assume the op amp is not ideal, but has an open-loop gain

$$A(j\omega) = \frac{A_o}{1 + \dfrac{j\omega}{\omega_c}}$$

(11L3.1)

where

$$
\begin{aligned}
A_o &\equiv \text{dc gain} \\
\omega &\equiv \text{frequency, rad/sec} \\
\omega_c &\equiv \text{simple pole}
\end{aligned}
$$

a. From the data sheet of the operational amplifier you will use in your lab, determine values for A_o and ω_c.

b. Determine the transfer function for the circuit of Figure 11L3.1 using Equation (11L3.1) for the open-loop gain of the op amp.

c. Given the transfer functions determined for the ideal and nonideal op amps, find the magnitude and phase plots for each case given in Table 11L3.1. Obviously, A_o and ω_c are not needed for the analysis ideal of the case.

Table 11L3.1 Component values for low-pass filter

Case	R_1 $K\Omega$	R_2 $K\Omega$	C μF	A_o	ω_c radians/sec
I	1	10	0.1	10^5	50
II	1	1,000	0.1	10^5	50
III	1	10	0.1	Values given for your op amp	
IV	1	1,000	0.1	Values given for your op amp	

d. Simulate cases III and IV using SPICE and the op amp model given in Figure 11L3.2 for the nonideal versions. Do a frequency analysis with amplitude 1 and let the frequency vary from 1 Hz to 1 MHz.

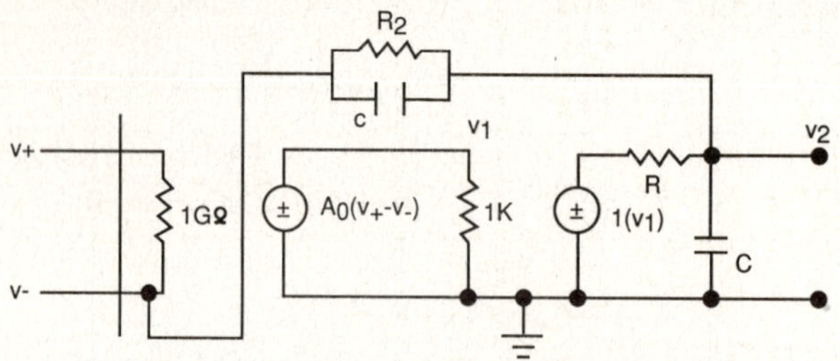

Figure 11L3.2: Op amp model for Cases III and IV

Pick R and C so that the RC time constant models ω_c for the nonideal cases. For Case I, assume the op amp is ideal and do a SPICE simulation. For this case use the model given in Figure 11L3.3.

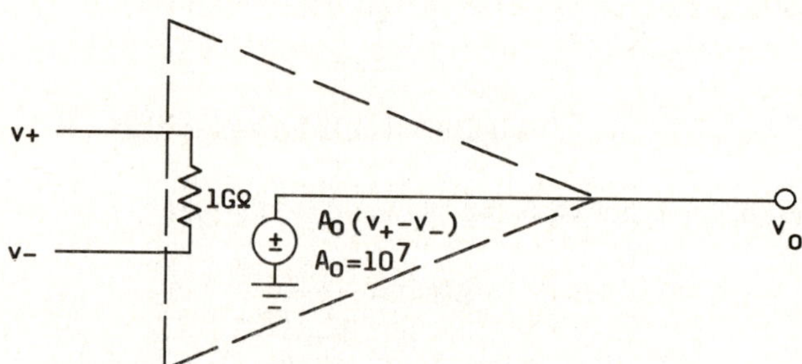

Figure 11L3.3: Op amp model for Cases I and II

2. Repeat the work of Part 1 using the circuit of Figure 11L3.4 and Table 11L3.2.

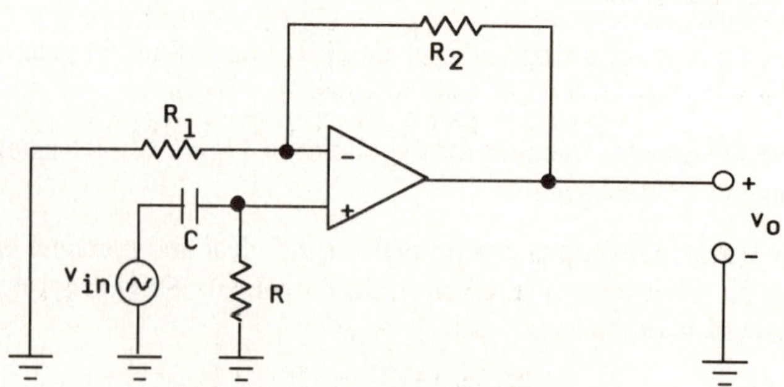

Figure 11L3.4: High-pass filter

Table 11L3.2

Case	R_1 KΩ	R_2 KΩ	C μF	R KΩ	A_o	ω_c
I	1	10	0.1	10	10^5	50
I	1	1000	0.1	1000	10^5	50
III	1	10	0.1	10	Values given for your op amp	
I	1	1000	0.1	1000	Values given for your op amp	

3. Cascade the circuits of Figure 11L3.1 and 11L3.4 as shown in Figure 11L3.5.

   a. If there is no loading effect between subcircuits 1 and 2, find the total transfer function $\dfrac{v_o(j\omega)}{v_{in}(j\omega)}$. Assume v_1 and v_2 are ideal.

   b. Assuming u_1 and u_2 are ideal op amps, pick values for R_1, R_2, C_1, C_2, R_3, R_4, and R_5 that yield the transfer function $\dfrac{v_o(j\omega)}{v_{in}(j\omega)}$ given in Figure 11L3.6.

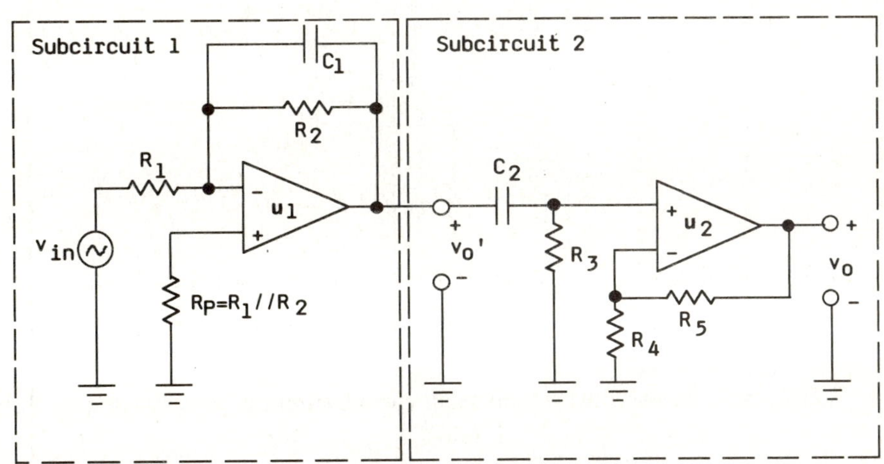

Figure 11L3.5: Band-pass filter

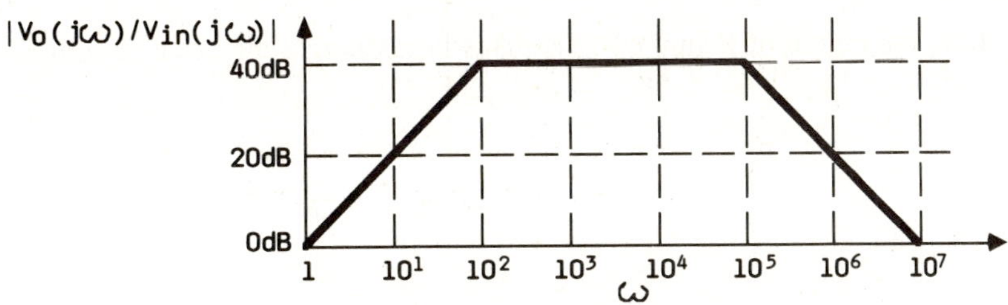

Figure 11L3.6: Transfer function

Verify your design using SPICE with the ideal op amp model given in Figure 11L3.3. Repeat the SPICE simulation using the nonideal op amp given in Figure 11L3.2. Compare the two simulations.

4. Use the open-loop gain as a function of frequency from your data sheet to determine the maximum bandwidth filter you could build described by the transfer function

$$\frac{v_o(j\omega)}{v_{in}(j\omega)} = \frac{j\omega \quad 10}{(1 + \dfrac{j\omega}{\omega_L})(1 + \dfrac{j\omega}{\omega_H})} \tag{11L3.2}$$

where

$$\omega_L = 100 \text{ rad/sec}$$

Use the circuit of Figure 11L3.5 to realize this filter. Determine values for all the components. Verify the design using SPICE.

Laboratory Procedure:

1. The low-pass filter.

   a. Build the circuit of Figure 11L3.7, which is essentially the same circuit as in Figure 11L3.1.

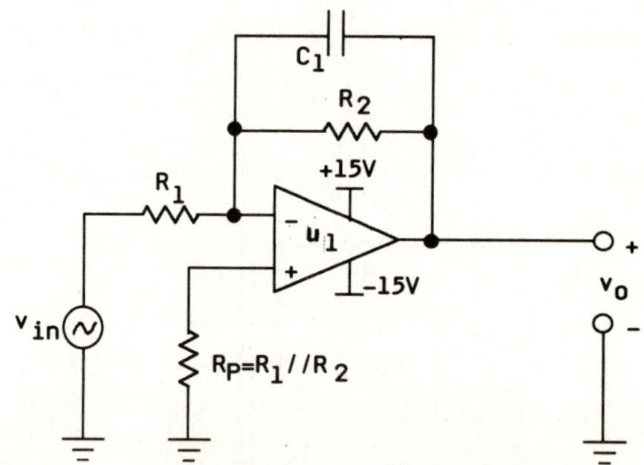

Figure 11L3.7: Low-pass filter

   b. Let v_{in} be a sinusoidal waveform small enough so that u_1 never saturates. Vary v_{in}'s frequency from dc to 1 MHz, and plot $\left|\dfrac{v_o(j\omega)}{v_{in}(j\omega)}\right|$ and $\dfrac{v_o(j\omega)}{v_{in}(j\omega)}$ against frequency on semilog paper. (Frequency is plotted on the log axis). Collect at least three points per decade.

2. The high-pass filter.

   a. Build the circuit of Figure 11L3.8. This is essentially the same circuit as that shown in Figure 11L3.4.

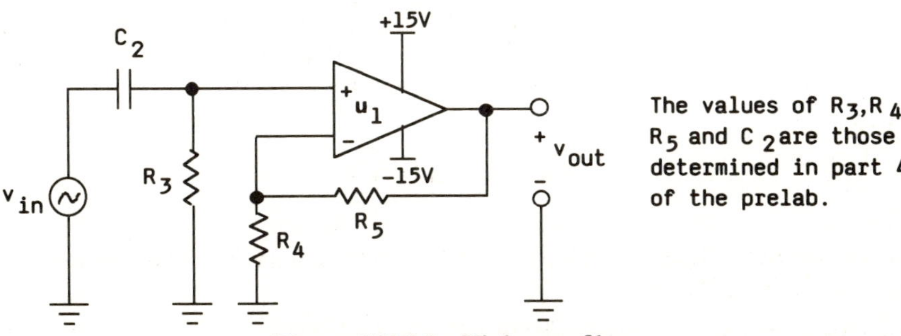

Figure 11L3.8: High-pass filter

b. Let v_{in} be a sinusoidal waveform small enough so that u_1 never saturates. Vary v_{in}'s frequency from dc to 1 MHz, and plot $\left| \dfrac{V_{out}(j\omega)}{V_{in}(j\omega)} \right|$ and $\dfrac{V_{out}(j\omega)}{V_{in}(j\omega)}$ against frequency on semilog paper. (Frequency is plotted on the log axis). Collect at least three points per power of 10.

3. The band-pass filter.

   a. Build the circuit of Figure 11L3.9. This is essentially the same circuit as that shown in Figure 11L3.5.

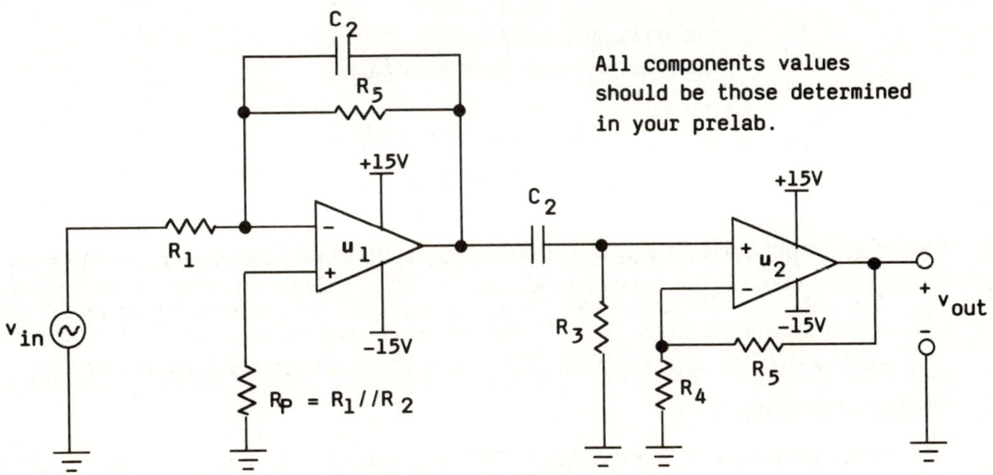

Figure 11L3.9: Band-pass filter

b. Let v_{in} be a sinusoidal waveform small enough so that u_1 never saturates. Vary v_{in}'s frequency from dc to 1 MHz, and plot $\left| \dfrac{V_{out}(j\omega)}{V_{in}(j\omega)} \right|$ and phase of $\dfrac{V_{out}(j\omega)}{V_{in}(j\omega)}$ on semilog paper. (Frequency is plotted on the log axis). Collect at least three points per decade.

c. Change R_3 so that ω_4 increases by a factor of 10. Repeat Part b for the larger value of ω_4.

d. Repeat Part c until you can no longer increase ω_4. Compare the last plot to that of Part 2.

DISCUSSION:

CONCLUSIONS: Compare your SPICE simulation to the experimental results.

LAB 11L.4

TITLE: Butterworth Filter Design

OBJECTIVE: In this laboratory exercise, the student must determine the order of the required Butterworth filter to meet the following design specifications:

> You are given a circuit that produces a square wave with a period of one millisecond. (You may use a function generator to produce the waves.) However, you need to generate a 1-kH sine wave. You will achieve this by designing a Butterworth filter to remove the harmonic content of the square wave so that the result is a sine wave with less than 1% harmonic distortion contributed by the first harmonic above the fundamental. Your design must be verified using an ac analysis in SPICE. The concepts of Fourier series, scaling (both impedance magnitude and frequency), and Bode plots are needed for this lab.

EQUIPMENT AND COMPONENTS:

Oscilloscope (dual trace)
Power supply ± 15 V
Function generator (up to 1 MHz)
Several unity-gain-compensated
  general purpose operational amplifiers

Capacitors $\Big\}$
Resistors $\quad$ determined by the student's design

PRELABORATORY:

1. Design the amplitude and frequency scale in your Butterworth filter at ω_c (cutoff frequency equal to 1 Hz and amplitude in the passband of 1 (or 0 dB)). Then scale them to the appropriate values. ($\frac{\omega_c}{2\pi}$ = 1KHz and amplitude A_o = 1). Using the concept of scaling up, determine values for the RC low pass filter given in Figure 11L4.1 that results in a 3-dB point at 10 KHz. The magnitude in the passband will remain unity.

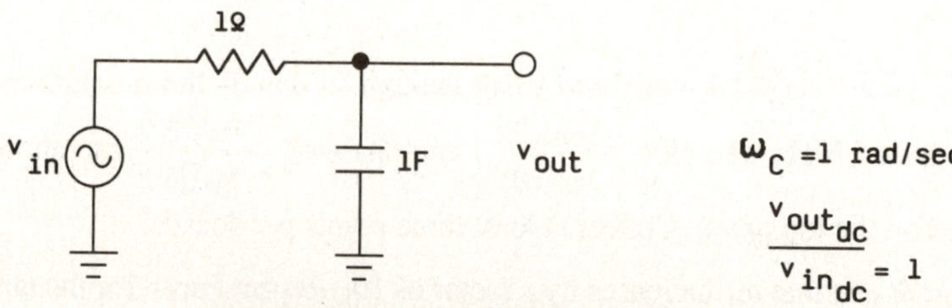

Figure 11L4.1: Circuit of section 1

2. Given a square wave as shown in Figure 11L4.2, find the Fourier series representation for this waveform and determine the amplitude of the first five harmonics. (Note that the wave is an odd function.)

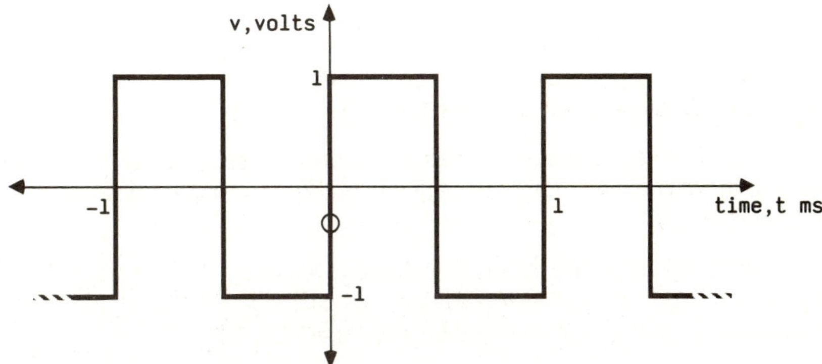

Figure 11L4.2: Square wave

3. Research the topic of Butterworth polynomials filters, and derive the Butterworth polynomial coefficients and factors for order n = 1, 2, 3, 4,.....,7.

4. Given the constant-K low pass filter of Figure 11L4.3, assume the op amp is ideal and show that the transfer function for the filter is given by

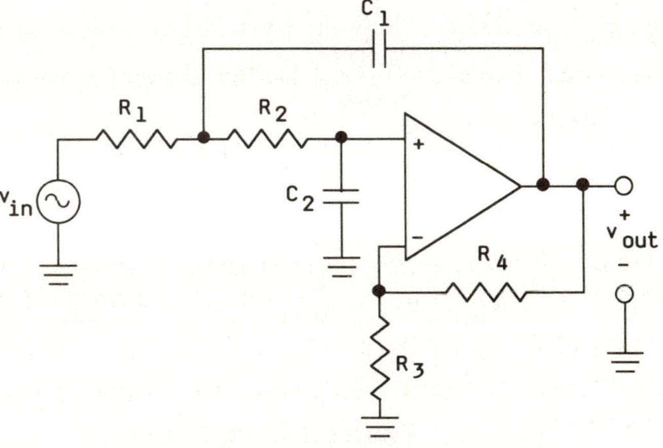

Figure 11L4.3: Constant-K low-pass filter

$$\frac{v_{out}(s)}{v_{in}(s)} = \frac{K\left(\dfrac{1}{R_1 R_2 C_1 C_2}\right)}{s^2 + s\left[\dfrac{1}{R_1 C_1} + \dfrac{1}{R_2 C_1} + \dfrac{1-K}{R_2 C_2}\right] + \dfrac{1}{R_1 R_2 C_1 C_2}} \tag{11L4.1}$$

where

$$K = 1 + \frac{R_4}{R_3}$$

Compare this result to the general form

$$H(s) = \frac{H_o \omega_c^2}{S^2 + \left[\dfrac{\omega_c}{Q}\right] S + \omega_c^2}$$

(11L4.2)

for a second-order low-pass filter, and identify $1/Q$, ω_C, and H_o. Finally, let $R_1 = R_2 = R$ and $C_1 = C_2 = C$, and represent H_o, ω_c, and Q in terms of K, R, and C.

5. From the initial design requirements, determine the minimum order of the needed Butterworth polynomial so that the second, third, and higher harmonics are smaller than 1% of the amplitude of the fundamental frequency.

6. Using the normalized frequency of $\omega_c = 1$ rad/sec, determine the normalized value of components ($R_1 = R_2 = R$, $C_1 = C_2 = C$ and R_3, R_4) so that each quadratic term in the Butterworth polynomial can be represented by the constant-K low-pass filter of Figure 11L4.3.

    a. What order of filter is required?

    b. If the order is odd, can the low-pass filter of Figure 11L4.1 represent the (s+1) term?

    c. Using a simple op amp equivalent circuit ($A = 10^5$, $R_{in} = 1$ G Ω, and $R_{out} = 10$ Ω), compose a SPICE command file to verify the performance of the normalized filter. Then, using a frequency scale factor $K_f = 6{,}280$ and a magnitude scale factor, K_m, of 10^4, determine all the R's and C's. Finally, analyze the denormalized circuit with SPICE and verify your design.

Laboratory Procedure:

1. Build the circuit of Figure 11L4.4. Vary the input frequency from 10 Hz to 100 kHz. Plot the magnitude of the transfer function $\left| \dfrac{v_{out}(j\omega)}{v_{in}(j\omega)} \right|$ in dB on the vertical axis and frequency in rad/sec. on the horizontal log axis of semilog paper.

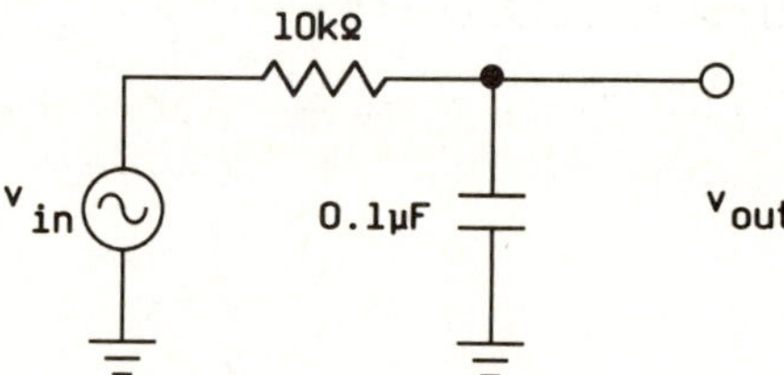

Figure 11L4.4: R-C circuit

    a. Normalize the magnitude of the components by $K_m = 10^4$ and the frequency by $K_f = 10^4$. Replot your data now that the frequency has been normalized.

    b. Does your plot in part IA, represent a Butterworth Filter of order n=1? Why?

2. Connect a 1-V-amplitude and 1-millisecond-period square wave to a spectrum analyzer. Measure and record the various frequency components of the square wave. Compare the amplitude ratios with those determined in your prelab. That is, normalize all harmonics to the fundamental in both the prelab and this section of the laboratory, and compare your results.

3. Build your circuit design from part 6 of the prelab.

a.   Qualitatively, verify the circuit's performance using the circuit of Figure 11L4.5. Scan the fundamental frequency of v_{in} from 10 Hz to 10 kHz. Observe and sketch the resulting display from the oscilloscope. If the output of your circuit is saturated due to the gain of the circuit, reduce the magnitude of the input sine wave.

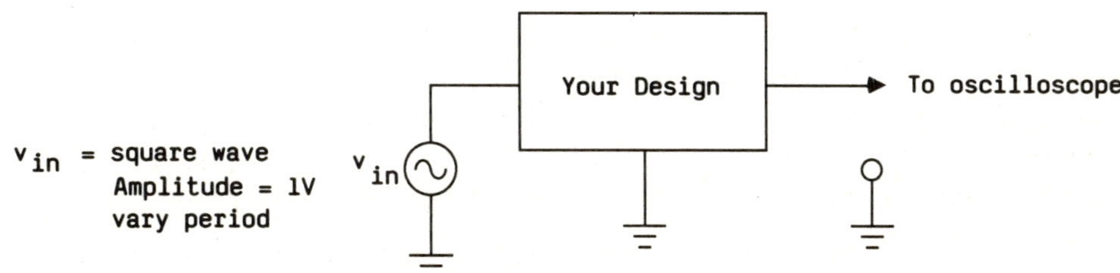

Figure 11L4.5: Test circuit for filter design

b.   Qualitatively, verify the circuit's performance using the circuit of Figure 11L4.6.

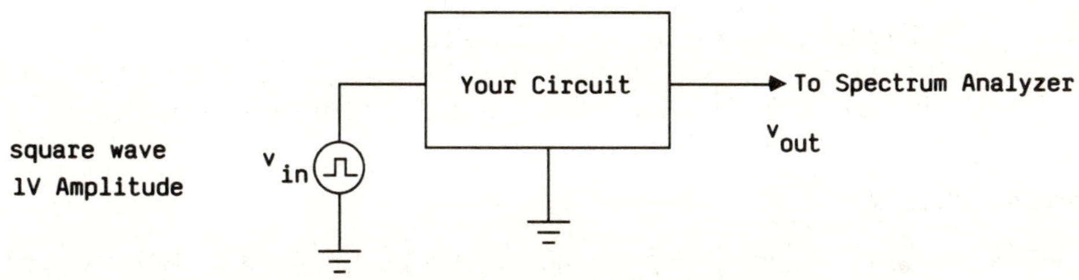

Figure 11L4.6: Test circuit for filter design

(1)   Record the amplitudes of the harmonics.

(2)   Normalize the harmonics to those of the fundamental.

(3)   Compare these results to those of Part 2.

DISCUSSION:

CONCLUSIONS:

LAB 11L.5

TITLE: The Gyrator and Negative Impedance Converter

OBJECTIVE: The objective of this laboratory exercise is to introduce the theory and implementation of the gyrator and negative impedance converter (NIC) using op amps. Several applications for these devices are given, and their performance is evaluated.

The op amp-based circuits are evaluated using SPICE, and the results are compared with those found in the laboratory.

EQUIPMENT AND COMPONENTS:
>Oscilloscope
>Function generator
>Power supply, ±15 V and 0 V-10 V
>Multimeter
>1/8-W resistors
>Capacitors
>General-purpose operational amplifiers,
>   unity gain compensated

PRELABORATORY:

1. The symbols and equations of Figures 11L5.1(a) and (b) represent the gyrator and NIC, respectively.

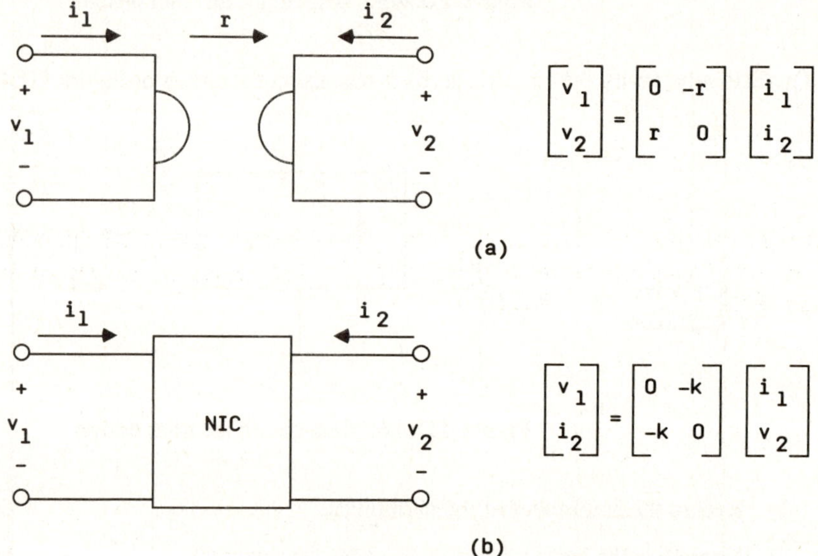

(a)

(b)

Figure 11L5.1: Symbols for gyrator and NIC

If a resistor R is connected across port two, where i_2 and v_2 are defined for the gyrator, the impedance seen looking into port one is determined as follows (the resulting circuit is given in Figure 11L5.2):

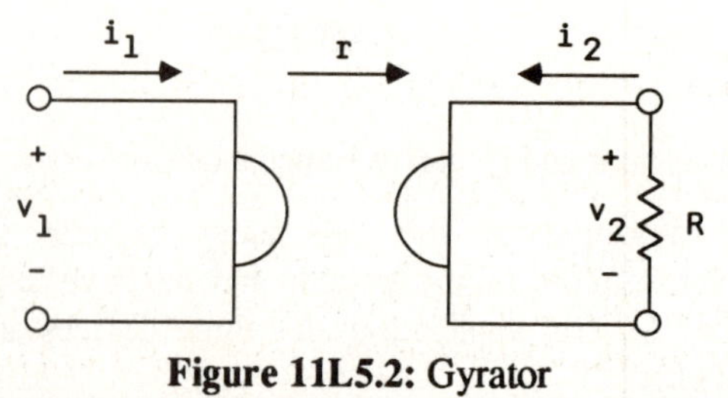

Figure 11L5.2: Gyrator

$$v_1 = -ri_2 \qquad\qquad (11L5.1)$$

and $\qquad i_2 = -v_2 / R \qquad\qquad (11L5.2)$

Combining Equations (11L5.1) and (11L5.2) yields

$$v_1 = rv_2 / R \qquad (11L5.3)$$

Since $v_2 = ri_1$, we obtain

$$v_1 = r^2 i_1 / R \qquad (11L5.4)$$

Therefore, the input impedance is

$$Z_{in} \equiv \frac{v_1}{i_1} = r^2 \left(\frac{1}{R}\right) \qquad (11L5.5)$$

The impedance is then proportional to the conductance of resistor R.

Now repeat the foregoing procedure for the gyrator and NIC with the following loads connected at port two:

$$\text{a. } Z_2 = R$$

$$\text{b. } Z_2 = \frac{1}{sC}$$

$$\text{c. } Z_2 = sL$$

2. Consider the circuit of Figure 11L5.3. Assume the op amp is ideal, and suppose that a virtual short exists between the input terminals of the op amp. In other words, $v_+ = v_-$, and no current flows into the input terminals.

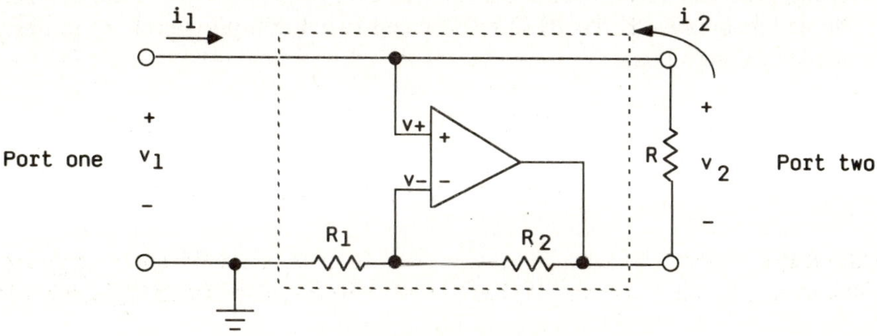

Figure 11L5.3: NIC

a. Find the impedance $\dfrac{v_1}{i_1}$ seen looking into port one.

b. Repeat the procedure of Part a for a load inductor L in place of R. Repeat for a load capacitor C.

c. Referring to Part b, replace the circuit inside the box of Figure 11L5.3 with the appropriate equivalent circuit of Figure 11L5.1. In terms of the circuit of Figure 11L5.3, what is the value of the constant in the matrix of Figure 11L5.1 for both cases of Part b?

3. Consider the circuit of Figure 11L5.4. Assume the op amps are ideal and a virtual short exits between the input terminals of each op amp.

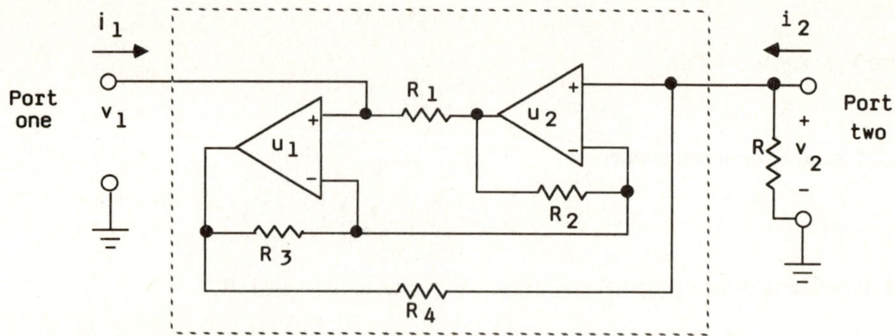

Figure 11L5.4: Generalized impedance converter (GIC)

a. Find the impedance $\dfrac{v_1}{i_i}$ seen looking into port one.

b. Replace R_4 with the a capacitor C_4, and find port one's input impedance.

c. Replace R_1 with a capacitor C_1, and find port one's input impedance.

d. Referring to Parts b and c, replace the circuit inside the box of Figure 11L5.4 with the appropriate equivalent circuit of Figure 11L5.1. What is the value of the constant in the matrix of Figure 11L5.1 if R_4 is replaced by a 1-μF capacitor and all other resistors are 1 KΩ?

4. LRC design problem

a. You are given a 10-Ω resistor and two 1,000-pF capacitors. You are to design an LRC circuit with Q = 20 and a resonance at 5 kHz. Use the two-port devices (gyrator, NIC, etc.) described in the prelaboratory and the 10-Ω resistor and two 1,000-pF capacitors to design the equivalent circuit of Figure 11L5.5.

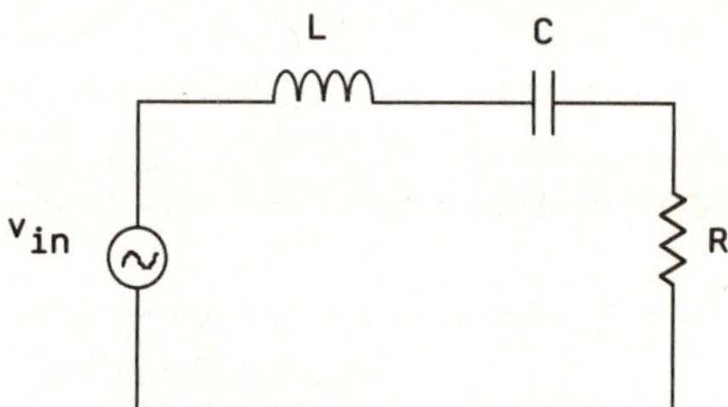

Figure 11L5.5: RLC circuit

b. Simulate your design using SPICE, and plot the frequency response $\left| \dfrac{v_2(j\omega)}{v_1(j\omega)} \right|$. Verify the resonance frequency ω_o and the quality factor Q with your paper design and experimental results. Use the op amp model of Figure 11L5.6 in your simulation.

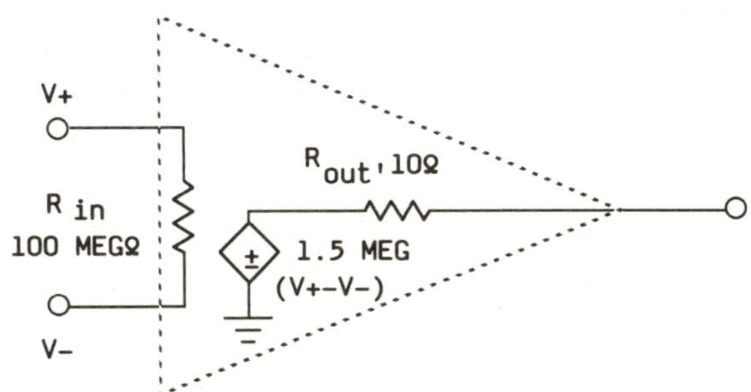

Figure 11L5.6: Op amp equivalent circuit

Laboratory Procedure:

1. Build the circuit of Figure 11L5.7. This is essentially the same circuit as that shown in Figure 11L5.3. Connect a dc test voltage to port one and a resistive load R to port two.

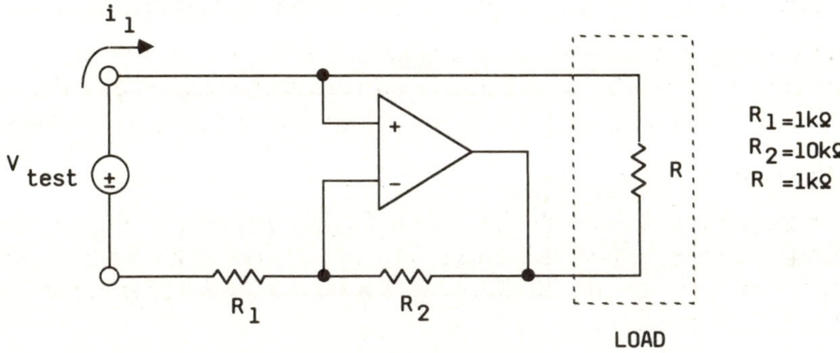

Figure 11L5.7: NIC

    a. Vary v_{test} from -1 V to 1 V in 0.1-V increments, measure the current i_1, and record the data.

    b. Plot the impedance, $\dfrac{v_{test}}{i_1}$ versus v_{test} on linear graph paper.

    c. Place a 5-kΩ resistor in series with v_{test} and repeat Parts a and b. Explain the results.

2. Simulated LR.

    a. Build the circuit of Figure 11L5.8. The device at the output is a generalized impedance converter. Let $R_s = 1$ kΩ, $Z_1 = Z_2 = Z_3 = Z_5 = 1$ kΩ and $Z_4 = 1/sC$ where $C = 1,000$ pF.

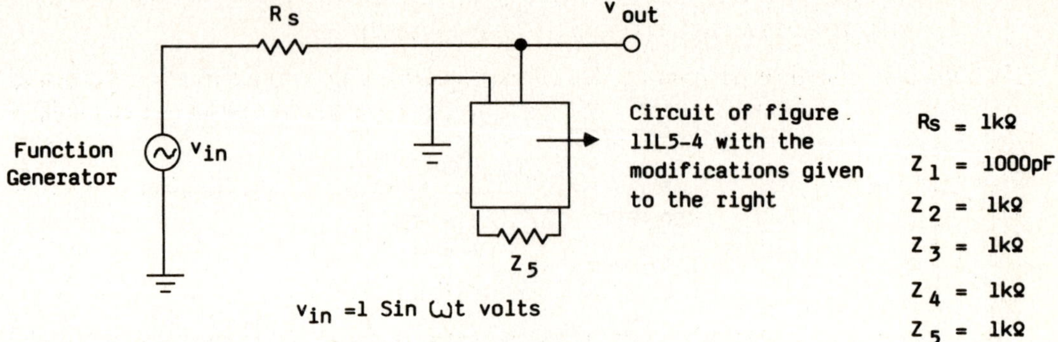

Figure 11L5.8: GIC

b. Vary v_{in} from 100 Hz to 1 MHz. Record five data points per decade. Plot $\left| \dfrac{v_{out}}{v_{in}} \right|$ versus frequency on semilog paper.

c. Compare the results of Part b with those theoretically generated by the circuit of Figure 11L5.9.

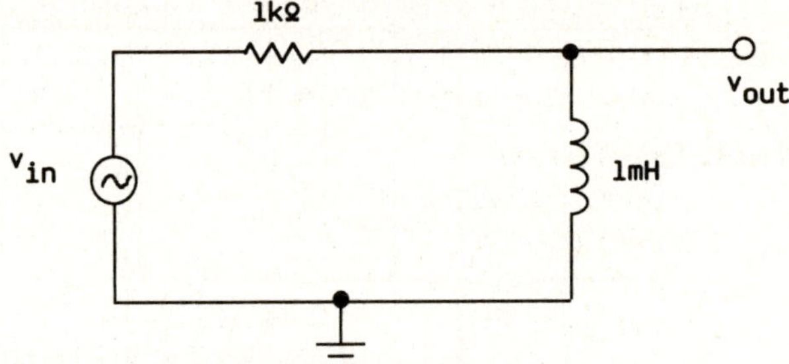

Figure 11L5.9: RL circuit

3. Design Problem.

    a. Build the circuit you designed and verified using SPICE in Part 4 of the prelab. The circuit, with simulated components, is shown in Figure 11L5.10.

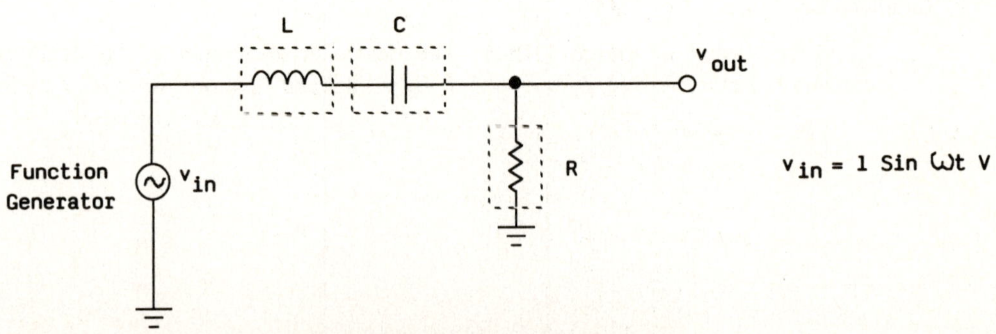

Figure 11L5.10: RLC design problem

b. Vary the frequency of v_{in} from 500 Hz to 50 kHz. Collect at least five data points of v_{out} versus frequency for each decade. Plot the data on semilog paper, and measure the bandwidth, resonance frequency, and quality factor for the circuit.

DISCUSSION: Include a discussion of why the simulation of an inductor is important to integrated circuit design. Is there a way to make the inductance variable with an external voltage or current?

CONCLUSIONS:

LAB 11L6

TITLE: A Voltage Regulator

OBJECTIVE: This laboratory exercise examines the use of an op amp in designing a voltage regulator. The sensitivity of the circuit to disturbance inputs, along with errors due to variation of parameters is evaluated. This evaluation is a combination of SPICE simulation, hand calculation and experimentation.

The prelab circuit is broken down into subcircuits, which constitute a simple voltage regulator. Then a design problem is presented that requires all of these subcircuits. Finally, the performance of a commercial monolithic 7800 series voltage regulator is tested and compared to the discrete design.

EQUIPMENT AND COMPONENTS:
Oscilloscope (dual trace)
Function generator
Power supply, ±15 V   0 V-10 V floating supply
Resistors, 2x1 kΩ,   1/8 W, 1% tolerance
General-purpose operational amplifier (frequency compensated)
Zener diode
Power transistor (2N1711)
7800 series monolithic voltage regulator

PRELABORATORY:

1. Voltage reference. The circuit of Figure 11L6.1 serves as a simple voltage reference.

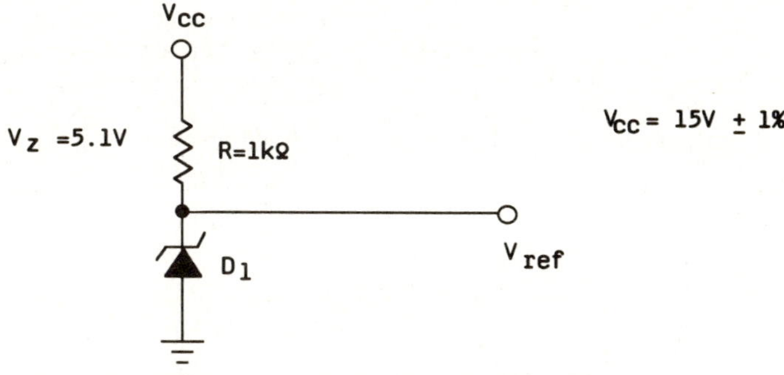

Figure 11L6.1: Simple voltage regulator

The following information describes the components used:

Diode, D_1	
Nominal zener voltage	5.1 V
Tolerance	±10%
Maximum zener impedance	17 Ω
Maximum zener current	70 mA
Temperature coefficient	4 mV/°C

Resistor, R	
Resistor value	1 kΩ
Resistor tolerance	1%
Resistor temperature coefficient	500 ppm/°C

From the information given, what are the maximum and minimum values of V_{ref}? (Assume a 25°C temperature change.)

2. Error Amplifier. The circuit of Figure 11L6.2 shows the op amp used as an error amplifier to detect the difference between the reference voltage V_{ref} and the output voltage V_{out}.

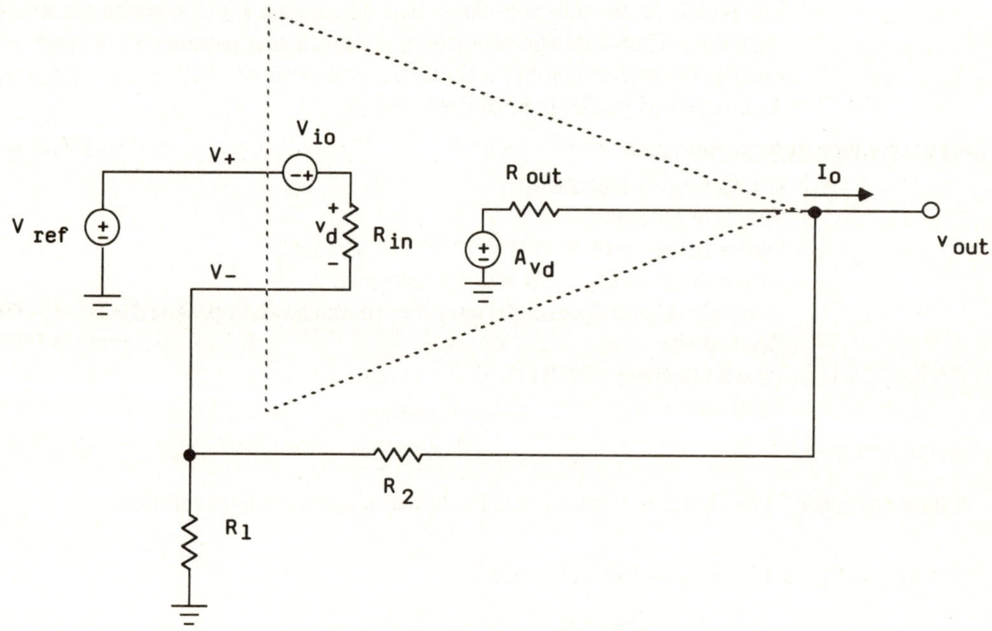

Figure 11L6.2: Circuit of Section 2

a. If we let R_{in} approach infinity, show that v_{out} is given by

$$V_{out} = \frac{(V_{ref} \pm V_{io}) - \dfrac{R_{out}}{A} I_o}{\dfrac{R_1}{R_1 + R_2} + \dfrac{1}{A}}$$

(11L6.1)

b. If R_1 and R_2 = 1 kΩ with a tolerance of 1% and a 500 ppm, temperature coefficient temperature $|V_{io} = 3mV|$ with a temperature coefficient of 10 μV/°C, and $10^4 \leq A \leq 10^5$, what is the maximum variation in v_{out} due to the parameter tolerance and a 25°C change in temperature? Assume $I_o = 0$ and $V_{ref} = 5.1V$.

c. Calculate the output impedance of the circuit of Figure 11L6.3 if R_{out} = 70 Ω and A = 10^5.

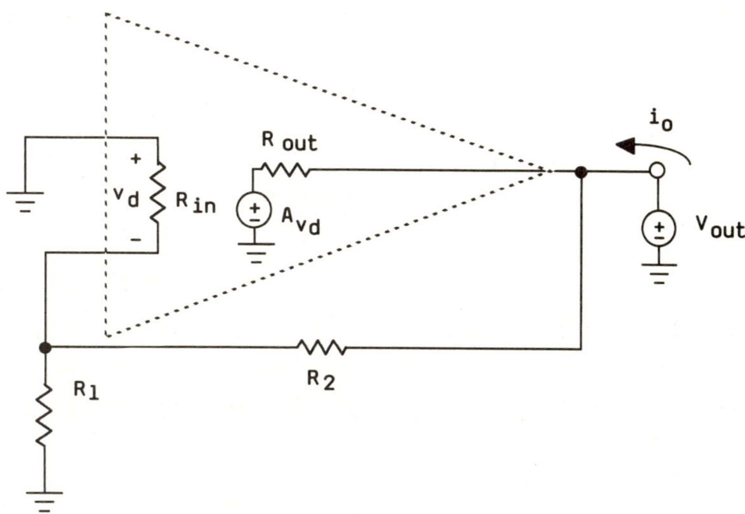

Figure 11L6.3: Circuit of Section 2.C

d. Why is a very low output impedance desirable?

3. Error amplifier common mode sensitivity. An error amplifier with V_{cc_+} = 15 V and V_{cc_-} = -15 V has a common-mode voltage

$$V_{cm} = \frac{V_+ + V_-}{2} \tag{11L6.2}$$

where V_+ and V_- are the noninverting and inverting input voltages, respectively. The effect of the common mode voltage is modeled in Figure 11L6.4. CMRR is the common-mode rejection ratio. The model is included in Figure 11L6.5. Show that

$$V_{out} = \frac{V_{ref} - \dfrac{\pm V_{cm}}{CMRR}}{\dfrac{R_1}{R_1 + R_2} + \dfrac{1}{A}} \tag{11L6.3}$$

Let $V_{CM} = V_{ref}$ and let $R_{out} \to 0$ and $R_{in} \to \infty$.

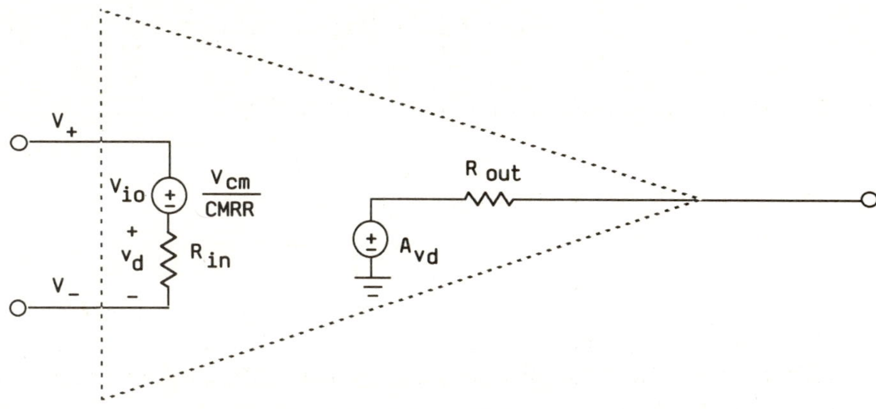

Figure 11L6.4: Circuit of Section 3

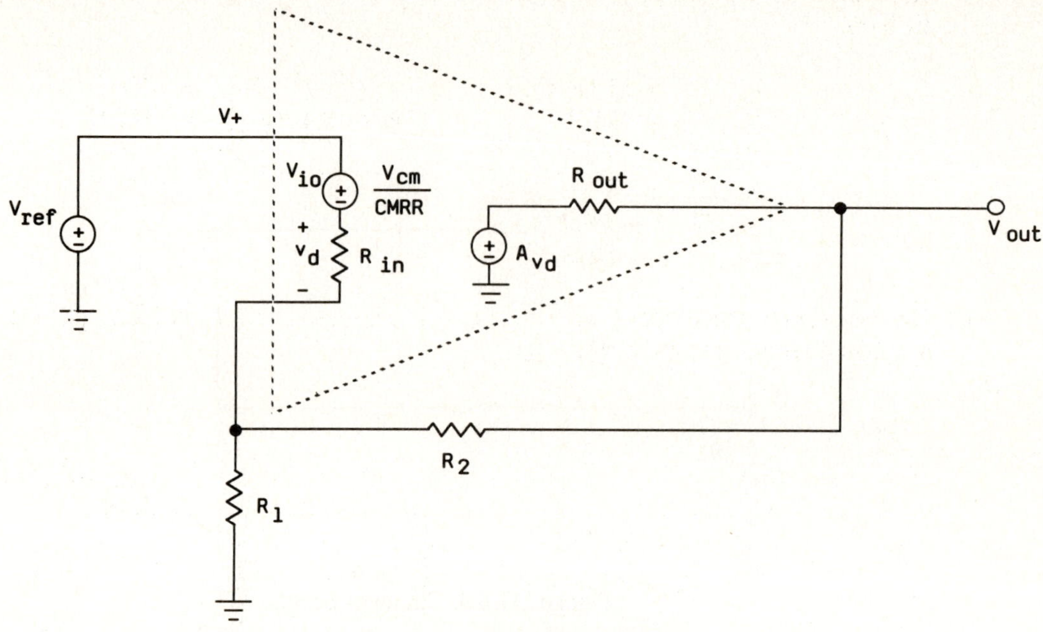

Figure 11L6.5: Error amplifier

4. Increasing the regulator current drive.

   a. If the op amp used in Figure 11L6.2 is an LM741, what is the maximum load (minimum resistance) the regulator can handle and still maintain the desired value for v_{out}?

   b. The transistor in Figure 11L6.6 has two positive effects on the performance of the circuit shown. First, it reduces the output impedance of the regulator and, second, it allows greater values of I_o to be delivered to the load. One disadvantage is that line regulation may not be as good due to the coupling of noise from V_{cc} to the output through Q_1. Show that R_{out}, the output resistance of the circuit of Figure 11L6.2, is reduced by $(1 + h_{FE}) \approx h_{FE}$ of Q_1.

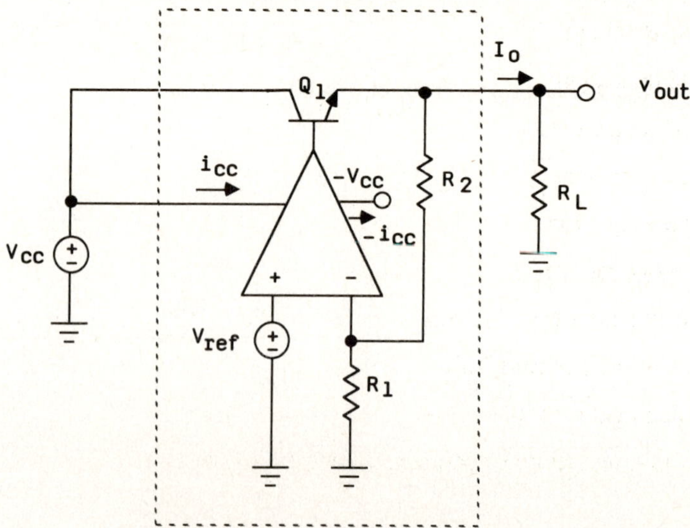

Figure 11L6.6: Circuit of Section 4

c.   In any power supply, power dissipation is a major consideration. For the following values, determine the maximum power dissipation of the regulator of Figure 11L6.6 (everything inside the box):

$$
\begin{array}{llll}
i_{cc} & = & 10 \text{ mA} \\
V_{cc_+} & = & 15 \text{ V} \\
V_{cc_-} & = & -15 \text{ V} \\
V_{ref} & = & 5.1 \text{ V} \\
R_1 = R_2 & = & 1 \text{ K}\Omega \\
R_L & = & 50 \text{ }\Omega
\end{array}
\qquad
\begin{array}{llll}
h_{FE} & = & 100 \\
\text{Early voltage } (Q_1) & = & 100 \text{ V}
\end{array}
$$

d.   Use the hybrid-π model for Q_1 given in Figure 11L6.7 and the results of Parts 1, 2, and 3 of the prelab to determine the variation in v_{out}.

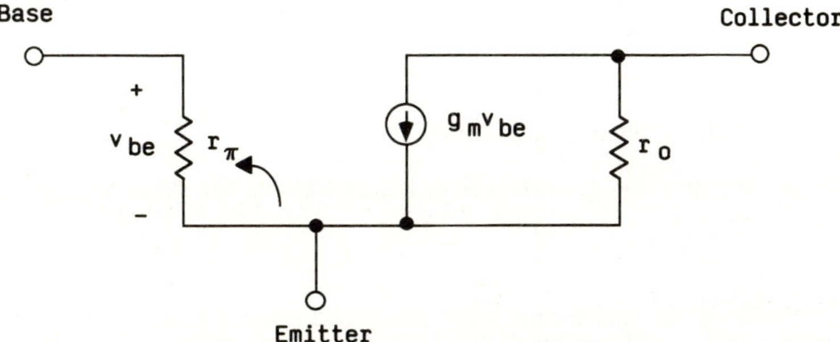

Figure 11L6.7: Hybrid-π model for Q_1

e.   Using SPICE, model the circuit of Figure 11L6.6 and place an ac source in series with V_{cc}. Determine the variation in v_{out} due to this ac input. Do a sensitivity analysis and determine those parameters that have the greatest effect on v_{out}. Assume the ac source has an amplitude of 1 V and a frequency of 60 Hz. Use the op amp model of Figure 11L6.2 with $R_{in} = 1 \text{ G}\Omega$.

5.   Evaluation of a commercial regulator (MC7800 series). Research the 7805 fixed-voltage regulator. Define the following characteristics and determine their values for the 7805 regulator:

   a.   Output voltage

   b.   Line regulation

   c.   Load regulation

   d.   Quiescent current

   e.   Ripple rejection

   f.   Dropout voltage

   g.   Output resistance

   h.   Short-circuit current limit

   i.   Peak output current

   j.   Average temperature coefficient of output voltage

Laboratory Procedure:

1. Voltage reference.

    a. Build the circuit of Figure 11L6.8.

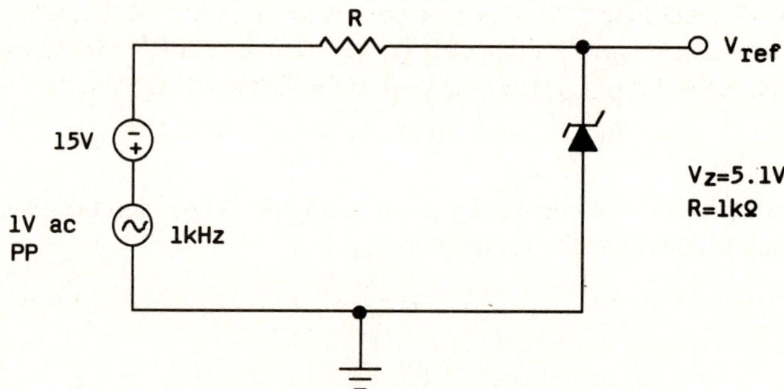

Figure 11L6.8: Voltage reference

       (1)   Measure the nominal voltage at V_{REF}.

       (2)   Measure the ac current flow and determine the small-signal resistance of the Zener diode, $\left(\dfrac{V_{ref(ac)}}{i_{ac}}\right)$.

   b.   Place the Zener diode on a curve tracer and observe where your circuit biases the diode. Would $R = 100\text{-}\Omega$ be better? Why?

2. Error amplifier.

    a. Build the circuit of Figure 11L6.9. Move the 1-V, 1-kHz ac source to each of the three locations shown, and measure the resulting ac signal at the output, v_{out}.

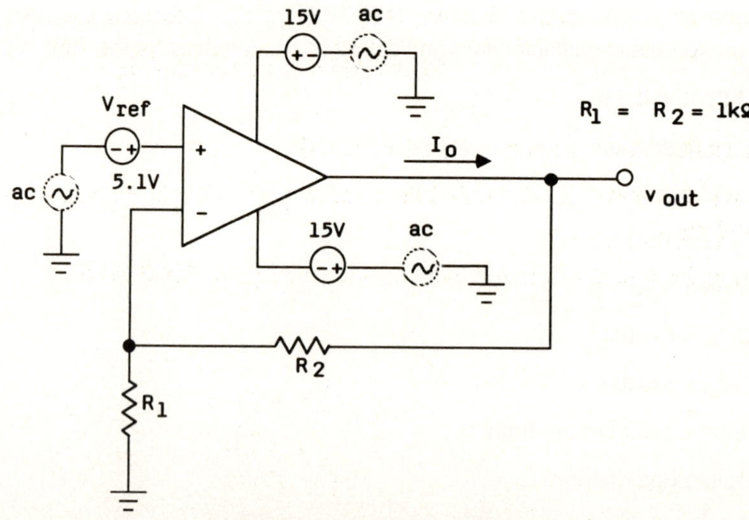

Figure 11L6.9: Error amplifier

b. Remove all ac sources from the circuit of Figure 11L6.9, and connect a load resistor R_L across the output v_{out}. Measure the maximum output current that can be sustained without causing v_{out} to change.

c. AC-couple a 10-kHz, 1 $V_{peak-to-peak}$ sine wave to the output v_{out} with all other ac signals set to zero. Measure the ac current and calculate the output resistance r_{out}. Increase the frequency to 100 kHz and repeat the measurement. Is the impedance the same? Why? Should the frequency be increased? Use the circuit of Figure 11L6.10 to couple the ac signal. If you are using a scope and a voltmeter or ammeter, remember the difference between rms and peak-to-peak voltage.

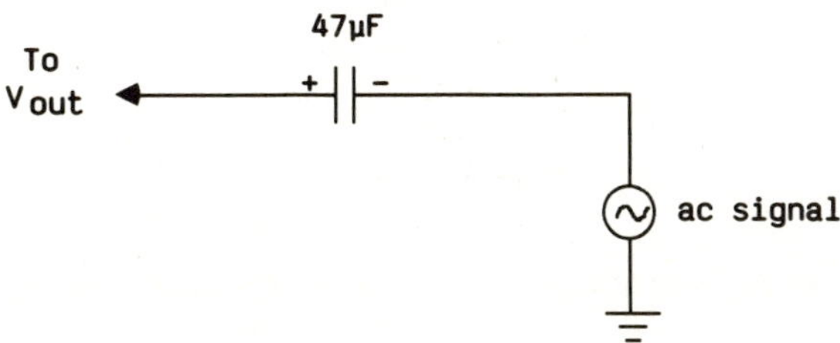

Figure 11L6.10: Circuit to complete the ac signal

3. Error amplifier common-mode sensitivity. Build the circuit of Figure 11L6.11. Use the same op amp as in Part 2.

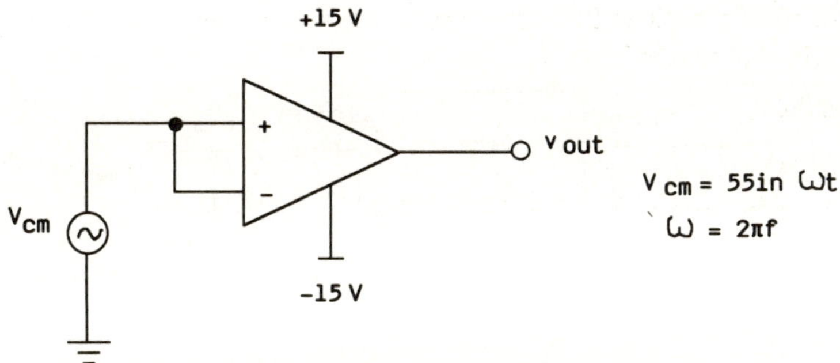

Figure 11L6.11: Error amplifier

a. Determine the common-mode rejection ratio.

b. Use Equation (11L6.3) to find the error in v_{out} due to the common-mode gain. (Let A approach infinity.)

4. Increasing the regulator current drive. Build the circuit of Figure 11L6.12.

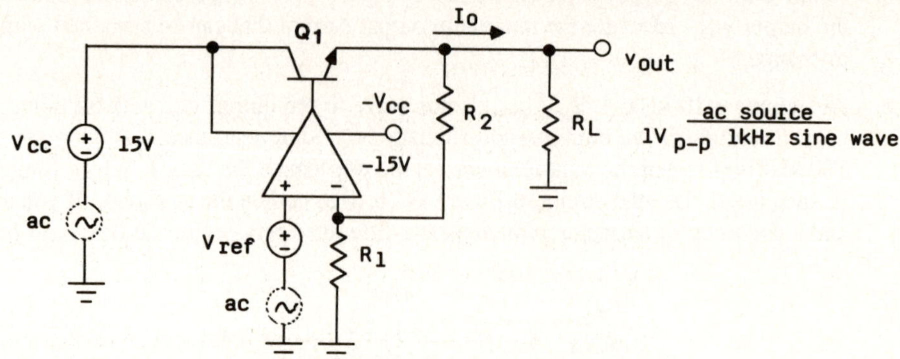

Figure 11L6.12: Circuit of Section 4

a. Measure the response at v_{out} due to the ac source applied at the indicated positions one at a time.

b. Again, capacitively couple an ac waveform to the output, as in Part 2, and measure r_{out}, i.e., $(\dfrac{v_{ac}}{i_{ac}})$.

c. Measure the minimum load resistance that the voltage regulator can drive. Care must be taken not to exceed the power dissipation rating of R_L or Q_1.

d. Place Q_1 on a curve tracer, and measure h_{FE} and the Early voltage.

5. Evaluation of the MC7805. Build the circuit of Figure 11L6.13.

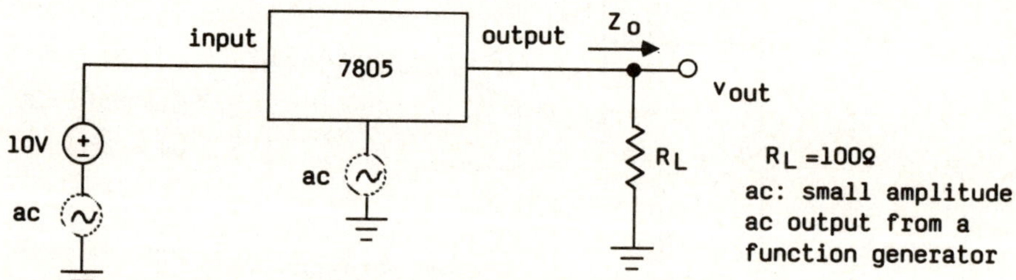

Figure 11L6.13: Circuit of Section 5

a. Separately measure the variation in v_{out} due to the ac disturbance inputs, and compare it with the variation in the discrete design.

b. As in the discrete circuit, capacitively couple an ac signal to the output and remove all other ac signals. Then measure the ac output, i_o, and calculate $r_o = \dfrac{v_{out}}{i_o}$. Compare this value with that found in the discrete design. Let $v = \dfrac{1}{2}\sin(2\pi ft)$, with $f = 10$ kHz, and $C = 47$ μf.

c. Reduce the output load resistance until v_{out} is no longer maintained. Watch the power dissipation in R_L!

DISCUSSION:

CONCLUSIONS:

LAB 11L7

TITLE:	Sinusoidal Oscillators Using Op Amps
OBJECTIVE:	This laboratory exercise introduces the concept of instability in terms of loop gain (magnitude and phase) and root locations of the characteristic equation. The condition is demonstrated with a nonunity-gain compensated op amp. Next, a Wien-bridge oscillator is analyzed and the frequency of oscillation is determined. A SPICE analysis aids in determining the sensitivity of the oscillator to variations in the component parameters.

Finally, a nonlinear clamping circuit is used to control amplitude. This circuit is analyzed by hand and by a SPICE simulation.

EQUIPMENT AND COMPONENTS:

AD 301 (Analog Devices 301 or equivalent), operational amplifier
Resistors, 1/8 W: 10 kΩ, 1 kΩ, 20.3 kΩ, and
   others determined by the student.
Capacitors (nonpolar)
Function generator (sinusoidal)
Power supply, ±15 V
Diodes (signal)

PRELABORATORY:

1. Consider the circuit of Figure 11L7.1.

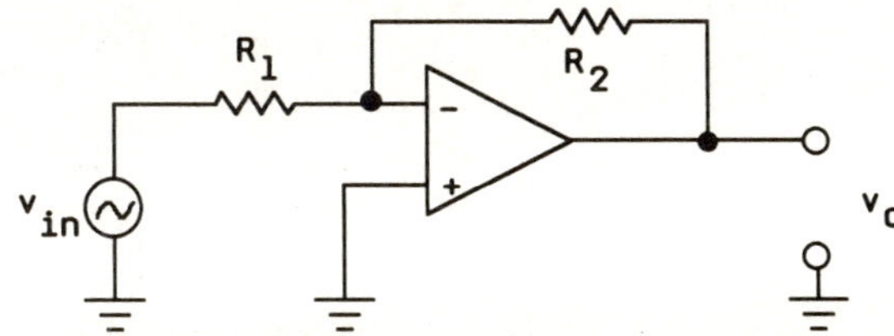

Figure 11L7.1: Uncompensated op amp

   a. Analyze the circuit, assuming that the op amp's open-loop gain A ($v_o = A\,(v_+ - v_-)$) is finite. Show that the transfer function v_o/v_{in} is given by

$$\frac{v_o}{v_{in}} = \frac{\dfrac{-AR_2}{(R_1 + R_2)}}{1 + \dfrac{R_1}{R_1 + R_2}A} \tag{11L7.1}$$

   b. Assume that A is now a function of frequency given by

$$A(s) = \frac{A_o}{(1 + \dfrac{s}{P_1})(1 + \dfrac{s}{P_2})(1 + \dfrac{s}{P_3})} \tag{11L7.2}$$

      where

$$\frac{P_1}{2\pi} = 10 \text{ Hz}; \quad \frac{P_2}{2\pi} = 1 \text{ kHz}; \quad \frac{P_3}{2\pi} = 100 \text{ kHz};$$

and

$$A_o = 100 \text{ dB}$$

c. Combine Equations (11L7.1) and (11L7.2) and find the characteristic equation for the circuit of Figure 11L7.1. If the term $\dfrac{R_1 A}{R_1 + R_2}$ is a function of frequency, then the term could be a complex quantity and possibly have a magnitude of 1 and a phase angle of 180°. This would result in the denominator being equal to zero and the transfer function going to infinity. This condition is one criterion of instability. The quantity $\dfrac{R_1\, A}{R_1 + R_2}$ is referred to as the **loop gain**.

d. On a magnitude-phase plot of the loop gain, instability can be inferred by seeing at what frequency the phase angle is -180°. If at that angle the magnitude is unity or greater, the system will be unstable and oscillate or saturate.

e. Plot the magnitude and the phase angle of the loop gain, $\left| \dfrac{R_1}{R_1 + R_2} A(j\omega) \right|$ and determine what range of values of $\dfrac{R_1}{R_1 + R_2}$ would result in a stable amplifier. What happens in terms of stability as the closed-loop gain is increased?

f. Use the op amp model given in Figure 11L7.2 to simulate the op amp frequency response. Try several iterations of R_1 and R_2 to identify the maximum closed-loop gain that results in instability. Compare this result with your hand calculations. (Hint: Fix $R_1 = 1 \text{ k}\Omega$ and let $R_2 = 100 \text{ k}\Omega, 50 \text{ k}\Omega, 40 \text{ k}\Omega,...$)

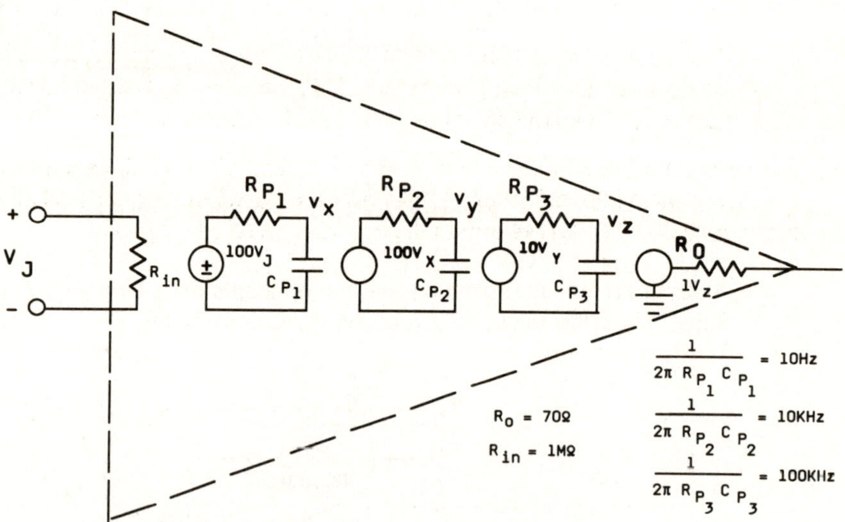

Figure 11L7.2: Op amp model

2. Consider the circuit of Figure 11L7.3

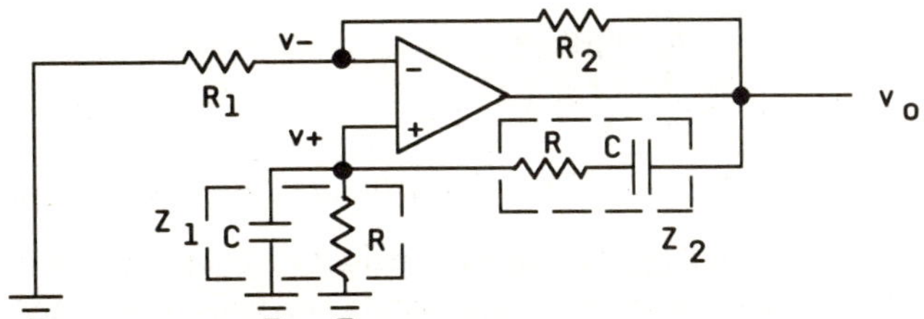

Figure 11L7.3: Wien-bridge oscillator

a.   Analyze the circuit. Assume the op amp is ideal. In writing a node equation at the inverting input, and knowing that $v_+ = v_-$, derive the Equation

$$v_o\left[\left[\frac{Z_1}{Z_1+Z_2}\right]\left[\frac{R_1+R_2}{R_1}\right]+1\right]=0 \qquad (11L7.3)$$

If we assume $v_o \neq 0$, then the term in the brackets must be zero. This requires that the quantity $\left[\dfrac{Z_1}{Z_1+Z_2}\right]\left[\dfrac{R_1+R_2}{R_1}\right]=-1$. The resonant frequency ω_o and the required value of R_2/R_1 are $\omega_o = \dfrac{1}{RC}$ and $\dfrac{R_2}{R_1}=2$, respectively, in order to ensure a loop gain of 1. Why should $\dfrac{R_2}{R_1}$ be made slightly greater than 2?

b.   Determine values for R, C, R_1, and R_2 that result in oscillations at 1 kHz.

c.   Do a SPICE simulation on the circuit using the op amp model from Section 1 and the component value from Section 2b. Do a transient analysis and determine the oscillation frequency. Compare this result with that of Section 2b.

3.   The circuit of Figure 11L7.4 is the same as that of Figure 11L7.3, except that of a limiting circuit is added. The additional circuitry determines the amplitude of the oscillation by decreasing the loop gain below 1 when the amplitude exceeds a specified magnitude.

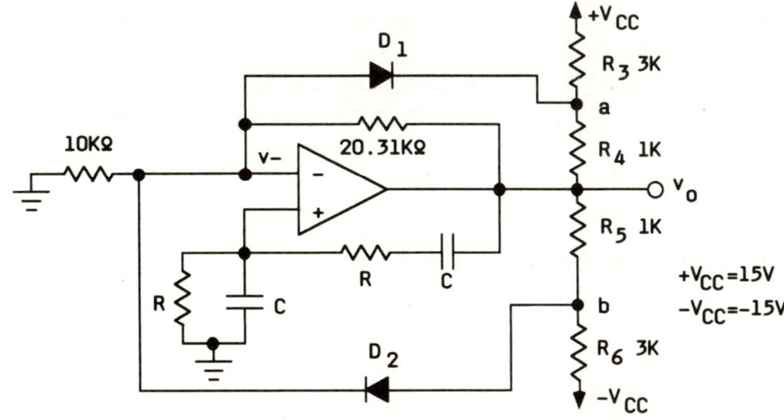

Figure 11L7.4: Wien-bridge oscillator with limiting circuit

To see how this limiting circuit works, assume the circuit will oscillate and do the following steps. Use the ideal diode model, and when current flows in the diode, assume it is negligible.

    a. Assume that the output voltage v_o is increasing in amplitude. The magnitude of the voltage at v_- is given by

$$v_- = \frac{v_o \, 10 \text{ k}\Omega}{30.31 \text{ k}\Omega} \simeq \frac{1}{3} v_o \tag{11L7.4}$$

Determine the voltage at node v_o when the voltage at node b is large enough to cause D_2 to conduct. What is the new loop gain of the circuit now that D_2 is conducting? (Hint: D_2 and R_5 are in parallel with the 20.31-kΩ resistor.) Is the new loop gain large enough to sustain oscillations? Do the same procedure for negative values of the voltage at v_o, this time looking at D_1 and node a.

    b. Determine values for R_3, R_4, R_5, and R_6 so that $\mid v_o \mid_{max} = 10$ V

4.   The phase shift oscillator. Consider the circuit of Figure 11L7.5.

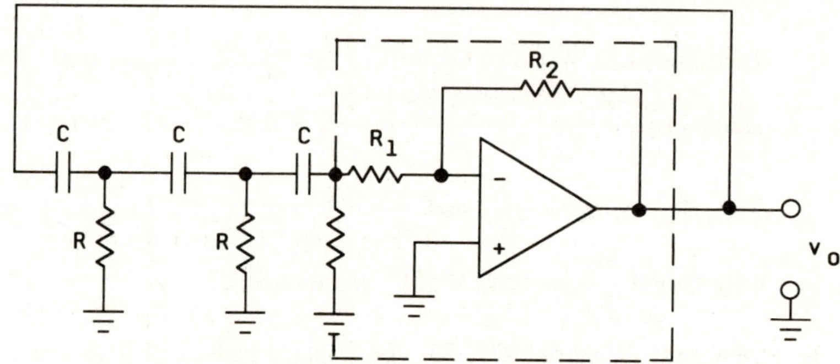

Figure 11L7.5: Phase shift oscillator

    a. Since the circuit inside the dashed box has a gain of $-R_2/R_1$, the remaining RC network will meet the criteria for producing oscillations if it has a magnitude of $\frac{1}{K}$ and a phase shift of 180°. Let $R = 1$ kΩ and $C = 2{,}200$ pF. Find the loop gain of the circuit, and determine the gain $-R_2 / R_1$ that makes the circuit oscillate. Also, determine at what frequency this occurs. Why should you set the gain R_2/R_1 slightly larger than the value you calculated for R_2/R_1?

    b. Incorporate the amplitude limiting circuit of Section 3 into the phase shift oscillator and draw the circuit. Do a SPICE simulation of the circuit of Figure 11L7.5 with and without the clamping circuit and discuss the results. Using SPICE, verify the clamping feature of this amplitude-limiting circuit. Let $R_{IN} = 16\Omega$, $A = 10^6$, and $R_{out} = 10\Omega$

    c. Verify your results from part 4a and b using SPICE.

Laboratory Procedure:

1.   Stability analysis of an uncompensated op amp.

    a. Build the circuit of Figure 11L7.6, which is essentially a detailed version of that shown in Figure 11L7.1.

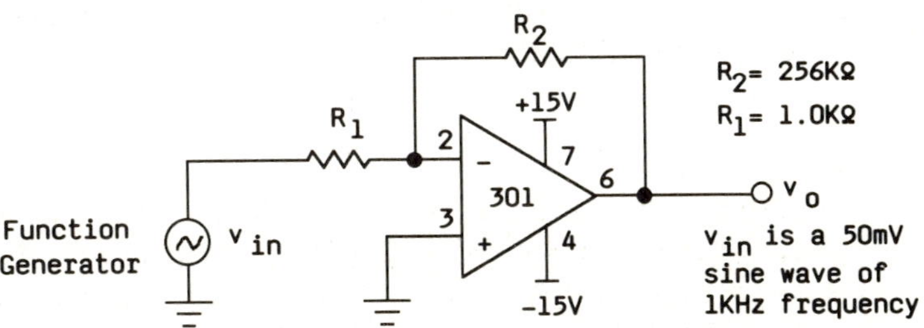

Figure 11L7.6: Uncompensated op amp

If the amplitude of the function generator cannot be set to a sine wave of 0.05 V amplitude, then the attenuator of Figure 11L7.7 can be connected between the function generator and R_1.

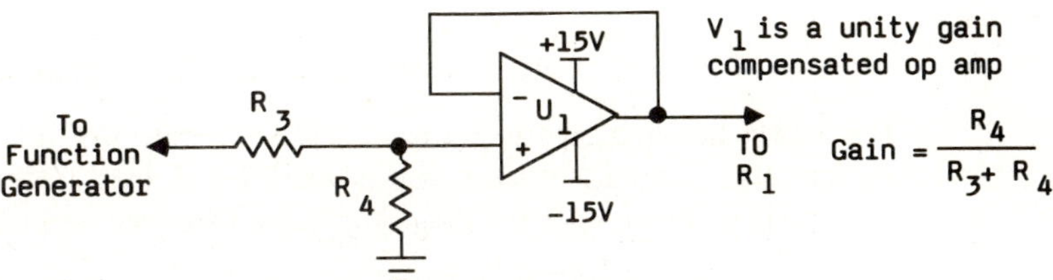

Figure 11L7.7: Attenuator circuit

Adjust the values of R_3 and R_4 to yield enough attenuation of the function generator signal so that a 50 mV sine wave can drive R_1.

b. Use a dual-trace oscilloscope and observe the function generator signal (or its attenuated version) and the output of the 301 op amp.

(1) Sketch both signals, and record the gain of the circuit of Figure 11L7.6. Reduce the closed-loop gain by changing the value of R_2. For best flexibility R_2 should be a resistor substitution box.

(2) Reduce the gain of the circuit by a factor of two, and observe the input and output waveforms. Sketch the waveforms and note the gain. Continue to reduce the gain by a factor of two and observe the output. For your op amp, find the minimum closed-loop gain that can be used without causing oscillations. At what frequency does the phase margin go to zero?

(3) Add a 30-pf capacitor between pins 1 and 8 of the 301 op amp. What effect does this have? Leave the capacitor connected to the op amp for the remainder of the lab.

2. The Wien-bridge oscillator.

a. Build the circuit of Figure 11L7.8, which is a version of that shown in Figure 11L7.4.

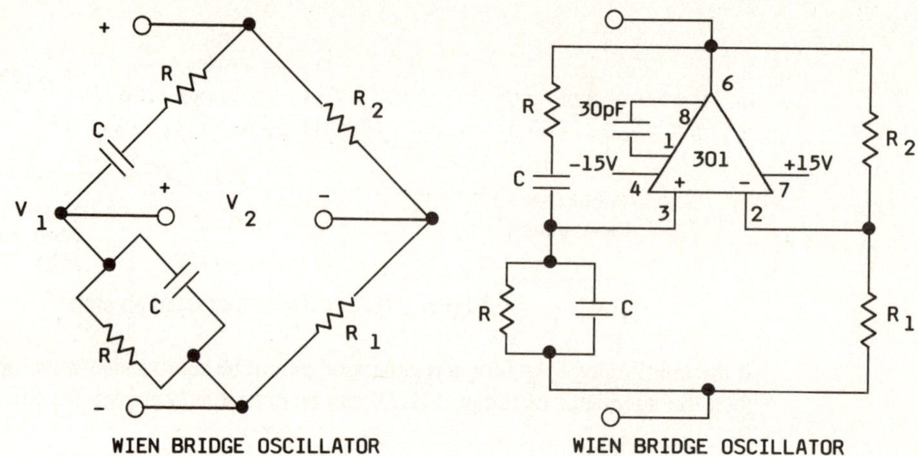

Figure 11L7.8: Wien-bridge oscillator

The values of R and C are determined in the prelab such that the circuit oscillates at 1 kHz.

(1)   Sketch the voltage waveforms at pins 2, 3, and 6 of the 301 op amp. Be sure to preserve the phase relationship of the three signals. Describe the operation of the circuit in terms of these voltage signals and how the op amp responds.

(2)   Change the value of R_1 to 20.31 kΩ. What effect does this have? Change R_1 to 1 kΩ and observe the effect on v_o.

b.   Add the amplitude-limiting circuit given in Section 3b of the prelab with the resistor values you determined to the circuit of Figure 11L7.8. The new circuit is shown in Figure 11L7.9. Be sure that R_1 is a 10-kΩ resistor.

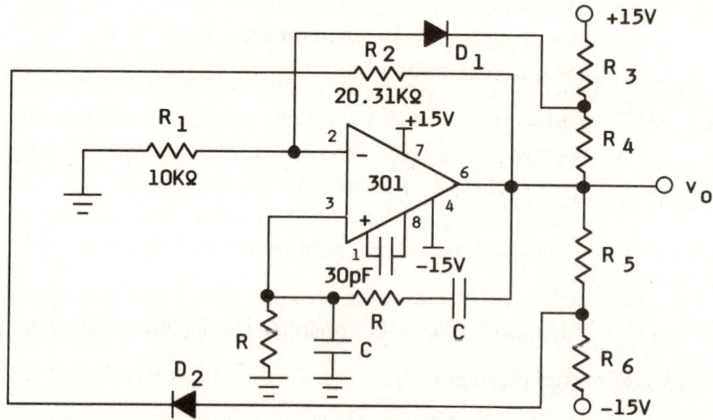

Figure 11L7.9: Wien bridge oscillator with limiting circuit

Sketch the voltage waveforms at pins 2, 3, and 6 of the 301 op amp in Figure 11L7.9. Be sure to preserve the phase relationship of the three signals. Compare these waveforms with those from Section A(1).

Now change the value of R_1 to 20.31 kΩ. What effect does this have? (Sketch v_o.) Change R_1 to 1 kΩ and sketch the resulting v_o. What are the differences in v_o for the three values of R_1?

3. The phase shift oscillator.

    a. Build the circuit of detail in Figure 11L7.10, which is essentially a detailed version of that shown in Figure 11L7.5. Use the 301 op amp discussed previously.

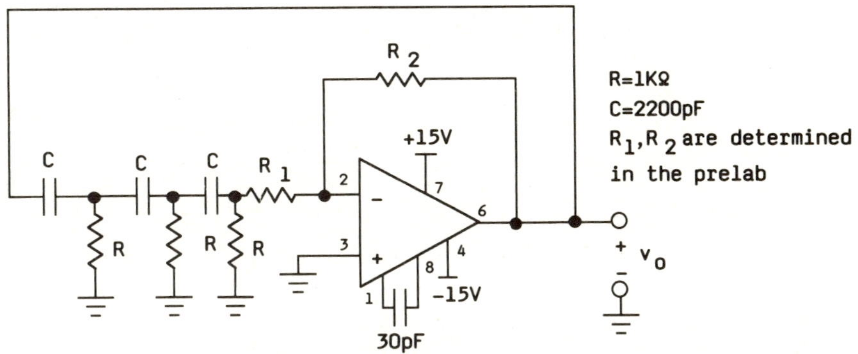

Figure 11L7.10: Phase shift oscillator

Sketch the voltage waveform at v_o. Increase the value of R_1 by a factor of five. What effect does this have on v_o? Decrease R_1 from its original value by a factor of five. What effect does this have on v_o?

    b. Add the amplitude-limiting circuit to the circuit of Figure 11L7.10. Observe and sketch the results. Be sure that R_1 is back to its original value. Repeat the variation of R_1 from Section 3a(1) for this circuit, and sketch the results.

DISCUSSION: Include a comparison of the SPICE results with those found in the laboratory.

CONCLUSIONS:

REFERENCES

[1] A. S. Sedra and K. C. Smith, *Microelectronic Circuits,* 2d ed. (New York: Holt, Rinehart and Winston, 1987), Chapter 3.

[2] C. J. Savant, Jr., M. S. Roden, and G. L. Carpenter, *Electronic Circuit Design* (Menlo Park, CA: Benjamin/Cummings, 1987), Chapter 8 and 9.

[3] E. J. Kennedy, *Operational Amplifier Circuits: Theory and Applications* (New York: Holt, Rinehart and Winston), 1988.

[4] S. G. Burns and P. R. Bond, *Principles of Electronic Circuits* (St. Paul: West, 1987), pp. 723-7.

[5] P. R. Gray and R. G. Meyer, *Analysis and Design of Analog Integrated Circuits,* 2d ed. (New York: John Wiley, 1984), pp. 515-9.

[6] Sedra and Smith, Chapter 12.

[7] Burns and Bond, Chapter 15.

[8] Kennedy, Chapter 6.

[9] Savant, Roden, and Carpenter, Chapter 13.

[10] Kennedy, pp. 380-8.

DIGITAL SYSTEMS AND DIGITAL ELECTRONICS

12.1 Introduction

This chapter is divided into two sections: digital systems and digital electronics. The digital systems section is in turn divided into three parts. The first part is a brief review of such topics as the binary system, Boolean algebra, logic gates, and the simplification of Boolean functions through the use of Karnaugh maps. The second part is a review of combinational logic. Topics covered are adders, subtractors, decoders, multiplexers, and gate arrays. The third part consists of a review of sequential logic, including such topics as latches, flip-flops, registers, and counters.

The section on digital electronics covers four logic circuit families. Two of the families are BJT logic families and include transistor-transistor logic (TTL) and emitter-coupled logic (ECL). The other two families are MOS families and include NMOS and CMOS. The first of these utilizes n-channel MOSFETs exclusively, while the second uses both n-channel and p-channel transistors in complementary symmetric circuit configurations. Discussed are such concepts as "high" and "low" or logic 1 and logic 0, the basic inverter, noise margins, power dissipation in a logic circuit, fan-in and fan-out, propagation delay, and delay-power product. All of these terms are discussed for each of the four logic families.

The laboratory experiments included in this chapter are designed to illustrate most of the concepts discussed. For example, Laboratory 12L.1 examines the concepts of fan-out and propagation delay, Laboratory 12L.2 is an introduction to combinational digital systems, Laboratory 12L.3 introduces the TTL gate, and Laboratory 12L.4 deals with sequential digital systems. The remaining six experiments are designed to illustrate other digital concepts discussed in the chapter.

352

12.2 Digital Systems

12.2.1 Binary Logic[1]

The definitions of the basic logical operations AND, OR, and NOT can be expressed in the form of truth tables. A **truth table** is a table of all possible combinations of the variables showing the relation between the values that the variables may take and the result of the operation. The truth tables for the AND, OR, and NOT operations are shown in Table 12.1.

Table 12.1: Truth tables for the AND, OR, and NOT logical operations

	AND			OR		NOT	
X	Y	X Y	X	Y	X+Y	X	$\overline{X}$
0	0	0	0	0	0	0	1
0	1	0	0	1	1	1	0
1	0	0	1	0	1		
1	1	1	1	1	1		

12.2.2 Boolean Algebra[2]

The basic identities of Boolean algebra are shown in Table 12.2. They are listed in two columns to demonstrate the property of duality of Boolean algebra.

Table 12.2: Basic identities of Boolean algebra

1.	$X + 0 = X$		2.	$X \cdot 1 = X$	
3.	$X + X' = 1$		4.	$X \cdot \overline{X} = 0$	
5.	$X + X = X$		6.	$X \cdot X = X$	
7.	$X + 1 = 1$		8.	$X \cdot 0 = 0$	
9.	$\overline{\overline{(X)}} = X$				
10.	$X + Y = Y + X$	Commutativity	11.	$XY = YX$	
12.	$X + (Y + Z) = (X + Y) + Z$	Associativity	13.	$X(YZ) = (XY)Z$	
14.	$X (Y + Z) = XY + XZ$	Distributivity	15.	$X + YZ = (X + Y)(X + Z)$	
16.	$\overline{(X+Y)} = \overline{X} \cdot \overline{Y}$	DeMorgan's laws	17.	$\overline{X \cdot Y} = \overline{X} + \overline{Y}$	
18.	$X + XY = X$		19.	$X(X + Y) = X$	

The dual of an algebraic expression is obtained by interchanging OR and AND and replacing 1's by 0's and 0's by 1's. Thus, a formula that appears in one column of the table can be obtained from the formula in the other column by taking the dual of the expressions on either side of the equal sign.

12.2.3 Karnaugh Maps[3]

The Karnaugh map method provides a procedure for simplifying Boolean functions of up to four variables. The map is a diagram made up of squares, with each square representing one minterm of the function. A Boolean function is a sum of minterms that can be displayed graphically on the map. A minterm is the product of binary variables, viz., $x\overline{y}z$. The Boolean function forms patterns on the map, and these patterns can be used to derive the simplest Boolean expression in the function. The simplified expressions are in the form of a sum of products or a product of sums of minterms. The simplest algebraic expression usually contains the minimum number of terms and the fewest possible number of literals in each term. The result is a circuit with both a minimum number of gates and a minimum number of inputs to the gates.

Two Variable MAP

A Boolean function with two variables has four minterms. The Karnaugh map, therefore, has four squares, one for each minterm. The map for a function with two variables is shown in Figure 12.1.

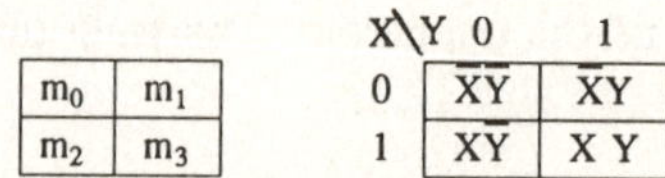

Figure 12.1: Two-variable Karnaugh Map

The values marked on the left side and the top of the map designate the values of the variables. The result of combining minterms m_0, m_1, and m_2 is

$$m_0 + m_1 + m_2 = \overline{X}\,\overline{Y} + X\overline{Y} + \overline{X}Y = \overline{X} + \overline{Y} \tag{12.1}$$

which is displayed in Figure 12.2.

X\Y	0	1
0	1	1
1	1	0

Figure 12.2: Map for combining minterms $m_0 + m_1 + m_2$

A 1 is placed inside the square for each minterm present. The area enclosed by the three squares is that for $\overline{X} + \overline{Y}$. Simple manipulation of Equation (12.1) will confirm this result.

12.2.4 Digital Logic Gates[4]

The standard gates used in digital design consist of AND, OR, the inverter, buffer, NAND, NOR, exclusive OR, and exclusive NOR. The symbols and truth tables for these eight gates are shown in Table 12.3. Each gate has one of two binary input variables specified by x and y and one binary output variable specified by F. The inverter circuit inverts the logic sense of a binary variable. It produces the NOT, or complement, function. The circle in the output of the symbol of an inverter designates the logic complement. The triangle symbol by itself designates a buffer circuit. A buffer produces the transfer function, but does not produce any logic operation because the binary value of the output is equal to the binary value of the input. The circuit is used for power amplification of the signal.

A gate can be extended to multiple inputs if the binary operation it represents is commutative and associative. Both the AND and OR operations have that property. The OR gate can also be extended to three or more variables.

The NAND and NOR functions are commutative, but not associative. To overcome this, the multiple NOR (or NAND) gate is defined as a complemented OR (or AND) gate. The symbols for the three input gates are shown in Figure 12.3.

Table 12.3 Digital logic gates

Name	Symbol	Algebraic function	Truth table

AND — $F = xy$

X	Y	F
0	0	0
0	1	0
1	0	0
1	1	1

OR — $F = x + y$

X	Y	F
0	0	0
0	1	1
1	0	1
1	1	1

Inverter — $F = \bar{x}$

X	F
0	1
1	0

Buffer — $F = x$

X	F
0	0
1	1

NAND — $F = (\overline{xy})$

X	Y	F
0	0	1
0	1	1
1	0	1
1	1	0

NOR — $F = (\overline{x + y})$

X	Y	F
0	0	1
0	1	0
1	0	0
1	1	0

Exclusive OR (XOR) — $F = x\bar{y} + \bar{x}y$, $x \oplus y$

X	Y	F
0	0	0
0	1	1
1	0	1
1	1	0

Exclusive NOR or equivalence — $F = xy + \overline{xy}$, $x \odot y$

X	Y	F
0	0	1
0	1	0
1	0	0
1	1	1

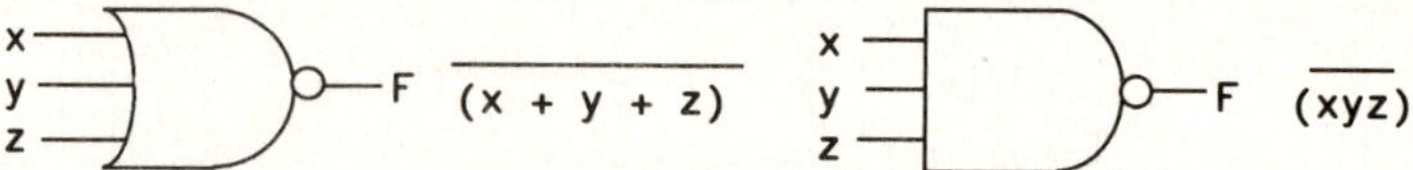

Figure 12.3: Multiple-input NOR and NAND gates

The exclusive OR and equivalence gates are both commutative and associative and, hence, can be extended to more than two inputs.

12.3 Combinational Logic

12.3.1 Adders[5]

A combinational circuit that performs the addition of two bits is called a **half-adder**. A circuit that performs the addition of three bits, the third bit being the carry, is a **full-adder**. A full adder can be implemented by employing two half-adders.

Half–Adder

Let x and y represent the two inputs, and S and C represent the sum and carry or the outputs. The truth table for the half-adder is then shown in Table 12.4.

Table 12.4 Truth table for the half-adder

Inputs		Outputs	
x	y	Carry	Sum
0	0	0	0
0	1	0	1
1	0	0	1
1	1	1	0

The simplified Boolean functions for the two outputs can be obtained directly from the truth table. The sum of products representation is

$$S = \overline{x}y + x\overline{y} \text{ or } x \oplus y \tag{12.2}$$

$$C = xy \tag{12.3}$$

and the product of sums representation is

$$S = (x + y)\overline{(x + \overline{y})} \tag{12.4}$$

$$C = xy \tag{12.5}$$

Full Adder

The full adder is a combination circuit that forms the arithmetic sum of three input bits. Two of the inputs represent the bits to be added and the third input represents any carry from a previous lower significant position. The outputs are represented by S, the value of the least significant bit of the sum, and C, the carry. Table 12.5 gives the truth table for the full adder.

Table 12.5: Truth table for the full-adder

x	y	z	C	S
0	0	0	0	0
0	0	1	0	1
0	1	0	0	1
0	1	1	1	0
1	0	0	0	1
1	0	1	1	0
1	1	0	1	0
1	1	1	1	1

The simplified Boolean expressions representing the Karnaugh map from Table 12.5 are

$$S = \bar{x}\,\bar{y}\,z + \bar{x}y\bar{z} + x\bar{y}\,\bar{z} + xyz \tag{12.6}$$

and

$$C = xy + xz + yz \tag{12.7}$$

The implementation of Equations (12.6) and (12.7) in the full adder is shown in Figure 12.4.

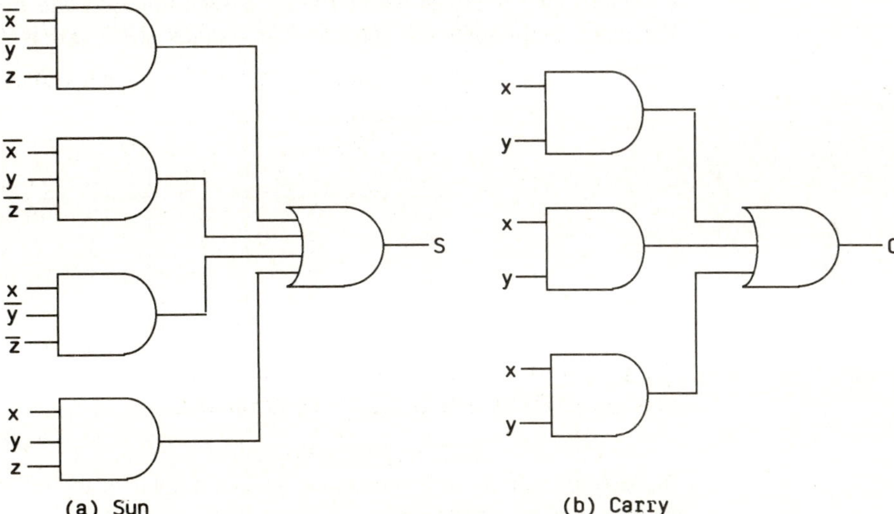

Figure 12.4: Full adder as the sum of products

Equations (12.6) and (12.7) can also be expressed as

$$S = z \oplus (x \oplus y) \tag{12.8}$$

and

$$C = z(x \oplus y) + xy \tag{12.9}$$

Figure 12.5 displays the implementation of a full-adder using two half-adders and an OR gate.

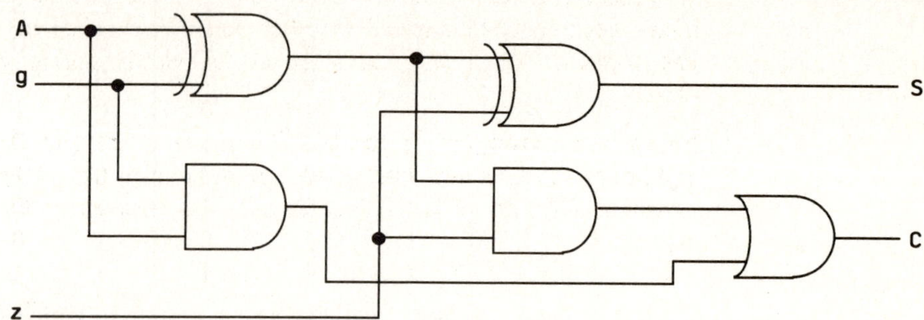

Figure 12.5: Implementation of the full adder using two half-adders and an OR gate

12.3.2 Subtractors[6]

Full–Subtractor

A full subtractor is a combinational circuit that performs a subtraction between two bits, but takes into account that a 1 may be borrowed by a lower significant number. The circuit has three inputs and two outputs.

In Table 12.6, the inputs x and y represent the minuend and subtrahend respectively. The input z represents the borrow. The outputs are D and B, the difference and the borrow.

Table 12.6: Truth table for the full subtractor

x	y	z	B	D
0	0	0	0	0
0	0	1	1	1
0	1	0	1	1
0	1	1	1	0
1	0	0	0	1
1	0	1	0	0
1	1	0	0	0
1	1	1	1	1

The simplified Boolean functions for the two outputs of the full subtractor, taken from the Karnaugh map, are

$$D = \overline{x}\overline{y}z + \overline{x}y\overline{z} + \overline{y}\overline{z} + xyz \qquad (12.10)$$

and

$$B = \overline{x}y + \overline{x}\overline{z} + yz \qquad (12.11)$$

Comparison of the full subtractor with the full adder reveals that the logic functions are the same for output D and output S. And output B resembles output C, except that the input variable x is complemented. Therefore, it is easy to convert a full adder into a full subtractor.

12.3.3 Decoders[7]

A decoder is a combinational circuit output that converts binary information from n input lines into a maximum of 2^n unique output lines. If the n-bit decoded information has unused or "don't care" combinations, the decoder output will have fewer than 2^n outputs.

Figure 12.6 shows a three- to eight-line decoder. The three inputs are decoded into eight outputs, with each output representing one of the minterms of the three-input variables. Each one of the eight AND gates generates one of the minterms.

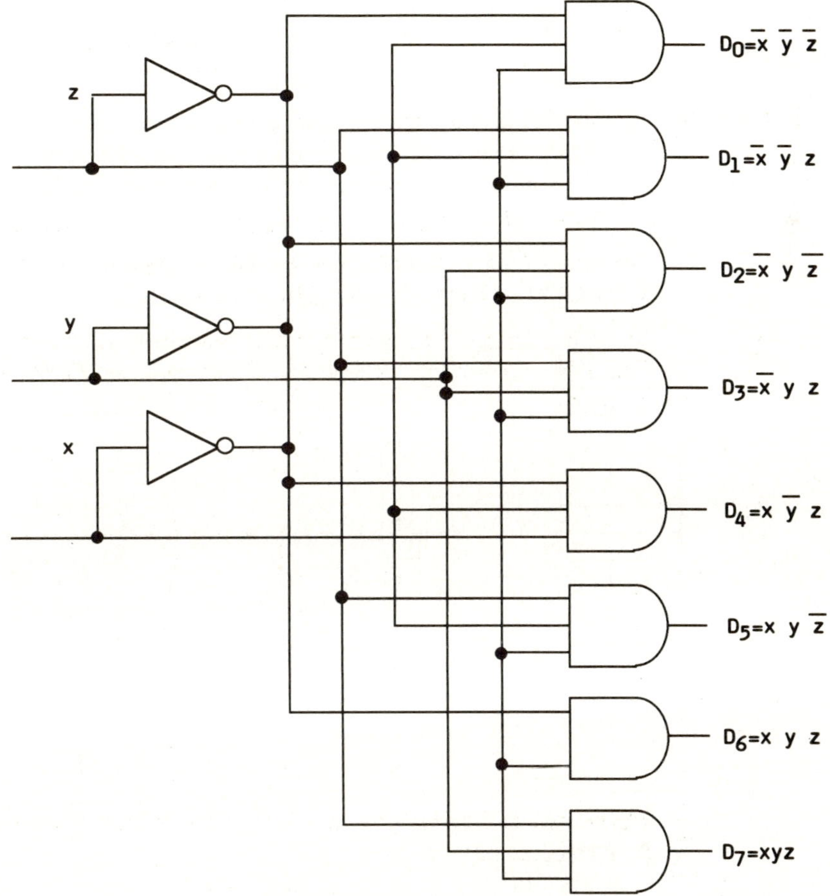

Figure 12.6: A 3 to 8 line decoder

An example of the use of a three- to eight-line decoder would be a binary-to-octal conversion. The input variables would represent a binary number, and the outputs would represent an eight-digit octal number.

The truth table for the three-to eight line decoder is shown in Table 12.7.

From the table, the output variables are mutually exclusive because only one output can be equal to 1 at any one time. The output line whose value is equal to 1 represents the minterm equivalent of the binary number of the input lines.

Table 12.7: Truth table for a 3 to 8 line decoder

Inputs			Outputs								
x	y	z	D_0	D_1	D_2	D_3	D_4	D_5	D_6	D_7	D_8
0	0	0	1	0	0	0	0	0	0	0	0
0	0	1	0	1	0	0	0	0	0	0	0
0	1	0	0	0	1	0	0	0	0	0	0
0	1	1	0	0	0	1	0	0	0	0	0
1	0	0	0	0	0	0	1	0	0	0	0
1	0	1	0	0	0	0	0	0	1	0	0
1	1	0	0	0	0	0	0	0	0	1	0
1	1	1	0	0	0	0	0	0	0	0	1

12.3.4 Multiplexers[8]

A multiplexer is a combinational circuit that selects binary information from one of many input lines and directs the information to a single output line. The selection of a particular input line is controlled by a set of selection lines. There are 2^n input lines and n selection lines whose bit combinations determine which input is selected.

An example of a multiplexer is shown in Figure 12.7. Each of the four input lines, I_0 through I_3, is applied to one input of an AND gate.

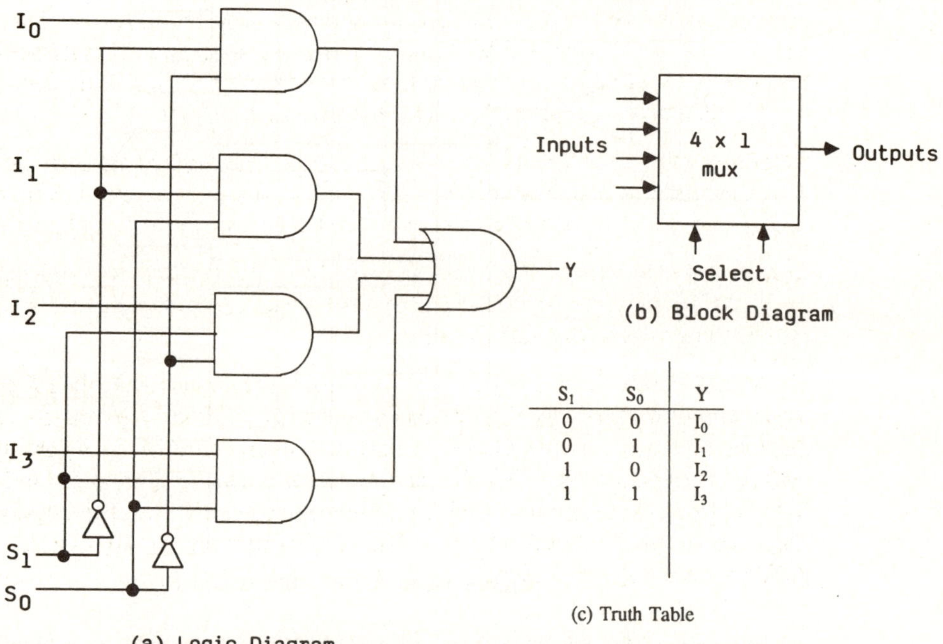

Figure 12.7: 4-to-1 line multiplexer

For example, when S_1 and $S_0 = 10$, the AND gate associated with input I_2 has two of its inputs equal to 1 and the third input connected to I_2. The other three AND gates have at least one input equal to 0, which means that their output is equal to 0. The output of the OR gate is then equal to the value of I_2. The multiplexer, then, provides a path from the selected input to the output.

Multiplexer ICs may have an enable input to control the operation of the circuit. When the enable input is in a particular binary state the outputs are disabled, and when it is in the enable state the circuit functions as a multiplexer.

12.3.5 Gate Arrays[9],[10]

A gate array consists of a pattern of gates, fabricated in an area of silicon, that is repeated until the entire chip is covered with identical gates. The design with gate arrays requires that the designer specify the layout of the chip and the way the gates are routed and interconnected. The first few levels of fabrication are common and do not depend on the final logic function. Additional fabrication levels are required to interconnect the gates in order to realize the desired function. There is enough complexity so that this task is usually performed by means of computer-aided design (CAD) methods.

12.4 Sequential Logic[11]

A sequential circuit consists of a combinational circuit and storage elements that form a feedback system. The sequential circuit receives binary information from external inputs. Then, together with the present state of the storage elements, these inputs determine the binary value of the outputs.

The outputs in a sequential circuit are a function of both external inputs, and the present state of the storage elements. The next state of the storage elements is also a function of the inputs and the present state. In effect then, a sequential circuit is specified by a time sequence of inputs, outputs, and internal states.

There are two types of sequential circuits: a **synchronous** sequential circuit and an **asynchronous** sequential circuit. The synchronous circuit is a system whose behavior can be determined from a knowledge of its signals at discrete instants of time. The behavior of an asynchronous circuit depends on the order in which the inputs change, and the state of the circuit can be affected at any instant of time.

Synchronization is achieved by a timing device called a **clock generator** that produces a periodic train of clock pulses. The clock pulses are distributed throughout the system in such a way that storage elements are affected only upon the arrival of each pulse.

The outputs of storage elements change only when clock pulses are present. Synchronous sequential circuits that use clock pulses in the input of storage elements are called **clocked sequential circuits**.

The storage elements employed in clocked sequential circuits are called **flip-flops**. A flip-flop is a binary storage device capable of storing one bit of information. The block diagram of a synchronous clocked sequential circuit is shown in Figure 12.8. The outputs can come from either the combinational circuit or from the flip-flops. The flips-flops receive their inputs from the combinational circuit and from a train of pulses that occur at fixed intervals of time. The next state of the flip-flops can change only during a clock pulse transition.

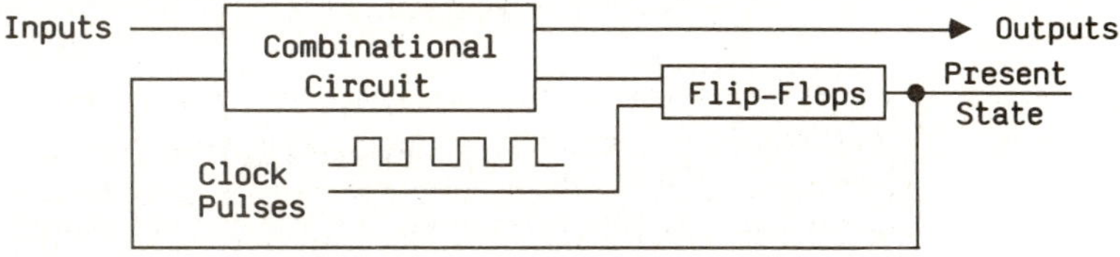

Figure 12.8: Synchronous clocked circuit

A flip-flop has two outputs, one for the normal value and a second for the complemented value of the bit that is stored.

The major differences among the various types of flip-flops are in the number of inputs they have and the ways they affect the binary state. The basic flip-flop called the **latch** operates with signal levels.

12.4.1 Latches[12]

The SR latch is a circuit with two cross-coupled NOR gates or two cross-coupled NAND gates (see Figures 12.9 and 12.10). It has two inputs labeled S for set and R for reset.

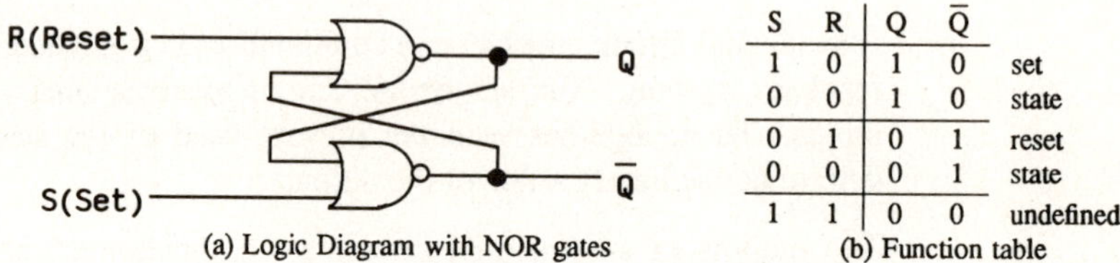

S	R	Q	$\overline{Q}$	
1	0	1	0	set
0	0	1	0	state
0	1	0	1	reset
0	0	0	1	state
1	1	0	0	undefined

(a) Logic Diagram with NOR gates   (b) Function table

Figure 12.9: SR Latch with NOR gates

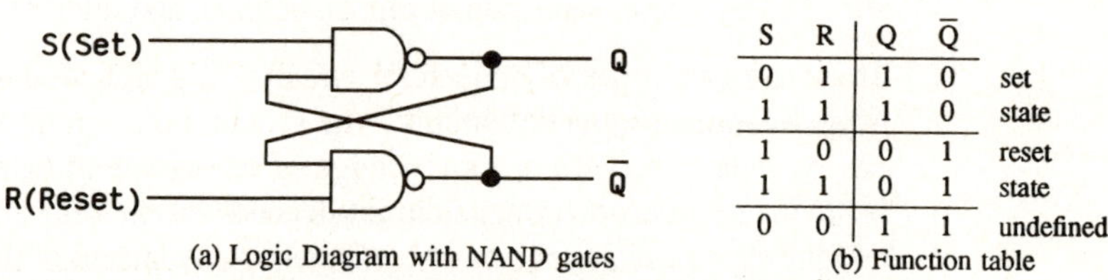

S	R	Q	$\overline{Q}$	
0	1	1	0	set
1	1	1	0	state
1	0	0	1	reset
1	1	0	1	state
0	0	1	1	undefined

(a) Logic Diagram with NAND gates   (b) Function table

Figure 12.10: SR Latch with NAND gates

For the NOR gate, when output Q=1 and $\overline{Q}$=0, the latch is in the set state, and when Q=0 and $\overline{Q}$=1, the latch is in the reset state. An undefined state occurs when both outputs are equal to 0. Normally, output Q and $\overline{Q}$ are complements of each other.

The D latch (Figure 12.11) eliminates the indeterminate state of the SR latch. In the D latch, inputs S and R are never equal to 1 at the same time.

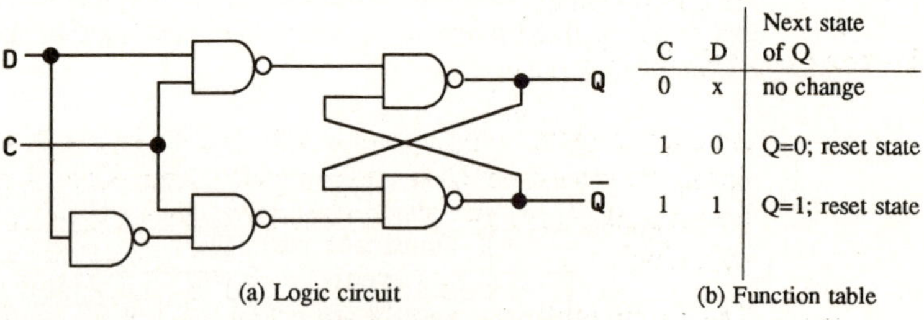

C	D	Next state of Q
0	x	no change
1	0	Q=0; reset state
1	1	Q=1; reset state

(a) Logic circuit   (b) Function table

Figure 12.11: D Latch

The D latch has only two inputs, D (data) and C (control). The D input goes to the S input and its complement goes to the R input. When C=0, inputs to the SR latch are 1 and the circuit cannot change states. When C=1 and D=1, the Q output goes to 1, placing the circuit in the set state. When C=1 and D=0, the output goes to 0 and the circuit is then in the reset state.

The D latch serves as temporary storage for binary information. The binary information present in the D input is transferred to the Q output when the control input is enabled. The output follows the input as long as the C enable remains active. When the control is disabled, the binary information present at the D input at the time the control is disabled is retained at the Q output until the control is activated again.

12.4.2 Master-Slave Flip-Flop[13]

The master-slave flip-flop consists of two latches and an inverter. A D-type master-slave flip-flop is shown in Figure 12.12, along with timing diagrams.

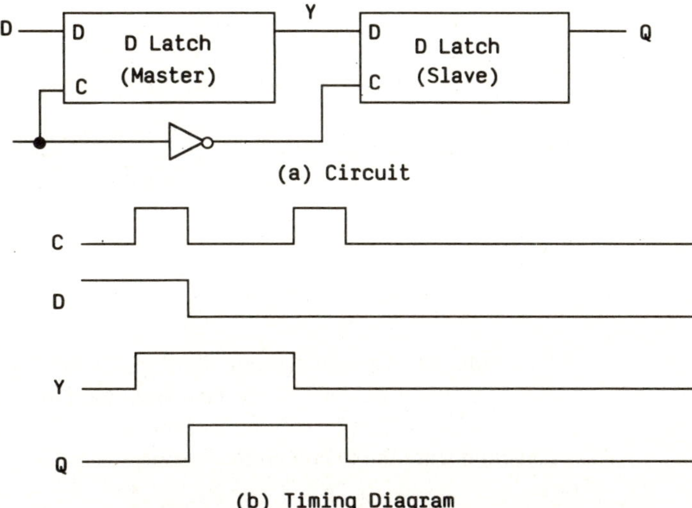

Figure 12.12: Master-Slave D-type flip-flop

When the clock pulse input C is 0, the output of the inverter is 1. The slave latch is enabled, and its output Q is equal to the master output Y. The master latch is disabled because C=0. When the input pulse to C changes to the logic-1 level, the data in the external D input is transferred to the master. The slave now is disabled as long as the pulse to C remains at the 1 level because its C input is equal to 0. Changes in the external D input change the master output Y, but cannot affect the slave output Q. However, when the pulse at C returns to 0, the master is disabled and isolated from the D input. But now the slave is enabled, and the value of Y is transferred to the output of the flip-flop at Q.

12.4.3 Edge-triggered Flip-Flop[13]

An edge-triggered flip-flop ignores the pulse while it is at a constant level, but triggers during the transition of the clock signal. Some edge-triggered flip-flops trigger on the positive edge (a 0-to-1 transition), and others trigger on the negative edge (a 1-to-0 transition).

The timing of the response of a flip-flop to input data and the clock must be taken into account when using edge-triggered flip-flops. There is a minimum time called **setup time** in which the input must be maintained at a constant value prior to the occurrence of the clock transition. Similarly, there is a definite time called the **hold time** when the input must not change after the application of the positive transition of the pulse. The **propagation delay time** of the flip-flop is defined as the time interval between the trigger edge and stabilization of the output to the new state.

12.4.4 Flip-Flop Excitation Tables[14]

The flip-flop excitation table lists the required inputs for a given change of state. That is, if we know the transition from the present state to the next state, the excitation table is used to find the flip-flop input conditions.

Excitation tables for the RS, JK, D, and T flip-flop are shown in Table 12.8.

Table 12.8: Flip-flop excitation tables

(a) RS Flip-Flop				(b) JK Flip-Flop			
$Q(t)$	$Q(t+1)$	S	R	$Q(t)$	$Q(t+1)$	J	K
0	0	0	x	0	0	0	x
0	1	1	0	0	1	1	x
1	0	0	1	1	0	x	1
1	1	x	0	1	1	x	0

(c) D Flip-Flop			(d) T Flip-Flop		
$Q(t)$	$Q(t+1)$	R	$Q(t)$	$Q(t+1)$	T
0	0	0	0	0	0
0	1	1	0	1	1
1	0	0	1	0	1
1	1	1	1	1	0

As an example, consider the RS flip-flop. If the flip-flop is in the 0-state, and it is desired to have it go the 1-state, then, from the table, the transition from the 0-state to the 1-state is obtained by making S=1 and R=0. Similarly, if the JK flip-flop is to have a transition from the 0-state to the 1-state, J must be equal to 1 and K may be either 0 or 1.

12.5 Registers and Counters[15]

12.5.1 Register

A register is a group of storage cells that can hold binary information. A group of flip-flops can be a register, since each flip-flop is a binary cell capable of storing one bit of information. An n-bit register has a group of n flip-flops and is capable of storing any binary information containing n bits. A register also has gates that perform certain data-processing tasks. The flip-flops hold the binary information, and the gates control when and how the information is transferred into the register.

12.5.2 Asynchronous or Ripple Counter

A counter is a register that goes through a predetermined sequence of states upon application of clock pulses.

In a ripple counter, the flip-flop output transition serves as a source for triggering other flip-flops. This means that the clock pulse inputs of all flip-flops except the first are triggered by the transition in other flip-flops. Figure 12.13 shows a JK four-bit binary ripple counter. The flip-flop that contains the least significant bit receives the incoming

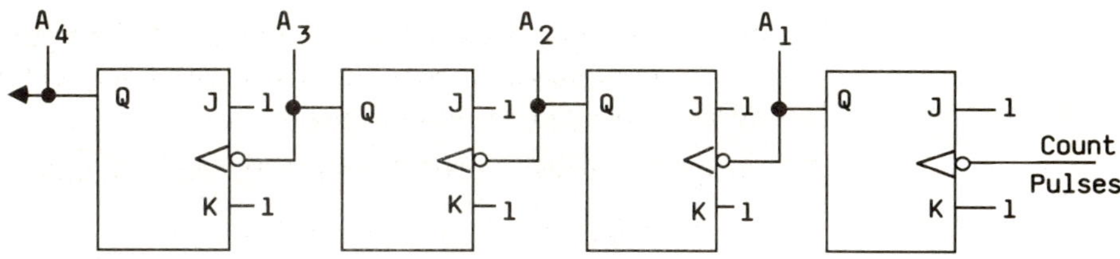

Figure 12.13: JK four-bit binary ripple counter

count pulses. All the inputs are equal to 1. The small circle in the clock pulse input indicates that the flip-flop complements during a negative-going transition. That is, an output that goes from 1 to 0 complements the flip-flop.

12.5.3 Synchronous Counter

Synchronous counters differ from ripple counters in that clock pulses are applied to the clock pulse inputs of all flip-flops. That is, clock pulse triggers all the flip-flops simultaneously, rather than one at a time in succession as in a ripple counter. Whether a flip-flop is complemented depends on the values of the J and K inputs at the time of the pulse. If J=K=0, the flip-flop remains unchanged. If J=K=1, the flip-flop complements.

12.6 Digital Electronics

12.6.1 Basic Concepts[16]

In binary digital circuits, two distinct voltage levels can represent the two values of binary variables. But, because of component tolerances and other effects, two distinct voltage **ranges** are defined.

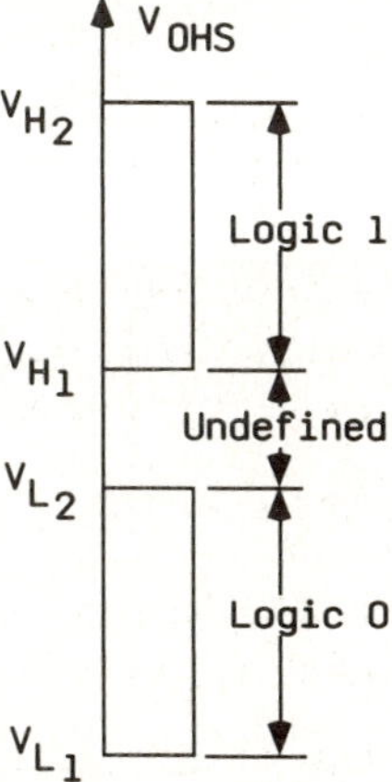

Figure 12.14: Two distinct voltage ranges represent the two values of binary variables.

As shown in Figure 12.14, if the signal lies in the range V_{L_1} to V_{L_2}, the signal is read as a logic 0. If the signal amplitude falls in the range V_{H_1} to V_{H_2}, it is interpreted as a logic 1. The forbidden band represents an undefined region. When logic-1 voltages are higher than logic-0 voltages, the system is said to use **positive logic**. If the band assignment is reversed, the system is said to be a **negative-logic** system.

12.6.2 Logic Families

In this chapter, we shall consider two MOS families: NMOS and CMOS. The NMOS family uses n-channel MOSFETs, while the CMOS family uses both n- and p-channel transistors in complementary symmetric circuit configurations. We shall also study two BJT logic families: transistor-transistor logic (TTL) and emitter-coupled logic (ECL). CMOS, TTL, and ECL are available as off-the-shelf components for conventional logic design, as well as for special VLSI circuits. NMOS, however, is used only in the design of VLSI circuits such as microprocessors and memory chips.

TTL is the most popular logic family. ECL is used in systems requiring high-speed operations. NMOS is used in circuits requiring high component density, and CMOS is used in systems requiring low power consumption.

TTL ICs are numerically coded as the 5400 (military) and 7400 (industrial) series.

ECL is designated as the 10,000 series. 10102 provides four two-input NOR gates, while 10107 provides three exclusive-OR gates.

CMOS circuits are designated as the 4000 series. The 4050 type provides six buffer gates.

12.6.3 Basic Inverter

The logic inverter is basically a voltage-controlled switch. In Figure 12.15, the input voltage is taken across nodes 1 and 3 and the output across nodes 2 and 3. When V_I is low (around 0 V) the switch is open and the output voltage V_0 is high. When V_I is high (above a certain threshold value), the switch is closed and the output voltage is low. This circuit realizes the logic inversion operation.

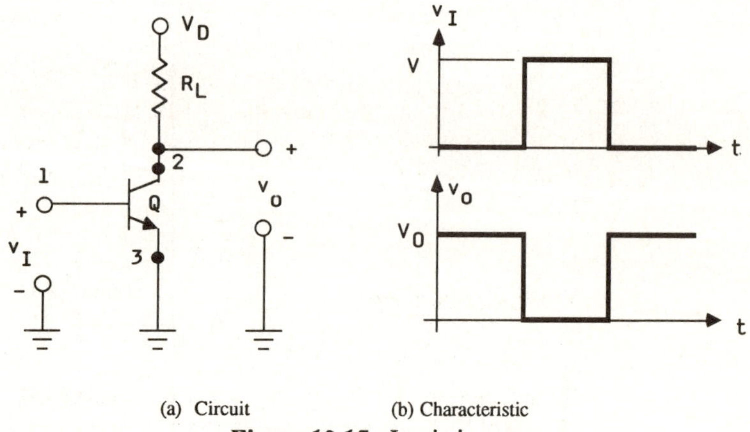

(a) Circuit   (b) Characteristic

Figure 12.15: Logic inverter

12.6.4 Inverter Transfer Characteristic

Figure 12.16 shows a typical inverter transfer characteristic.

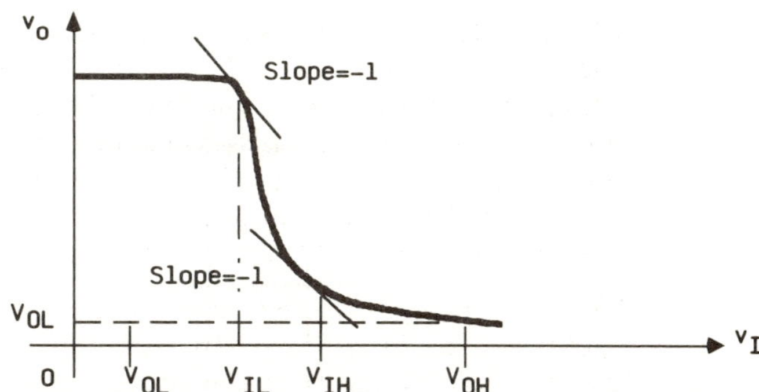

Figure 12.16: Typical transfer characteristic of a logic inverter

There are three distinct regions:

1. The low-input region; $V_I < V_{IL}$
2. The transition region; $V_{IL} < V_I \leq V_{IH}$
3. The high-input region; $V_I > V_{IH}$

Since the transitions between one region and the next might not be sharp, V_{IL} and V_{IH} are defined as the points at which the slope of the voltage transfer curve is -1. The slope in the transition region is much greater than 1. Input voltages less than V_{IL} are defined as representing logic 0, and V_{IL} is taken as the maximum allowable logic-0 value. Values of input voltage greater than V_{IH} are defined as representing logic 1, and V_{IH} is the minimum allowable logic-1 value.

12.6.5 Noise Margins

As long as the input signal is correctly perceived as low or high, the operation is not affected by variations in the signal level, or **noise**. Noise is extraneous signals that can be capacitively or inductively coupled into the digital circuit. If noise were superimposed on the output signal of the driving gate (V_{OH}), the gate would not be bothered as long as the amplitude of the noise voltage was lower than the safety margin, ($V_{OH} - V_{IH}$). This difference is called a **high** noise margin and is represented by

$$NM_H \equiv V_{OH} - V_{IH} \qquad (12.12)$$

The **low** noise margin is

$$MM_L \equiv V_{IL} - V_{OL} \qquad (12.13)$$

Manufacturers usually specify worst case values for the four parameters V_{OA}, V_{IH}, V_{OL}, and V_{IL}. These parameters and the noise margins are shown in Figure 12.17. In the figure,

$$V_{OH} \equiv \text{the minimum voltage that will be}$$
available at a gate output when the
output is supposed to be high

$$V_{IH} \equiv \text{the minimum gate input voltage that will be}$$
clearly recognized by the gate as corresponding
to a high

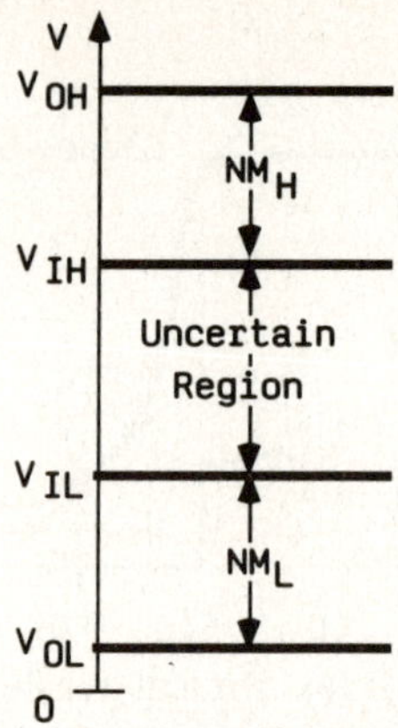

Figure 12.17: Logic band diagram indicating noise margins

$$V_{OL} \equiv \text{the maximum voltage that will be available at a gate output when the output is supposed to be low}$$

$$V_{IL} \equiv \text{the minimum gate input voltage that will be clearly recognized by the gate as a low}$$

To maximize the voltage swing, V_{OL} should be as close to 0 V as possible and V_{OH} should be as high as possible, or close to the power supply voltage.

12.6.6 Power Dissipation

The power dissipated in a logic circuit is composed of a static and a dynamic component. The static power is the power dissipated while the circuit is not changing states. If the inverter is switched on and off f times per second, the dynamic power dissipation will be

$$\text{Dynamic power dissipation} = fC_L(V^+)^2 \tag{12.14}$$

where C_L is the load capacitance. The load capacitance could be the input capacitance of another logic gate or the capacitance of the interconnections.

12.6.7 Fan-in and Fan-out

The fan-in of a gate is the number of gate inputs. Fan-out is the maximum number of similar gates or standard loads that a gate can drive while remaining within specifications. A standard load is defined as the amount of current needed by an input of another gate. The terms **loading** and **fan-out** are sometimes interchanged. The term **overload** comes from the fact that the output of a gate can supply a limited amount of current, and above this current it ceases to operate properly. Fan-out is expressed as a number.

12.6.8 Propagation Delay

Figure 12.18 shows the typical response of an inverter to an input pulse with finite rise and fall times. In addition to the finite rise and fall times, there is a delay time between the input and output pulses. The usual way of expressing this propagation delay is to specify the times between the 50% points of the input and output waveforms at both the leading and trailing edges. These times are called t_{PHL} (HL indicates high-to-low transition of the output) and t_{PLH} (LH means the low-to-high transition of the output). The propagation delay, t_P, is then defined as the average of the two times:

$$t_p = \frac{1}{2}\,(t_{PHL} + t_{PLH}) \tag{12.15}$$

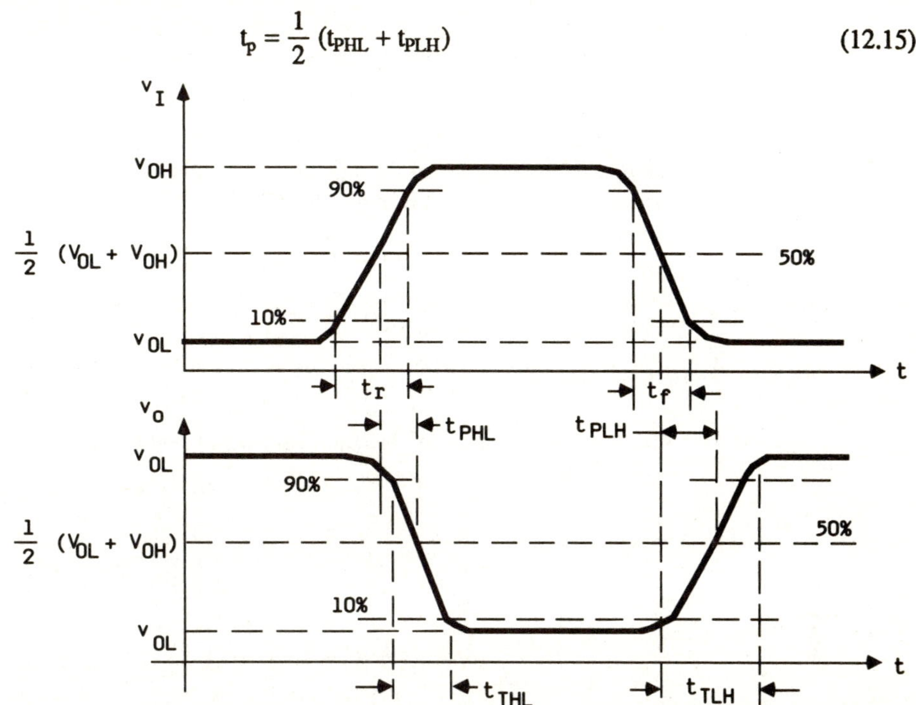

Figure 12.18: Illustration of propagation delay and transition times of logic gates

12.6.9 Power-Delay Product

The power-delay product is a figure of merit for comparing logic families and is defined as

$$PDP \equiv t_p P_D \quad \text{(joules)} \tag{12.16}$$

where P_D is the power dissipation of the gate. PDP may be interpreted as energy consumed per logic decision. Therefore, the lower the PDP for a logic family, the more effective is the family.

12.6.10 Positive and Negative Logic

Except during transitions, the binary signal takes on one of two possible values: logic 1 or logic 0. Accordingly, there are two distinct logic-signal systems.

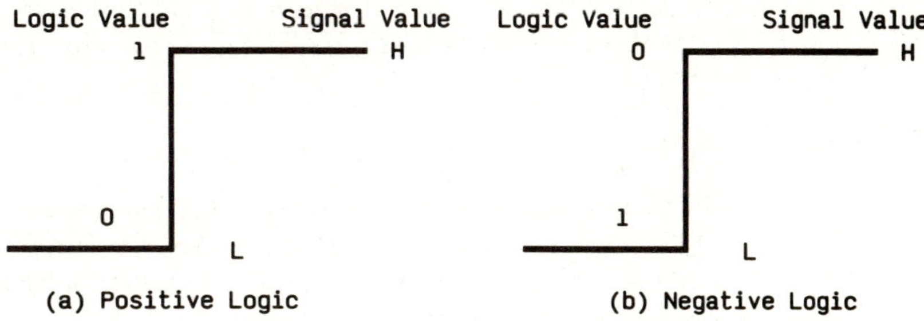

Figure 12.19: Logic-value assignment

When the high level, H, represents logic 1, the system is referred to as a positive-logic system; When the low level, L, represents logic 1, the system is called a negative-logic system. Figure 12.19 shows both types of system.

What determines the type of a logic-signal system is the assignment of logic values by the relative amplitudes of the signals, not whether the signal is positive or negative. An examination of manufacturers' data sheets shows that digital functions are defined in terms of H (high) or L (low) levels. It is the user's prerogative to choose a positive logic or negative logic assignment. The data sheets show a range of voltage values that the circuit will recognize as a high or low level. Also listed are typical values and the voltage supply requirements.

12.7 *Transistor-Transistor Logic (TTL)*[17],[18]

The circuit diagram of a TTL gate is shown in Figure 12.20.

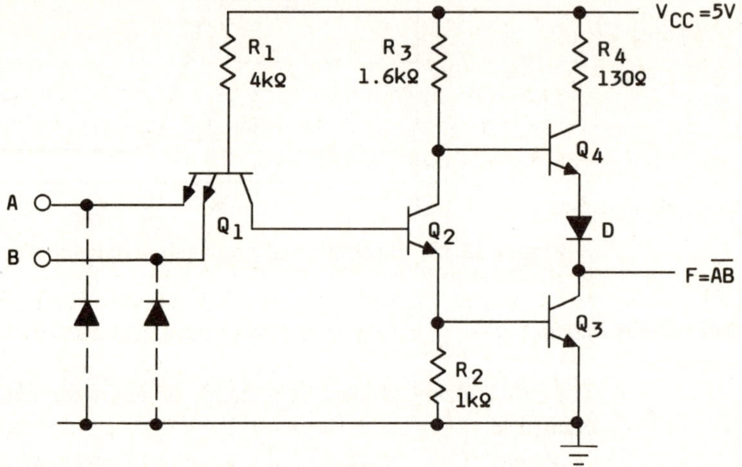

Figure 12.20: Standard two-input TTL NAND gate circuit (series 54/74)

The voltage transfer characteristic for the standard TTL gate is given in Figure 12.21.

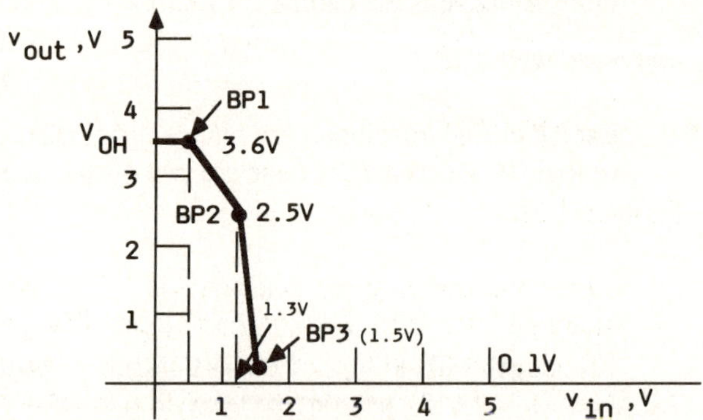

Figure 12.21: Voltage transfer characteristic (VTC) of the standard TTL gate

The noise margins of the gate can be determined from the critical points as follows:

$$
\begin{aligned}
V_{OH} &= 3.6\ \text{V} \\
V_{IL} &= 0.6\ \text{V} \\
V_{OL} &= 0.1\ \text{V} \\
V_{IH} &= 1.5\ \text{V} \\
NM_H = V_{OH} - V_{IH} &= 2.1\ \text{V} \\
NM_L = V_{IL} - V_{OL} &= 0.5\ \text{V}
\end{aligned}
$$

The values have been computed on the assumption that the gate is not loaded and without considering power-supply or temperature variations.

12.7.1 Fan-out

The gate input current in the low state, I_{IL}, is equal to the current that flows out of the emitter of Q_1 (see Figure 12.20). If the TTL gate is driven by another TTL gate, the output transistor Q_3 should sink this current I_{IL} while remaining in saturation and while maintaining a saturation lower than a guaranteed maximum. Since the output current that a TTL gate can sink is limited to a certain maximum value, the maximum fan-out of the gate is determined directly by the value of I_{IL}. That is,

$$N \le \frac{I_{C3(EOS)}}{I_{IL}} \tag{12.17}$$

where EOS means "edge of saturation."

12.7.2 Propagation Delay Time

The propagation delay of TTL gates is defined as the time between the 50% points of corresponding edges of the input and output waveforms (see Section 12.6.8). For standard TTL, t_p is typically about 10 ns.

12.7.3 Power Dissipation

It can be shown that when the gate output is high, the gate dissipates 4 mW of power, and when the output is low, the dissipation is 16.7 mW. The average dissipation is then 11 mW. This results in a delay-power product of about 100 pJ.

The electrical characteristics of standard TTL are shown in Table 12.9.

Table 12.9: Standard TTL (54/74) typical electrical characteristics at T_A-25°C (Reprinted by permission from *Analyses and Design of Digital Integrated Circuits* by D. A. Hodges and H. G. Jackson. Copyright © 1988 by McGraw-Hill.)

V_{OH}/V_{OL}	3.5 V/0.2 V	Fan-out	10
V_{IH}/V_{IL}	1.5 V/0.5 V	Supply voltage	+5.0 V
NM_H/NM_L	2.0 V/0.3 V	Power dissipation per gate	10 mW
Logic swing	3.3 V	Propagation delay time	10 ns

12.7.4 Schottky-Clamped TTL

Improvements in speed and reductions in power dissipation have been achieved for standard TTL. The speed of the standard TTL gate is limited by the fact that transistors Q_1, Q_2, and Q_3 saturate and, therefore, we must consider their storage time. Further, Q_3 discharges slowly through a 1-kΩ resistor in its base current. The resistances in the circuit, along with the various transistor and wiring capacitances, form relatively long time constants. To counter these problems, we would like to prevent the transistors from going into saturation and to reduce the values of all resistances.

In Schottky-clamped TTL, transistors are prevented from going into saturation by connecting a low-voltage-drop diode between the base and the collector (Figure 12.22).

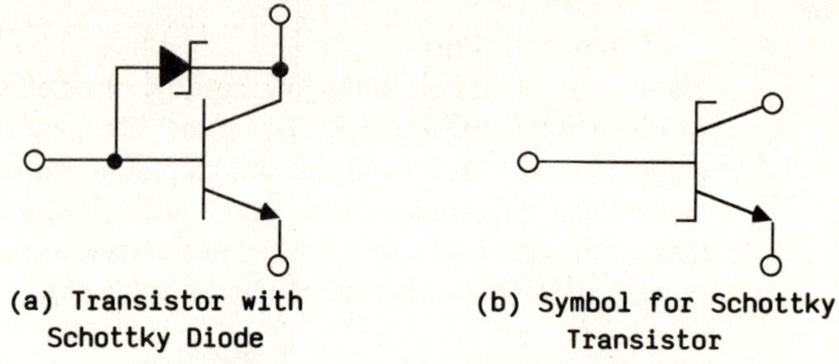

Figure 12.22: Schottky clamped transistor

A Schottky-clamped TTL NAND gate is shown in Figure 12.23.

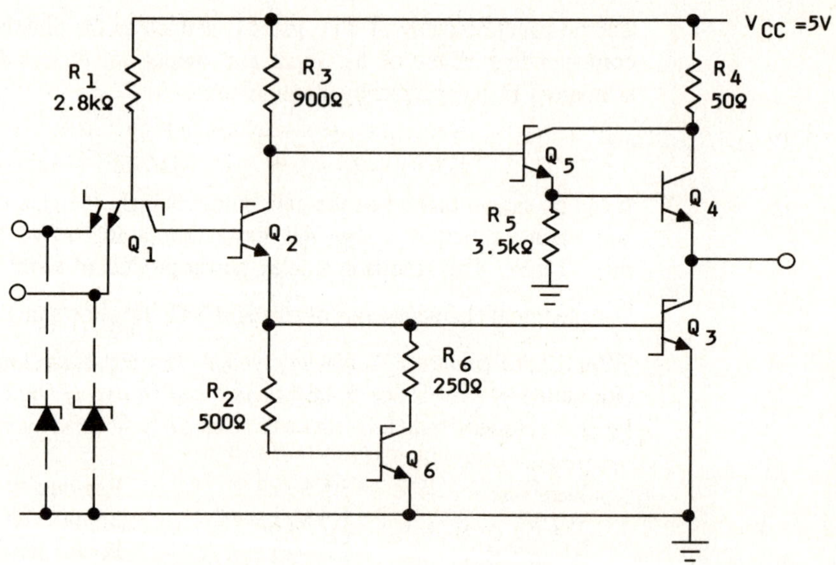

Figure 12.23: Schottky-clamped TTL NAND gate (54S/74S)

The performance characteristics for Schottky TTL (74S) are specified to have the following worst case parameters:

$$
\begin{array}{llll}
V_{OH} &=& 1.7\,\text{V} \qquad & V_{OL} &=& 0.5\,\text{V} \\
V_{IH} &=& 2.0\,\text{V} & V_{IL} &=& 0.8\,\text{V} \\
t_p &=& 3\,\text{ns} & P_D &=& 20\,\text{mW}
\end{array}
$$

The delay-power product is 60 pJ, compared to 100 pJ for standard TTL.

12.7.5 Low-power Schottky TTL

The low-power Schottky-clamped TTL [TTL(LS)] is shown in Figure 12.24. Typical propagation delay time for this gate is 10 ns, but the power dissipation is only 2 mW. To obtain this low power, it is necessary to increase the resistance values by a factor of five compared with those in the standard TTL gate.

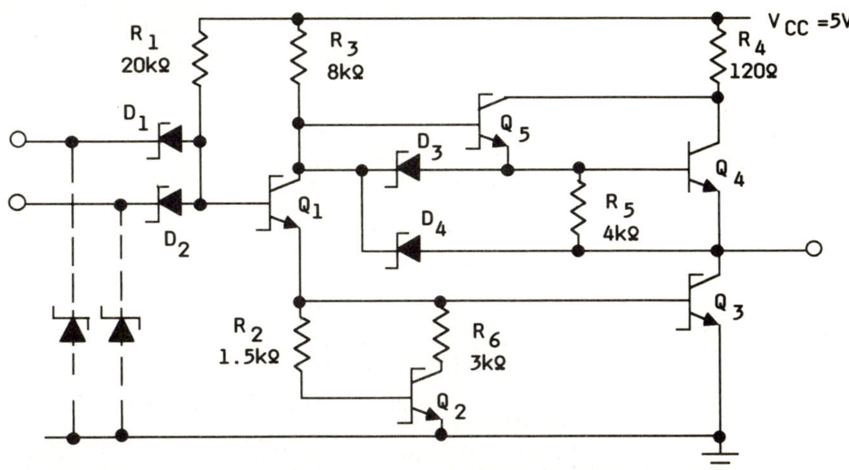

Figure 12.24: Low-power Schottky-clamped TTL NAND gate (54LS/74LS)

Table 12.10 compares the principal characteristics for the standard TTL, Schottky TTL and low-power Schottky.

Table 12.10: TTL Logic Characteristics at $T_A = 20°C$ (Reprinted by permission from *Analyses and Design of Digital Integrated Circuits* by D. A. Hodges and H. G. Jackson. Copyright © 1988 by McGraw-Hill.)

	Series 74	Series 74S	Series 74LS
min V_{OH}/max V_{OL}	2.4 V/0.4 V	2.7 V/0.5 V	2.7 V/0.5 V
min V_{IH}/max V_{IL}	2.0 V/0.8 V	2.0 V/0.8 V	2.0 V/0.8 V
min I_{OH}/max I_{OL}	-0.4 mA/16 mA	-1.0 mA/20 mA	-0.4 mA/8 mA
min I_{IH}/max I_{IL}	40 µA/-1.6 mA	50 µA/-2.0 mA	20µA/-0.4mA
typical propagation delay time	10 ns	3 ns	10 ns
typical power dissipation per gate	10 mW	20 mW	2 mW

From the table, the worst case logic swing and transition width can be determined. Also, from the minimum and maximum values of the input and output currents, the fan-out can be determined. The negative sign on I_{OH} and I_{IL} indicates that the current is out of the gate.

12.7.6 Advanced Schottky-clamped TTL

The advanced Schottky (54AS/74AS) and advanced low-power Schottky (54ALS/74ALS) circuits make use of the latest developments in bipolar IC fabrication technology. The characteristics of advanced Schottky TTL are given in Table 12.11. The AS series aims for the highest speed. The propagation delay time of 1.5 ns is one-half that of the early Schottky series, but the power dissipation is the same at 20 mW. The ALS series aims for the lowest power. The power dissipation is only 1 mW, but, in addition, the propagation delay time has been reduced from 10 ns to 4 ns. The F series is intermediate between the AS and the ALS, with a propagation delay time of 2.5 ns and a power dissipation of 4 mW.

Table 12.11: Advanced Schottky TTL characteristics at $T_A = 25°C$ (Reprinted by permission from *Analyses and Design of Digital Integrated Circuits* by D. A. Hodges and H. G. Jackson. Copyright © 1988 by McGraw-Hill.)

	Series 74F	Series 74AS	Series 74ALS
min V_{OH}/max V_{OL}	2.7 V/0.5 V	same as 74S	same as 74S
min V_{IH}/max V_{IL}	2.8 V/0.8 V	same as 74S	same as 74S
min I_{OH}/max I_{OL}	-1.0 mA/20 mA	-2.0 mA/20 mA	-0.4 mA/4.0mA
min I_{IH}/max I_{IL}	20 µA/-0.6 mA	0.2 mA/-2.0 mA	20 µA/-0.2 mA
typical propagation delay time	2.5 ns	1.5 ns	4 ns
typical power dissipation per gate	4 mW	20 mW	1 mW

12.7.7 TTL Gate Circuits

All the basic combinational logic forms (AND, OR, NAND, NOR, INVERT, exclusive-OR and AND-OR-INVERT) are available in each of the TTL circuits. The number of inputs to these gates is from one to eight.

12.8 Emitter-coupled Logic[19],[20]

Emitter-coupled logic (ECL) is the fastest logic circuit family because all of the transistors are operated out of saturation and the logic signal swings are kept small. These techniques avoid storage time delays and reduce the time required to charge and discharge the load and parasitic capacitances.

The two forms of ECL are the ECL 10K and ECL 100K. Typical electrical characteristics for the ECL 10K are shown in Table 12.12.

Table 12.12: ECL 10K typical electrical characteristics at $T_A=25°C$ (Reprinted by permission from *Analyses and Design of Digital Integrated Circuits* by D. A. Hodges and H. G. Jackson. Copyright © 1988 by McGraw-Hill.)

V_{OH}/V_{OL}	-0.9 V/-1.7 V	Fan out	10
V_{IH}/V_{IL}	-1.2 V/-1.4 V	Supply voltage	-5.2 V
NM_H/NM_L	0.3 V/0.3 V	Power dissipation per gate	25 mW
Logic swing	0.8 V	Propagation delay time	2 ns

For an ECL 100K, the gate delay is approximately 0.75 ns and the power dissipation per gate is about 40 mW. The result is a delay-power product of 30 pJ. For the ECL 10K, these numbers are 2ns, 25mW and 50pJ for gate delay, power dissipation, and delay-power product, respectively. While the ECL 10K series is slightly slower than the 100K, it is easier to use.

12.8.1 ECL 10K Basic Circuit

The basic gate circuit of the ECL is shown in Figure 12.25.

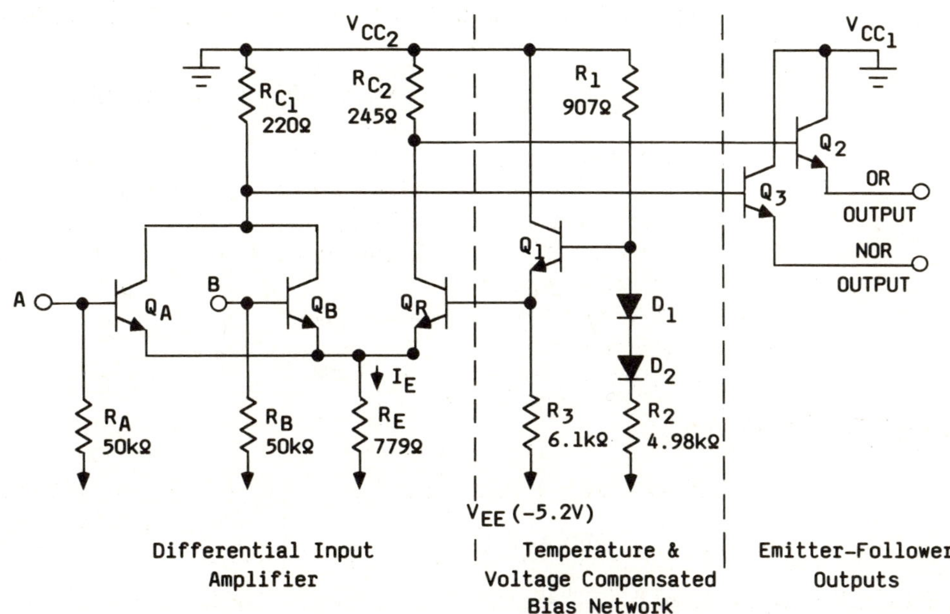

Figure 12.25: Basic ECL gate circuit

12.8.2 Voltage Transfer Characteristics

The simplified version of the ECL gate is shown in Figure 12.26.

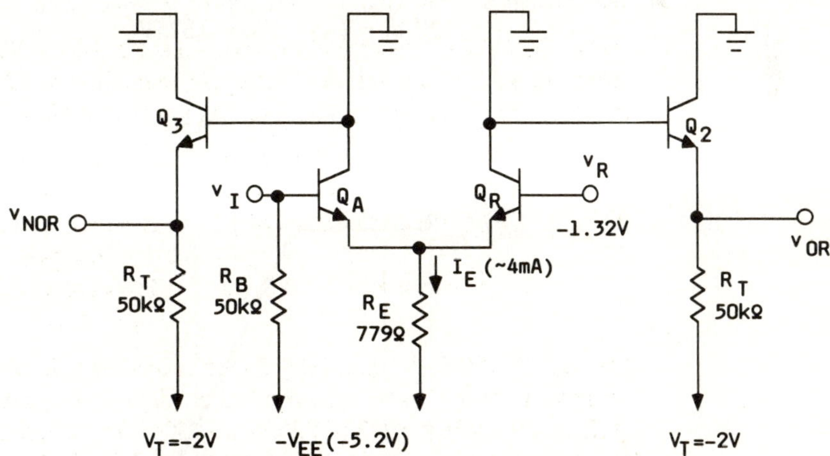

Figure 12.26: Simplified version of the ECL gate

The simplified schematic of Figure 12.26 aids in determining the OR and NOR transfer curves. It is assumed that the outputs are terminated as shown; that the B input is low, and therefore Q_B (see Figure 12.25) is off; that $V_{BE_{(ON)}} = 0.75$ V; and that β is high so that base currents are neglected. The result is the OR transfer characteristic of Figure 12.27 and the NOR transfer characteristic of Figure 12.28.

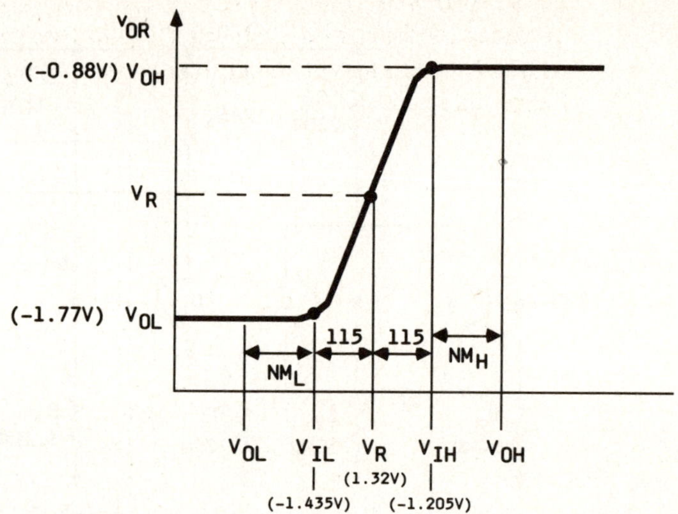

Figure 12.27: The transfer characteristic for the circuit of Figure 12.26

From the OR transfer characteristic, the output voltage corresponding to $V_I=V_R$ is approximately V_R. This is also approximately the midpoint of the logic voltage swing, since

$$\frac{V_{OL} + V_{OH}}{2} = -1.325 \text{ V} \tag{12.18}$$

This means that the output logic levels are centered around the midpoint of the input transition band. Mathematically,

$$NM_H = V_{OH} - V_{IH} = 0.325V \tag{12.19}$$

and

$$NM_L = V_{IL} - V_{OL} = 0.33V$$

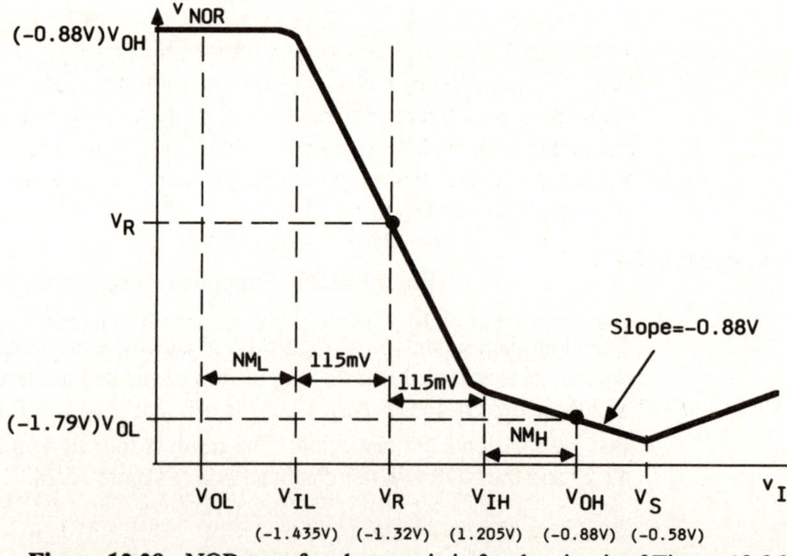

Figure 12.28: NOR transfer characteristic for the circuit of Figure 12.26

For the NOR transfer characteristic, the values of V_{IL} and V_{IH} are identical to those for the OR characteristic. For $V_I<V_{IL}$, Q_A is off and v_{NOR} can be determined by analyzing the circuit consisting of R_{C_1}, Q_3, and the 40-Ω termination resistor. For $V_I>V_{IH}$, Q_A is on and is conducting the entire bias current. The circuit reduces to that shown in Figure 12.29.

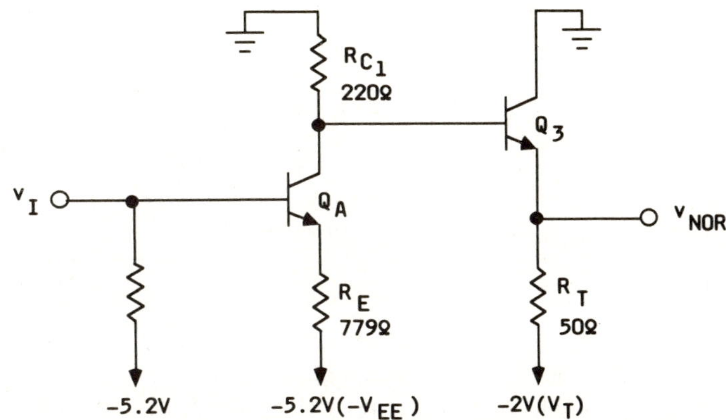

Figure 12.29: Circuit for finding V_{NOR} versus V_I for $V_I>V_{IH}$

Because R_{C_1} is slightly smaller than R_{C_2}, when $V_I=V_{TH}$, the result is an output voltage slightly higher than V_{OL}. However, when V_I equals the logic-1 value, i.e., V_{OH} = -0.888 V, V_{OL} will be equal to -1.77 V. As V_I exceeds V_{IH}, transistor Q_A operates in the active mode and the circuit of Figure 12.29 can be analyzed to determine the gain of the amplifier. The gain of the amplifier is the slope of the segment $V_{IL}<V_I<V_{IH}$. At $V_I>V_S$ Q_A saturates, and any further increases in V_I cause the collector voltage and V_{NOR} to increase.

Determination of the various voltages values specified in Figure 12.28 can be carried out by analyzing the circuit of Figure 12.29.

12.8.3 Fan-out

At low clock rates, the fan-out is on the order of I_C/I_B, or the current gain $\beta_F(\approx 100)$. Here, the output current of the gate is from emitter-followers, and the input current to the load gates is the base current of the nonsaturating current switch transistors. But, associated with each load gate is a finite load capacitance that must be charged and discharged as the driving gate changes state. This voltage change takes time and therefore limits the fan-out to about 10.

12.8.4 Propagation Delay Time

The small logic swing, together with the nonsaturation of the transistors in the normal mode of operation, leads to a propagation delay of 2 ns for the ECL 10K.

12.8.5 Signal Transmission

Because of the high speed of operation of the ECL, attention should be paid to the method of interconnecting the various logic gates. The connecting wires should be very short, in terms of the signal rise time. For the ECL 10K, 10 cm from end to end is recommended, but for the ECL 100K, 5 cm from end to end is suggested. If greater lengths are needed, transmission lines, possibly including twisted pairs or ribbon wires, will be necessary.

12.8.6 Power Dissipation

The power dissipation consists of the power consumed by the current switch and the reference voltage supply. The result is a power dissipation per gate of about 25 mV.

12.8.7 Wired-OR Capability

The emitter-follower output stage of the ECL allows an additional level of logic by wiring the outputs of several gates in parallel (see Figure 12.30).

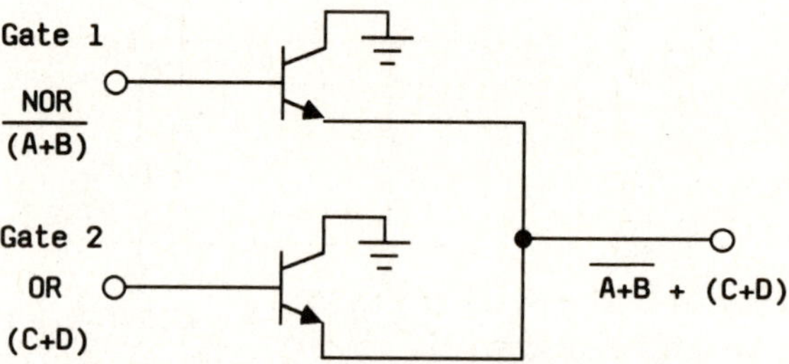

Figure 12.30: Wired-OR capability of ECL

12.8.8 ECL 100K Series

In the ECL 10K series, the reference voltage and output levels increase with temperature. Also, changes in the power supply voltage (-5.2 V) cause changes in V_R and V_{OL}. The ECL 100K series was designed to make the transfer characteristic almost independent of supply voltage and temperature variations. The schematic of a two-input OR/NOR gate is shown in Figure 12.31.

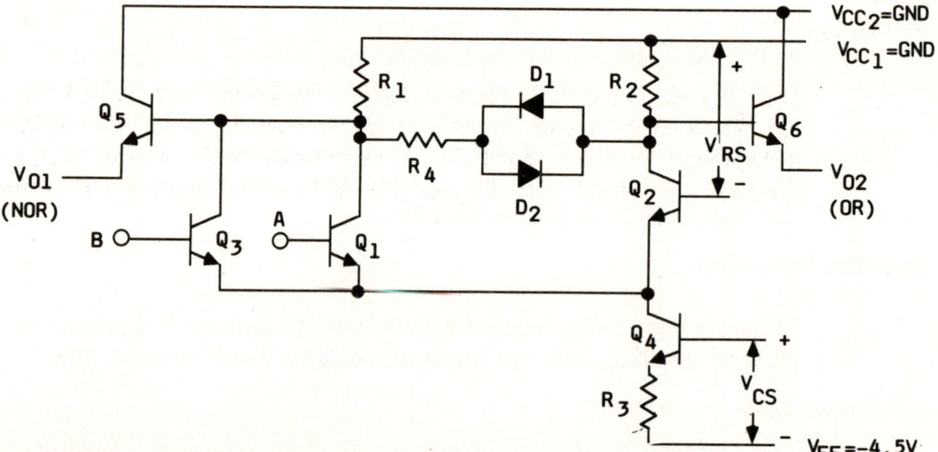

Figure 12.31: Schematic of ECL 100K two-input OR/NOR gate (current switch)

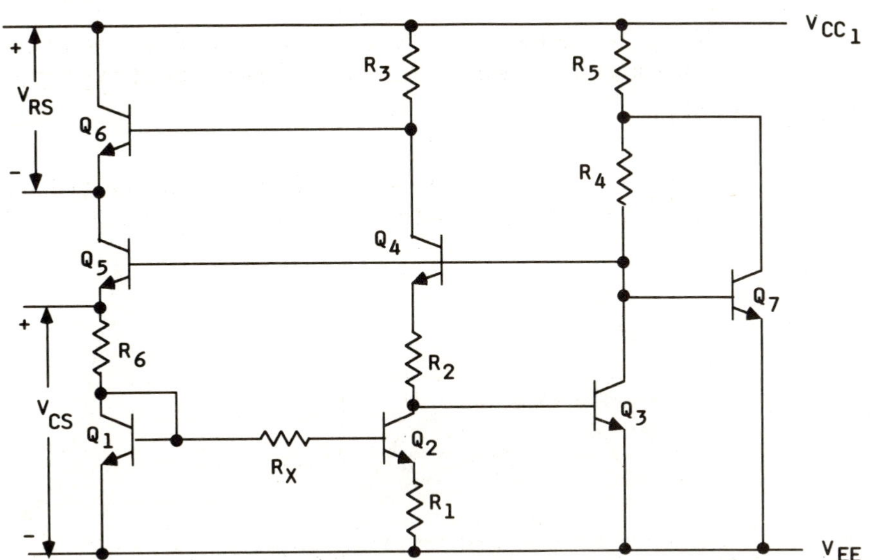

Figure 12.32: Bias Network for V_{RS} and V_{CS} in ECL 100K two-input OR/NOR Gate
(Reprinted by permission from *Analyses and Design of Digital Integrated Circuits*
by D. A. Hodges and H. G. Jackson. Copyright © 1988 by McGraw-Hill.)

The bias network for the reference voltages V_{RS} and V_{CS} is shown in Figure 12.32.
V_{RS} and V_{CS} are designed to be invariant to supply and temperature changes. The
voltage supply is reduced from -5.2 V to -4.5 V to reduce the power dissipation of the
circuit. In one work,[27] analysis of the circuits of Figures 12.31 and 12.32 shows that
for supply voltage variations, typical changes for V_R and V_{OL} are 10 mV/V and 15
mV/V, respectively. This compares with 150 mV/V and 250 mV/V for V_R and V_{OL},
respectively, for the ECL 10K. The temperature coefficients for the 10K series are 1.1
mV/°C for V_R, 0.6 mV/°C for V_{OL}, and 1.4 mV/°C for V_{OH}. For the 100K series, the
changes are less than 0.1 mV/°C for each of the parameters.

12.9 NMOS[21],[22]

12.9.1 NMOS Inverter

The inverter is the basic building block from which most MOS logic circuits are
developed. The inverter displays all of the essential features of MOS logic gates
except logic functions. The techniques developed for the inverter can be extended to
the NOR and NAND gates. Figure 12.33(a) shows the depletion-load inverter.

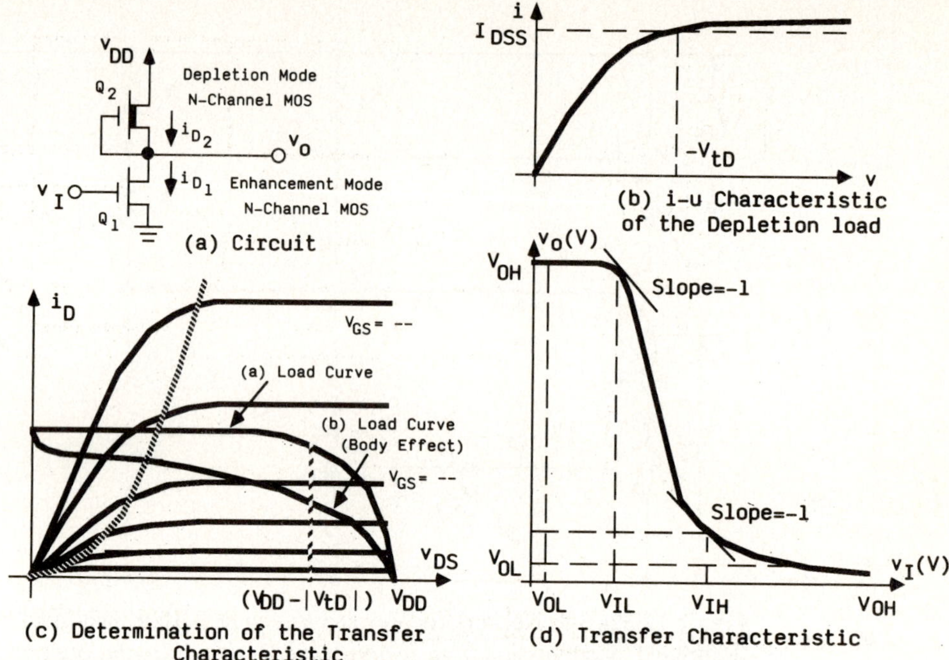

Figure 12.33: NMOS depletion-load inverter

Figure 12.33(b) shows the i-v characteristic of the depletion load. In Figure 12.33(c), the i-v load curve is superimposed on the $i_D - v_{DS}$ characteristic of the enhancement transistor Q_1. We locate the point V_{DD} on the v_{DS} axis and draw a mirror image of the i-v characteristic of the load device. Also shown in the figure is the load curve of the depletion load with the body effect (see Chapter 10 section 4 for a discussion of body effect) taken into account. Without the body effect, the depletion device behaves as a constant-current source over a wide range of v_{DS}. This means that the inverter transfer characteristic will be very sharp in the transition region. It also implies that a relatively large current is available to charge a load capacitance, resulting in a short t_{PLH}. However, the body effect causes the depletion load to depart from constant-current operation. Therefore, the inverter characteristics are not as good in practice as they are based on calculations alone. The noise margins can be calculated as

$$NM_L = V_{IL} - V_{OL} \tag{12.20}$$

and

$$NM_H = V_{OH} - V_{IH}$$

and the average static power is given by

$$P_D = \frac{1}{2} V_{DD} (I_{D_{2\text{HIGH STATE}}} + I_{D_{2\text{LOW STATE}}}) \tag{12.21}$$

Figure 12.33(d) shows the transfer characteristic of the inverter.

12.9.2 Dynamic Operation

Figure 12.34 is used to determine the dynamic operation of the depletion-load inverter in the presence of a load capacitance C. The propagation delay times t_{PLH} and t_{PHL}, shown in part (b), can be calculated by finding the average currents available to charge and discharge C.

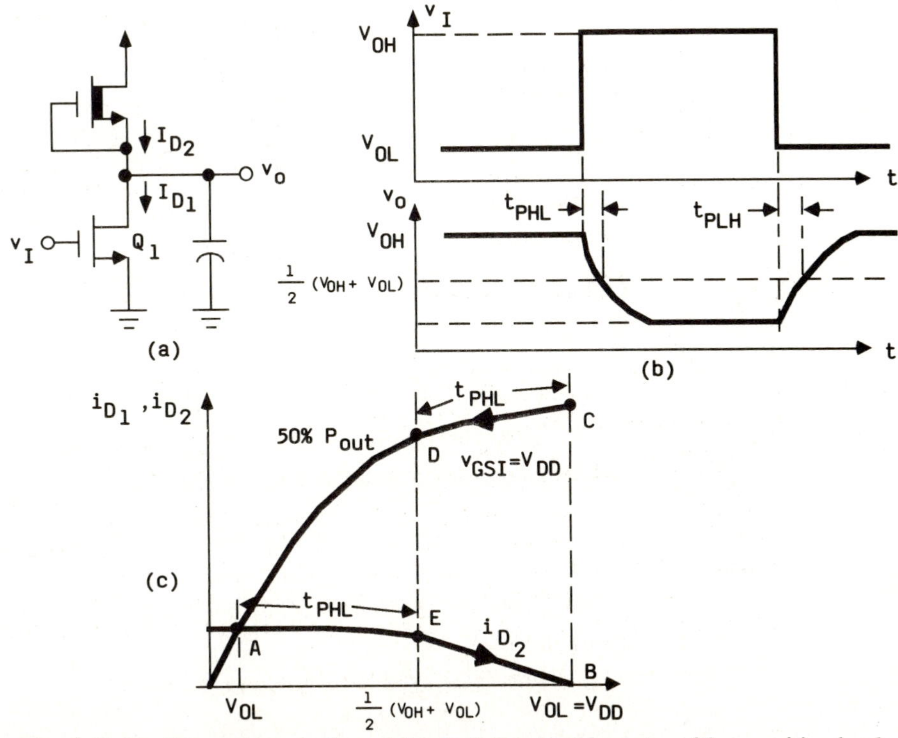

Figure 12.34: Dynamic operation of the depletion-load inverter with capacitive load
(Reprinted by permission from *Microelectronic Circuits*, Second Edition,
by A. S. Sedra and K. C. Smith. Copyright © 1987 by Holt, Rinehart & Winston.)

The total propagation delay is

$$t_p = \frac{1}{2}(t_{PHL} + t_{PLH}) \tag{12.22}$$

12.9.3 Power-Delay Product

Measured values of average power consumption ($P_{D(av)}$) and average delay (t_p) may be
used to calculate the PDP for any gate:

$$PDP = P_{D(av)}t_p = (\text{watts})\,(\text{sec}) = \text{joules} \tag{12.23}$$

Average power is the product of the supply voltage V_{DD} and the average supply current
$I_{D(max)}/2$. The delay times are inversely proportional to $I_{D(max)}$ and proportional to
$C(V_{OH} - V_{OL})$. Thus,

$$PDP = \frac{C(V_{OH} - V_{OL})V_{DD}}{2} \tag{12.24}$$

12.9.4 NMOS Logic Circuits

Figure 12.35 shows a two-input NOR gate in depletion-load NMOS technology.

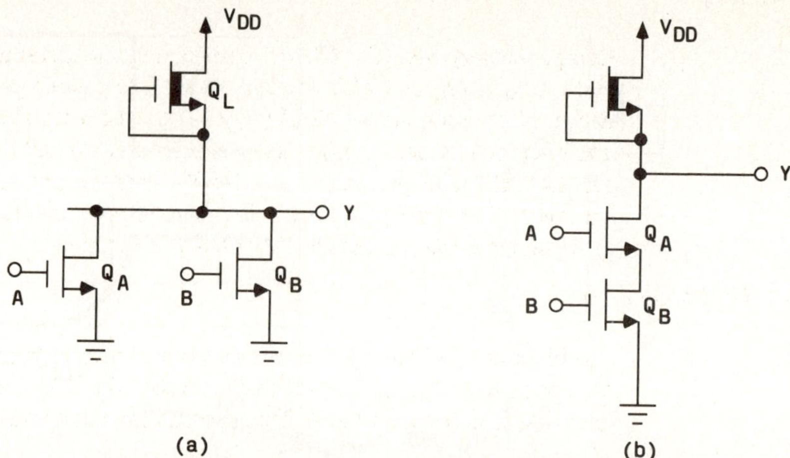

(a)   (b)

Figure 12.35: Logic gates in depletion-load NMOS logic

If the voltage at any of the input terminals of the gate is high (V_{DD}), then the associated transistor will be on and the output voltage will be low (V_{OL}). The output voltage will be high if the two inputs are low. Then, both input transistors will be off and $V_Y = V_{DD}$. The operation is described by the Boolean expression

$$Y = \overline{A}\,\overline{B} \tag{12.25}$$

or

$$Y = \overline{A+B} \tag{12.26}$$

The input transistors are matched to the inverter transistor. If both inputs are simultaneously high, then the output voltage will be lower than V_{OL}. The gate fan-in can be increased by adding additional input transistors.

For the NAND gate, the output will be low only when both Q_A and Q_B are on or when both inputs are high. Therefore,

$$\overline{Y} = AB \tag{12.27}$$

or

$$Y = \overline{AB} = \overline{A} + \overline{B} \tag{12.28}$$

When both Q_A and Q_B of the NAND gate are conducting, the effective channel length between the output node and ground is twice that of the inverter transistor. In order to keep the output voltage at the value of V_{OL} obtained in the inverter, each of the input transistors in the NAND gate should have double the width of the inverter transistor. Then the two series conducting transistors will have the same effective W/L ratio as the inverter transistor. If we have N inputs, the width of each of the input transistors should be N times that of the inverter transistor. The silicon area required by a NAND gate is greater than that required by a NOR gate having the same number of inputs.

The NMOS logic gate can be very densely packed on an IC chip, and this permits very high levels of integration. The most important application of NMOS is in the design of VLSI circuits such as microprocessors and RAMS. The NMOS low load-driving capability makes it impractical for conventional digital system design, and thus, NMOS logic is not available as SSI or MSI "off-the-shelf" components, as is CMOS or TTL.

12.10 CMOS[23],[24]

Complementary MOS, or CMOS, logic circuits are available as standard SSI (small-scale integration) and MSI (medium-scale integration) packages and are used in the design of general-purpose VLSI (very large scale integration) circuits such as memory and microprocessors. CMOS is favored for custom and semicustom VSLI. We shall see that all static parameters of CMOS inverters are superior to those of NMOS inverters. The price paid for the improved characteristics is increased complexity in the manufacturing process and increased area per circuit.

12.10.1 The CMOS Inverter

The inverter (see Figure 12.36) consists of two matched enhancement-type MOSFETS. One transistor, Q_N, has an n-channel and the other, Q_P, has a p-channel. The body of each device is connected to its source; therefore, there is no body effect.

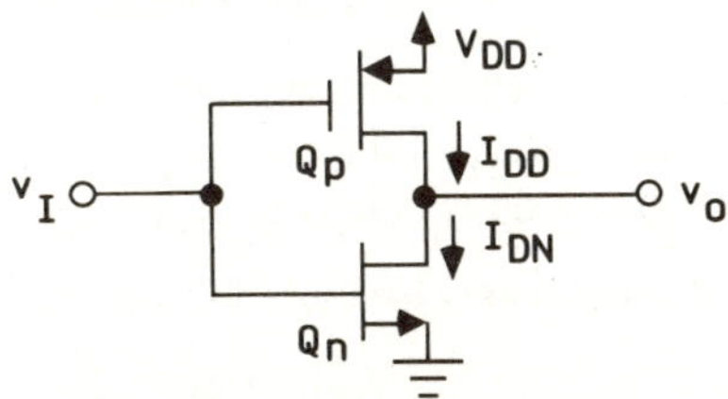

Figure 12.36: CMOS inverter

12.10.2 DC Analysis

For circuit operation, the n-channel device Q_n is considered the driving transistor and the p-channel device Q_p is considered to be the load. Since the circuit is symmetric, the reverse case would yield the same results.

The two extremes of operation, i.e., when V_I is V_{DD} and when V_I is 0, are shown in Figure 12.37.

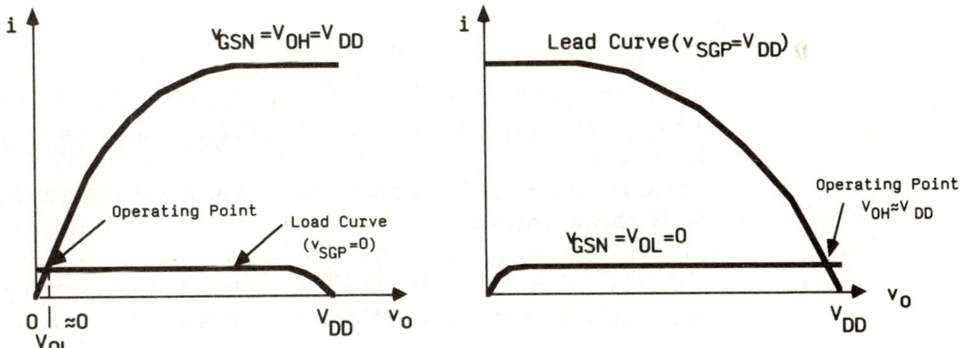

Figure 12.37: Operation of the CMOS inverter

The voltage transfer characteristic (VTC) for the CMOS inverter is shown in Figure 12.38.

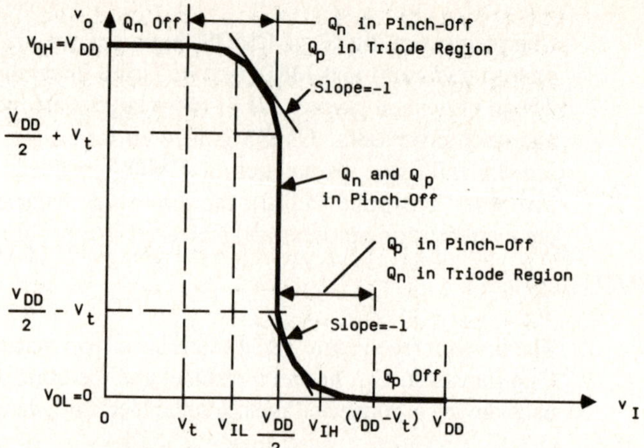

Figure 12.38: VTC for the CMOS inverter

The transfer characteristic has five distinct regions corresponding to Q_n off, Q_n in pinch-off and Q_p in the triode region, Q_n and Q_p both in pinch-off, Q_p in pinch-off and Q_n in the triode region, and Q_p off.

The critical points for the VTC are as follows:

$$V_{OH} = V_{DD} \tag{12.29}$$

$$V_{OL} = 0 \tag{12.30}$$

$$V_{IH} = 1/8(5V_{DD} - 2V_t) \tag{12.31}$$

$$V_{IL} = 1/8(3V_{DD} + 2V_t) \tag{12.32}$$

The noise margins can then be determined as

$$\begin{aligned} NM_H &= V_{OH} - V_{IH} \\ &= 1/8(3V_{DD} + 2V_t) \end{aligned} \tag{12.33}$$

and

$$\begin{aligned} NM_L &= V_{IL} - V_{OL} \\ &= 1/8(3V_{DD} + 2V_t) \end{aligned} \tag{12.34}$$

The symmetry of the VTC (Q_n and Q_p match) yields equal noise margins.

When the inverter, loaded by a capacitor C, is switched, the dynamic power dissipation can be shown to be

$$P_D = fCV_{DD}^2 \tag{12.35}$$

where f is the frequency at which the inverter is switched.

12.10.3 Dynamic Operation

In Section 12.9.1 we observed that the NMOS inverter circuit has unequal turn-on and turn-off times. In contrast, the CMOS inverter features equally fast turn-on and turn-off times in the presence of capacitor loads.

The delay-power product for CMOS is the product of the dynamic power dissipation and the propagation delay time. The delay-power is directly proportional to the rate of switching and can be reduced by operating at low rates. The delay-power can also be reduced by reducing the load capacitance and power-supply voltage. Typical values for the delay-power product are 1 pJ for VLSI circuits to approximately 10 pF for SSI circuits.

12.10.4 CMOS Gate Circuits

NOR and NAND gates in CMOS are shown in Figure 12.39. Two transistors must be added for each additional input. This requirement results in a lower circuit density for CMOS as compared to NMOS.

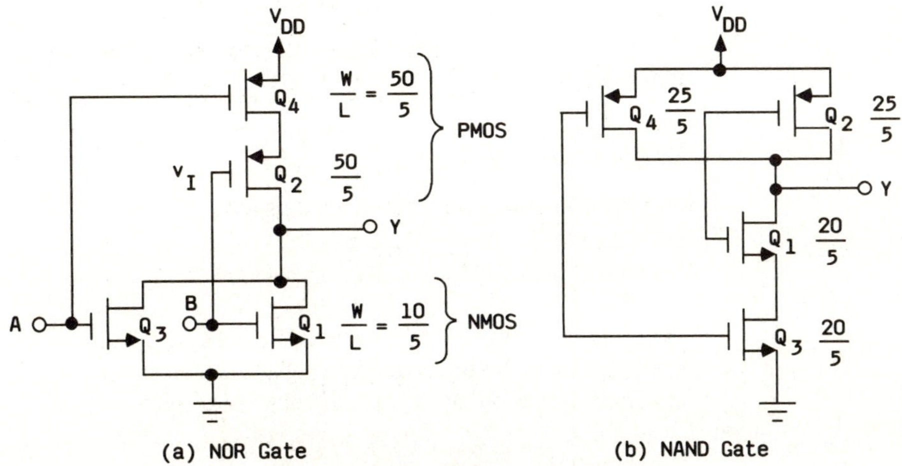

Figure 12.39: CMOS gate circuits

The output of the circuit in Figure 12.39(a) will be high (V_{DD}) if both Q_2 and Q_4 are simultaneously on. This means that A and B must be simultaneously low. We have

$$Y = \overline{A}\,\overline{B} \tag{12.36}$$

or

$$Y = \overline{A + B} \tag{12.37}$$

This is the NOR logic function.

The output of the circuit in Figure 12.39(b) will be low (0 V) only when Q_1 and Q_3 are both on. This occurs only when A and B are both high.

The result is

$$\overline{Y} = AB \tag{12.38}$$

or

$$Y = \overline{AB} \tag{12.39}$$

The result, then, is a NAND logic function.

Table 12.13 contains a summary of the advanced CMOS logic performance characteristics, along with a summary of the same characteristics for the 74LS series of TTL circuits for comparison.

Table 12.13 Bipolar and CMOS Logic Performance Characteristics @$T_A = 25°C$ (Reprinted by permission from *Analyses and Design of Digital Integrated Circuits* by D. A. Hodges and H. G. Jackson. Copyright © 1988 by McGraw-Hill.)

		Series			
Parameter	74LS	74HC	74HCT	74AC	74ACT
min V_{OH}/max V_{OL} CMOS load	2.7/0.5	4.4/0.1	4.4/0.1	4.4/0.1	4.4/0.1
TTL load	2.7/0.5	4.0/0.3	4.0/0.3	3.9/0.3	3.9/0.3
min V_{IH}/max V_{IL}	2.0/0.8	3.1/0.9	2.0/0.8	3.1/1.3	2.0/0.8
min I_{OH}/min I_{OL}, μA	-0.4/8	±4	±4	±24	±24
min I_{IH}/min I_{IL}, μA	20/-400	±0.1	±0.1	±0.1	±0.1
typical t_p, ns	10	10	10	5	5
typical dc P_D/gate	2 mW	2.5 μW	2.5 μW	2.5 μW	2.5 μmW

For 74LS: $V_{CC} = 5$ V, $C_L = 15$ pF
For CMOS Series: $V_{CC} = 4.5$ V, $C_L = 50$ pF

Maximum fan-out of a CMOS gate is limited by the deterioration in the gate's dynamic response due to the increased load capacitance in the circuit. For TTL, the maximum fan-out is limited by the degradation in noise margins.

12.11 Summary

The two general topics of digital systems and digital electronics were discussed in this chapter. Under digital systems, we considered binary logic and gave the truth tables for the logical operations AND, OR, and NOT. Also given was a table of digital logic gates along with the algebraic function and truth table for each gate. Karnaugh maps to simplify logic functions were introduced, and we reviewed the combinatorial logic concepts of the adder, subtractor, decoder, multiplexer and gate array. Finally, we considered sequential logic, reviewing storage elements such as flip-flops and discussing methods of triggering them. Registers and counters were introduced and the properties of synchronous and asynchronous counters were examined.

Under the topic of digital electronics, we reviewed four kinds of digital logic gates: TTL, ECL, NMOS, and CMOS. For TTL and ECL gates, we analyzed each circuit completely by studying their voltage transfer characteristics. NMOS and CMOS gates were studied by analyzing the basic inverter circuit. The noise margin, power dissipation, fan-in and fan-out, propagation delay, and power-delay product of each of the four kinds of gate were discussed, and tables summarizing typical electrical characteristics at room ambient temperature for TTL, ECL, and CMOS were presented. Bipolar and CMOS logic performance were compared.

TTL has an extensive list of digital functions and is currently the most popular logic family. ECL is used in systems requiring high-speed operations. NMOS is used in circuits requiring high component density, and CMOS is used in systems requiring lower power consumption.

LAB 12L1

TITLE: Fan-out and Propagation Delay for TTL (Black Box Approach)

OBJECTIVE: This laboratory exercise examines the concepts of fan-out and propagation delay for several versions of TTL from a black box perspective. Fan-out and noise margin can be defined in terms of the following eight parameters: V_{OH}, V_{OL}, V_{IH}, V_{IL}, I_{OH}, I_{OL}, I_{IH}, and I_{IL}. In the prelab, these parameters are determined from data sheets on the two-input quad NAND gate (74XX00) for several versions of TTL. The propagation delays for several TTL versions of the 74XX00 are also determined in the prelab. In the laboratory procedure section these parameters are measured and compared to the worst-case numbers from the data sheets. Finally, a short design problem that is worked out in the prelab is verified in the laboratory.

EQUIPMENT AND COMPONENTS:
Oscilloscope   (dual channel)
Power supply (2 0-V to 5-V variable)
Pulse generator (0-V to 5-V variable)
Breadboard
Multimeter
TTL two-input quad NAND gate
  7400
  74L00
  74LS00
  74ALS00

PRELABORATORY:

1.   From the data sheets, answer the following questions concerning the 74XX00. (XX indicates all four versions).

   a.   Draw the symbol for the 74XX00 two-input NAND gate in accordance with the ANSI/IEEE standards. Also, draw the function table for the device.

   b.   List the eight voltage and current parameters and two propagation delay parameters for all four of the 74XX00 gates. Be sure to include any special operating conditions that may be required when making these measurements.

2.   Use the propagation delay information from Question 1 to answer the following question:

   a.   Complete the timing diagram for the circuit of Figure 12L1.1. Assume worst case propagation delays for t_{PLH} and t_{PHL}. Also, assume that U_1 is a 7400, U_2 is a 74L00, U_3 is a 74LS00, and U_4 and U_5 are 74ALS00s. Define the logic function performed by the circuit.

3.   A digital system was designed many years ago using standard TTL (7400 series). Suppose that, because this unit has always sold well and because you have a limited engineering staff, the system was never upgraded. However, now you want to design a new board for the system using advanced low-power Schottky (ALS) parts. One of your ALS NAND gates must interface with five standard TTL NAND gate inputs. Will the circuit of Figure 12L1.2 work? If not, how could the problem be solved using the ALS and standard TTL parts? Note that the schematic of Figure 12L1.2 shows only the relevant components.

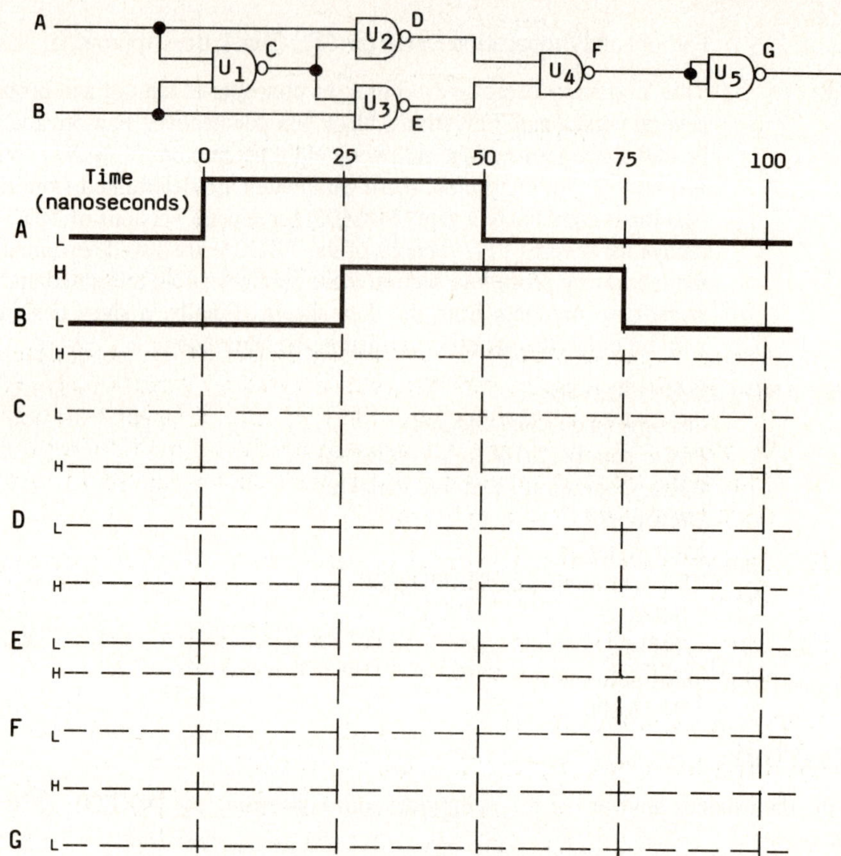

Figure 12L1.1: Circuit and timing diagram of Part 2

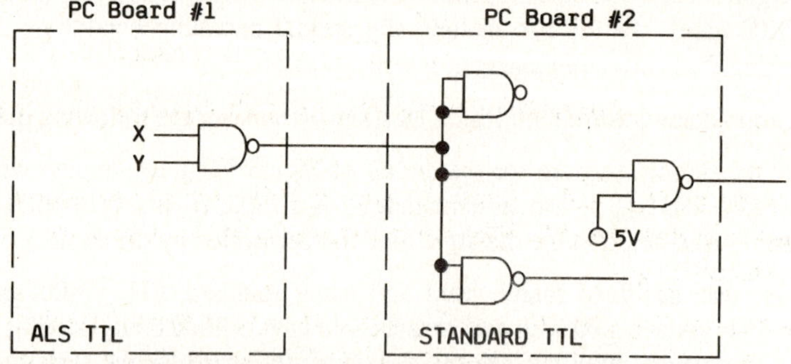

Figure 12L1.2: Circuit for design problem

Laboratory Procedure:

1. Measure the eight voltage and current parameters and two propagation delay parameters for the four versions of TTL 74XX00 NAND gates.

    a. Measure I_{IH} and I_{IL}.

       (1) Build the circuit of Figure 12L1.3 using a 74ALS00.

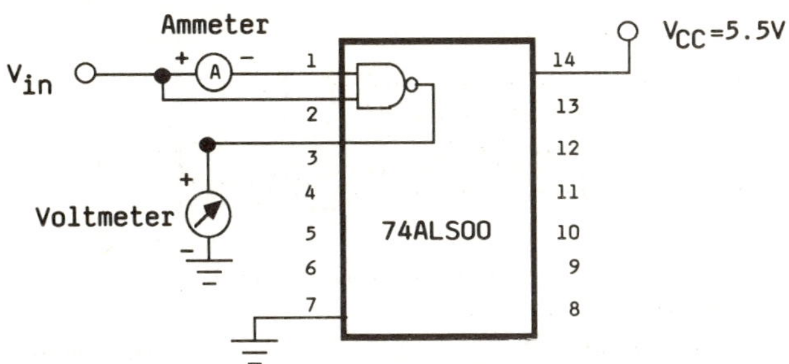

Figure 12L1.3: Circuit to measure I_{LH} and I_{IL}

(2)   To measure I_{IH}, set V_{in} to 2.7 V and measure the current flow through the ammeter. Verify that the voltage at pin 3 is less than or equal to 0.5 V. Record the current flowing into pin 1. Then move the ammeter to pin 2 so that the value of I_{IH} can be measured and recorded. To measure I_{IL}, set V_{IN} to 0.4 V and measure the current flow in pin 1 and then pin 2. Also, verify that the voltage at pin 3 is greater than or equal to 3.5 V. Record these values of current as two measurements of I_{IL}.

(3)   Compare your measured values of I_{IL} and I_{IH} to the worst case values given in the data sheets.

(4)   Repeat Parts (1)-(3) for the three other versions of TTL, with the adjustment that for standard TTL, L, and LS, the voltage at pin 14 is to be set to 5.25 V.

b.   Measure V_{OH}, I_{OH}, V_{OL}, and I_{OL}.

(1)   Build the circuit of Figure 12L1.4 using a 74ALS00.

Figure 12L1.4: Circuit to measure V_{OH}, I_{OH}, V_{OL}, and I_{OL}

(2)   To measure V_{OH} and I_{OH}, set V_{in} equal to 0 V. Adjust the value of R until the current flowing in R is -0.4 mA. Measure the voltage at pin 3. Record the value of v_o as V_{OH} and the current as I_{OH}.

(3)   To measure V_{OL} and I_{OL}, set V_{in} equal to 3 V. Adjust R so that I is equal to 8 mA. Measure and record the voltage at pin 3 as V_{OL} and the current I as I_{OL}. These indicate the drive capability of the NAND gate.

(4)   Compare these values to worst case values for V_{OH}, I_{OH}, and I_{OL}, on the data sheet.

(5)   Repeat Parts (1)-(4) for the other three versions of TTL, with the adjustment that for standard TTL and L, R for an I_{OL}, is to be set equal to 16 mA when measuring V_{OL}.

c.   Measure V_{IH} and V_{IL}.

(1)   Build the circuit of Figure 12L1.5.

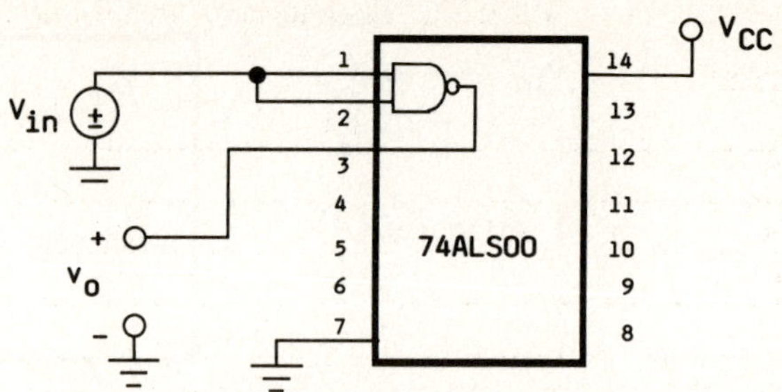

Figure 12L1.5: Circuit to measure V_{IH} and V_{IL}

(2)   Start with $v_{in} = 0$ V, and measure V_o. Set V_{CC} equal to 4.5 V. Increase the value of v_{IN} slowly until v_o changes to a low state. The value of V_{in} corresponds to $V_{IH}(min)$ for your component.

(3)   Start with $V_{in} = 5$ V and set $V_{CC} = 5.5$ V. Slowly reduce the value of V_{in} until the output goes from a low to a high. When this occurs the value of V_{in} is $V_{2L(max)}$ for your device.

(4)   Compare these results with the worst case values listed on the data sheet.

(5)   Repeat Parts (1)-(4) for the three other versions of TTL, with the adjustment that for standard TTL, L, and LS, V_{CC} is to be kept at 5 V.

d.   Measure the propagation delays (t_{PLC} and t_{PHL}). Wire the circuit of Figure 12L1.6. The diagram represents 16 74ALS00 gates connected in series. The four gates contained within each of the four packages are connected in series, and the four packages themselves are connected in series. The result is a delay of 16 gates from input to output. Connect a pulse generator to the input NAND gate as shown, and display the output and input signals on the oscilloscope. Divide the measured delays (t_{PLH} and t_{PHL}) by 16 to get an average delay per gate and compare the figure with that given on the data sheets.

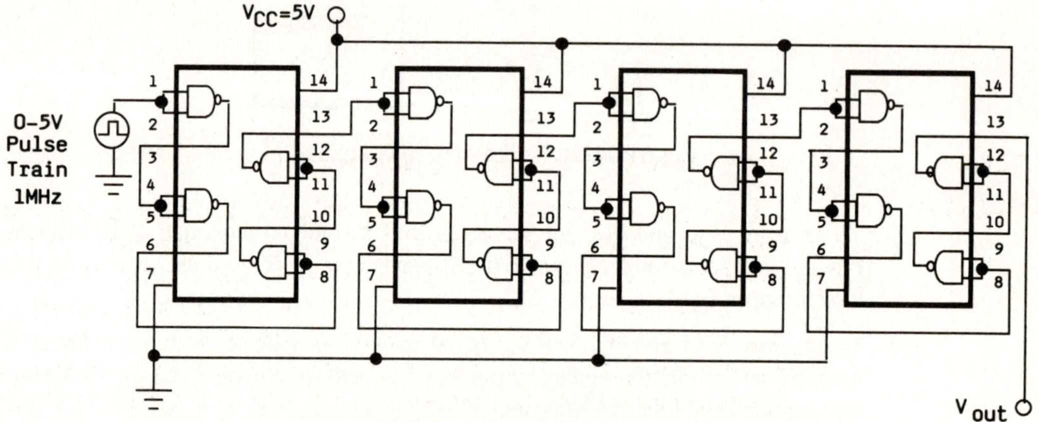

Figure 12L1.6: Circuit to measure t_{PLC} and t_{PHC}

e.   Compile all the data.
Complete Table 12L1.1 with your data.

Table 12L1.1: Indicate Which Family

Parameter	Max	Min	Units
V_{IH} (V)			
V_{IL} (V)			
V_{OH} (V)			
V_{OL} (V)			
I_{IH} (mA)			
I_{IL} (mA)			
I_{OH} (mA)			
I_{OL} (mA)			
t_{PLH} (ns)			
t_{PHL} (V)			
NM_{H} (V)			
NM_{L} (V)			

   f. Design problem.

      (1) Build the circuit shown in Figure 12L1.2. Test to see whether the outputs are the ideal values.

      (2) Would the circuit work if all the devices were at their worst case values?

      (3) Test your redesign. Compare the noise margins of both versions. Which is better? Why?

DISCUSSION:

CONCLUSIONS:

LAB 12L2

TITLE: Introduction to Combinational Digital System

OBJECTIVE: This laboratory exercise serves as an introduction to the design and analysis of combinational digital circuits using simple AND, OR, and NOT logic functions. Truth tables and K-maps are needed to express the functionality of some medium-scale integration (MSI) devices. The student will use these tools to analyze existing MSI functions and determine their Boolean expressions. Also, the student will receive a verbal description of an MSI function and use truth tables and K-maps to express this function and then implement the function with specified SSI gates.

The laboratory exercise will adhere to the 1984 ANSI/IEEE symbology adopted by industry, government, and professional organizations to represent logic functions. The availability of some version of TTL gates is assumed; however, these functions are also available in other logic families (e.g., CMOS). Any version is acceptable, as long as appropriate supply voltage levels are used. Avoid mixing logic families, however, unless you confirm compatibility between parts.

EQUIPMENT AND COMPONENTS:
> Digital multimeter (DVM)
> 5-V power supply
> Breadboard

PRELABORATORY:

1. Referring to a TTL data book, determine the functionality and DIP (dual-in-line package) pinouts for the following parts (include the truth table for each):

   a. 74LS00

   b. 74LS02

   c. 74LS04

   d. 74LS10

   e. 74LS20

   f. 74LS30

   g. 74LS40

   h. 74LS86

2. Consult a TTL data book, and record the following information about the 74LS139:

   a. Draw the 1984 ANSI/IEEE standard diagram for this device.

   b. Draw the truth table or function table for the device.

   c. Draw and label the pinout for this device when packaged in a 16-pin DIP.

   d. What is the difference between the military temperature range and the commercial temperature range?

   e. Using AND, OR, and NOT gates (2-input, 2-input, and 1-input functions, respectively), design a circuit that performs the operation of a 74LS139. Assume your device is always enabled, so that an enable input is not necessary. Since the 74LS139 is a dual device, you need only implement half of it.

   f. Repeat Part (e) using only NAND gates.

3. A four-bit binary adder with carry-in and carry-out accepts two four-bit binary numbers along with a carry-in and outputs the four-bit sum and a carry-out. A full adder accepts two binary variables and a carry-in, if necessary (say X, Y, and C_{IN}, respectively), and produces two outputs. One output variable is the arithmetic sum of the three inputs and is called S. The remaining output, C, is set if the arithmetic sum of the three inputs results in a carry into the next binary digit.

   a. Draw a truth table describing the full adder.

   b. Design a full adder using only NAND gates. **Hint:** Analyze and use the circuit of Figure 12L2.1 as part of your design.

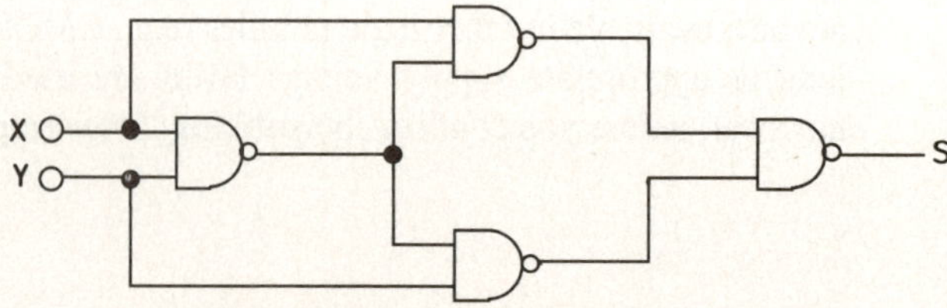

Figure 12L2.1: Use in full adder design

Laboratory Procedure:

1. Identification of SSI components. Build a circuit to verify the truth table for each device listed in Part 1 of the prelab. For each device, a high input should be greater than 3 V and less than 5 V. Record not only a high or low at the outputs, but also the actual voltage. If a DIP has more than one of a particular function, you need only look at one device within the package. See Figure 12L2.2 as an example.

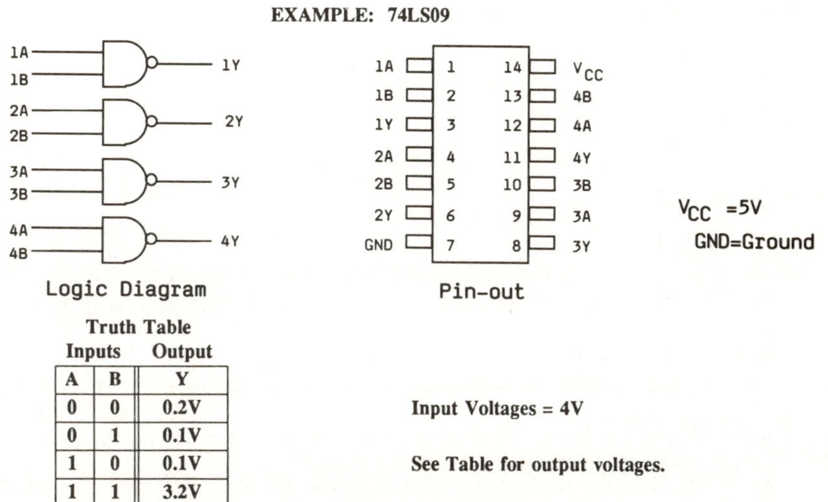

Truth Table

Inputs		Output
A	B	Y
0	0	0.2V
0	1	0.1V
1	0	0.1V
1	1	3.2V

Figure 12L2.2: Part 1a

2. Analysis of the 74LS139. Build the circuit of Figure 12L2.3. Set the switches to the eight possible combinations, and record the voltages of the four outputs in Table 12L1.1.

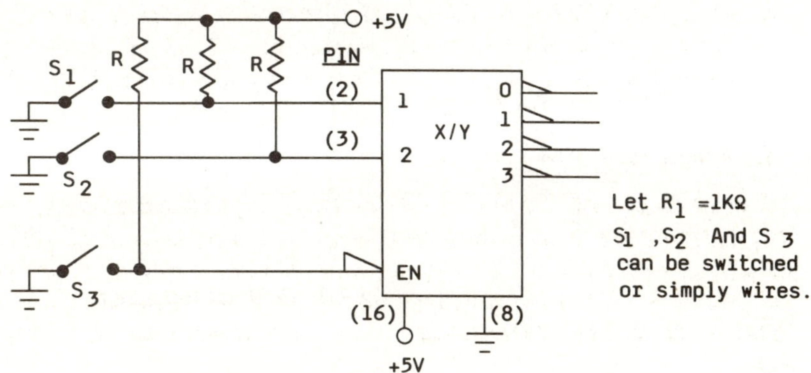

Figure 12L2.3: Circuit for analysis of 74LS139

Table 12L2.1: Data for analysis of 74LS139

Inputs			Outputs			
S_1	S_2	S_3	0	1	2	3
0	0	0				
0	0	1				
0	1	0				
0	1	1				
1	0	0				
1	0	1				
1	1	0				
1	1	1				

Record the actual voltage at the output, and determine the response as high or low. What happens at the output when S_3 is open? Build your design using AND, OR, and NOT gates.

(1) Set the inputs of your design in a way similar to that of Part 2. Do not let any gate input float. That is, inputs should be connected to ground for a low or connected to 5 V through a 1 kΩ resistor for a high.

(2) Compare your discrete design with the 74LS139.

3. Three-bit full adder.

   a. Build a full adder using only NAND gates. The block diagram in Figure 12L2.4 represents the full adder. Build this circuit and complete the accompanying table.

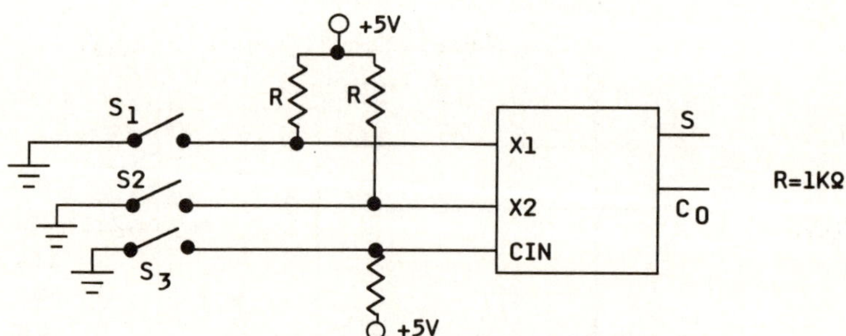

Figure 12L2.4: Three-bit full adder

Table 12L2.2: Data for three-bit full adder

	Inputs		Outputs	
X_1	Y_1	C_{IN}	S	C_0
0	0	0		
0	0	1		
0	1	0		
0	1	1		
1	0	0		
1	0	1		
1	1	0		
1	1	1		

b. Build and verify a three-bit full adder by building three full adders and connecting them as shown in Figure 12L2.5.

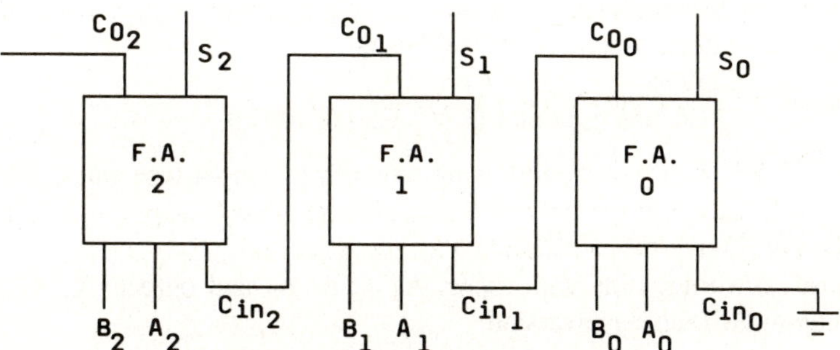

Figure 12L2.5: Three-bit full adder

DISCUSSION:

CONCLUSIONS:

---

*   The 0's and 1's in the table represent ground and 5 V, respectively, applied to the inputs. A "0" corresponds to a positive logic low and a "1" to a positive logic high.

LAB 12L3

TITLE: Designing with Digital Function Blocks (Implementation in TTL SSI Gates)

OBJECTIVE: During the past 10 years, development tools and the increased understanding of certain fabrication techniques have led to the proliferation of programmable logic arrays (PLAs) and gate arrays. These monolithic devices contain simple building blocks that are similar in complexity and function to an SSI type of gate. Both PLAs and gate arrays get the designer into the mode of minimizing gate count rather than package count. The major difference in designing with gate arrays rather than discrete gates or MSI packages is that the entire circuit is contained on a single-chip, resulting in a smaller, less expensive circuit board and a much more reliable design.

This laboratory addresses this design approach, but uses SSI and TTL gates to implement the combinational MSI blocks required to design a large digital system. TTL gates are used rather than gate arrays because of their popularity, easy access, and low cost.

EQUIPMENT AND COMPONENTS:
Power supply (2, 0 V-5 V, variable)
Breadboard
Oscilloscope (5 V, 500 mA)
DVM
Function generator (5-V pulses at 1 kHz)

PRELABORATORY: In each of the following, use only two input-type gates. All designs should be minimized.

1. The Binary-to-BCD code converter.

    a. Construct a truth table with inputs $A_3 A_2 A_1 A_0$ and outputs $X_4 X_3 X_2 X_1 X_0$ that describe a binary-to-BCD code conversion.

    b. Draw a schematic of this converter implemented with AND, OR, and NOT gates.

    c. Compare you design with that of the 74ALS485 in terms of function features and gate count.

2. The full adder.

    a. Construct a truth table with inputs $A_1 A_0 B_1 B_0$ and a carry-in, C_{IN}, and outputs $D_1 D_0$ and carry-out, C_{OUT}.

    b. Draw a schematic of this two-bit full adder using AND, OR, and NOT gates.

    c. How could your design be expanded to be a four-bit adder such as the 74ALS83?

3. The multiplier.

    a. Construct a truth table with inputs $A_1 A_0$ and $B_1 B_0$ and outputs $D_3 D_2 D_1 D_0$ such that the output is the product of $A_1 A_0$ and $B_1 B_0$.

    b. Draw a schematic of this two-by-two bit multiplier using AND, OR, and NOT gates.

    c. Assuming you are using LS types of gates (low-power Schottkys). What is the maximum amount of time between inputs and stable outputs for your circuit? Compare this to the 74LS261.

4. The three-to-eight line decoder.

    a. Construct a truth table with inputs A_2, A_1, A_0 and outputs $Y_7, Y_6, Y_5, Y_4, Y_3, Y_2, Y_1, Y_0$ such that Y_7 is low only when $A_2 = A_1 = A_0 =$ high with all the other Y's high. Also, Y_6 is low only when $A_2 = A_1 =$ high and $A_0 =$ low with all the other Y's high. This binary pattern continues until $Y_0 =$ low when $A_2 = A_1 = A_0 =$ low with all the other Y's high.

    b.   Implement this design in NAND gates and draw the schematic.

    c.   Calculate the maximum power consumption of your design, assuming it was implemented using ALS technology and compare to that of the 74ALS138.

5.   The four-to-one line multiplexer.

    a.   Construct the truth table for a device with inputs A_0 B_0 C_0 D_0 and select lines S_1 S_0. The device has one output, Y_0, which equals one of the input lines, depending on the bit pattern of $S_1 S_0$. Each bit pattern corresponds to selecting one of the four inputs to appear at the output.

    b.   Draw a schematic of your design using NOR gates.

    c.   Compare your design with the 74ALS153 in terms of propagation delay, assuming your design is done using ALS technology.

Laboratory Procedure:

1.   Binary-to-BCD code converter.

    a.   Build the converter you designed in Part 1 of the prelab.

    b.   Verify your design by completing the truth table of Table 12L3.1.

Table 12L3.1: Truth table for the binary-to-BCD code converter

A_3	A_2	A_1	A_0		X_4	X_3	X_2	X_1	X_0
0	0	0	0						
0	0	0	1						
0	0	1	1						
0	1	0	0						
0	1	0	1						
0	1	1	0						
0	1	1	1						
1	0	0	1						
1	0	1	0						
1	0	1	1						
1	1	0	0						
1	1	0	1						
1	1	1	0						
1	1	1	1						

2.   The full adder.

    a.   Build the two-bit full adder you designed in Part 2 of the prelab.

    b.   Confirm your design by showing the sum of the following numbers:

        (1)   2, 2

        (2)   0, 0

        (3)   3, 3

        (4)   2, 3

3. The multiplier.

    a. Construct the two by two multiplier you designed in Part 3 of the prelab.

    b. Verify the design by showing the product of the following numbers:

        (1) 2, 2

        (2) 0, 0

        (3) 3, 3

        (4) 2, 3

4. The three-to-eight line decoder.

    a. Construct the decoder you designed in Part 4 of the prelab.

    b. Complete Table 12L3.2 to confirm your design. Also include in the table the power consumption of the circuit. This can be determined by multiplying $V_{CC} = 5V$ by I_{CC} (the current flow from the supply for each condition).

Table 12L3.2: Data for the three-to-eight line decoder

Power ($V_{CC} \times I_{CC}$)	A_2	A_1	A_0	Y_0	Y_1	Y_2	Y_3	Y_3	Y_5	Y_6	Y_7
	0	0	0								
	0	0	1								
	0	1	0								
	0	1	1								
	1	0	0								
	1	0	1								
	1	1	0								
	1	1	1								

5. The four-to-one line multiplexer.

    a. Build the circuit you designed in Part 5 of the prelab.

    b. Construct a table and confirm the operation of your design.

DISCUSSION: Include a comparison of the performance of your design and that of the TTL MSI functions for the three-to-eight decoder and four-to-one line multiplexer in terms of speed and power consumption. Use the worst case times and powers from the data sheets for the SSI parts you used in your design and for a comparable MSI-to-TTL part.

CONCLUSIONS:

LAB 12L4

TITLE: The Flip-Flop (A Storage Device)

OBJECTIVE: Without the ability to retain the values of digital information, one could neither store intermediate values of a mathematical operation nor electronically store a program that is ready for execution. In other words, an electronic computer is not possible without a storage element.

The basic storage device that holds digital data is the flip-flop. This laboratory exercise examines the NAND and NOR gate implementation of the flip-flop and the limitations of such a simple device.

In virtually all digital systems, feedback around a flip-flop is required to perform various functions (e.g., the function of a counter or state machine). This laboratory examines the latch or level-triggered flip-flop with feedback and demonstrates the problems with this design. The master-slave and edge-triggered flip-flops are used as solutions to the problems.

EQUIPMENT AND COMPONENTS:
Power supply (2 0 V-5 V, variable)
Breadboard
Oscilloscope
DVM
Function generator
3 x 74LS00
1 x 74LS10

PRELABORATORY:

1.   The NAND gate flip-flop. For the circuit of Figure 12L4.1:

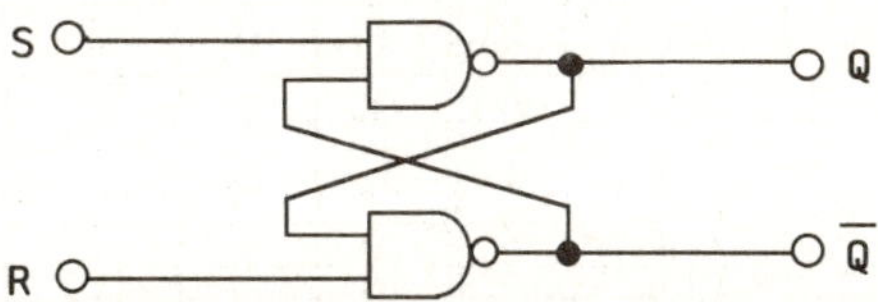

Figure 12L4.1: NAND gate flip-flop

a.   Complete the truth table in Table 12L4.1. Q_n and $\overline{Q}_n$ represent the initial guess at the values of the Q and $\overline{Q}_n$ outputs. Q_{n+1} and $\overline{Q}_{n+1}$ represent the future stable values of Q and $\overline{Q}$ due to the present state of Q, $\overline{Q}$, and the inputs, S and R.

Table 12L4.1: NAND gate truth table

S	R	Q_n	$\overline{Q}_n$	Q_{n+1}	$\overline{Q}_{n+1}$
0	0	1	0		
0	0	0	1		
0	1	1	0		
0	1	0	1		
1	0	1	0		
1	0	0	1		
1	1	1	0		
1	1	0	1		

b.   SR=00 is not a valid input for this circuit. Why?

c.   When SR=11, can the output change state?

2.   **The NOR gate flip-flop.** For the circuit of Figure 12L4.2:

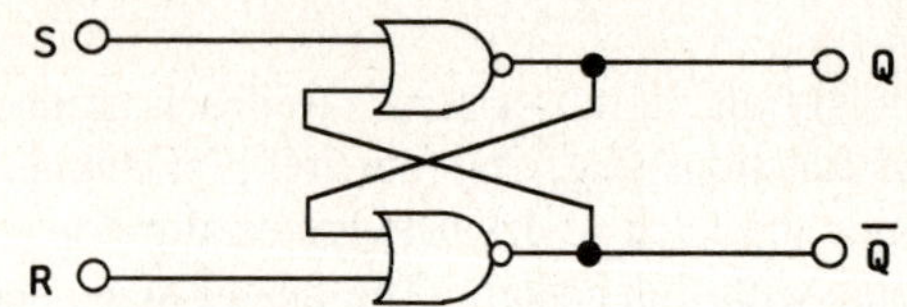

Figure 12L4.2: NOR gate flip-flop

a. Complete the truth table in Table 12L4.2. Again, Q_n and $\overline{Q}_n$ can be thought of as initial guesses at the values of Q and $\overline{Q}_n$. Q_{n+1} and $\overline{Q}_{n+1}$ are the future stable values of Q and $\overline{Q}$ due the present states of S, R, Q, and $\overline{Q}$.

Table 12L4.2: NOR gate truth table

S	R	Q_n	$\overline{Q}_n$	Q_{n+1}	$\overline{Q}_{n+1}$
0	0	1	0		
0	0	0	1		
0	1	1	0		
0	1	0	1		
1	0	1	0		
1	0	0	1		
1	1	1	0		
1	1	0	1		

b. SR=11 is not a valid input for this circuit. Why?

c. When SR=00, can the output change state?

3. The NOR gate flip-flop with a control input.

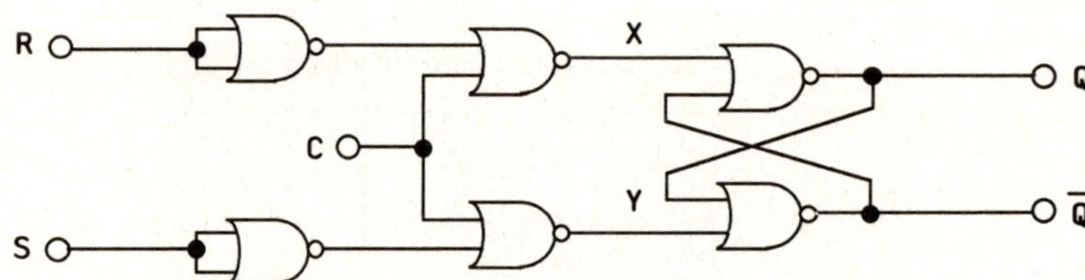

Figure 12L4.3: NOR gate flip-flop with a control input

a. Analyze the circuit of Figure 12L4.3 and complete Table 12L4.3. It is up to you to pick initial values for Q and $\overline{Q}$ and determine the value, given the inputs R, S, and C.

Table 12L4.3: NOR gate truth table

C	R	S	X	Y	Q	$\overline{Q}$
0	0	0				
0	0	1				
0	1	0				
0	1	1				
1	0	0				
1	0	1				
1	1	0				
1	1	1				

   b.   The input variable C is considered a control signal. What happens to X, Y, Q and $\overline{Q}$ when C=1?

   c.   Why is CRS=000 considered an invalid input?

4.   The NAND gate flip-flop with a control input.

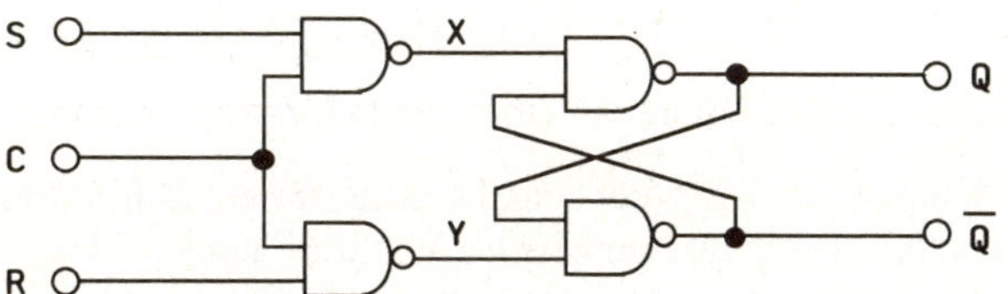

Figure 12L4.4: NAND gate flip-flop with a control input

   a.   Analyze the circuit of Figure 12L4.4 and complete Table 12L4.4. It is up to you to pick initial values for Q and $\overline{Q}$ and determine the actual values given the inputs R, S, and C.

Table 12L4.4: Truth table for NAND gate

C	S	R	X	Y	Q	$\overline{Q}$
0	0	0				
0	0	1				
0	1	0				
0	1	1				
1	0	0				
1	0	1				
1	1	0				
1	1	1				

   b.   The input variable C is considered a control signal. What happens to X, Y, Q, and $\overline{Q}$ when C=0?

   c.   Why is CSR=111 considered an invalid input?

   d.   What is an advantage of the circuit of Figure 12L4.4 over the circuit of Figure 12L4.3?

5.   The D flip-flop (latch).

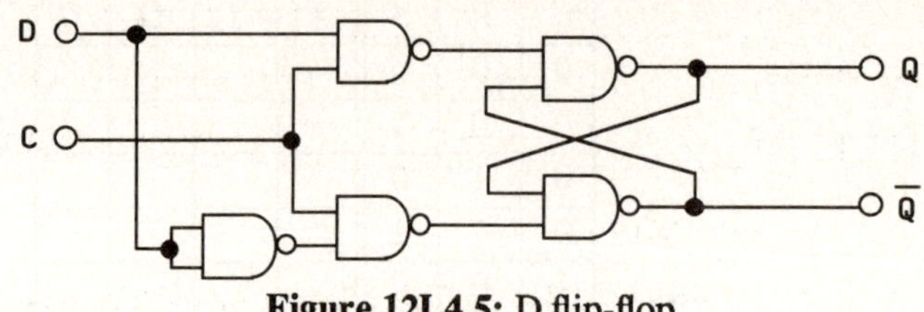

Figure 12L4.5: D flip-flop

    a.   Analyze the circuit of Figure 12L4.5 and complete Table 12L4.5.

Table 12L4.5: Truth table for D flip-flop

C	D	Q	$\overline{Q}$
0	0		
0	1		
1	0		
1	1		

    b.   Compare Table 12L4.5 to the truth table of a D flip-flop. Are they the same?

6.   The T flip-flop. A toggle, or T flip-flop, can be made from a D flip-flop by connecting the $\overline{Q}$ output to the D input. The output should change with each control input.

    a.   Analyze the circuit of Figure 12L4.6 and complete Table 12L4.6.

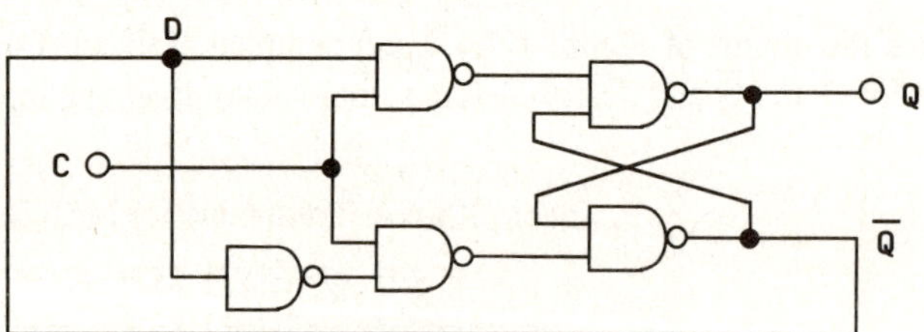

Figure 12L4.6: T flip-flop

Table 12L4.6: Truth Table for T flip-flop

C	D	Q_n	$\overline{Q}_n$	Q_{n+1}	$\overline{Q}_{n+1}$
0	0				
0	1				
1	0				
1	1				

    b.   Does the circuit of Figure 12L4.6 act as a toggle flip-flop, where C is the toggle input? Why?

7.   The D flip-flop (master-slave type). Consider the circuit of Figure 12L4.7. Each block in Figure 12L4.7 represents the circuit of Figure 12L4.5.

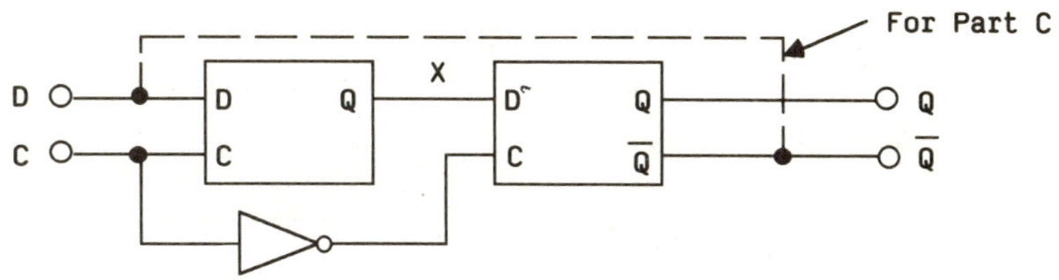

Figure 12L4.7: D flip-flop (master-slave type)

a.   Analyze the circuit of Figure 12L4.7 and complete Table 12L4.7.

Table 12L4.7: Truth table for D flip-flop

D	C	X	Q_n	Q_{n+1}
0	0	0	0	
0	0	0	1	
0	0	1	0	
0	0	1	1	
0	1	0	0	
0	1	0	1	
0	1	1	0	
0	1	1	1	
1	0	0	0	
1	0	0	1	
1	0	1	0	
1	0	1	1	
1	1	0	0	
1	1	0	1	
1	1	1	0	
1	1	1	1	

b.   Complete the timing diagram of Figure 12L4.8.

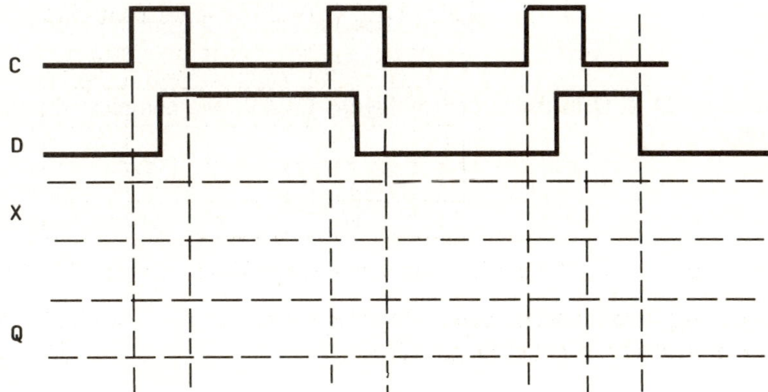

Figure 12L4.8: Timing diagram for D flip-flop

   c.   Connect the $\overline{Q}$ output of Figure 12L4.7 to the D input, and repeat the analysis of Part a. Note that $\overline{Q}$ now equals D.

   d.   Does the circuit of Part c act as a toggle flip-flop?

8.   The edge-triggered D flip-flop.

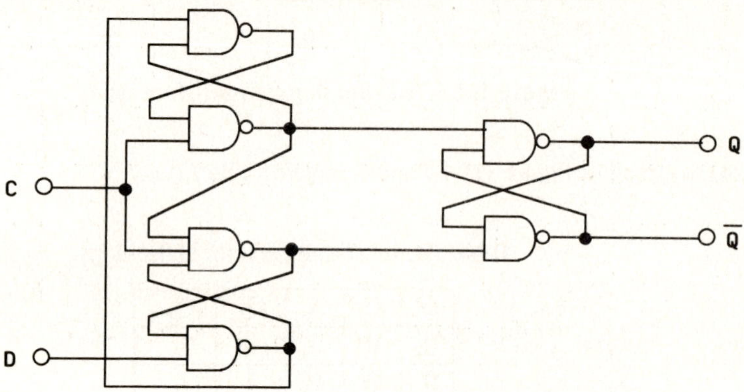

Figure 12L4.9: Edge triggered D flip-flop

   a.   Analyze the circuit of Figure 12L4.9. Complete the timing diagram of Figure 12L4.10.

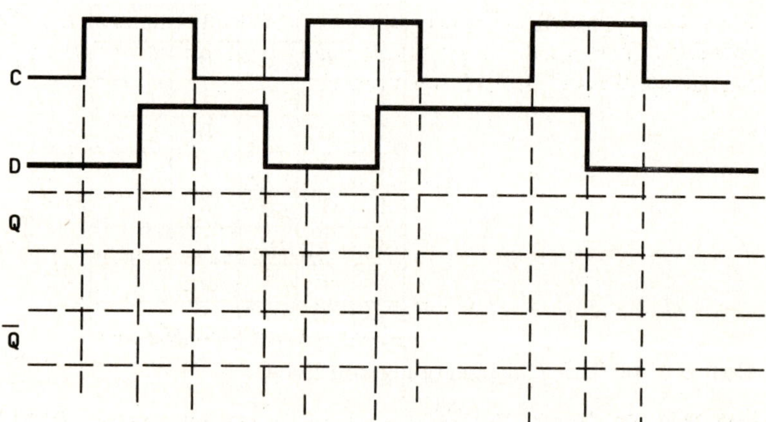

Figure 12L4.10: Timing diagram for Part a

   b.   Connect $\overline{Q}$ to D in the circuit of Figure 12L4.9, and complete the timing diagram of Figure 12L4.11.

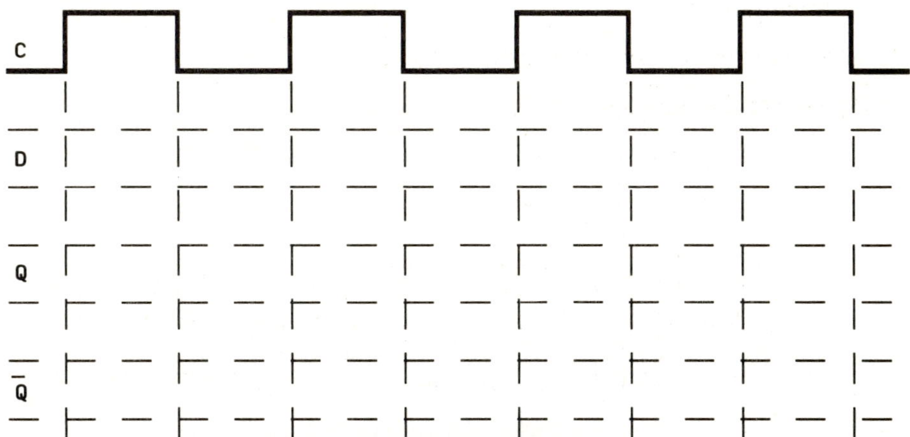

Figure 12L4.11: Timing diagram for Part b

    c.   Does the circuit described by part b act as a toggle flip-flop?

    d.   If the period of C was 100 ns, what is the period of Q?

Laboratory Procedure:

1.   The NAND gate RS flip-flop.

    a.   Use the two-input NAND gates given in the list of equipment and components, and build the circuit of Figure 12L4.12. Be sure to connect 5 V and ground to the DIP package. (Note: The 74LS00 has four-two-input NAND gates.)

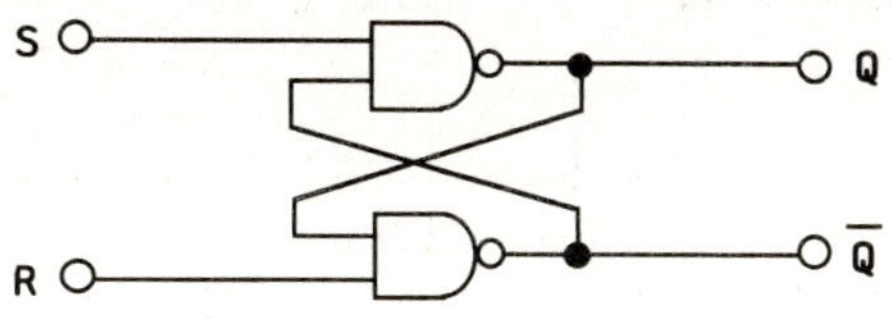

Figure 12L4.12: NAND gate

    b.   Complete the truth table of Table 12L4.8, where a 1 corresponds to a high and is a voltage greater than 2 V and a 0 corresponds to a voltage less than 0.7 V. When applying highs and lows to the S and R inputs, use 5 V and 0 V, respectively.

Table 12L4.8: NAND gate truth table

S	R	Q_n
0	0	
0	1	
1	0	
1	1	

    c.   Add two more NAND gates to the circuit of Figure 12L4.12 as shown in Figure 12L4.4 and repeated in Figure 12L4.13.

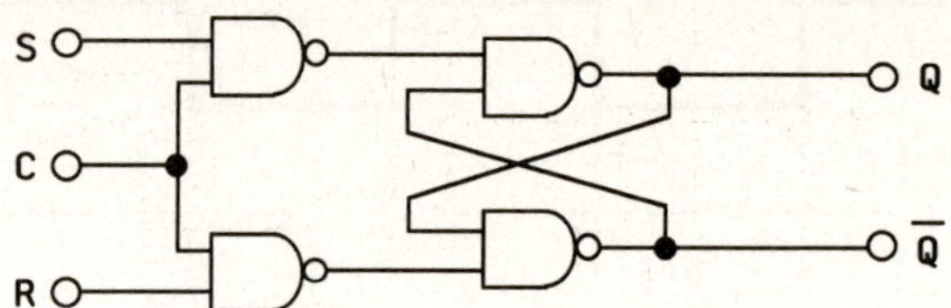

Figure 12L4.13: NAND gate for Part c

d. Complete the truth table of Table 12L4.9 where a 1 corresponds to a high and is a voltage greater than 2 V. A 0 corresponds to a low and is a voltage less than 0.7 V. When applying highs and lows to the S and R inputs, use 5 V and 0 V, respectively.

Table 12L4.9: Truth table for Part d

S	R	Q_n
0	0	
0	1	
1	0	
1	1	

e. Add one NAND gate to the circuit of Figure 12L4.13 to form the circuit of Figure 12L4.14, with the +5-V input shown explicitly.

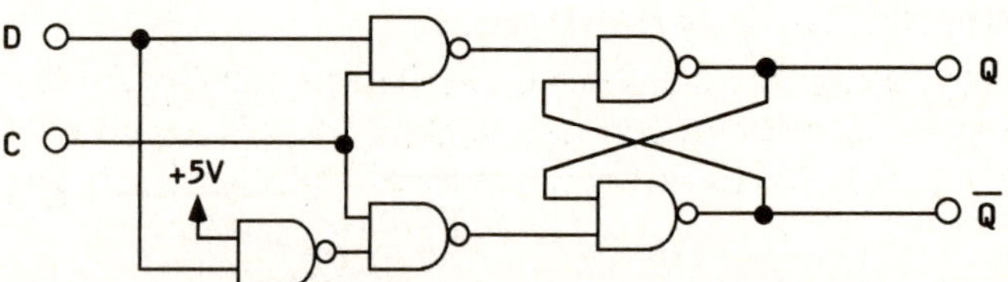

Figure 12L4.14: NAND gate for Part e

f. Complete Table 12L4-10 where a 1 corresponds to a high and is a voltage greater than 2 V. A 0 corresponds to a low and is a voltage less than 0.7 V. When applying highs and lows to the D and C inputs, use 5 V and 0 V, respectively. Also, record the actual voltages at Q and $\overline{Q}$.

Table 12L4.10: Truth table for Part f

C	D	Q	$\overline{Q}$
0	0		
0	1		
1	0		
1	1		

g. Use the circuit of Part e and connect the $\overline{Q}$ output to the D input as shown in in Figure 12L4.15, with the +5-V input shown explicitly.

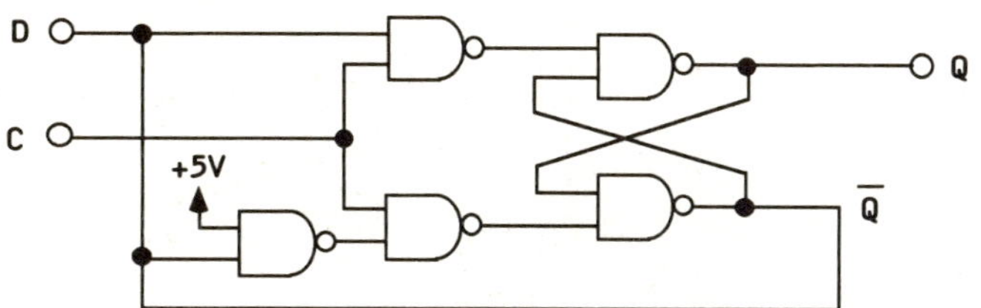

Figure 12L4-15: Circuit for Part g

h.   Toggle the input C from low to high using a low-frequency (100 Hz) 0- to 5-V pulse train from a function generator. Observe and sketch Q and $\overline{Q}$. How does this result compare with your prelab results?

i.   Add a NAND gate to the circuit of Figure 12L4.14, as shown in Figure 12L4.16.

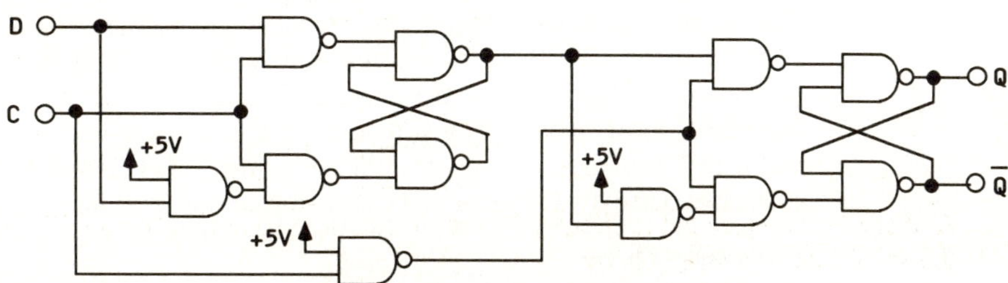

Figure 12L4.16: Circuit for Part i

j.   Connect the Q output to the D input. Demonstrate that this circuit is a toggle flip-flop, where C is the T input. Connect a 0.2-ms period, 0- to 5-V pulse train from a function generator to the C input, and sketch the output Q. What is the period of Q? Is this what was predicted in the prelab?

k.   Build the circuit of Figure 12L4.17. (Note: You need a three-input NAND gate.)

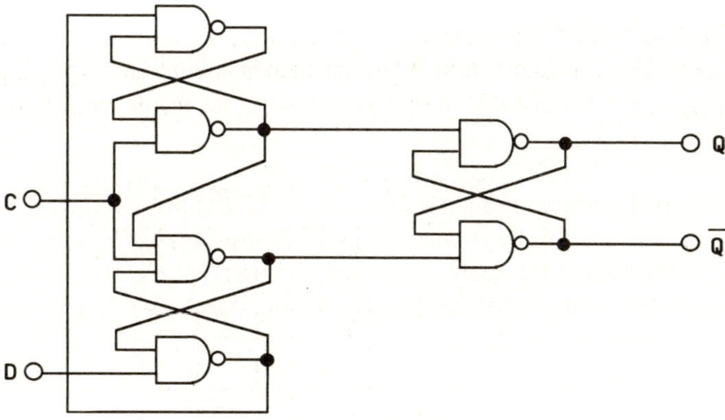

Figure 12L4.17: Circuit for Part k

l.   Devise a way to demonstrate that Q follows the D input only on the positive edge of C. Describe your set up and sketch the results.

DISCUSSION:

CONCLUSIONS:

LAB 12L5

TITLE: State Machine Design

OBJECTIVE: State machines are found in virtually all large digital systems that require a repetitive sequence of outputs. A state machine may or may not be influenced by external inputs. A counter is an example of a state machine that is not influenced by any external input variable. The controller in a digital computer can be a state machine that outputs the microinstructions necessary to execute a given machine instruction. In this case, the value of a state machine's output is influenced by some external inputs, specifically, the machine instruction pulled from memory during a fetch cycle.

The design of a state machine is a straightforward, though regimented, procedure. The prelab walks you through the procedure to design a four-bit up-and-down counter. In this case your state machine will count up from 0000 to 1111 and repeat, or count down from 1111 to 0000 and then repeat, depending on whether the input variable X equals 1 or 0. You will do two paper designs, one with JK flip-flops and one with D flip-flops. However, in the laboratory, you will only build and verify one of the two designs. Then you will compare your design to that of the 74LS169.

EQUIPMENT AND COMPONENTS:

Digital multimeter or logic probe

Breadboard

TTL logic

$2 \times$ 74LS74 dual D flip-flop, which triggers
   on the positive-edge of the clock

$2 \times$ 74LS112 dual JK flip-flop, which triggers
   on the negative edge of the clock

Combinational logic gates (minimize your design)

The 74 and 112 are available in other versions of TTL and are acceptable for this laboratory.

One 74LS169   four-bit up-and-down counter.

PRELABORATORY:

As described in the objective, you will design two up-and-down binary counters, one using JK flip-flops and one using D flip-flops. The external input X_1 determines the direction of the count. $X_1=1$ means count up; $X_1=0$ means count down. X_0 is a second external input that disables the count. $X_0=1$ means allow the count, and $X_0=0$ means disable the count.

1. Determine the number of flip-flops required for the counter based on the D flip-flop. Also, determine the number of flip-flops based on JK flip-flop.

2. Draw the excitation tables for the JK and D flip-flops.

3. Draw a state table and state diagram for the JK-based counter and the D-based counter. Your tables and diagrams should include the states as well as the external variables. Then minimize the input expressions to the flip-flops using K-maps.

4. Using the 74LS74 and any other required combinational logic, draw the schematic for the up-and-down counter. If you are not using the PRESET and CLEAR inputs connect them to their inactive state. (Note: the PRESET and CLEAR inputs are asynchronous.)

5. Using the 74LS112 and any other required combinational logic, draw the schematic for the up-and-down counter. If you are not using the PRESET and CLEAR inputs connect them to their inactive state. (Note: the PRESET and CLEAR inputs are asynchronous.)

6. Using the 74LS169, draw a schematic of the up-and-down counter.

7. Compare the performance of each design in these terms:

   a. maximum clock speed

   b. maximum propagation delay from the triggering clock edge to a change in state

   c. power consumption

   d. chip count

Laboratory Procedure:

1. The D-flip-flop counter.

   a. Build the counter you designed in part 4 of the prelab. Be sure that all power supplies are connected before you connect any input signals.

   b. Use a function generator to produce the clock signal. Make sure that the voltages of the clock are compatible with your chip's requirements.

   c. Verify that the control inputs (disable and up and down) work.

2. The JK flip-flop counter

   a. Build the counter you designed in Part 5 of the prelab. Be sure that all power supplies are connected before you connect input signals.

   b. Use a function generator to produce the clock signal. Make sure that the voltages of the clock are compatible with your chips' requirements.

DISCUSSION:

CONCLUSIONS:

LAB 12L6

TITLE:	The MOS Inverter
OBJECTIVE:	This laboratory exercise is an introduction to digital electronics using the metal-oxide-semiconductor field-effect transistor, or MOSFET. Most devices currently available have polysilicon gates and not metal, as the original devices did. However, the devices are still referred to as MOSFETs.

In this laboratory, the static voltages associated with a logic gate will be measured and calculated for the inverter. These voltages are V_{OL}, V_{OH}, V_{IL}, and V_{IH}. From these four voltages, the following characteristics are found: output voltage swing, high-noise margin, low-noise margin, and transition width.

Next, an additional inverter stage is added and the change in the preceding characteristics is determined. Then the concept of fan-out is used to determine the total number of loads the inverter can drive when the noise margins go to zero.

Finally, a two-MOSFET gate is examined and its logic function determined. Before starting the prelaboratory, the student should receive two switching n-channel enhancement MOSFETS. The student should use a curve tracer and determine K (A/V^2) and the threshold voltage, V_t for each device and label them Q_1 and Q_2. The student should also obtain a family of curves for i_D versus v_{DS} as a function of v_{GS}.

EQUIPMENT AND COMPONENTS:
> 0- to 5-V variable power supply
> fixed supply at 5 V/100 mA is required
> 2 DVMs (only one is necessary)
> 2 100 kΩ 1/8-W resistors
> 3 3N169 n-channel enhancement-mode MOSFET

PRELABORATORY:

1. Given the circuit of Figure 12L6.1, find the following information:

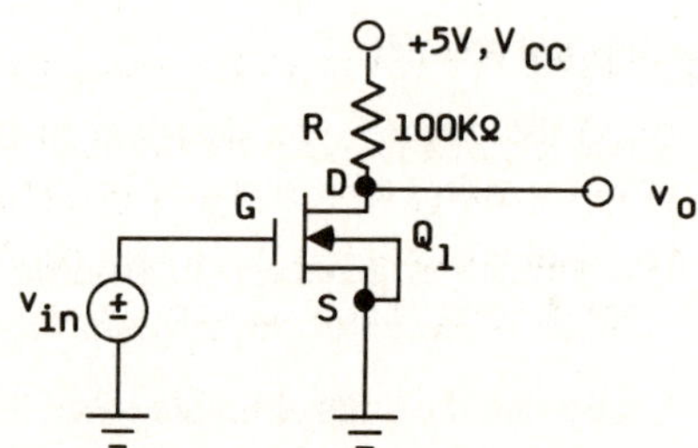

Figure 12L6.1: MOS inverter

    a. Plot the static transfer function between v_{in} and V_o. This can be done by varying v_{in} from 0 V to 5 V in 0.5-V increments, calculating v_o at each point.

    b. Place the voltages V_{OH}, V_{OL}, V_{IH}, and V_{IL} on your graph from Part a. Also, graphically show NM_H, NM_L, the output voltage swing, and the transition width.

2. By adding a load to the output of Q_1 as shown in the circuit of Figure 12L6.2, find the following:

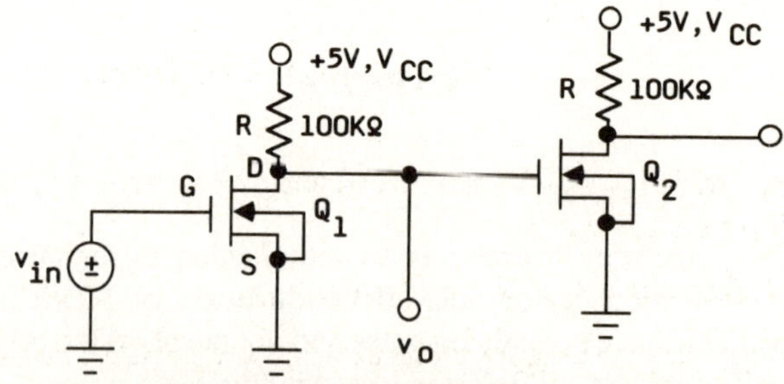

Figure 12L6.2: MOS inverter and load

    a. Determine new values for V_{OL}, V_{OH}, V_{IH}, and V_{IL}.

    b. Also determine NM_H, NM_L, output voltage swing, and transistor width.

    c. Determine how many loads could be added to Q_1 before NM_H or NM_L goes to zero. Use whichever value of NM_H or NM_L is smaller.

3. Consider the circuit of Figure 12L6.3.

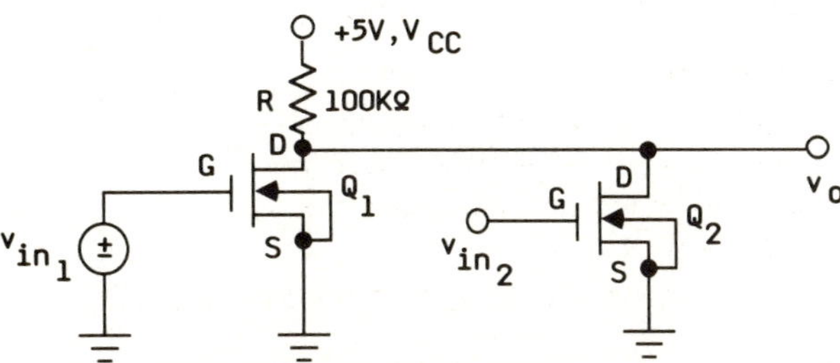

Figure 12L6.3: Circuit of Part 3

    a. What logic function is performed by the circuit?

    b. Design an RS flip-flop with MOSFETS using the gate shown in the figure as a building block.

4. Do a SPICE simulation for all three prelaboratory circuits shown as well as the flip-flop design. Do a transient analysis for each circuit, with all input signals taken as 5-V pulses with 500-ns widths and vertical edges. For the flip-flop and two-input gate, all four combinations of input should be analyzed. For the flip-flop, initial values for the output should be set.

Laboratory Procedure:

1. Single inverter stage.

    a. Build the circuit of Figure 12L6.4.

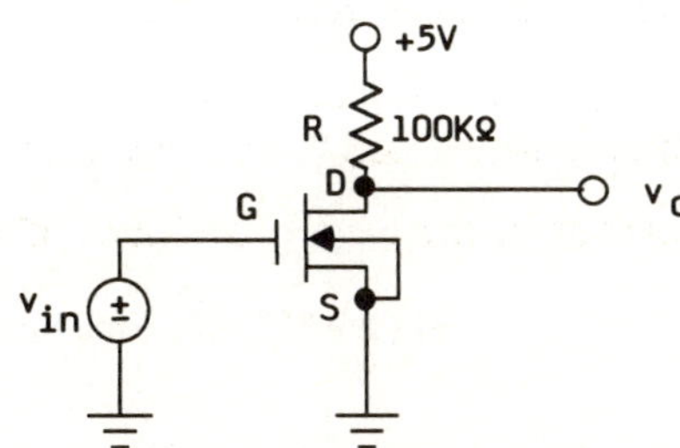

Figure 12L6.4: MOS inverter

    b. Vary V_{in} from 0 to 5 V in 0.5-V increments, and plot v_o versus V_{in} for each setting of V_{in}. (Plot v_{in} on the independent x-axis.)

    c. From this plot, determine V_{OH}, V_{OL}, V_{IH}, and V_{IL}.

    d. Determine NM_H, NM_L, the output voltage swing, and the transition width.

2. Single inverter stage with one gate load.

    a. Build the circuit of Figure 12L6.5.

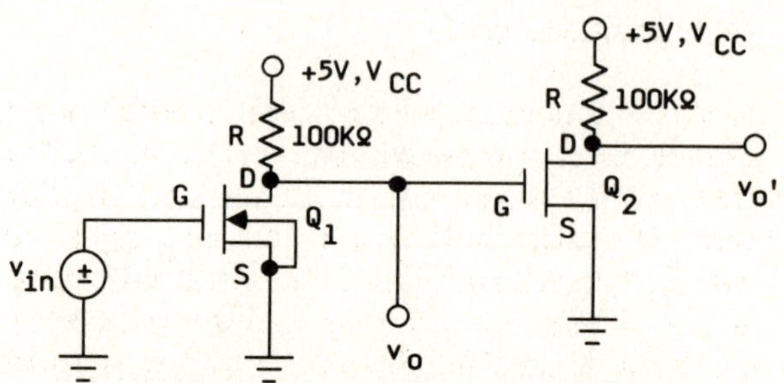

Figure 12L6.5: MOS inverter with load

b. Vary V_{in} from 0 to 5 V in 0.5-V increments, and plot V_o versus V_{in}, (Plot V_{in} on the independent x-axis.)

c. Label V_{OH}, V_{OL}, V_{IH}, and V_{IL}. How many loads could be added to Q_1 and have its output considered valid, if the noise margins could equal zero? (Remember, this is only the static case.)

3. The two-input gate.

a. Build the circuit of Figure 12L6.6.

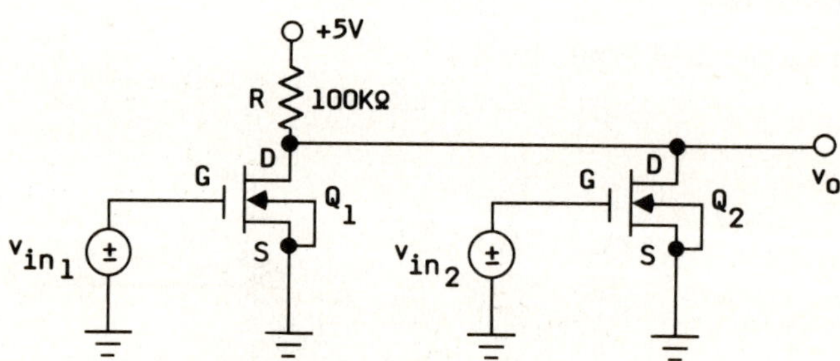

Figure 12L6.6: Two input gate

b. Verify the logic function from the prelab.

c. Build and verify the operation of the RS flip-flop you designed in the prelab.

DISCUSSION:

CONCLUSIONS:

LAB 12L7

TITLE: The BJT Inverter

OBJECTIVE: This laboratory exercise is an introduction to digital electronics using the bipolar junction transistor as an inverter. During the lab, the four static voltages V_{OL}, V_{OH},

V_{IL}, and V_{IH} associated with a logic gate are measured. From these four voltages, you will determine the following characteristics for the inverter:

1. Output voltage swing

2. High-noise margin, NM_H

3. Low-noise margin, NM_L

4. Transition width

Next, an additional inverter stage is added and the change in the foregoing characteristics is measured. You are then expected to use these results to determine the maximum fan-out for high and low inputs by allowing the noise margins to go to zero. The student will receive three small signal transistors (i.e., 2N2222As), and will need to characterize these transistors on a curve tracer and label them Q_1, Q_2 and Q_3. The information required is the base-emitter voltages with the transistor in cutoff and saturation, $V_{BE(EOC)}$, and $V_{BE(EOS)}$ respectively. Also needed is the forward current gain, i_c/i_b or β_F.

EQUIPMENT AND COMPONENTS:

Function generator (0-V to 5-V square wave)
Power supply, (0-5 V at 100 mA)
2, 1-kΩ, 1/8-W resistors
2, 10-kΩ, 1/8-W resistors
3, small-signal transistors, i.e., 2N222A
Oscilloscope with delayed sweep

PRELABORATORY:

1. Given the circuit of Figure 12L7.1, with $R_1 = 10$ kΩ and $R_2 = 1$ kΩ, find the following information:

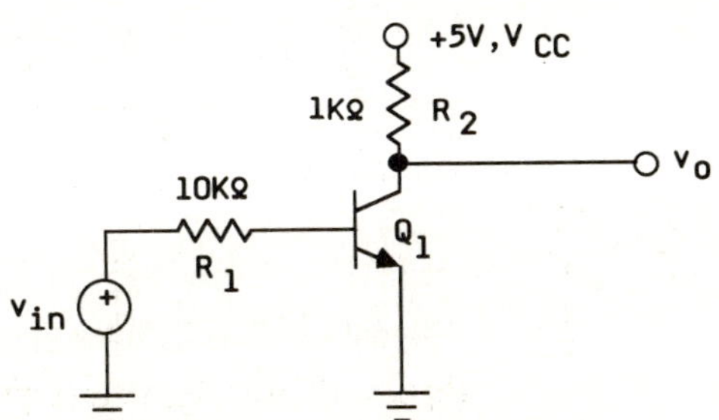

Figure 12L7.1: BJT inverter

    a. Plot the static transfer function between V_{in} and V_o. This can be done by varying V_{in} from 0 to 5 V in 0.5-V increments and calculating V_o at each point.

    b. Place the voltages V_{OH}, V_{OL}, V_{IH}, and V_{IL} on your graph from Part a. Also, graphically show NM_H, NM_L, the output voltage swing, and the transition width.

2. By adding a load to the output of Q_1 ($R_1 = R_4 = 10$ kΩ and $R_2 = R_3 = 1$ kΩ) as shown in the circuit of Figure 12L7.2, find the following:

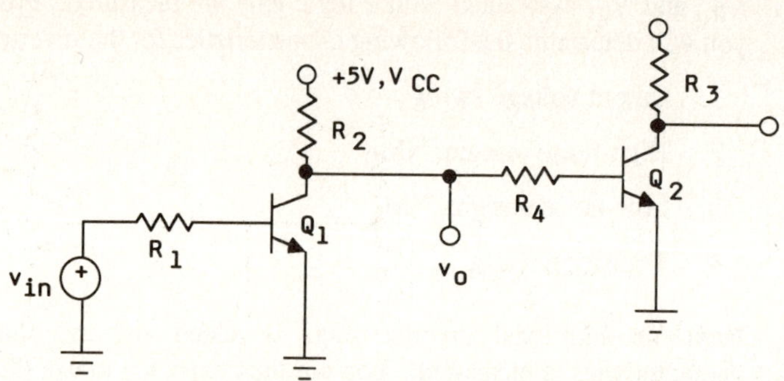

Figure 12L7.2: BJT inverter with one load

a. Determine new values for V_{OL}, V_{OH}, V_{IH}, and V_{IL}.

b. Repeat Part b of Question 1 with the new circuit. The inverter formed by Q_2 is considered the load on the Q_1 inverter.

c. Determine how many loads could be added to Q_1 before NM_H or NM_L go to zero. Use whichever value is smaller.

Laboratory Procedure:

1. Single Inverter Stage.

   a. Build the circuit of Figure 12L7.3, which is the same as that of Figure 12L7.1 with the power supply shown explicitly. (let $R_1 = 10\ k\Omega$ and $R_2 = 1\ k\Omega$)

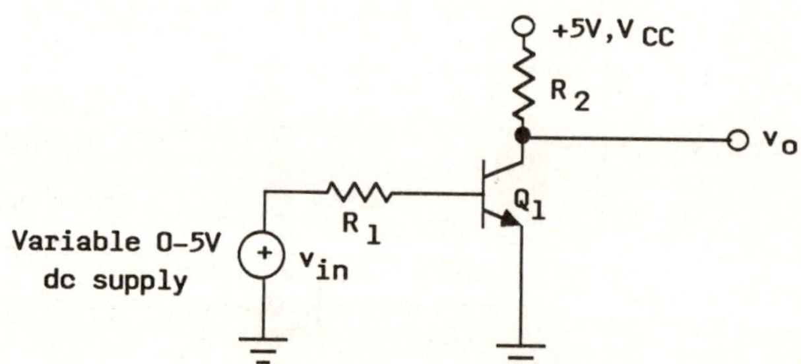

Figure 12L7.3: BJT inverter

   b. Vary V_{in} from 0 V to 5 V in 0.5-V increments, and plot V_o versus V_{in} for each setting of V_{in}. (Plot V_{in} on the independent x-axis.)

   c. From this plot, determine V_{OH}, V_{OL}, V_{IH}, and V_{IL}.

   d. Determine NM_4 and NM_L, the output voltage swing, and the transition width.

2. Single inverter with one gate load.

   a. Build the circuit of Figure 12L7.4, which is essentially the same as that shown in Figure 12L7.2. (Let $R_1 = R_3 = 10\ k\Omega$ and $R_2 = R_4 = 1\ k\Omega$)

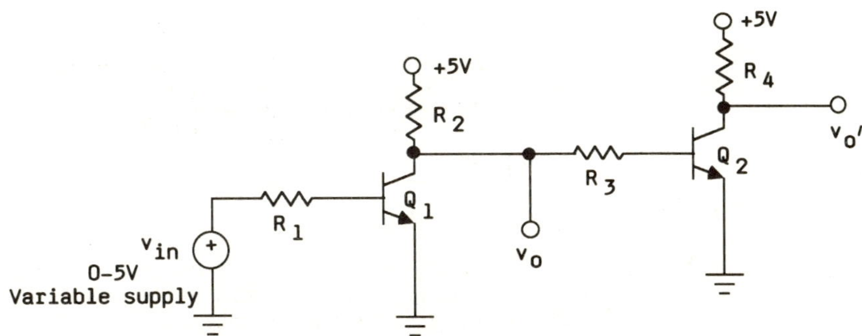

Figure 12L7.4: Single inverter with one gate load

b. Vary V_{in} from 0 V to 5 V in 0.5-V increments, and plot v_o versus V_{in}. (Plot V_{in} on the independent x-axis.)

c. Label V_{OH}, V_{OL}, V_{IH}, and V_{IL}.

d. Add another load to Q_1, shown in Figure 12L7.5. Measure V_o, and determine how many gate loads could be added to Q_1 and still allow V_o to be a valid input. (Let $R_1 = R_3 = 10$ kΩ and $R_2 = R_4 = 1$ kΩ)

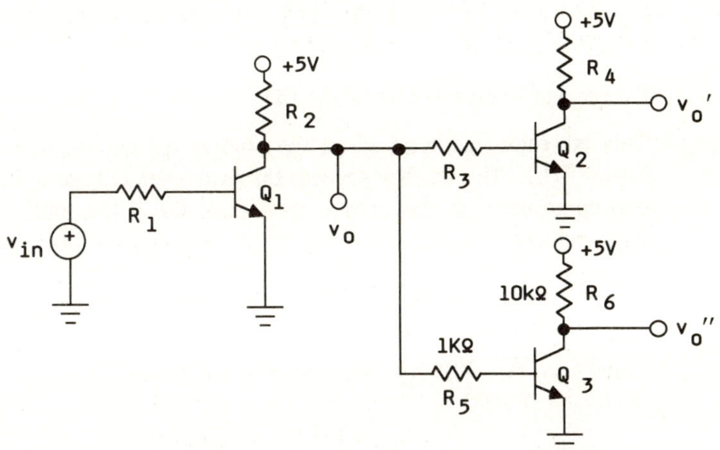

Figure 12L7.5: Single inverter with two gate loads

3. SPICE Simulation.

   a. Simulate the circuit of Figure 12L7.1 using SPICE. The following parameters must be included for Q_1:

$$C_{jeo} = 0.4 \text{ pF}$$
$$\phi_e = 0.9 \text{ V}$$
$$m_e = 0.5$$
$$\tau_f = 0.3 \text{ ns}$$
$$\tau_{BF} = 14 \text{ ns}$$
$$\tau_s = 21 \text{ ns}$$
$$C_{jco} = 0.2 \text{ pF}$$
$$\phi_c = 0.7 \text{ V}$$
$$m_c = 0.3$$

$$\left.\begin{array}{c} V_{BE(EOC)} \\ V_{BE(EOS)} \\ V_{CE(SAT)} \end{array}\right\} \text{your device}$$

b. Let the input be a 0-V to 5-V pulse with a width of 1 μs and vertical edges. Do a transient analysis, and determine the same information about the inverter as in Part 1. Also, determine the propagation delays τ_{PLH} and τ_{PHL}.

DISCUSSION:

CONCLUSIONS:

LAB 12L8

TITLE: Design and Analysis of Standard TTL

OBJECTIVE: This laboratory exercise gives the student an introduction to transistor-transistor logic (TTL). The lab begins with the basic elements of a TTL gate with a passive pull-up resistor on the output, then adds the active pull-up transistor for speed improvement.

Hand analysis of the inverter circuit is done along with a SPICE simulation. The transient response of the circuit is performed so the eight parameters V_{OH}, V_{OL}, V_{IH}, V_{IL}, I_{OH}, I_{OL}, I_{IH}, and I_{IL} characterizing a digital logic function can be determined.

Next, matched transistors are used as the input stage so that a two-input NAND gate can be built.

And finally, the performance of the NAND gate is compared with that of a current version of TTL. For this part of the lab, you may use LS, AS, ALS, or whatever version of TTL you have available.

EQUIPMENT AND COMPONENTS:
Oscilloscope
Power Supply (5 V) and 0 V-5 V variable)
Function generator (square waves with variable dc offset)
Q_1, Q_2 are often a matched pair where indicated (see appendix)
All other Q's are 2N2222
R's as shown, 1%, 1/8 W
1 74LS169   four-bit up-and-down counter

PRELABORATORY:

1. Consider the circuit of Figure 12L8.1. Assume $\beta_F = 100$, $V_{BE(ON)} = 0.7$ V, $V_{D(ON)} = 0.7$ V, $V_{BC(ON)} = 0.5$ V, $\beta_R = 0.5$, $V_{CE(SAT)} = 0.1$ V, and $V_{BE(SAT)} = 0.8$ V.

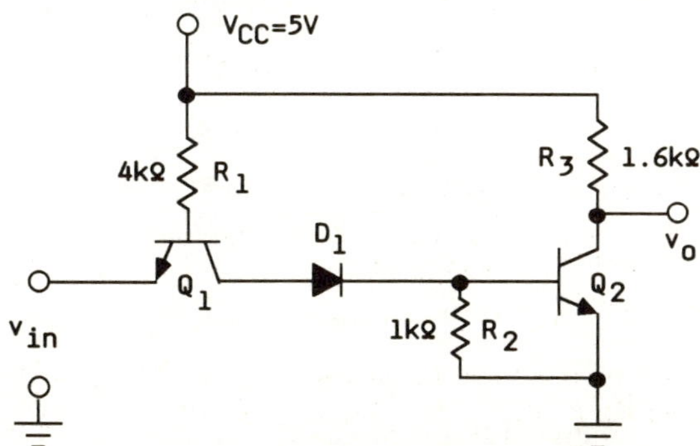

Figure 12L8.1: TTL gate

   a. Let $V_{in} = 0.2$ V. Determine all node voltages and branch currents. (Hint: Assume various states for the transistor Q_1, and then verify that the node voltages support your assumption.)

   b. Let $V_{in} = 5$ V. Determine all node voltages and branch currents. (Hint: Assume Q_1 is in the inverse active mode, and then verify that the voltages support your assumption.)

   c. What logic function is performed by this circuit?

   d. Use SPICE, and do a transient analysis of the circuit assuming that the collector-base junction capacitance at zero bias is 1 pF for each transistor and the emitter-base junction capacitance with zero bias is 10 pF for each transistor. Also assume the diode junction capacitance is 10 pF. Your analysis should give you the eight voltages and currents that characterize the logic gate. Use as your input the 0- to 5-V pulse, illustrated in Figure 12L8.2.

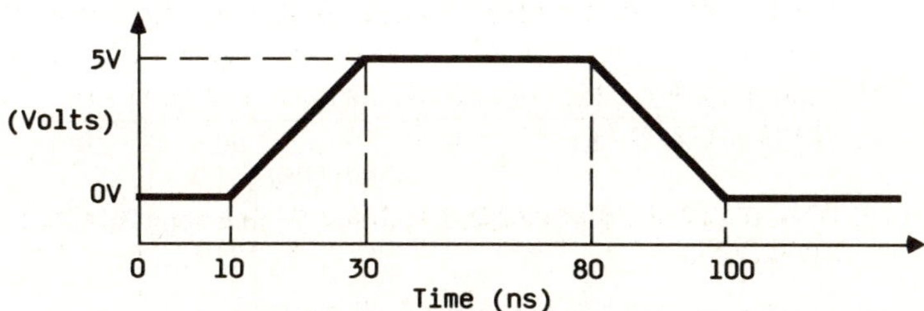

Figure 12L8.2: Pulse for Part 1d

2. Consider the modified TTL gate of Figure 12L8.3. Let $\beta_F = 100$, $V_{BE(ON)} = 0.7$ V, $V_{BC(ON)} = 0.5$ V and $\beta_R = 0.5$.

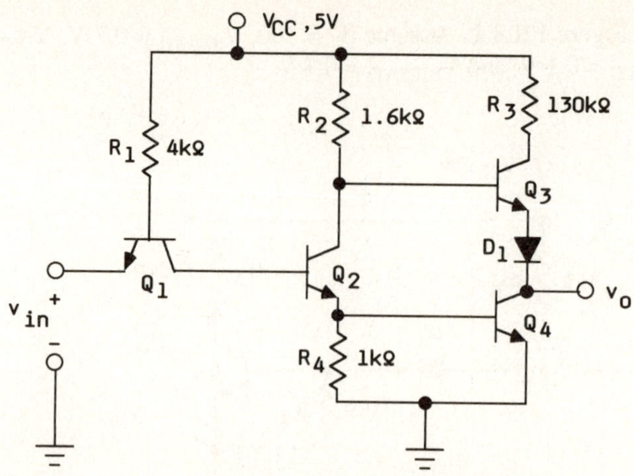

Figure 12L8.3: Modified TTL gate

a. Let $V_{in}=0.2$ V. Determine all node voltages and branch currents. The same considerations for Q_1 apply as in part 1.

b. Let $V_{in} = 5$ V. Determine all node voltages and branch currents. Again, assume that Q_1 is in the inverse active mode, and then verify that the voltages support your assumption.

c. What logic function is performed by this circuit?

d. Use SPICE and do a transient analysis. Your analysis should give you the eight voltages and currents that characterize your logic gate. Use the 0- to 5-V pulse shown in Figure 12L8.4 as your input. Use the same capacitor values as used in section 1d.

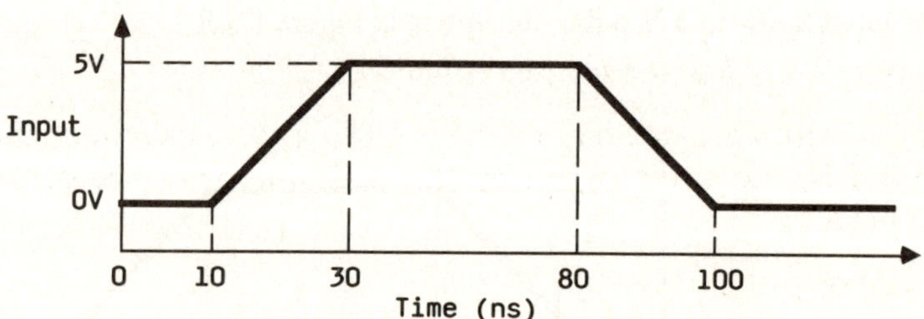

Figure 12L8.4: 0- to 5-V pulse for Part 2

3. Modify the circuit of Figure 12L8.3 with matched transistors as shown in Figure 12L8.5. Assume $\beta_F = 100$, $V_{BE(ON)} = 0.7$ V, $V_{BC(ON)} = 0.5$ V and $\beta_R = .5$.

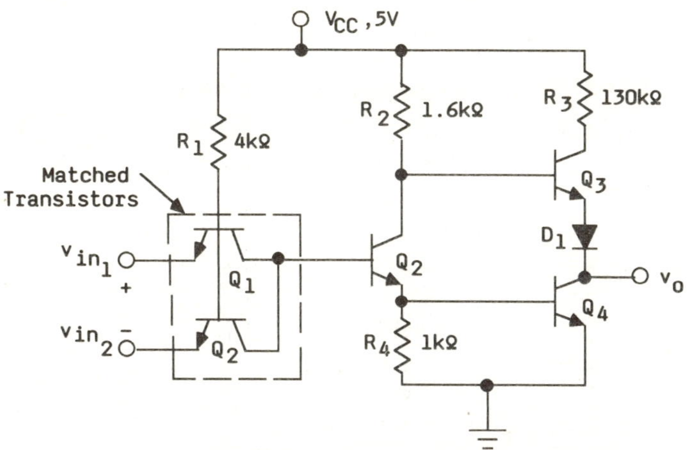

Figure 12L8.5: Two input TTL gate

a. Analyze the input stage, Q_1 and Q_2, and determine what happens at the output, V_o for the four combinations of v_{in_1}, and v_{in_2}. Use 0.2 V and 5 V as low and high values, respectively, for v_{in_1} and v_{in_2}.

b. What logic function does this circuit perform?

c. Use SPICE to do a transient analysis of the circuit. Your analysis should yield the eight voltages and currents that characterize your logic gate. Use the 0- to 5-V pulse shown in Figure 12L8.4 as your input. Let $V_{in_1} = V_{in_2} = V_{in}$ for this analysis. Use the same capacitor values as used in Section 1d.

4. Evaluation of the 74XX00 Two-Input NAND gate.

a. Draw the pinout and internal logic circuit for a version of the 74XX00 available in your laboratory. (XX indicates any version of the TTL gate.)

b. Find the worst case values of the eight voltages and currents that characterize the logic gate. Also find the worst-case propagation delay and power consumption. Compare these results with those of Part 3.

Laboratory Procedure:

1. Simplified TTL gate.

a. Build the circuit of Figure 12L8.6.

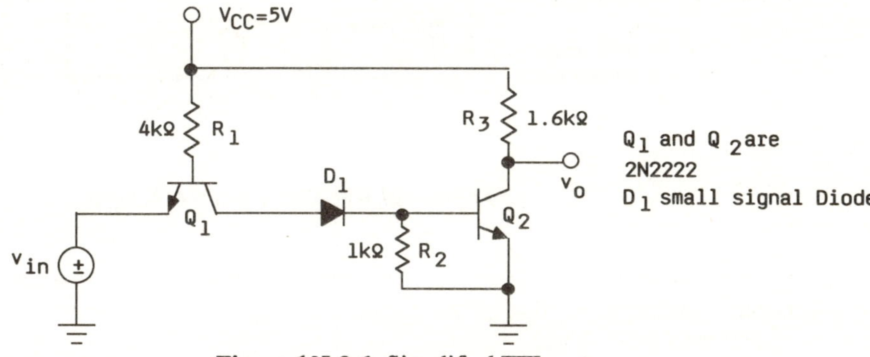

Figure 12L8.6: Simplified TTL gate

b. Vary V_{in} from 0 V to 5 V in 0.2 V increments. Record and plot V_o as a function of V_{in}. Determine V_{OH}, V_{OL}, V_{IH}, and V_{IL}.

c. Use a function generator to produce $v_{in} = 0$ V to 5 V square waves. Can you use the delay sweep of the oscilloscope to measure the propagation delay of the circuit?

2. Modified TTL gate.

  a. Build the circuit of Figure 12L8.7.

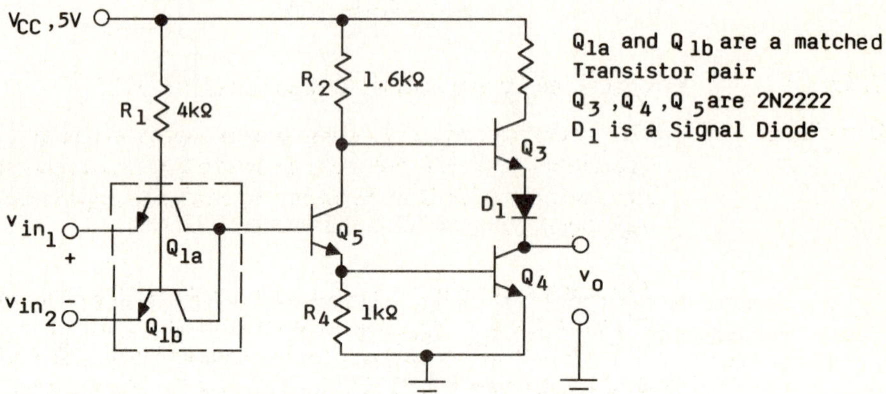

Figure 12L8.7: Modified TTL gate

b. Verify the performance of this circuit by applying a high imput of 5 V through a 1-kΩ pull-up resistor. Ground is taken as the low input. Generate the four combinations of v_{in_1} and v_{in_2}.

c. Use a function generator to produce v_{in_1} while $v_{in_2} = 5$ V through a 1-kΩ pull-up resistor. The function generator should produce and input voltage, v_{in_1}, previously described in section 1c of the laboratory procedure.

3. The 74XXOO two-input NAND gate.

  a. Build the circuit of Figure 12L8.8.

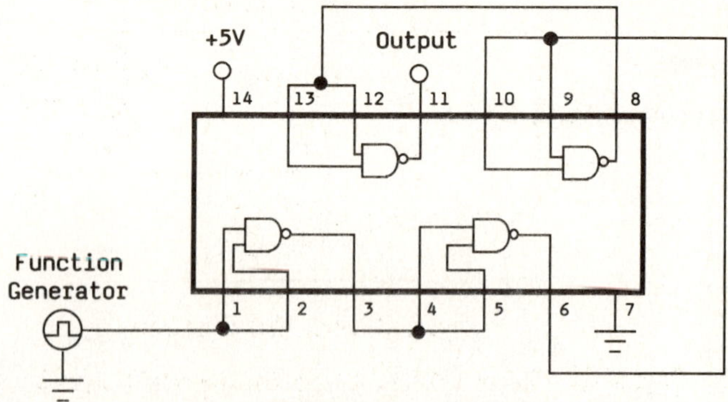

Figure 12L8.8: Two-input NAND gate

b. Observe the output and input on a dual-trace oscilloscope with delayed sweep. Observe the positive and negative edge of the output. Can you measure an average propagation delay for your NAND gate? Why?

    c. Using a DVM, devise a circuit to measure the eight voltages and currents that characterize a logic gate.

DISCUSSION:

CONCLUSIONS:

LAB 12L9

TITLE: Design and Analysis of Emitter-Coupled Logic (ECL)

OBJECTIVE: This laboratory gives the student an introduction to emitter-coupled logic, or ECL. The advantage of ECL over other logic families is in the switching speed parameter. However, the price is an increase in power consumption as compared with other logic families.

This laboratory exercise begins with a basic differential pair and current mirror circuits to show the operation of an ECL circuit. Next, the NOR/OR functions of the 10K series of ECL is analyzed with hand calculations and SPICE simulation. These hand calculations and SPICE simulation will yield the eight voltages and currents that characterize a digital logic gate: V_{OH}, V_{OL}, V_{IH}, V_{IL}, I_{OH}, I_{OL}, I_{IH}, and I_{IL}.

Finally, the performance of a series 10K OR/NOR gate is evaluated and compared with that of a discrete gate.

EQUIPMENT AND COMPONENTS:
Power supply, ± 5 V and -5 V to + 5 V, variable
Function generator with dc offset
Q_1, Q_2 are a matched pair where indicated; All other Q's are 2N2222.
R's (as indicated) 1%, 1/8 W

PRELABORATORY:

1. Consider the circuit of Figure 12L9.1.

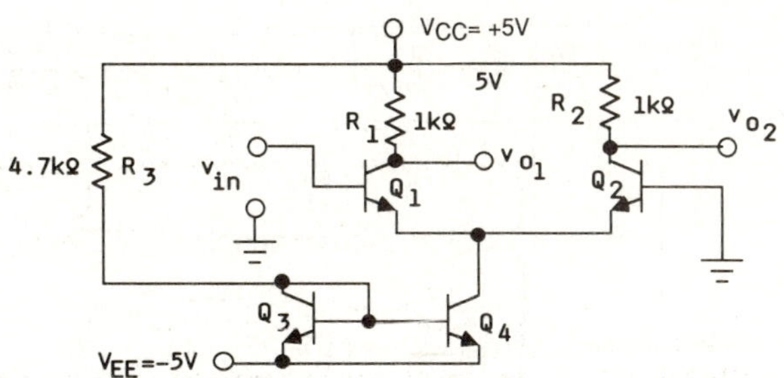

Figure 12L9.1: Circuit of Part 1

(Assume that Q_1–Q_2 and Q_3–Q_4 are matched pairs; $V_{BE(ON)} = 0.7$; $I_{C4} = 2$ mA; $V_{CB(sat)} = 0.5$ V; β's = 100; and $V_{CC} = +5$ V)

    a. Determine all node voltages for

        (1) $v_{in} = $ -1-V

        (2) $v_{in} = $ +1-V

b. What is the maximum value of v_{in} that will not cause Q_1 to saturate?

c. Determine the port voltages and currents that characterize the logic function between output V_{o_1} and the input, and output v_{o_2} and the input. What logic functions are defined by this circuit?

d. Use SPICE, and do a transient analysis for this circuit, assuming that a collector-base junction capacitance, with zero bias, is 1 pF for each transistor and a base-emitter junction capacitance, with zero bias, is 10 pF for each transistor. Use a 0- to 5-V pulse, shown in Figure 12L9.2, as your input.

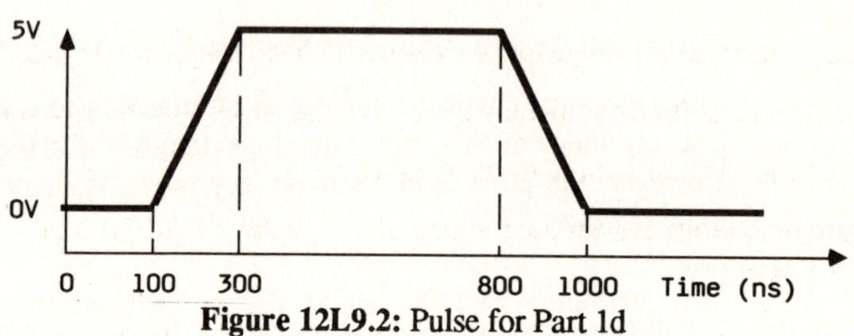

Figure 12L9.2: Pulse for Part 1d

Also, determine the propagation delay, both low to high and high to low.

e. Are the output voltage levels compatible with the input voltage levels? Why?

2. Consider the circuit of Figure 12L9.3. Assume $\beta_F = 100$ and $V_{BE(ON)} = 0.75$ V.

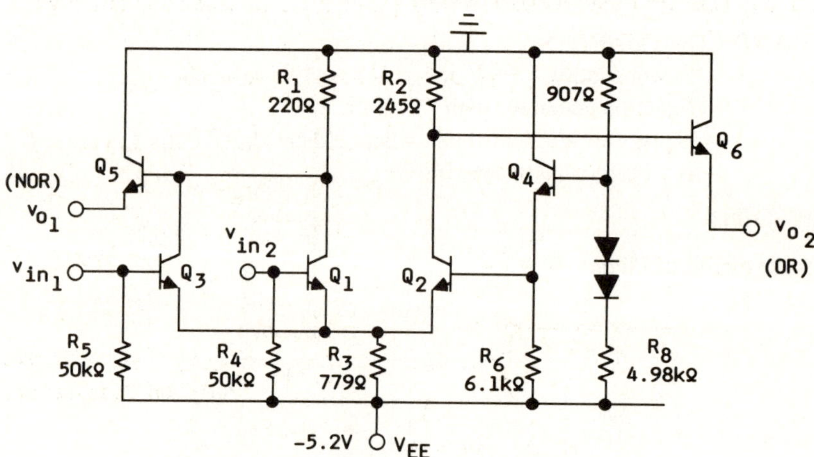

Figure 12L9.3: Emitter couple logic circuit

a. Determine all node voltages for two values of V_{IN}. Let $V_{in_2} = 0$.

    (1) $V_{in_1} = -1.4$ V

    (2) $V_{in_1} = -1.2$ V

b. Determine the port voltages and currents that characterize the logic function between outputs $V_{o(OR)}$ and $V_{o(NOR)}$ and inputs 1 and 2. What logic function defines this circuit?

c. Use SPICE, and do a transient analysis for this circuit, assuming the same values as in Part 1d of the prelab. Use a 0- to 5-V pulse, shown in Figure 12L9.4, as your input.

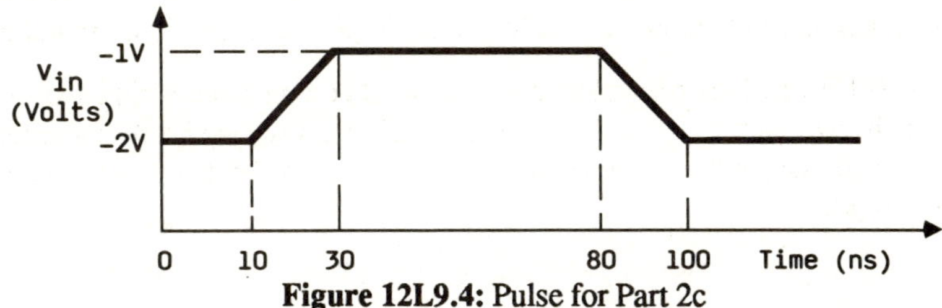

Figure 12L9.4: Pulse for Part 2c

Also, determine the propagation delay, both low to high and high to low through the circuit.

    d.   Are the output levels compatible with the input levels? What is the value of the noise margin?

3.   Evaluate the MC10H101 quad two-input NOR/OR gate.

    a.   Draw the pinout and internal logic circuit for the MC10H101 or a comparable part available in your laboratory.

    b.   From the data sheets, find the worst case values for the eight voltages and currents that characterize the logic gate. Also, find the worst case propagation delays and power consumption. Compare these results with those of Part 2.

Laboratory Procedure:

1.   A simplified ECL gate.

    a.   Build the circuit of Figure 12L9.5. $Q_1 Q_2$ and $Q_3 Q_4$ are matched transistors.

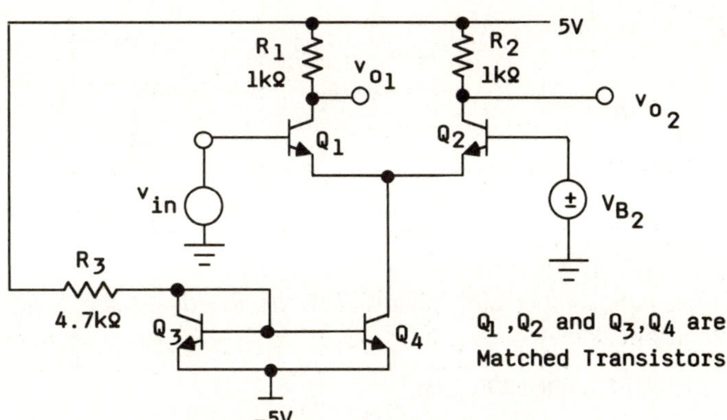

Figure 12L9.5: Simplified ECL gate

    (1)   Vary v_{in} from -1 V to 1-V in 0.1 V increments. Plot V_{01} and V_{02} versus V_{in}. Let $V_{B_2} = 0$ volts.

    (2)   Determine what happens when V_{in} exceeds the linear range for the transistor.

    (3)   Set V_{B_2} equal to 1 V and repeat Parts (1) and (2) with V_{in} ranging from 0 to 2 V.

2.   Discrete version of a 10K ECL Gate.

    a.   Build the circuit of Figure 12L9.6, which is essentially the same as that shown in Figure 12L9.3. $Q_1 Q_2$ are matched transistors, and Q_3, Q_4, Q_5 and Q_6 are 2N2222 transistors.

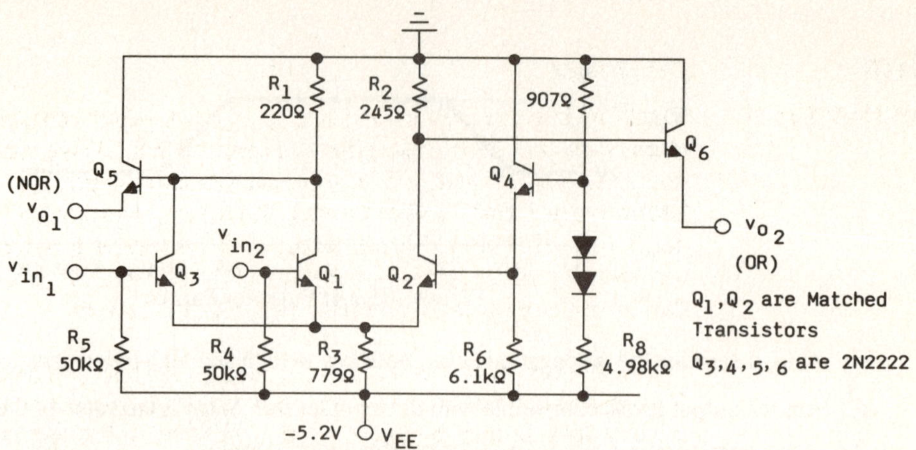

Figure 12L9.6: 10K ECL gate

(1) Let V_{in_2} be as shown in the circuit, floating, and connect V_{in_1} to a voltage ranging from -2 V to -1.4 V in 0.1 V increments. Measure, record and plot V_{02} and V_{01} as a function of V_{in_1}. Then switch the connections for V_{in_1} and V_{in_2}, and repeat this procedure. Determine V_{OH}, V_{OL}, V_{IH}, and V_{IL}.

(2) Let V_{in_2} be as shown in the circuit, and connect a function generator with a dc offset to produce a square wave as shown in Figure 12L9.7.

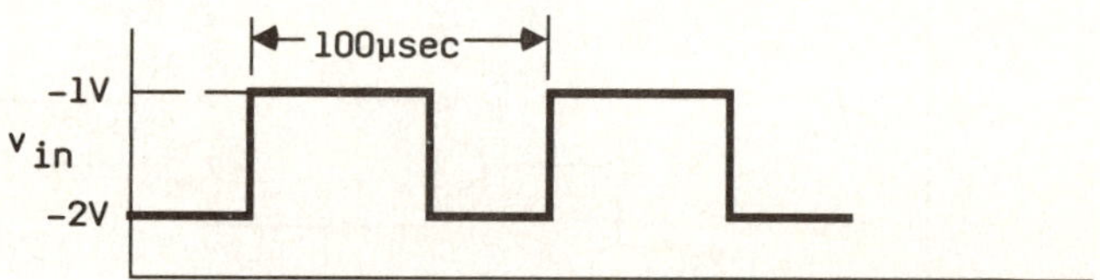

Figure 12L9.7: Square wave with dc offset

Display the output and input signals on the dual-trace oscilloscope, and measure the propagation delay.

3. Evaluation of a 10-K ECL OR/NOR gate.

   a. Connect the quad NOR gates so that they are in series.

   (1) Connect a square wave to the input of the first gate.

   (2) Is it possible to measure the propagation delay? Why?

   b. Port parameters

   (1) Design a circuit to measure V_{OH}, V_{OL}, V_{IH}, V_{IL}, I_{OH}, I_{OL}, I_{IH}, and I_{IL}.

   (2) Compare these values to Part 2 of the laboratory procedure.

DISCUSSION:

CONCLUSIONS:

LAB 12L10

TITLE: CMOS Logic

OBJECTIVE: CMOS logic is the most used technology for low power consumption applications. One can obtain CMOS logic gates as commercially available parts. Also, CMOS is used to design LSI and VLSI devices such as microprocessors. Another benefit of CMOS is that it lends itself to the early fabrication of analog and digital devices on a single chip of silicon. This is useful in the design of a microprocessor with an analog-to-digital or digital-to-analog converter on the same piece of silicon. Such a design is referred to as a **monolithic analog digital system**.

This laboratory starts with the CMOS inverter. The transfer function for the inverter is reasoned out quantitatively in the prelab and found experimentally in the lab. The inverter idea is then expanded to the two-input NOR and NAND gates. The laboratory finishes with a comparison between discrete CMOS logic and the 1400-series commercially available CMOS logic.

EQUIPMENT AND COMPONENTS:
          DVM power supplies: +5 V and 0-V to 5-V, adjustable
          3 x n-channel enhancement MOSFETS
          3 x p-channel enhancement MOSFETS
          1400UB dual three-input NOR gate and inverter (Motorola)
          Breadboard

PRELABORATORY:

1. Consider the circuit of Figure 12L10.1. Both Q_1 and Q_2 are enhancement-mode p- and n-channel, devices, respectively. Describe the operation of the circuit as V_{in} goes from 0 V to 5 V. Assume that the devices are matched and each has a threshold voltage of 1.5 V.

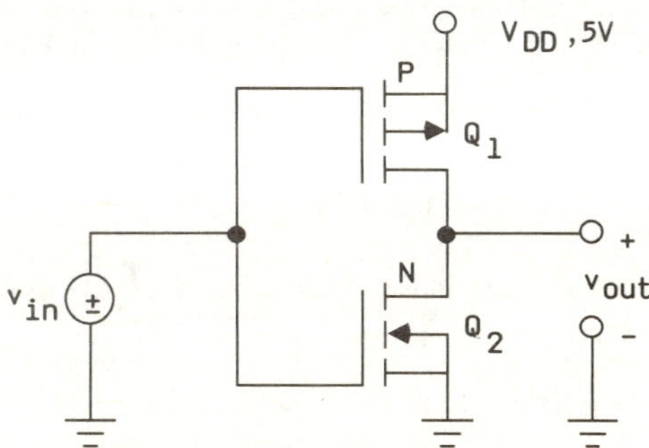

Figure 12L10.1: CMOS inverter

    a.   If v_{in} is a 1 msec 0 to 5 volt square wave with a 50% duty cycle, at what value of v_{in} does current flow through: Q_1 and/or Q_2?

    b.   Determine the high and low voltages at the output by applying the appropriate high (5 V) or low (0 V) at the input.

2. Consider the logic circuits of Figure 12L10.2. One circuit represents a two-input NOR gate, and the other a two-input NAND gate.

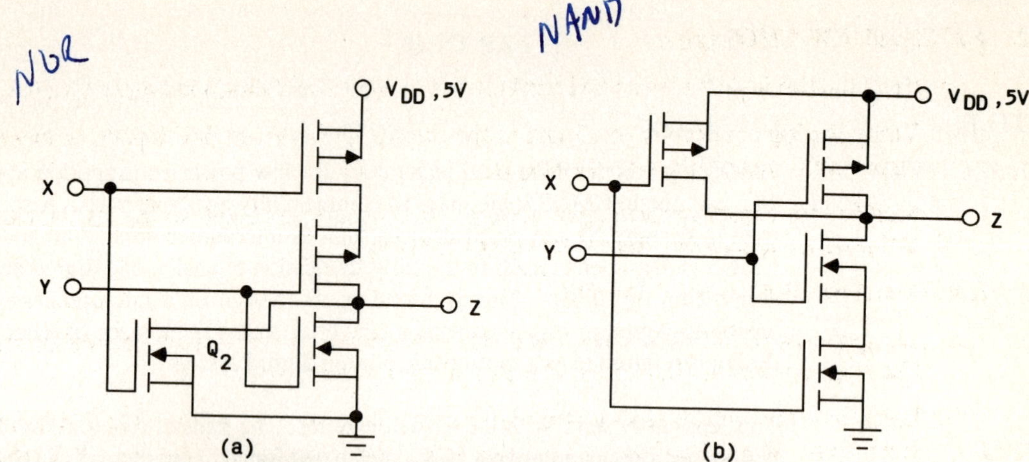

Figure 12L10.2: 2-input NOR and 2-input NAND gate

Describe the operation of both circuits. Assume that all MOSFET are matched (n and p channels are complementary matched devices). Which circuit is the NOR gate and which is the NAND gate?

a. How can the design be expanded to a three-input device?

b. Show a three-input NAND gate.

Consider the MC1400UB CMOS device.

a. Show the pinouts and describe the operation of this device.

b. Find V_{OH}, V_{OL}, V_{IN}, and V_{IL} for this device from the data sheets.

Laboratory Procedure:

1. CMOS inverter.

a. Build the circuit of Figure 12L10.3. This is essentially the same circuit as that shown in Figure 12L10.1.

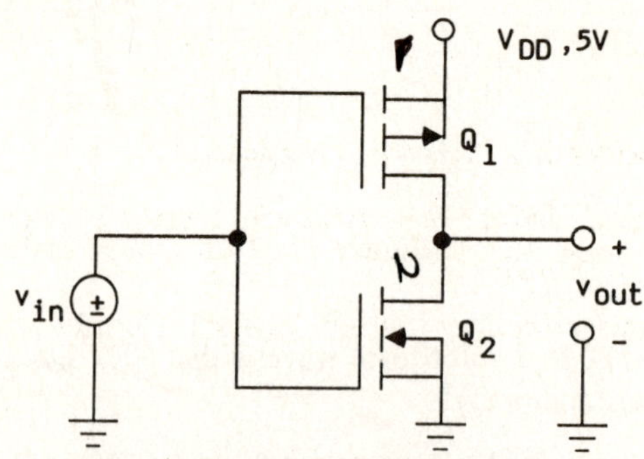

Figure 12L10.3: CMOS inverter

b. Vary V_{in} from 0 to 5 V in 0.5-V increments. Measure and record V_{out}. Plot the transfer function, v_{out} versus v_{in}. Determine the threshold voltage for each device from these data.

c. Compare the values of V_{IH}, V_{IL}, V_{OH}, and V_{OL} with the levels described for these quantities on the 1400UB data sheets. Also, compare them with the values predicted in Part 1 of the prelab.

2. NAND and NOR CMOS gates.

    a. Build the two logic functions given in Part 2 of the prelab and shown in Figure 12L10.2.

    b. Verify the logic functions you found in the prelab. Use 0-V and 5-V inputs for low and high, respectively. Record the output voltages.

    c. Modify your circuit for three-input devices. Verify that your circuit works. Again, use 0-V and 5-V inputs, respectively. Record the output voltages.

3. Commercially available CMOS 1400UB.

    a. Set up the 1400UB CMOS device, and verify the truth table as listed on the 1400UB data sheet. Use $V_{DD}=5$ volts and 0 and 5 volts for logical 0 and 1, respectively.

    b. Using only the inverter section of the device, apply v_{in} and vary it from 0 to 5 volts in 0.5 volt increments. Measure and record v_{out}. Plot your results and identify V_{OH}, V_{OL}, V_{IH}, and V_{IL}.

    c. Compare the results obtained in Part b with the values for these quantities listed for your discrete inverter. Include a SPICE simulation of your discrete inverter.

DISCUSSION:

CONCLUSIONS:

REFERENCES

[1] M. M. Mano, *Digital Logic and Computer Design* (Englewood Cliffs, NJ Prentice-Hall, 1979) pp. 25-7.

[2] M. M. Mano, *Computer Engineering Hardware Design* (Englewood Cliffs, NJ Prentice-Hall, 1988), pp. 32-8.

[3] Ibid., pp. 43-58.

[4] M. M. Mano, *Digital Logic and Computer Design*, (Englewood Cliffs, NJ: Prentice-Hall, 1979), pp. 56-60.

[5] Ibid., pp. 119-123.

[6] Ibid., pp. 123-5.

[7] Ibid., pp. 167-172.

[8] Ibid., pp. 175-180.

[9] Mano, *Computer Engineering Hardware Design*, p. 209.

[10] D. A. Hodges and H. G. Jackson, *Analyses and Design of Digital Integrated Circuits*, 2d ed. (New York: McGraw-Hill, 1988), pp. 388-393.

[11] Mano, *Computer Engineering Hardware Design*, pp. 120-122.

[12] Ibid., pp. 122-5.

[13] Ibid., pp. 126-132.

[14] Ibid., pp. 144-5.

[15] Mano, *Digital Logic and Computer Design*, pp. 257-89.

[16] A. S. Sedra and K. C. Smith, *Microelectronic Circuits*, 2d ed. (New York: Holt, Rinehart and Winston, 1987), pp. 845-853.

[17] Ibid., pp. 915-942.

[18]   Hodges and Jackson, pp. 241-57.

[19]   Ibid., pp. 257-69.

[20]   Sedra and Smith, pp. 942-956.

[21]   Ibid., pp. 854-869.

[22]   Hodges and Jackson, pp. 56-73, pp. 82-5, pp. 91-2.

[23]   Ibid., 85-91, 92-8.

[24]   Sedra and Smith, pp. 869-83.

INDEX